Lecture Notes in Computer Science 6859

Commenced Publication in 1973
Founding and Former Series Editors:
Gerhard Goos, Juris Hartmanis, and Jan van Leeuwen

Andrea Corradini Bartek Klin
Corina Cîrstea (Eds.)

Algebra and Coalgebra in Computer Science

4th International Conference, CALCO 2011
Winchester, UK, August 30 – September 2, 2011
Proceedings

Springer

Volume Editors

Andrea Corradini
Università di Pisa, Dipartimento di Informatica
56127 Pisa, Italy
E-mail: andrea@di.unipi.it

Bartek Klin
University of Warsaw, Faculty of Mathematics, Informatics, and Mechanics
ul. Banacha 2, 02-097 Warszawa, Poland
E-mail: klin@mimuw.edu.pl

Corina Cîrstea
University of Southampton, School of Electronics and Computer Science
Southampton, SO17 1BJ, UK
E-mail: cc2@ecs.soton.ac.uk

ISSN 0302-9743 e-ISSN 1611-3349
ISBN 978-3-642-22943-5 e-ISBN 978-3-642-22944-2
DOI 10.1007/978-3-642-22944-2
Springer Heidelberg Dordrecht London New York

Library of Congress Control Number: 2011933920

CR Subject Classification (1998): F.4, F.3, F.2, I.1, G.2, G.4, D.2.1

LNCS Sublibrary: SL 1 – Theoretical Computer Science and General Issues

Typesetting: Camera-ready by author, data conversion by Scientific Publishing Services, Chennai, India

Printed on acid-free paper

Springer is part of Springer Science+Business Media (www.springer.com)

Preface

CALCO, the International Conference on Algebra and Coalgebra in Computer Science, is a high-level, bi-annual event formed by joining CMCS (the International Workshop on Coalgebraic Methods in Computer Science) and WADT (the Workshop on Algebraic Development Techniques). CALCO aims to bring together researchers and practitioners with interests in foundational aspects, and both traditional and emerging uses of algebras and coalgebras in computer science. The study of algebra and coalgebra relates to the data, process and structural aspects of software systems.

Previous CALCO editions took place in Swansea (UK, 2005), Bergen (Norway, 2007) and Udine (Italy, 2009). CALCO 2011, the fourth conference in the series, took place in the city of Winchester (UK), from August 30 to September 2, 2011.

CALCO 2011 received 41 submissions, out of which 21 were selected for presentation at the conference. The standard of submissions was generally very high. The selection process was carried out by the Program Committee, taking into account the originality, quality and relevance of the material presented in each submission, based on the opinions of expert reviewers, three or four for each submission. The selected and revised papers are included in this volume, together with the contributions from the invited speakers Vincent Danos, Javier Esparza, Philippa Gardner and Gopal Gupta.

CALCO 2011 was co-located with two workshops. The CALCO Young Researchers Workshop, CALCO-jnr, was dedicated to presentations by PhD students and young researchers. CALCO-jnr was organized by Corina Cîrstea, Magne Haveraaen, John Power, Monika Seisenberger and Toby Wilkinson. The CALCO-tools Workshop, organized by Dorel Lucanu, provided presentations of tools. The papers presenting the tools also appear in this volume.

We wish to thank all the authors who submitted their papers to CALCO 2011, the Program Committee and the external reviewers who supported the Committee in the evaluation and selection process.

We are grateful to the University of Southampton for hosting CALCO 2011 at Winchester and to the Organizing Committee, chaired by Corina Cîrstea, for all the local arrangements. We also thank the London Mathematical Society, the British Logic Colloquium, and the University of Southampton for their financial support. At Springer, Alfred Hofmann and his team supported the publishing process. We gratefully acknowledge the use of EasyChair, the conference management system by Andrei Voronkov.

June 2011

Andrea Corradini
Bartek Klin
Corina Cîrstea

Organization

CALCO Steering Committee

Jiří Adámek	Technical University Braunschweig, Germany
Michel Bidoit	CNRS and ENS de Cachan, France
Corina Cîrstea	University of Southampton, UK
Andrea Corradini	University of Pisa, Italy
José Luiz Fiadeiro	University of Leicester, UK
H. Peter Gumm (Co-chair)	Philipps University Marburg, Germany
Rolf Hennicker	Ludwig Maximilians University Munich, Germany
Bart Jacobs	Radboud University Nijmegen, The Nertherlands
Bartek Klin	University of Warsaw, Poland
Hans-Jörg Kreowski	University of Bremen, Germany
Alexander Kurz	University of Leicester, UK
Marina Lenisa	University of Udine, Italy
Ugo Montanari	University of Pisa, Italy
Larry Moss	Indiana University, Bloomington, USA
Till Mossakowski (Co-chair)	DFKI Lab Bremen and University of Bremen, Germany
Fernando Orejas	Polytechnical University Catalonia, Barcelona, Spain
Francesco Parisi-Presicce	University of Rome La Sapienza, Italy
John Power	University of Bath, UK
Horst Reichel	Technical University of Dresden, Germany
Jan Rutten	CWI and Radboud University Nijmegen, The Netherlands
Andrzej Tarlecki	University of Warsaw, Poland
Hiroshi Watanabe	AIST, Japan

Program Committee

Jiří Adámek	Technical University Braunschweig, Germany
Lars Birkedal	IT University of Copenhagen, Denmark
Filippo Bonchi	University of Pisa, Italy
Corina Cîrstea	University of Southampton, UK
Andrea Corradini (Co-chair)	University of Pisa, Italy
Maribel Fernández	King's College London, UK

José Luiz Fiadeiro	University of Leicester, UK
H. Peter Gumm	Philipps University Marburg, Germany
Ichiro Hasuo	RIMS, Kyoto University, Japan
Bart Jacobs	Radboud University Nijmegen, The Netherlands
Bartek Klin (Co-chair)	University of Warsaw, Poland
Alexander Kurz	University of Leicester, UK
Barbara König	University of Duisburg-Essen, Germany
Marina Lenisa	University of Udine, Italy
José Meseguer	University of Illinois at Urbana-Champaign, USA
Ugo Montanari	University of Pisa, Italy
Larry Moss	Indiana University, Bloomington, USA
Till Mossakowski	DFKI Lab Bremen and University of Bremen, Germany
Catuscia Palamidessi	INRIA and LIX, Ecole Polytechnique, France
Dusko Pavlovic	Kestrel Insitute, USA
John Power	University of Bath, UK
Jan Rutten	CWI and Radboud University Nijmegen, The Netherlands
Lutz Schröder	DFKI and University of Bremen, Germany
Sam Staton	University of Cambridge, UK
Andrzej Tarlecki	University of Warsaw, Poland
Yde Venema	University of Amsterdam, The Netherlands
Martin Wirsing	Ludwig Maximilians University Munich, Germany

Additional Reviewers

Thorsten Altenkirch
Giorgio Bacci
Marcello Bonsangue
Roberto Bruni
Carlos Caleiro
Vincenzo Ciancia
Mihai Codescu
Jácome Cunha
Giovanna D'Agostino
Pietro Di Gianantonio
Jörg Endrullis
Egbert Fohry
Nicola Gambino
Neil Ghani
Daniel Gorin
Helle Hvid Hansen
Claudio Hermida
Peter Jipsen
Jan Komenda
Oliver Kutz
Raul Leal
Sergueï Lenglet
Paul Blain Levy
Gérard Ligozat
James Lipton
Alberto Lluch Lafuente
Paulo Mateus
Marino Miculan
Stefan Milius
Rasmus Møgelberg
Andrzej Murawski
Jorge A. Perez
Grigore Roşu
Vlad Rusu
Katsuhiko Sano
Ivan Scagnetto
Christoph Schubert
Michael Shulman
Paweł Sobociński
Ana Sokolova
Nicolas Tabareau
Carolyn Talcott
Nikos Tzevelekos
Tarmo Uustalu
Frank Valencia
Jiří Velebil
Jamie Vicary
Uwe Waldmann
Hans Zantema

Organizing Committee

Corina Cîrstea	University of Southampton, UK
Dirk Pattinson	Imperial College London, UK
Toby Wilkinson	University of Southampton, UK

Table of Contents

Invited Talks

Contributed Papers

CALCO Tools Workshop

On the Statistical Thermodynamics of Reversible Communicating Processes

Giorgio Bacci[1], Vincent Danos[2,⋆], and Ohad Kammar[2]

[1] DiMI, University of Udine
[2] LFCS, School of Informatics, University of Edinburgh

Abstract. We propose a probabilistic interpretation of a class of reversible communicating processes. The rate of forward and backward computing steps, instead of being given explicitly, is derived from a set of formal energy parameters. This is similar to the Metropolis-Hastings algorithm. We find a lower bound on energy costs which guarantees that a process converges to a probabilistic equilibrium state (a grand canonical ensemble in statistical physics terms [19]). This implies that such processes hit a success state in finite average time, if there is one.

1 Introduction

Regardless of the task a distributed algorithm is trying to complete, it has to deal with the generic problem of escaping deadlocks. One can solve this problem in a general fashion for CCS, π-calculus, and similar message-passing models of concurrency by equipping communicating processes with local memories. The role of memories is to record what local information will allow processes to backtrack, if needed. Backtracking preserves the granularity of distribution and can be seen as a special kind of communication [1,2,8]. A somewhat similar idea, based on resets rather than memories, can be found in early work from K.V.S. Prasad [17], and more recently in the context of higher-order π-calculus [11].

Following this approach, it is enough for the programmer to design a process that *may* succeed according to its natural forward semantics, and the same process, wrapped in a reversible operational semantics, *must* succeed. Separating the concern of deadlocks from the fundamentally different one of advancing locally towards a distributed consensus - makes the code simpler to understand and amenable to efficient automated verification [3,9]. What this method does not do, however, is to generate efficient code - that the code must eventually succeed does not say anything about the time it will take to do so. In fact, the general framework of non-deterministic transition systems is ill-equipped to even discuss the issue, as the worst time of arrival will often be infinite.

In the following, we provide a reinterpretation of our earlier construction of a reversible CCS in probabilistic terms. This sets the question in a well-understood quantitative framework. In particular, we can define a notion of exhaustive search

⋆ Corresponding author: vdanos@inf.ed.ac.uk

A. Corradini, B. Klin, and C. Cîrstea (Eds.): CALCO 2011, LNCS 6859, pp. 1–18, 2011.

for our distributed reversible processes, namely that the search reaches a *probabilistic equilibrium*. Better, we can show that under suitable constraints on the parameters arbitraging between forward and backward moves, a process will reach such an equilibrium. In other words, when such constraints are met, a particular event that may happen in a basic forward process, not only must happen in its reversible form, but must do so in finite average time. Obviously, this is a stronger guarantee.

1.1 Energy Landscaping

To achieve this, we build an energy function, or simply a *potential*, on the state space of a reversible process. This potential generates convergent search behaviours, on the transition graph that the process generates. The actual grammar of processes that we use slightly generalizes CCS [14] in that synchronizations can be many-way (more than two processes can synchronize together; eg, as in Petri nets), and can also be non-exclusive (a channel can be used to synchronize with many others, possibly at different rates; eg, as in Kwiatkowski-Stark's continuous π-calculus [10]). Most importantly, the potential we zero in on constrains the continuous-time Markov chain semantics without altering the granularity of distribution.

Not any potential works, and some energy policies will be too liberal and allow some processes to undergo an explosive growth which never reaches an equilibrium (despite reversibility). Our key finding is that, in order to obtain the existence of a general equilibrium, processes must have a probability to fork that decreases exponentially as a function of the size of their local memory. We prove that any rate of decrease larger than $\alpha^2\beta\log(4(\beta+1))$, where α is the maximal number of processes that can synchronize at once, and β the maximal number of successive forks that a process can undergo before communicating again, is sufficient.

This lower bound is reasonably sharp.

1.2 Related Work

Beyond the technical result, this paper borrows from a fundamental physical intuition, that of driving and deriving the dynamics by shaping the limit probability distribution it should converge to. A stochastic semantics is simpler in this respect, but the idea can be adapted to deterministic semantics as well. This is, in essence, the celebrated Metropolis algorithm which rules supreme in search and sample problems [13,7]: one first describes the energetic landscape of the state space, equivalently the probabilistic equilibrium, and then one defines the probabilistic transitions used to travel this landscape. Transition probabilities are chosen in a way that guarantees that the dynamics converge to the said equilibrium (this is explained in more details in §2).

Our work also borrows from Ref. [4], where the question of the existence of an equilibrium for a recursive Markov chain is proved to be undecidable. The proof uses a sequential reduction of an archetypical search problem, namely the Post correspondence problem, into the equilibrium existence problem. To obtain an equilibrium, even in the absence of a solution to the underlying Post problem instance, one introduces an energy penalty on searches. The class of potential that we use in the present paper is a distributed version of the former (although, here, we only deal with finding a potential that converges, as opposed to checking whether a given dynamics admits of one).

Note that the simple syntax of reversible CCS is merely used here for the sake of a simple reduction of the idea to practice. As for our earlier work on the qualitative aspects of reversible CCS, other syntaxes could do just as well.

Finally, and despite its concurrent-theoretic nature, the present work also draws inspirations from practical applications. In a modeling context, Ollivier's et al. recent paper [16] proposes a method to design models of biomolecular networks of allosteric interactions that is entirely based on the systematic usage of a certain grammar of local potentials. Among other things, this guarantees the thermodynamic consistency of the models. This has been a major source of inspiration for this work, as for the earlier Ref [5] where we prove that the thermodynamic consistency of mass action Petri nets is decidable (and find a definite shape for the associated potentials).

1.3 Outline

The paper is organised as follows. The next section (§2) is a reminder of the basics of discrete-space/continuous-time Markov chains, and the (in the context of this paper) central notion of equilibrium - in essence, a special kind of fixed point for the action of the Markov chain. Next, in §3, we turn to the (slightly generalized) syntax of reversible CCS and discuss the important property of simplicity which the transition graph underlying a reversible process enjoys. Roughly, this means that the transition graph of a reversibilized process is acyclic because it incorporates its own history, and this has consequences on the construction of equilibria. As in both cases, we are dealing here with simple or well-understood objects, §2-3 will be a bit concise, but still, hopefully, reasonably self-contained.

In §4, we introduce and compare different candidate potentials which one could think of using to lansdcape the reversible CCS state space. The §5 investigates an example of explosive growth which shows that no equilibrium can be obtained if energy penalties are purely based on the number of synchronizations of processes; this prepares the ground for a general approach. Finally, in §6, which is the technical core of the paper, we obtain our convergence result, which gives sufficient lower bounds on energy costs for communication to ensure the existence of a probabilistic equilibrium. The main technicality has to do with finding lower bounds for the potential of a process as a function of its number of synchronizations (Lemma 4). The conclusion returns to some of the issues discussed above, and touches on likely future research and intersections with more traditional concurrent-theoretic work on rewriting and termination.

2 Probabilistic Reminders

Throughout the paper we write $\log(x)$ for the *natural* logarithm; we manipulate multisets as vectors, that is to say additively, and write $|\boldsymbol{a}|$ for the size of a multiset (equivalently its L_1-norm); eg $\boldsymbol{a} = a + 2b$, and $|\boldsymbol{a}| = 3$.

We start with a quick reminder on CTMCs.

2.1 Timers and Chains

A *random exponential time* of parameter $\lambda > 0$ is an $[0, +\infty)$-valued random variable T such that $p(T > t) = \exp(-\lambda t)$. Thus, the density of T is $\lambda \exp(-\lambda t)$, for $t \geq 0$; and T's mean is $\int_0^{+\infty} \lambda \exp(-\lambda t) t \, dt = \lambda^{-1}$.

Suppose given a set X which is at most countably infinite, and a rate function $q(x, y) \in \mathbb{R}^+$, for x, y in X, and $x \neq y$.

The *transition graph* or the support of q, written $|q|$, is the binary relation, or the directed graph, on X which contains (x, y) iff $q(x, y) > 0$.

We suppose $|q|$ has finite out-degree (this also called being image-finite).

We can define a continuous-time Markov chain over X in the following way. When the chain is at x in X, for each of the finitely many ys such that $q(x, y) > 0$, draw a random exponential time $\tau(x, y)$ with parameter $q(x, y)$; advance time by $\tau = \min \tau(x, y)$, and jump to the (almost surely) unique y such that $\tau(x, y) = \tau$.

The idea is that all possible next states compete, and the higher the rate of $q(x, y)$, the more likely it is that y will be the next state. It is easy to calculate that the probability to jump to y is actually $q(x, y) / \sum_z q(x, z)$; and that for small ts, the probability to jump to x within t is equivalent to $q(x, y)t$, hence one can think of $q(x, y)$ as the rate at which one jumps from x to y.

Note that for the above definition to make sense it is important to suppose as we have done that $|q|$ is image-finite. We will also suppose thereafter that $|q|$ is symmetric (not to be confused with the much stronger assumption that q is a symmetric function, ie $q(x, y) = q(y, x)$), and define $\rho(x, y) = q(y, x)/q(x, y)$ when either (equivalently both) of $q(x, y)$ and $q(y, x)$ are > 0.

2.2 Equilibrium

Now, on to the definition of an equilibrium that will be our central concern here.

Consider a function p defined on X and with values in $\mathbb{R}^+$. One says p is an equilibrium for q if p is not everywhere zero, and:

- [*detailed balance*] for all $(x, y) \in |q|$, $p(x)q(x, y) = q(y, x)p(y)$
- [*convergence*] $Z = \sum_X p(x) < +\infty$

If such a p exists, we can obtain a probability on X by normalizing p as p/Z. Naturally, if X is finite the second condition always holds.

The detailed balance condition implies that p, construed as a probabilistic state of the system, is a fixed point of the action of the chain q, and as $|q|$ is symmetric, regardless of the initial state, the chain will converge to p (see Ref. [5, §2] for more details, or for a comprehensive textbook explanation, see Ref. [15]).

2.3 Potentials

Suppose given a real-valued function V on X (the energy landscape), together with a symmetric graph G on X (the moves one can make to travel the landscape). One can always define a rate function $q(x, y)$ over G for which $p_V(x) = \exp(-V(x))/Z$ is an equilibrium, when X is finite.

For instance, set $q(x, y) = 1$ if $V(x) \geq V(y)$ (one is always willing to travel downhill), $q(x, y) = \exp(V(x) - V(y))$ else (one is increasingly reluctant to travel uphill).

We can readily see that, with these settings, we have detailed balance:

$$p_V(x)/p_V(y) = e^{V(y)-V(x)} = q(y, x)/q(x, y) \tag{1}$$

When X is finite, this is enough to define an equilibrium, and this particular choice of a rate function q together with the choice of G is the *Metropolis algorithm*. In the case which interests us, when X can be countably infinite, the idea still applies, but one has to make sure that the potential V defines a finite $Z_V = \sum_X \exp(-V(x))$.

The converse problem of, given q, finding a potential for q, is also interesting. In general, we can pick an origin x_0 arbitrarily and within the connected component of x_0 in $|q|$, define the potential V as:

$$V(x) = \sum_{(x,y)\in\gamma} \log \rho(x, y) \tag{2}$$

for some path γ leading from x_0 to x.

Such an assignment is correct, meaning that detailed balance holds for the pair (q, V), iff the assignment does not depend on the choice of γ. In this case, V is defined uniquely on x_0's component up to an additive constant. If there *is* a dependency, then there is no solution, ie the dynamics on x_0's component is not describable by a potential. One says then that the chain is *dissipative*. Even for simple CTMCs this property is undecidable [4].

If the transition graph $|q|$ is acyclic, seen as an *un*directed graph, there is only one choice for γ, so independence trivially holds.

In the next section, X will be the state space of some reversible communicating process p_0. Reversibility will be obtained by equipping threads with memories which collectively capture the history of the computation (up to causal equivalence of computation paths, as we will see). Hence, the set of states reachable from the initial state p_0 is nearly equivalent to the space of its computations, and in particular, the underlying transition graph is (nearly) acyclic. This means that detailed balance will be for free, and only convergence will be an issue.

We mention in passing that in the case of stochastic mass action Petri nets, we find the opposite situation. Namely, verifying detailed balance might be involved as one needs to compute reactions invariants (the Petri net "loops"), whereas the convergence automatically follows [5].

3 Qualitative Semantics

We start with a reminder of CCS and its reversible form. In this minimalistic model, communication is devoid of any content, that is to say no value or name changes hands in a communication event; hence we will talk rather about synchronizations (synchs for short).

We assume a countable set of *channels* A, and a finite set of *synchronizable* (non-empty) multisets of channels $A^\star$.

A process p can be a product $p_1, \ldots, p_n$, or a guarded sum $a_1 p_1 + \ldots + a_n p_n$ with coefficients $a_i \in A$.

Products and sums are considered associative and commutative.

When $p = (p_1, \ldots, p_n)$, p forks into the siblings p_is which then run in parallel. When $p = (a_1 p_1 + \ldots + a_n p_n)$, p waits for an opportunity to synch on any of the channels a_i with a set of processes willing to synchronize on a tuple $\boldsymbol{a}$ such that $a + \boldsymbol{a} \in A^\star$. When that happens, p runs p_i.

Recursive definitions are allowed only if *guarded*, meaning that a recursively defined process variable only appears prefixed (aka guarded) by a channel; such definitions are considered to unfold silently.

As the reader may have noticed, to enhance readability (as we need to compute some examples in §4-5) we use lightfooted notations. Specifically, we use the comma to denote the product, and juxtaposition for prefixing.

An example of process (which we analyze from close in §5) is $p_0 = p, p'$, with $p = a(p, p)$, $p' = a'(p', p')$. Assuming $a + a' \in A^\star$, the two top threads p, p' can synchronize, after what they will fork into two copies of themselves, which can synchronize too, etc. Clearly p_0 has countably many computation traces, therefore we do need to deal with countable state spaces.

3.1 Memories and Transitions

Reversibility is obtained by adjoining memory stacks to processes in order to record transitions. One pushes a sibling identifier on the memory, when forking, and information about the synch partners, when synching.

Thus we have *fork* transitions (with $n > 0$):

$$\Gamma \cdot (p_1, \ldots, p_n) \rightarrow^f \Gamma 1 \cdot p_1, \ldots, \Gamma n \cdot p_n$$

where the memory Γ is copied over to each sibling, with a unique integer identifier for each one of them.

And we also have *synch* transitions (with $m > 0$):

$$\begin{array}{l} \Gamma_1 \cdot (a_1 p_1 + q_1), \ldots, \Gamma_m \cdot (a_m p_m + q_m) \rightarrow^s_{\boldsymbol{a}} \\ \Gamma_1(\boldsymbol{\Gamma}, a_1, q_1) \cdot p_1, \ldots, \Gamma_m(\boldsymbol{\Gamma}, a_m, q_m) \cdot p_m \end{array}$$

where $\boldsymbol{\Gamma}$ is short for $\Gamma_1 a_1, \ldots, \Gamma_m a_m$. This means that each of the threads taking part in the synch records the memory and channel of all participants, its own channel a_i, and its sum remainder q_i (preempted by the synch).

We have labelled the transition arrows for convenience, where $\rightarrow^{s}_{\boldsymbol{a}}$ means synch on a multiset $\boldsymbol{a} \in A^\star$. Naturally, the above transitions can happen in any product context.

(NB: Memories are reversed in this notation compared to Ref. [1].)

Communicating processes commonly offer additional constructs: name creation, restrictions, value- or name-passing, etc, which are not considered here. None should make a difference to our main argument, but this has to be verified.

Consider a process with an empty memory $\varnothing \cdot p_0$, and define $p \in \Omega(p_0)$ if p is reachable from $\varnothing \cdot p_0$ by a sequence of transitions, also known as a computation trace, as defined above. It is easy to see that within any $p \in \Omega(p_0)$, memories uniquely identify their respective processes. Thus, both types of transitions store enough information to be reversed unambiguously. In particular, adding the symmetric backward transitions leaves $\Omega(p_0)$ unchanged.

Hereafter, we will suppose that transitions are effectively symmetric.

3.2 Near Acyclicity and Simplicity

Now that we have our symmetric transition graph in place, we can return to the acyclicity property that we alluded to in §2.

Consider a computation trace γ, taking place in some $\Omega(p_0)$. If in γ we find a forward move followed immediately by its symmetric backward move, then we can cancel both and obtain a new (shorter) trace γ' with the same end points. Likewise, if in γ we find two synch moves in immediate succession which are triggered by entirely disjoint set of threads (one says the synchs are concurrent), then we can commute the two steps and obtain a new trace γ' (of equal length) with the same end points.

This defines a notion of *causal equivalence* on traces with common end points. We know from Ref. [1] that any two computation traces γ, γ' with the same end points are causally equivalent. We will refer to this as the *labeling property.* (The name is chosen in relation to Lévy's labeling for λ-calculus [12]; indeed, forward reversible CCS is a Lévy-labeling of CCS.) Essentially, this means that the transition graph on $\Omega(p_0)$ is nearly acyclic.

The labeling property implies a convenient property of *simplicity*, namely that, for p, p' in $\Omega(p_0)$, there is *at most one* transition from p to p'.

3.3 An Aside on Degenerate Sums

Actually, for the labeling and simplicity properties to be strictly true, one needs to be precise in the management of degeneracy in sums (as noticed in Ref. [8]). Consider a simple binary synch, with $\boldsymbol{a} = a + a' \in A^\star$:

$$x := \Gamma \cdot ap + r, \Gamma' \cdot a'p' + r' \rightarrow \Gamma(a, \boldsymbol{\Gamma}, r) \cdot p, \Gamma'(a', \boldsymbol{\Gamma}, r') \cdot p' =: y$$

Suppose ap and $a'p'$ occur with multiplicities $\mu(ap)$, $\mu(a'p')$, then there are $\mu(ap)\mu(a'p')$ distinct ways to jump from x to y.

To handle this additional cyclicity/lack of simplicity in the transition graph, one can forbid $p + p$ in sums altogether; or one can recover simplicity by memorising the particular term in the sum that was used (which introduces non-commutative sums that are a bit awkward); or simply incorporate such parallel edges in the causal equivalence - and handle the induced local symmetry factor in the quantitative part of the development (next section and onwards). The latter solution seems preferable, as it is more general.

Then, anticipating somewhat on the next section, the compound rate ratio $\rho(x, y)$ for the above parallel jumps on $a + a'$ is:

$$\rho(x, y) = \frac{k_{\boldsymbol{a}}^{-}}{k_{\boldsymbol{a}}^{+}} \cdot \frac{1}{\mu(ap)} \cdot \frac{1}{\mu(a'p')} \tag{3}$$

where $k_{\boldsymbol{a}}^{+}$, $k_{\boldsymbol{a}}^{-}$ are the forward and backward rates for a synch on $\boldsymbol{a}$ (which can depend on x and y in general).

The multiplicity factors appearing above are perfectly manageable, but they do make the treatment a little less smooth. Henceforth, we will assume such degeneracies do not happen, and simplicity holds as is.

3.4 Which Potential to Look for?

Returning to the main thrust, we are now looking for a quantitative version of the above calculus.

As the labeling property guarantees near-acyclicity of the underlying symmetrized transition graph, we know from §2 that any rate function q will lead to a potential V definable as in (2), provided that it respects the causal equivalence, ie the little cyclicity left in reversible CCS.

By which we mean, specifically, that the ratios ρ should verify: 1) $\rho(x, y)\rho(y, z)$ $= \rho(x, y')\rho(y', z)$ when y, y' are intermediate forms obtained by interleaving concurrent synchs in either order, and 2) should be equal when coming from degenerate sums as above. There is no need to say anything for trivial forward/backward loops, as by definition $\rho(x, y) = \rho(y, x)^{-1}$.

The fact that (almost) any rate assignment works notwithstanding, we would like to derive the rates from some potential, for reasons explained in the introduction. This raises the question of what a good potential is. Two requirements stand out. Firstly, the potential should be such that the implied dynamics is implementable concurrently and does not require any global knowledge of the state of the system. Secondly, it should converge. Besides, one might also want a potential that is invariant under natural syntactic isomorphisms (eg the numbering of siblings), and such that disjoint sums of processes implies independent dynamics.

The solution we will eventually home in on, ticks all of the above boxes (but we are not going to be sure until §6).

4 Concurrent Potentials

We examine now two potentials that seem natural in the light of the discussion right above, and establish that they are implementable concurrently (within the abstract model of concurrency on which CCS relies, that is). Convergence issues are investigated in the next two sections.

Throughout this section we fix an initial process $\varnothing \cdot p_0$, and a real-valued *energy vector* indexed over $A^\star$, and written $\boldsymbol{\epsilon}$.

4.1 Total Stack Size Potential

The first potential we consider is defined inductively on the syntax of a reversible process in $\Omega(p_0)$:

$$\begin{aligned} V_1(p_1,\ldots,p_n) &= V_1(p_1)+\ldots+V(p_n)\\ V_1(\Gamma\cdot p) &= V_1(\Gamma i) = V_1(\Gamma)\\ V_1(\Gamma(\boldsymbol{\Gamma},a,q)) &= V_1(\Gamma)+\epsilon_{\boldsymbol{a}}\end{aligned}$$

with $\boldsymbol{\Gamma} = \Gamma_1 a_1,\ldots,\Gamma_m a_m$ and $\boldsymbol{a} = a_1+\ldots+a_m$.

Equivalently, $V_1(p)$ is the inner product $\langle\boldsymbol{\epsilon},\tilde{\Gamma}(p)\rangle$, where $\tilde{\Gamma}(p)(\boldsymbol{a})$ is the number of occurrences of $\boldsymbol{a}$ in p; $\tilde{\Gamma}(p)$ can be seen as a forgetful and commutative projection of the memory structure of p.

Note that $V_1(\varnothing\cdot p_0) = 0$ with this definition; one can always choose a zero energy point in the (strongly) connected component of the initial state, and it is natural to choose the initial state itself.

For each of the two types of transitions, we can easily compute the energy balance:

$$\begin{aligned} \Delta V_1 &= (n-1)V_1(\Gamma) && n\text{-ary fork with memory } \Gamma\\ \Delta V_1 &= m\epsilon_{\boldsymbol{a}} && \text{synch on } \boldsymbol{a}\end{aligned}$$

Now, we need to understand how these constrain the rate function. This is analogous to what we have done earlier with (1) in §2.3.

Let us write k_f^-, k_f^+ for backward and forward forking rates, and $k_{\boldsymbol{a}}^-$, $k_{\boldsymbol{a}}^+$ for backward and forward synching rates. For a fork, and by the simplicity property, the constraint translates into $\log\rho(x,y) = \log(k_f^-/k_f^+) = (n-1)V_1(\Gamma)$. A possible solution is:

$$\begin{aligned} k_f^- &= 1\\ k_f^+ &= e^{-(n-1)V_1(\Gamma)}\end{aligned}$$

This is an entirely local solution, as the increasing reluctance to fork only depends on the local memory of the process of interest (and the number of siblings, but that can be statically controlled). Similarly, for a synch, the constraint is $\log\rho(x,y) = \log(k_{\boldsymbol{a}}^-/k_{\boldsymbol{a}}^+) = m\epsilon_{\boldsymbol{a}}$. A possible solution is:

$$\begin{aligned} k_{\boldsymbol{a}}^- &= 1\\ k_{\boldsymbol{a}}^+ &= e^{-m\epsilon_{\boldsymbol{a}}}\end{aligned}$$

not only this is local, but in contrast with the fork case, the assignment does not depend on the memories of the synching processes.

Note that there are many other solutions compatible with V_1.

4.2 Total Synch Potential

Perhaps the most natural potential is the following.

Given a path γ from $\varnothing \cdot p_0$ to p:

$$V_0(p) = \sum_{\boldsymbol{a} \in A^\star} \sum_{x \to_{\boldsymbol{a}}^{s} y \in \gamma} (-1)^{v(s)} \epsilon_{\boldsymbol{a}}$$

where $v(s) = \pm 1$ depending on whether the synch is forward or backward. This V_0 is based on the idea that only communication costs, and forking is done at constant potential. As for V_1, $V_0(\varnothing \cdot p_0) = 0$.

Clearly, this definition is independent of the choice of γ. Indeed, by the labeling property, any γ, γ' linking $\varnothing \cdot p_0$ to p are convertible by swaps of concurrent synchs, and trivial cancellations, both of which leave V_0 invariant. The corresponding constraints can be met by a locally implementable rate function, eg, $k_f^- = k_f^+$, and $k_{\boldsymbol{a}}^- / k_{\boldsymbol{a}}^+ = \exp(\epsilon_{\boldsymbol{a}})$. (We could add a term to V_0 to count the forks as well.)

Differently from V_1, there is no inductive formula for $V_0(p)$, as to compute it one essentially needs to replay a reduction to p.

4.3 V_1 vs. V_0

Let us compare the potentials on two simple examples. Below, we suppose $\boldsymbol{a} = a + a'$, $\boldsymbol{b} = b + b'$ in $A^\star$, and $\epsilon_{\boldsymbol{a}} > 0$; we do not represent the entirety of the memory elements, just what we need to compute the Vs.

Here is a first example:

$$\begin{aligned}\varnothing \cdot a(a, b, a', b'), a' &\to 0\boldsymbol{a} \cdot (a, b, a', b'), 1\boldsymbol{a} \cdot \_ \\ &\to 0\boldsymbol{a}0 \cdot a, 0\boldsymbol{a}1 \cdot b, 0\boldsymbol{a}2 \cdot a', 0\boldsymbol{a}3 \cdot b', 1\boldsymbol{a} \cdot \_ \\ &\to 0\boldsymbol{a}0\boldsymbol{a} \cdot \_, 0\boldsymbol{a}1\boldsymbol{b} \cdot \_, 0\boldsymbol{a}2\boldsymbol{a} \cdot \_, 0\boldsymbol{a}3\boldsymbol{b} \cdot \_, 1\boldsymbol{a} \cdot \_ = p\end{aligned}$$

and we get:

$$V_0(p) = 2\epsilon_{\boldsymbol{a}} + \epsilon_{\boldsymbol{b}} < 7\epsilon_{\boldsymbol{a}} + \epsilon_{\boldsymbol{b}} = V_1(p)$$

We can use the expansion law, replacing a, b with $ab + ba$, and similarly for a', b' in p_0, and get a variant of the above trace:

$$\begin{aligned}\varnothing \cdot a(ab + ba, a'b' + b'a'), a' &\to 0\boldsymbol{a} \cdot (ab + ba, a'b' + b'a'), 1\boldsymbol{a} \cdot \_ \\ &\to 0\boldsymbol{a}0 \cdot (ab + ba), 0\boldsymbol{a}1 \cdot (a'b' + b'a'), 1\boldsymbol{a} \cdot \_ \\ &\to 0\boldsymbol{a}0\boldsymbol{ab} \cdot \_, 0v1\boldsymbol{ab} \cdot \_, 1\boldsymbol{a} \cdot \_ = p'\end{aligned}$$

with:

$$V_0(p) = V_0(p') = 2\epsilon_{\boldsymbol{a}} + \epsilon_{\boldsymbol{b}} < 4\epsilon_{\boldsymbol{a}} + 2\epsilon_{\boldsymbol{b}} = V_1(p') < V_1(p)$$

We see that V_1, unlike V_0, is truly concurrent in the sense that it is sensitive to sequential expansions. In fact, according to V_1, an expanded form using a sum is cheaper by an amount of $V_1(\Gamma)$; a sequentialized version is bolder in its search (and the backward options are fewer).

In general, $V_0 \leq V_1$, as a synch performed on the way to p is visible at least once in a memory in p (in fact, at least $|\boldsymbol{a}|$ for a synch on $\boldsymbol{a}$); and $V_0(p) = V_1(p)$ if p has only forks with $n \leq 1$.

So which potential should one prefer? Both seem equally good, but the next section will tell a very different story.

5 Explosive Growth

Any potential V partitions $\Omega(p_0)$ into finite level sets $\Omega_v(p_0)$, defined as the set of reachable ps such that $V(p) = v$.

Among other things, to address the convergence issue, we will need to control the cardinality of $\Omega_v(p_0)$, which by the labeling property, is the number of traces (up to causal equivalence) γ leading to $\Omega_v(p_0)$.

Let us try to see how this plays out with our earlier example, $p_0 = p, p'$, with $p = a(p,p)$, $p' = a'(p',p')$ (or isomorphically $q_0 = aq_0, a'q_0$).

Call ϕ_k the following trace (synch partners not represented in memories):

$$\begin{array}{rl} \varnothing \cdot p_0 \rightarrow^{f} & 0 \cdot p, 1 \cdot p' \\ \rightarrow^{fs} & 0\boldsymbol{a}0 \cdot p, 0\boldsymbol{a}1 \cdot p, 1\boldsymbol{a}0 \cdot p', 1\boldsymbol{a}1 \cdot p' \\ = & 0\boldsymbol{a}0 \cdot a(p,p), 0\boldsymbol{a}1 \cdot a(p,p), \\ & 1\boldsymbol{a}0 \cdot a'(p',p'), 1\boldsymbol{a}1 \cdot a'(p',p') \\ \rightarrow^{fs} & 0\boldsymbol{a}0\boldsymbol{a}0 \cdot p, 0\boldsymbol{a}0\boldsymbol{a}1 \cdot p, 0\boldsymbol{a}1\boldsymbol{a}0 \cdot p, 0\boldsymbol{a}1\boldsymbol{a}1 \cdot p, \\ & 1\boldsymbol{a}0\boldsymbol{a}0 \cdot p', 1\boldsymbol{a}0\boldsymbol{a}1 \cdot p', 1\boldsymbol{a}1\boldsymbol{a}0 \cdot p', 1\boldsymbol{a}1\boldsymbol{a}1 \cdot p' \\ \ldots & \\ \rightarrow^{fs} & \prod_{w \in \{0,1\}^k} 0w(\boldsymbol{a}) \cdot p, \prod_{w \in \{0,1\}^k} 1w(\boldsymbol{a}) \cdot p' = p_k \end{array}$$

where $w(x)$, for $w \in \{0,1\}^k$, is defined as the fair interleaving of w and x^k - where x begins, eg $01(\boldsymbol{a}) = \boldsymbol{a}0\boldsymbol{a}1$.

Note that ϕ_k is maximally synchronous, in the sense that synchronizations are all intra-generational. In this respect it is a very peculiar trace, and one which is easy to compute with.

As the computation unfolds symmetrically, ϕ_k has $2^k - 1$ synchs, and its end process p_k has 2^{k+1} threads, and each has a memory where $\epsilon_{\boldsymbol{a}}$ occurs k times.

Hence process p_k has respective potentials:

$$V_0(p_k) = (2^k - 1)\epsilon_{\boldsymbol{a}} \leq k2^{k+1}\epsilon_{\boldsymbol{a}} = V_1(p_k)$$

Non-causally equivalent realizations of ϕ_k, can be obtained by picking different intra-generational matchings. Each choice leads to distinct end processes p_k, all with the same V_1 and V_0 potentials - and thus all in the intersection of $\Omega_{V_0(p_k)}$ and $\Omega_{V_1(p_k)}$.

There are $\prod_{0 \leq h < k} 2^h!$ distinct such ϕ_ks, hence, using $n^n e^{-n} \leq n!$, we get the following lower bound on the cardinality of $\Omega_{V_0(p_k)}$ and $\Omega_{V_1(p_k)}$:

$$(2^{k-1})^{2^{k-1}} e^{-2^{k-1}} \leq 2^{k-1}! \leq \prod_{0 \leq h < k} 2^h! \tag{4}$$

Thus $\log |\Omega_{V_0(p_k)}|$ and $|\log \Omega_{V_1(p_k)}|$ grow asymptotically faster than $k2^{k-1} \log 2$.

This entropic term will trump the term opposed by V_0 which is asymptotically equivalent to $-2^k \epsilon_{\boldsymbol{a}}$. The inescapable conclusion is that, no matter how costly a synch on $a + a'$ is made to be, V_0 will diverge, and, concretely, the process p_0 will undergo an infinite growth if it follows this potential. So, we can forget V_0 for infinite state spaces.

On the other hand, V_1's term is $-k2^{k+1}\epsilon_a$ which can control our lower bound of the entropic term, if $4\epsilon_a > \log 2$. So V_1 might still work.

Now (4) only provides a lower bound. As there are many other traces, using extra-generational matchings, that might end up in the same level set, it is hard to know how sharp it is. And, anyway, this is just an example. We have yet to prove that for all p_0, there are suitable choices of ϵ, that will make V_1 converge. This is what we do in the next section.

6 Main Statement

We need a couple of combinatorial lemmas.

Lemma 1. *Suppose $k > 0$ and $\sum_{i=1}^{i=k} n_i = n$, then $\sum_i n_i \log n_i \geq n \log(n/k)$.*

Proof.

$$\begin{aligned}\textstyle\sum_i n_i \log n_i &= -n \textstyle\sum_i -(n_i/n) \log n_i/n + n \log n \\ &= -nS(n_i/n) + n \log n \\ &\geq -n \log k + n \log n\end{aligned}$$

where, in the last step, we use $S(n_i/n) \leq \log k$, the usual upper bound on the entropy over a finite set $\{1, \ldots, k\}$. □

Lemma 2 (lower bound on tree depth). *Consider the set of trees t with maximal branching β, then for some $c > 0$, $\|t\| := \sum_{u \in t^\circ} d(u) \geq c \cdot n \log n$ with $n = |t^\circ|$ the number of internal nodes of t. Specifically, $c = 1/\log 4\beta$ works.*

Proof. For $n = 0, 1$, the lower bound holds for any c as the rhs is 0.

Suppose $n \geq 2$. We partition t into its k immediate subtrees with non-zero internal nodes, $n_i > 0$; as $n \geq 2$, we know that $k > 0$.

$$\begin{aligned}\|t\| &= \textstyle\sum_i \sum_{u_i \in t_i^\circ} d_i(u) + 1 & \\ &= \textstyle\sum_i n_i + \|t_i\| & \\ &= n - 1 + \textstyle\sum_i \|t_i\| & \\ &\geq n - 1 + c \textstyle\sum_i n_i \log n_i & \text{by induction} \\ &\geq n - 1 + c(n-1) \log((n-1)/k) & \text{by Lemma 1} \\ &= (n-1)(1 - c \log k) + c(n-1) \log(n-1) &\end{aligned}$$

So we need to find c such that for $n \geq 2$:

$$(n-1)(1 - c \log k) + c(n-1) \log(n-1) \geq cn \log n$$

Set for $x \geq 1$:

$$g(x) = (x-1)(1 - c \log k) + c(x-1) \log(x-1) - cx \log x$$

We have $g(1^+) = 0$, $g'(x) = 1 - c \log(xk/(x-1)) \geq 0$ as soon as $c \leq 1/\log(xk/(x-1)) =: h(x)$. $(1/h)'(x) = -1/(x(x-1)) \leq 0$ for $x \geq 1$, so h is increasing on $(1, \infty)$, and if we take $c \leq 1/\log 2k$, $c \leq h(x)$ for $x \geq 2$, and g increases for $x \geq 2$.

Now $g(2) = 1 - c \log 4k \geq 0$ as soon as $c \leq 1/\log 4k$.

Set $c = 1/\log 4\beta$ where β is the maximum branching of t.

Clearly $\beta \geq k$, so we have $g(x) \geq 0$ for $x \geq 2$. □

Note that the proof gives explicit control in terms of the maximal branching degree β, namely $\|t\| \geq n \log n / \log 4\beta$. If one allows arbitrary branching, any tree t with all $n-1$ nodes right below the root of t verifies $\|t\| = n-1$. While the (best) inequality says $n-1 \geq n \log n/(2\log 2 + \log(n-1))$, which is true indeed. Clearly the inequality is more interesting if one imposes a maximal branching degree.

It is possible to specialize the inequality (which is central to the main convergence result below) to the case of balanced trees. As these minimize depth for a given number of internal nodes (else one can always move groups of sibling leaves upwards, and in so doing decrease the potential), they should be a good test of the sharpness of our lower bound. We consider only binary trees to keep computations simpler.

Lemma 3 (balanced binary case). *Let t_k be the balanced binary tree with 2^k leaves, equivalently $n = 2^k - 1$ internal nodes, $k \geq 0$:*

$$\|t_k\| = \textstyle\sum_{1 \leq i < k} i2^i = (n+1)\log(n+1)/\log 2 - 2n$$

Proof. The formula holds for $k = 0, 1$, and $\|t_0\|=\|t_1\|=0$.

Suppose k, the number of 'generations' in t_k, is strictly positive.

As t_k has 2^k leaves, $2^k - 1$ internal nodes, we have by induction on the last generation of the tree:

$$\begin{aligned} \|t_1\| &= 0 \\ \|t_{k+1}\| &= k2^k + \|t_k\| \end{aligned}$$

Therefore $\|t_k\| = \sum_{1 \leq i < k} i2^i$.

Set $\phi_k(x) = (x^k - 1)/(x - 1) = \sum_{0 \leq i < k} x^i$. We have:

$$\phi_k'(x) = \textstyle\sum_{1 \leq i < k} ix^{i-1} = ((k-1)x^k - kx^{k-1} + 1)/(x-1)^2$$

Hence $\sum_{1 \leq i < k} i2^i = 2\phi'(2) = 2((k-1)2^k - k2^{k-1} + 1) = k2^k - 2^{k+1} + 2$. □

Therefore, in this case the inequality of Lemma 2 amounts to saying that $(n+1)\log(n+1)/\log 2 - 2n \geq n \log n / 3 \log 2$ (with $\beta = 2$) or equivalently:

$$(n+1)\log(n+1) \geq 1/3 \cdot n \log n + (2 \log 2)n$$

which is indeed true for $n \geq 0$ and a rather sharp estimate for small values of n. This means that the lower bound provided by Lemma 2 is good.

6.1 Lower Bound on the Potential

With Lemma 2 in place, we can bound below the energy of a process/trace with a given number of synchs. But first, we need to fix some notations.

As in §4-5, we suppose given a process p_0, and consider only computation traces starting from the initial state $\varnothing \cdot p_0$.

We write $T(n)$ for the set of traces containing n synchs, considered up to causal equivalence, and set $\epsilon_m := \min_{a \in A^\star} \epsilon_a$, for the minimal energetic cost of a synch. We also write $\Omega_n(p_0)$ for the set of processes reachable in n synchs.

As traces originating from the initial state are isomorphic to their end points, ie $T(n) \sim \Omega_n(p_0)$, we will treat traces from $\varnothing \cdot p_0$ and processes in $\Omega(p_0)$ as nearly synonymous.

We suppose given an upper bound α on the number of processes that can synchronize at once during an execution of $\varnothing \cdot p_0$. Eg $\alpha = \max_{\boldsymbol{a} \in A^\star} |\boldsymbol{a}|$, the maximal synch size in $A^\star$.

We write δ for the thread *increment* of a particular forking event in a trace. This means that one replaces one thread with $\delta + 1$ ones. We suppose also that we are given two numbers (eg obtained from a trivial syntactic analysis) such that $\beta_- \leq \delta + 1$ and $\delta \leq \beta_+$ always hold. If $\beta_- > 1$, then $\delta > 0$, which amounts to saying that the number of threads always increase under fork.

We can establish the following lower bound on the potential:

Lemma 4. *Suppose $\beta_- > 1$, $\epsilon_m > 0$, $p \in \Omega_n(p_0)$:*

$$\frac{\epsilon_m}{\log 4 + \log(\beta_+ + 1)} \cdot n \log n \leq V_1(p)$$

Proof. Consider the set $U(n)$ of trees with n internal nodes labeled in $A^\star$. It clearly makes sense to extend the definition of V_1 to such labeled trees.

Consider $t \in U(n)$, $n > 0$, and u an internal node in t with label $\boldsymbol{a}$, all the children of which are leaves (so that u is on the boundary of the set of internal nodes). Define $t \setminus u \in U(n-1)$ as the tree obtained from t by erasing the $\delta(u)+1$ leaves right below u (as well as u's label).

Write $\tilde{\Gamma}(u)$ for the multiset of occurences of labels above u, and $d(u)$ for the depth of u in t (as we have already done above). We can bound below the difference of potential incurred by erasing the $\delta(u) + 1$ children of u:

$$\begin{aligned} V_1(t) - V_1(t \setminus u) &= (\delta(u) + 1)\epsilon_{\boldsymbol{a}} + \delta(u)\langle \boldsymbol{\epsilon}, \tilde{\Gamma}(u) \rangle \\ &\geq \epsilon_m \delta(u) d(u) \\ &\geq \epsilon_m d(u) \end{aligned}$$

We have used $\delta(u) > 0$.

It follows that V/ϵ_m decreases by chunks of at least $d(u)$ for each deletion of a node on the internal boundary, therefore $V(t)/\epsilon_m \geq \sum_i d(u_i) =: \|t\|$, and we can apply Lemma 2, to obtain $V(t)/\epsilon_m \geq n \log n / \log 4(\beta_+ + 1)$.

As any p in $\Omega_n(p_0)$ projects to a labeled tree in $U(n)$, by forgetting the information on communication partners and remainders, and this forgetful operation leaves V_1 invariant, the statement follows. □

With the same notations as in the proof above, consider a leaf $v \in t$, and define $t(u, v)$ as the new tree obtained by moving the leaves below u, to below v; clearly, if $d(v) < d(u)$, $d(t(u, v)) < d(t)$. If no such move exists, by definition t is balanced. So, as alluded to earlier, the lower bound we get for the potential is obtained for balanced concurrent structures of execution - and as they have lower energies, they will be highly favoured by the dynamics. In other words, our potential V_1 penalizes depth - one could call it a *breadth-first* potential - and different threads will tend to stay synchronous.

We turn now to the other pending question, namely that of binding above the entropy (that is to say the logarithm of the cardinality) of the set of traces of a given number of synchs.

6.2 Upper Bound on the Number of Traces

Dually to Lemma 4, which a lower bound on potentials, we can derive an upper bound on entropies:

Lemma 5. *For large ns,* $\log|T(n)| \leq \beta_+\alpha^2 O(n \log n)$

Proof. By induction on n, there are at most $\delta_0 + n\beta_+\alpha$ threads in the end process of a trace in $T(n)$, as each synch adds at most $\delta\alpha$ new threads, where we have written δ_0 for the initial number of threads in p_0.

Any trace with $n+1$ synchs can be obtained (perhaps in many ways but we are looking for an upper bound) by synching one of length n, so $|T(n+1)| \leq |T(n)|(\delta_0 + n\beta_+\alpha)^\alpha$. As $T(0) = 1$, we get $\log|T(n)| \leq \alpha \log(\delta_0 + n\beta_+\alpha)!$.

Since:

$$1 - n + n\log n \leq \log n! \leq 1 - n + (n+1)\log n$$

it follows that $\log(\delta_0 + n\beta_+\alpha)! \sim \beta_+\alpha O(n\log n)$. □

The first inequality is sharp if all synchs are possible, and one has the maximal thread count, and no sums (as they decrease the number of matches), which is exactly the situation of the explosive example of §5.

As the arithmetic progression that gives rise to the factorial, samples the factorial only with frequency $1/\delta\alpha$ (this is sometimes called a shifted j-factorial [18, p.46], where $j = \alpha\delta$, and the shift is δ_0 in our example), it seems the upper bound above could be improved. But, if we return to the maximal synchronous traces computed in §5, we see that the bound above is quite sharp, so it seems unlikely.

6.3 Convergence

Now we can put both bounds to work to get the convergence of our potential.

Proposition 1. *Suppose* $1 < \beta_-$, *and* $\beta_+\alpha^2 \log(4(\beta_+ + 1)) < \epsilon_m$, *then:*

$$Z(p_0) := \textstyle\sum_{p\in\Omega(p_0)} e^{-V_1(p)} < +\infty$$

Proof. We can partition $Z(p_0)$ by number of synchs:

$$\begin{aligned} Z(p_0) &:= \textstyle\sum_n \sum_{p\in\Omega_n(p_0)} e^{-V_1(p)} \\ &\leq \textstyle\sum_n e^{-\epsilon_m n\log n/\log 4(\beta_+ + 1)} \cdot |T(n)| \quad \text{by Lemma 4} \end{aligned}$$

By Lemma 5, the logarithm of the general term of the upper series is equivalent to $-\epsilon_m n\log n/\log 4(\beta_+ + 1) + \beta_+\alpha^2 O(n\log n)$, so both series converge if $\epsilon_m > \delta\alpha^2\log(4(\beta_+ + 1))$. □

We can summarise our findings:

Theorem 1. *Consider a reversible process $\varnothing \cdot p_0$ equipped with a rate function q which satisfies the §4.1-constraints of the V_1 potential; suppose that for any fork event in $\Omega(p_0)$, the thread increment δ verifies $0 < \delta \leq \beta_+$ for some constant β_+, and assume that for $\boldsymbol{a} \in A^\star$, V_1 stipulates a synch cost $\epsilon_{\boldsymbol{a}}$ which is at least $\max_{\boldsymbol{a} \in A^\star} |\boldsymbol{a}|^2 \cdot \beta_+ \log(4(\beta_+ + 1))$.*

Then q has an equilibrium on $\Omega(p_0)$ defined as $\pi(p) \propto e^{-V_1(p)}$.

6.4 Discussion

This concludes the comparison of the potentials introduced in §4. Unlike the potential V_0 which is not enough to control growth (as we have seen in §5), V_1 which forces forking costs to increase with the size of the local memory will, when parametrized suitably, lead to an equilibrium.

Note that 1) the condition given above on the minimal energy cost is a sufficient one, and might not be necessary (we don't know at the time of writing), 2) in particular, to obtain refined effects on the equilibrium population of certain level sets, and therefore modulate the search, one might need more flexibility. Whether this is possible and useful remains to be seen.

As said in §2, for general reasons in the theory of continuous-time Markov chains, the symmetry of the underlying transition graph guarantees that the probabilistic state of the system converges to the invariant probability $\pi(p)$ defined above. This avoids co-Zenoid situations where the probability of return to a p is 1, while the mean return time is actually infinite. Here we are guaranteed finite mean return times. This form of probabilistic termination, which does not lose itself in infinite branches, is the technical definition of exhaustivity. It is easy to construct examples (and we have seen one earlier in §5) where one has an invariant measure, and almost certain returns, but infinite mean return times. This is why it is fundamental to prove the convergence of $Z(p_0)$.

The restriction to a minimum branching degree $\beta^- > 0$ is not very strong, as one can always use some padding with null processes to make the minimal forking degree higher. Nevertheless, it would be nice to have a more elegant way to deal with non-expansive forks, as surely they cannot seriously stand in the way of convergence.

7 Conclusion

There has been a lingering desire in concurrency theory for a metaphor of computation as a physical process. We present here evidence that one can promote this metaphor to an operational conceptualization of a certain, arguably rather abstract, type of distributed programming. Specifically, we have shown how a slightly generalized version of reversible CCS can be equipped with a distributed potential. This potential is parametrized by costs for different types of synchronizations, in a way that any process eventually reaches a probabilistic equilibrium over its reachable state space - provided that the rate at which processes fork decreases exponentially as a function of the size of the local history.

Much remains to be done.

It would be interesting to see how our findings can be adjusted to a less abstract model of distributed computing, and whether one can find examples where this technique solves problems. We intend to seriously pursue such examples in the future, perhaps in the field of multi-party and simultaneous transactions. To talk about efficiency, solutions, and examples, one needs to make room for the inclusion of irreversible synchronizations expressing success states. This, we have done already in the qualitative case [2], and the extension should be straightforward.

Another natural companion question is to optimize parameters for efficiency of the search mechanism. Instead of minimizing the time to irreversible synchs (aka commits), which begs the question, one could use as a proxy the following objective. Namely to maximize the *equilibrium residency time* of the search reflected on success states ps which live on the boundary ∂X of the fully reversible state space:

$$\text{argmax}\,\boldsymbol{\epsilon}.\ \sum_{p\in\partial X} \pi(\boldsymbol{\epsilon}, p) = \int \mathbf{1}_{\partial X}\, d\pi$$

(by reflected search we mean that irreversible synchs are de-activated, and, hence, the process bounces off the success boundary) where π is the equilibrium probability, and $\boldsymbol{\epsilon}$ its energy vector. Such quantities are nice optimization targets as they can estimated via the ergodic theorem by the averages $\frac{1}{n}\sum \mathbf{1}_{\partial X}(X_k)$, ie the empirical time of residence in a success state (under reflective regime). From there on, it seems one might be in a good position to interface with machine learning techniques to discover efficient parametrizations.

One can also think of this result as a termination one, more in line with the tradition of rewriting and proof-theory. Of course, it is a kind of termination, just as in actual physical systems, which does not mean the complete disappearance of any activity in the system, but rather the appearance of a steady or stable form of activity. As such, it introduces a discourse on resources which is not the one commonly offered in relation to termination proofs in the context of programming languages and rewriting systems, where one tries to limit copies, sizes, and iterations. There has been a thread of research studying termination by various typing systems in process languages (for a recent example, see Ref. [6]) - here we propose what seems a fundamentally different way to achieve the same, in a probabilistic setting, and one wonders if perhaps there is a fruitful relationship to be established.

Finally, it seems that one could squeeze more out of the statistical physical metaphor, and start thinking about the concepts of temperature (which here is degenerate, as it only measures the energy scale) as they are used in the context of sequential simulated annealing algorithms, where the potential can change over time.

References

1. Danos, V., Krivine, J.: Reversible communicating systems. In: Gardner, P., Yoshida, N. (eds.) CONCUR 2004. LNCS, vol. 3170, pp. 292–307. Springer, Heidelberg (2004)

2. Danos, V., Krivine, J.: Transactions in RCCS. In: Abadi, M., de Alfaro, L. (eds.) CONCUR 2005. LNCS, vol. 3653, pp. 398–412. Springer, Heidelberg (2005)
3. Danos, V., Krivine, J., Tarissan, F.: Self-assembling trees. Electr. Notes Theor. Comput. Sci. 175(1), 19–32 (2007)
4. Danos, V., Oury, N.: Equilibrium and termination. In: Cooper, S.B., Panangaden, P., Kashefi, E. (eds.) Proceedings Sixth Workshop on Developments in Computational Models: Causality, Computation, and Physics. EPTCS, vol. 26, pp. 75–84 (2010)
5. Danos, V., Oury, N.: Equilibrium and termination II: the case of *Petri Nets*. Mathematical Structures in Computer Science (to appear, 2011)
6. Demangeon, R., Hirschkoff, D., Kobayashi, N., Sangiorgi, D.: On the complexity of termination inference for processes. In: Barthe, G., Fournet, C. (eds.) TGC 2007. LNCS, vol. 4912, pp. 140–155. Springer, Heidelberg (2008)
7. Diaconis, P.: The Markov chain Monte-Carlo revolution. AMS 46(2), 179–205 (2009)
8. Krivine, J.: Algébres de Processus Réversible - Programmation Concurrente Déclarative. Ph.D. thesis, Université Paris 6 & INRIA Rocquencourt (November 2006)
9. Krivine, J.: A verification algorithm for declarative concurrent programming. CoRR abs/cs/0606095 (2006)
10. Kwiatkowski, M., Stark, I.: The continuous π-calculus: A process algebra for biochemical modelling. In: Heiner, M., Uhrmacher, A.M. (eds.) CMSB 2008. LNCS (LNBI), vol. 5307, pp. 103–122. Springer, Heidelberg (2008)
11. Lanese, I., Mezzina, C.A., Stefani, J.-B.: Reversing higher-order pi. In: Gastin, P., Laroussinie, F. (eds.) CONCUR 2010. LNCS, vol. 6269, pp. 478–493. Springer, Heidelberg (2010)
12. Lévy, J.J.: Réductions correctes et optimales dans le λ-calcul. Ph.D. thesis, Thèse de doctorat d'État, Université Paris 7 (1978)
13. Metropolis, N., Rosenbluth, A., Rosenbluth, M., Teller, A., Teller, E., et al.: Equation of state calculations by fast computing machines. The Journal of Chemical Physics 21(6), 1087 (1953)
14. Milner, R.: Communication and concurrency. International Series on Computer Science. Prentice Hall, Englewood Cliffs (1989)
15. Norris, J.: Markov chains. Cambridge University Press, Cambridge (1998)
16. Ollivier, J., Shahrezaei, V., Swain, P.: Scalable rule-based modelling of allosteric proteins and biochemical networks. PLoS Computational Biology 6(11) (2010)
17. Prasad, K.V.S.: Combinators and bisimulation proofs for restartable systems. Ph.D. thesis, University of Edinburgh (1987)
18. Schmidt, M.: Generalized j-factorial functions, polynomials, and applications. Journal of Integer Sequences 13(2), 3 (2010)
19. Streater, R.: Statistical dynamics. Imperial College Press, London (1995)

Solving Fixed-Point Equations by Derivation Tree Analysis*

Javier Esparza and Michael Luttenberger

Institut für Informatik, Technische Universität München, 85748 Garching, Germany
{esparza,luttenbe}@in.tum.de

Abstract. Systems of equations over ω-continuous semirings can be mapped to context-free grammars in a natural way. We show how an analysis of the derivation trees of the grammar yields new algorithms for approximating and even computing exactly the least solution of the system.

1 Introduction

We are interested in computing (or approximating) solutions of *systems of fixed-point equations* of the form

$$\begin{aligned} X_1 &= f_1(X_1, X_2, \ldots, X_n) \\ X_2 &= f_2(X_1, X_2, \ldots, X_n) \\ &\vdots \\ X_n &= f_n(X_1, X_2, \ldots, X_n) \end{aligned}$$

where $X_1, X_2, \ldots, X_n$ are variables and $f_1, f_2, \ldots, f_n$ are n-ary functions over some common domain S. Fixed-point equations are a natural way of describing the equilibrium states of systems with n interacting components (particles, populations, program points, etc.). Loosely speaking, the function f_i describes the next state of the i-th component as a function of the current states of all components, and so the solutions of the system describe the equilibrium states. In computer science, a prominent example of fixed-point equations are dataflow equations. In this case, the system is a program, the components are the control points of the program, the common domain is some universe of data facts, and the f_i's describe the dataflow to (or from) the i-th control point to all other control points in one program step (see e.g. [NNH99]).

Without further assumptions on the functions $f_1, \ldots, f_n$ and the domain S, little can be said about the existence and computability of a solution. In the last years we have studied *polynomial systems* (systems in which the f_i's are multivariate polynomials) in which S is an *ω-continuous semiring*, a well-known algebraic structure [Kui97]. This setting has the advantage that the system always has a least solution, a result usually known as Kleene's theorem [Kui97], which allows us to concentrate on the task of approximating or computing it.

This paper surveys recent results [EKL07a, EKL07b, EKL08a, EKL10, Lut10] and some work in progress [Lut]. The presentation emphasizes the connection between the

* This work was partially supported by the project "Polynomial Systems on Semirings: Foundations, Algorithms, Applications" of the Deutsche Forschungsgemeinschaft.

A. Corradini, B. Klin, and C. Cîrstea (Eds.): CALCO 2011, LNCS 6859, pp. 19–35, 2011.

algebraic study of equations and formal language theory. In fact, our main goal is to show how equations can be mapped to context-free grammars in a natural way,[1] and how an analysis of the derivation trees of the grammars yields new algorithms for approximating and even computing the least solution of the equations.

The paper is structured as follows. After some preliminaries (Section 2), we introduce a known result (Section 3): the least solution of a system is equal to the *value* of its associated grammar, where the value of a grammar is defined as the sum of the values of its derivation trees, and the value of a derivation tree is defined as the (ordered) product of its leaves. This connection allows us to approximate the least solution of a system by computing the values of "approximations" to the grammar. Loosely speaking, a grammar G_1 approximates G_2 if every derivation tree of G_1 is a derivation tree of G_2 up to irrelevant details. We show that Kleene's theorem, which not only proves the existence of the least solution, but also provides an algorithm for approximating it, corresponds to approximating G by grammars $G^{[1]}, G^{[1]}, \ldots$ where $G^{[h]}$ generates the derivation trees of G of height h. We then introduce (Section 4) a faster approximation by grammars $H^{[1]}, H^{[1]}, \ldots$ where $H^{[h]}$ generates the derivation trees of G of *dimension* h [EKL08a, EKL10]. We show that this approximation is a generalization of Newton's method for approximating the zero of a differentiable function, and present a new result about its convergence speed when multiplication is commutative [Lut][2]. In the final part of the paper (Section 5) we apply the insights obtained from Newton's and Kleene's approximation to different classes of *idempotent* semirings, i.e., semirings in which the law $a + a = a$ holds. We obtain approximation algorithms that actually provide the exact solution after a finite number of steps.

2 Polynomial Equations Over Semirings

For the definition of polynomial systems we need a set S and two binary operations on S, *addition* and *multiplication*, satisfying the usual associativity and distributivity laws:

Definition 1. *A* semiring *is a tuple* $\langle S, +, \cdot, 0, 1\rangle$, *where (i)* S *is a set with* $0, 1 \in S$ *called the carrier of the semiring, (ii)* $\langle S, +, 0\rangle$ *is a commutative monoid with neutral element* 0, *(iii)* $\langle S, \cdot, 1\rangle$ *is a monoid with neutral element* 1, *(iv)* 0 *is an annihilator, i.e.* $0 \cdot a = a \cdot 0 = 0$ *for all* $a \in S$, *and (v) multiplication distributes over addition from the left and from the right.*

When addition and multiplication are clear from the context, we identify a semiring $\langle S, +, \cdot, 0, 1\rangle$ with its carrier S . We also often write ab for $a \cdot b$. A polynomial over a semiring S is a finite sum of finite products of variables and semiring elements. For instance, if X, Y denote variables and $a, b, c \in S$ denote semiring elements, then $aYb + XYXc$ is a polynomial. Notice that multiplication is not required to be commutative, and so we cannot represent a single polynomial in *monomial form*, i.e. as a finite sum of products of the form $aX_1 \cdots X_m$, where $a \in S$ is a coefficient and $X_1 \cdots X_n$ is

[1] We do not claim to be the first to come up with this connection. See e.g. [BR82, Boz99].

[2] The proof has not yet been published, but we feel confident it is correct.

a product of variables. Things change for polynomial *systems*. In this case, we may introduce auxiliary variables following the procedure used to put a context-free grammar in Chomsky normal form; for instance, the univariate equation

$$X = aXb + XcX + e$$

which is not in monomial form, can be transformed into the multivariate system

$$X = aXY + XZ + e \qquad Y = b \qquad Z = cX$$

which *simulates* the original system w.r.t. the X-component. Although our results do not require systems to be in monomial form, for this survey we always assume it to simplify notation.

Polynomial systems over semirings may have no solution. For instance, $X = X + 1$ has no solution over the reals. However, if we extend the reals with a maximal element ∞ (correspondingly adapting addition and multiplication so that these operations still are monotone), we can consider ∞ a solution of this equation. We restrict ourselves to semirings with these "limit" elements.

Definition 2. *Given a semiring S, define the binary relation $\sqsubseteq$ by*

$$a \sqsubseteq b :\Leftrightarrow \exists d \in S : a + d = b.$$

A semiring S is ω-continuous *if (i) $\langle S, \sqsubseteq\rangle$ is a ω-complete partial order, i.e., the supremum $\sup_{i\in\mathbb{N}} a_i$ of any ω-chain $a_0 \sqsubseteq a_1 \sqsubseteq \ldots$ exists in S w.r.t. the partial order $\sqsubseteq$ on S; and (ii) both addition and multiplication are ω-continuous in both arguments, i.e., for any ω-chain $(a_i)_{i\in\mathbb{N}}$ and semiring element a:*

$$a + \sup_{i\in\mathbb{N}} a_i = \sup_{i\in\mathbb{N}}(a + a_i) \qquad \text{and} \qquad a \cdot \sup_{i\in\mathbb{N}} a_i = \sup_{i\in\mathbb{N}}(a \cdot a_i)$$

and symmetrically in the other argument.

We adopt the following convention:

If not stated otherwise, S denotes an ω-continuous semiring $\langle S, +, \cdot, 0, 1\rangle$.

In an ω-continuous semiring we can extend the summation operator $\sum$ from finite to countable families $(a_i)_{i\in I}$ by defining

$$\sum_{i\in I} a_i := \sup\left\{\sum_{i\in F} a_i \mid F \subseteq I, |F| < \infty\right\}.$$

It then can be shown that $\sum$ is still associative and multiplication distributes over $\sum$ from both the left and the right [DKV09]. Note that in a ω-continuous semiring S we have $0 \sqsubseteq a$ for all $a \in S$. Hence, the reals extended by ∞ do not constitute an ω-continuous semiring w.r.t. the canonical order $\leq$, but the *nonnegative* reals do.

It is easy to see that *Kleene's fixed-point theorem* applies to polynomial systems over ω-continuous semirings:[3]

[3] The theorem is often also attributed to Tarski. In fact, it can be seen as a slight extension of Tarski's fixed-point theorem for complete lattices [Tar55], or as a particular case of Kleene's first recursion theorem [Kle38].

Theorem 1 ([Kui97]). *Every polynomial system* $X = f(X)$ *over an* ω*-continuous semiring has a least solution* μf *w.r.t.* $\sqsubseteq$*, and* μf *is equal to the supremum of the* Kleene sequence*:*

$$0 \sqsubseteq f(0) \sqsubseteq f(f(0)) \sqsubseteq \ldots \sqsubseteq f^i(0) \sqsubseteq f^{i+1}(0) \sqsubseteq \ldots \tag{1}$$

Observe that Kleene's theorem not only guarantees the existence of the least fixed point, but also provides a first approximation method, usually called fixed-point iteration.

3 From Equations to Grammars

We illustrate by means of examples how Kleene's theorem allows us to connect polynomial systems of equations with context-free grammars and the derivation trees associated with them. For a formal presentation see e.g. [Boz99, EKL10, DKV09].

Consider the equation

$$X = \frac{1}{4}X^2 + \frac{1}{4}X + \frac{1}{2} \tag{2}$$

over the nonnegative reals extended by ∞, which is an ω-continuous semiring. The equation is equivalent to $(X-1)(X-2) = 0$, and so its least solution is $X = 1$. We introduce identifiers a, b, c for the coefficients, yielding the formal equation

$$X = f(X) := aX^2 + bX + c\,. \tag{3}$$

We say that (2) is an *instance* of (3). Formally, instances correspond to valuations. A *valuation* is a mapping $V\colon \Sigma \to S$, where Σ is the set of identifiers of the formal equation (in our example $\Sigma = \{a, b, c\}$), and S is an ω-continuous semiring. So (2) is the instance of (3) for the valuation where S are the nonnegative reals with ∞, $V(a) = V(b) = 1/4$, and $V(c) = 1/2$. We denote the instance for V by $X = f_V(X)$, and its least solution by μf_V.

We associate a context-free grammar with Equation (3) by reading every summand of the right-hand side as a production:

$$G\colon X \to aXX \mid bX \mid c\,, \tag{4}$$

We denote by $T(G)$ the set of derivation trees of G. We depict derivation trees in the standard way as ordered finite trees and say that a derivation tree $t \in T(G)$ *yields* a word $a_1 a_2 \ldots a_l \in \Sigma^*$ if the i-th leaf from the left of t is labeled by a_i. For instance, the following trees t_1, t_2, t_3, t_4 yield the words c, bc, acc, $abcc$, respectively:

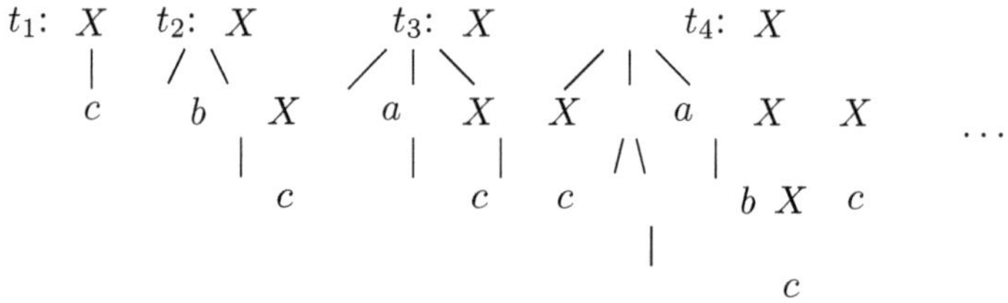

Note that the grammar G for an univariate equation has only a single nonterminal, and thus the axiom of G is clear. In the case of a multivariate polynomial system $X = f(X)$, we construct in the same way a context-free grammar G, but without an explicit axiom. $T(G)$ stands for the union of the sets $T_1(G), \ldots, T_n(G)$ of derivation trees corresponding to setting $X_1, \ldots, X_n$ as axiom. In the following, we do not explicitly distinguish between the univariate and the multivariate case, and adopt the convention:

> Given a grammar G without explicit axiom, a result regarding G or $T(G)$ is to be understood as holding for any possible choice of the axiom.

A valuation $V\colon \Sigma \to S$ extends naturally to the derivation trees of G: for a tree $t \in T(G)$ yielding $a_1 a_2 \ldots a_l$, we define

$$V(t) = V(a_1) \cdot V(a_2) \cdot \ldots \cdot V(a_l),$$

and for a set of trees $T \subseteq T(G)$, we define $V(T) = \sum_{t \in T} V(t)$. For instance, for the trees t_1, t_2, t_3, t_4 shown in the picture above and the valuation mapping a, b, c to $1/4$, $1/4$, and $1/2$, respectively, we get

$$V(\{t_1, t_2, t_3, t_4\}) = V(t_1) + V(t_2) + V(t_3) + V(t_4) = 1/2 + 1/8 + 1/16 + 1/64 = 45/64.$$

Now, as a last step, we can extend V to a valuation of the complete grammar.

Definition 3. *Let G be the grammar of a formal polynomial system $X = f(X)$, and let $V\colon \Sigma \to S$ be a valuation over some ω-continuous semiring S. We define $V(G) = V(T(G)) = \sum_{t \in T(G)} V(t)$ over S.*

The starting point of our paper is a well-known result stating that, given a formal polynomial equation $X = f(X)$ and a valuation V, the least solution of $X = f_V(X)$ is equal to $V(G)$ (see e.g. [Boz99], and, independently, [EKL10]; the essence of the result can be traced back to [BR82, Tha67, CS63]). In other words, the least solution can be obtained by adding the values under V of all its derivation trees.

Theorem 2 ([Boz99, EKL10]). *Let $X = f(X)$ be a formal polynomial system with a set Σ of formal identifiers, and let $V\colon \Sigma \to S$ be a valuation. Then:*

$$\mu f_V = V(G). \tag{5}$$

By our convention, for a multivariate system Theorem 5 states that for every variable X_i the X_i-component of μf_V is given by the infinite sum of all evaluated derivation trees derivable from X_i w.r.t. G.

We sketch a proof of this theorem for the particular case of equation (3). Let us "unfold" the grammar G of (4) by augmenting the nonterminal X with a counter keeping track of the height of a derivation:

$$
\begin{aligned}
X^{\langle 1\rangle} &\to c\\
X^{[1]} &\to X^{\langle 1\rangle}\\
X^{\langle 2\rangle} &\to aX^{\langle 1\rangle}X^{\langle 1\rangle} \mid bX^{\langle 1\rangle}\\
X^{[2]} &\to X^{\langle 2\rangle} \mid X^{[1]}\\
X^{\langle 3\rangle} &\to aX^{\langle 2\rangle}X^{\langle 2\rangle} \mid aX^{[1]}X^{\langle 2\rangle} \mid aX^{\langle 2\rangle}X^{[1]} \mid bX^{\langle 2\rangle}\\
X^{[3]} &\to X^{\langle 3\rangle} \mid X^{[2]}\\
&\ \vdots\\
X^{\langle h\rangle} &\to aX^{\langle h-1\rangle}X^{\langle h-1\rangle} \mid aX^{[h-2]}X^{\langle h-1\rangle} \mid aX^{\langle h-1\rangle}X^{[h-2]} \mid bX^{\langle h-1\rangle}\\
X^{[h]} &\to X^{\langle h\rangle} \mid X^{[h-1]}\\
&\ \vdots
\end{aligned}
$$

Let $G^{[h]}$ ($G^{\langle h\rangle}$) be the grammar consisting of those "unfolded" rules whose left-hand side is given by one of the variables of $\mathcal{X}^{[h]} = \{X^{\langle 0\rangle}, X^{[0]}, \ldots, X^{\langle h\rangle}, X^{[h]}\}$, taking $X^{[h]}$ ($X^{\langle h\rangle}$) as axiom.[4] An easy induction shows the existence of a bijection between $T(G^{[h]})$ ($T(G^{\langle h\rangle})$) and the trees of $T(G)$ of height at most (exactly) h. In fact, it is easy to see that $G^{[h]}$ ($G^{\langle h\rangle}$) and G are both unambiguous[5], and the bijection just assigns to a tree of $T(G^{[h]})$ the unique tree of G yielding the same word. For instance, the tree of $G^{[3]}$ shown on the left of the figure below is mapped to the tree of G of height 3 shown on the right:

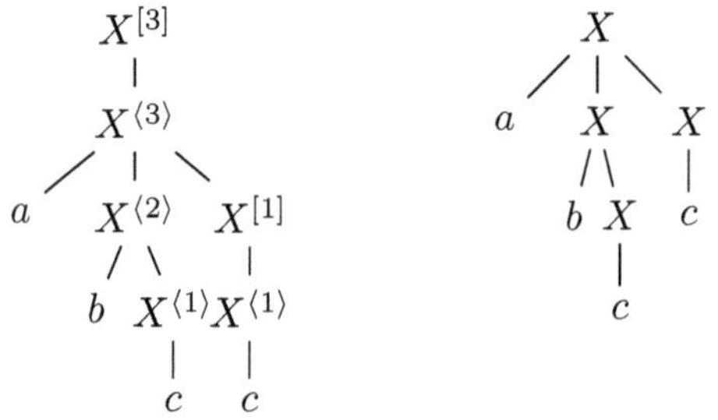

Hence, $V(G^{[h]})$ ($V(G^{\langle h\rangle})$) is the contribution to $V(G)$ of the derivation trees of height at most (exactly) h to $V(G)$. It therefore suffices to show that $f_V^h(0) = V(G^{[h]})$. Note that by the extension of V to derivation trees, $V(G^{[h]})$ and $V(G^{\langle h\rangle})$ can be computed recursively as follows (with $a_V := V(a), b_V := V(b), c_V := V(c)$):

$$
\begin{aligned}
V(G^{\langle h\rangle}) &= a_V V(G^{\langle h-1\rangle})^2 + a_V V(G^{[h-2]})V(G^{\langle h-1\rangle})\\
&\quad + a_V V(G^{\langle h-1\rangle})V(G^{[h-2]}) + b_V V(G^{\langle h-1\rangle})\\
V(G^{[h]}) &= V(G^{[h-1]}) + V(G^{\langle h\rangle})
\end{aligned}
$$

[4] In the multivariate case, for every choice Z of the axiom of G, define $G^{[h]}$ ($G^{\langle h\rangle}$) analogously with $Z^{[h]}$ ($Z^{\langle h\rangle}$) as axiom.

[5] A grammar G is *unambiguous* if for every word $w \in L(G)$ there is a unique derivation (tree) w.r.t. G.

where $V(G^{\langle 1 \rangle}) = c_V$ and $V(G^{[-1]}) := V(G^{[0]}) := 0$.

Now, an easy induction proves the stronger claim that

$$f_V^h(0) = V(G^{[h]}) \text{ and } f_V^h(0) = f_V^{h-1}(0) + V(G^{\langle h \rangle})$$

and by Kleene's theorem we get $\mu f_V = \sup_{h \in \mathbb{N}} f_V^h(0) = \sup_{h \in \mathbb{N}} V(G^{[h]}) = V(G)$.

Notice that this proof not only reduces the problem of computing the least solution of $X = f_V(X)$ to the problem of computing $V(G)$, it also shows that:

Kleene's approximation sequence is the result of evaluating the derivation trees of G by increasing height.

4 Newton's Approximation

In the last section we have constructed grammars $G^{\langle 1 \rangle}, G^{\langle 2 \rangle}, \ldots$ that, loosely speaking, "partition" the derivation trees of G according to height. Formally, there is a bijection between the derivation trees of $G^{\langle h \rangle}$ and the derivation trees of G of height exactly h. Using these grammars we can construct grammars $G^{[1]}, G^{[2]}, \ldots$ such there is a bijection between the derivation trees of $G^{[h]}$ and the derivation trees of G of height at most h. The grammars $G^{[h]}$ allow us to iteratively compute approximations $V(G^{[h]})$ to $V(G) = \mu f_V$.

We can transform this idea into a general principle for developing approximation algorithms. Given a grammar G, we say that a sequence $(G^{\langle i \rangle})_{i \in \mathbb{N}}$ of grammars *partitions* G if $T(G_i) \cap T(G_j) = \emptyset$ for $i \neq j$, and there is a bijection between $\bigcup_{i \in \mathbb{N}} T(G^{\langle i \rangle})$ and $T(G)$ that preserves the yield, i.e., the yield of a tree is equal to the yield of its image under the bijection.. Every sequence $(G^{\langle i \rangle})_{i \in \mathbb{N}}$ that partitions G induces another sequence $(G^{[i]})_{i \in \mathbb{N}}$, defined as in the previous section, such that $T(G^{[i]}) = \bigcup_{j \leq i} T(G^{\langle i \rangle})$. We say that $(G^{[i]})_{i \in \mathbb{N}}$ *converges* to G. The following proposition follows easily from these definitions.

Proposition 1. *Let $X = f(X)$ be a formal polynomial system with a set Σ of formal identifiers, and let G be the context-free grammar associated to it. If a sequence $(G_i)_{i \in \mathbb{N}}$ of grammars converges to G, then*

$$\mu f_V = \sup_{i \in \mathbb{N}} V(G_i) \, .$$

The unfolding of the last section assigns to *every* variable in the right-hand-side of a production a lower index (height) than the variable on the left-hand-side, which forbids any kind of unbounded recursion in the unfolded grammars. We now unfold the grammar G so that *nested-linear recursion* is allowed [EKL08b, GMM10]. Again we augment each variable X by a counter, yielding variables $X^{\langle i \rangle}, X^{[i]}$. A derivation starting from $X^{\langle i \rangle}$ ($X^{[i]}$) allows for exactly i (at most i) nested-linear recursions. For the grammar (4) we get:

$$
\begin{aligned}
X^{\langle 1\rangle} &\rightarrow c \mid bX^{\langle 1\rangle} \\
X^{[1]} &\rightarrow X^{\langle 1\rangle} \\
X^{\langle 2\rangle} &\rightarrow aX^{\langle 1\rangle}X^{\langle 1\rangle} \mid aX^{[1]}X^{\langle 2\rangle} \mid aX^{\langle 2\rangle}X^{[1]} \mid bX^{\langle 2\rangle} \\
X^{[2]} &\rightarrow X^{\langle 2\rangle} \mid X^{[1]} \\
&\vdots \\
X^{\langle i\rangle} &\rightarrow aX^{\langle i-1\rangle}X^{\langle i-1\rangle} \mid aX^{[i-1]}X^{\langle i\rangle} \mid aX^{\langle i\rangle}X^{[i-1]} \mid bX^{\langle i\rangle} \\
X^{[i]} &\rightarrow X^{\langle i\rangle} \mid X^{[i-1]} \\
&\vdots
\end{aligned}
$$

It is instructive to compare the productions of $X^{\langle h\rangle}$ in Kleene's approximation, and the productions of $X^{\langle i\rangle}$ as defined above:

$$
\begin{aligned}
X^{\langle h\rangle} &\rightarrow aX^{\langle h-1\rangle}X^{\langle h-1\rangle} \mid aX^{[h-2]}X^{\langle h-1\rangle} \mid aX^{\langle h-1\rangle}X^{[h-2]} \mid bX^{\langle h-1\rangle} \\
X^{\langle i\rangle} &\rightarrow aX^{\langle i-1\rangle}X^{\langle i-1\rangle} \mid aX^{[i-1]}X^{\langle i\rangle} \mid aX^{\langle i\rangle}X^{[i-1]} \mid bX^{\langle i\rangle}
\end{aligned}
$$

Let $H^{[i]}$ ($H^{\langle i\rangle}$) denote the grammar with axiom $X^{[i]}$ ($X^{\langle i\rangle}$) and consisting of those productions "reachable" from $X^{[i]}$ ($X^{\langle i\rangle}$) in the above unfolding. As in the case of Kleene approximation, we can easily show by induction that $H^{[i]}$ is unambiguous, and that the mapping assigning to a tree of $T(H^{[i]})$ the unique tree of G deriving the same word is a bijection. Since every word of $L(G)$ belongs to $L(H^{[i]})$ for some $i \in \mathbb{N}$, the sequence $(H^{[i]})_{i\in\mathbb{N}}$ converges to G.

Again, we can compute $V(H^{\langle i\rangle})$ and $V(H^{[i]})$ recursively where $\mu X.g(X)$ denotes the the least solution of the equation $X = g(X)$ (again $a_V := V(a), \ldots$):

$$
\begin{aligned}
V(H^{\langle i\rangle}) &:= \mu X.(\, a_V X V(H^{[i-1]}) + a_V V(H^{[i-1]})X + b_V X + a_V V(H^{\langle i-1\rangle})^2 \,) \\
V(H^{[i]}) &:= V(H^{\langle i\rangle}) + V(H^{\langle i-1\rangle})
\end{aligned}
$$

where $V(H^{\langle 1\rangle}) := \mu X.(\, b_V X + c_V \,)$.

At this point the reader may ask whether any progress has been made: instead of solving the polynomial system $X = f_V(X)$ we have to solve the polynomial systems $X = g_i(X)$. However, these systems are linear, while $X = f_V(X)$ may be nonlinear, and in ω-continuous semirings solving linear equations reduces to computing the Kleene star $a^* := \sum_{i\in\mathbb{N}} a^i$. So, for any ω-continuous semiring which allows for an efficient computation of a^*, this approximation scheme becomes viable. For instance, over the nonnegative reals we have $a^* = \frac{1}{1-a}$ if $a < 1$ and $a^* = \infty$ otherwise. Thus, if V is a valuation on the real semiring, then the solution of a linear equation can be easily computed. For the equations above elementary arithmetic yields

$$
V(G^{\langle i\rangle}) := \frac{a_V V(G^{\langle i-1\rangle})^2}{1 - 2a_V V(G^{[i-1]}) - b_V} \qquad V(G^{[i]}) := V(G^{\langle i\rangle}) + V(G^{[i-1]}) \tag{6}
$$

with $V(G^{[1]}) = V(G^{\langle 1\rangle}) := \dfrac{c_V}{1 - b_V}$.

The following table compares the first approximations obtained by using the approximation schemes derived in this and the previos section for our example (2):

Kleene	$V(G^{\langle i \rangle})$	$1/2$	$3/8$	$105/1024$	$\dots$
	$V(G^{[i]})$	$1/2$	$11/16$	$809/1024$	$\dots$
Newton	$V(H^{\langle i \rangle})$	$2/3$	$4/15$	$16/255$	$\dots$
	$V(H^{[i]})$	$2/3$	$14/15$	$254/255$	$\dots$

(7)

It is now time to explain why we call this scheme Newton's approximation. For every valuation V over the reals, the least solution of $X = f_V(X)$ is a zero of the polynomial $g(X) = f_V(X) - X = a_V X^2 + (b_V - 1)X + c_V$. Again, an easy induction shows:

$$V(H^{\langle i \rangle}) = -\frac{g(V(H^{[i-1]}))}{g'(V(H^{[i-1]}))} \qquad V(H^{[i]}) = -\frac{g(V(H^{[i-1]}))}{g'(V(H^{[i-1]}))} + V(H^{[i-1]})$$

starting now from $V(H^{\langle 0 \rangle}) = V(H^{[0]}) = 0$, where $g'(X)$ denotes the derivative of g – in our example: $g'(X) = 2a_V X + b_V - 1$. These equations are nothing but Newton's classical method for approximating the solution of $g(X) = 0$ starting at the point 0, and this is not a coincidence: we have recently shown that this relation holds for every polynomial equation $X = f_V(X)$ over the nonnegative reals [EKL10]. So this approximation scheme generalizes Newton's method to equations over arbitrary ω-continuous semirings.

Recall that Kleene's approximation corresponds to evaluating the derivation trees of G by increasing height. The question whether we can charaterize Newton's approximation in a similar way has been answered positively in [EKL10]. We need the notion of dimension of a derivation tree.

Definition 4. *Let t be a derivation tree. If t consists of a single node, then its dimension is* 1*. Otherwise, let d be the maximal dimension of the children of t. If two or more children have dimension d, then t has dimension* $d+1$*; otherwise, t has dimension d.*

For instance, the derivation tree of the grammar (4) shown below on the left has dimension 3 (its second and third child have dimension 2, because both of them have two children of dimension 1).

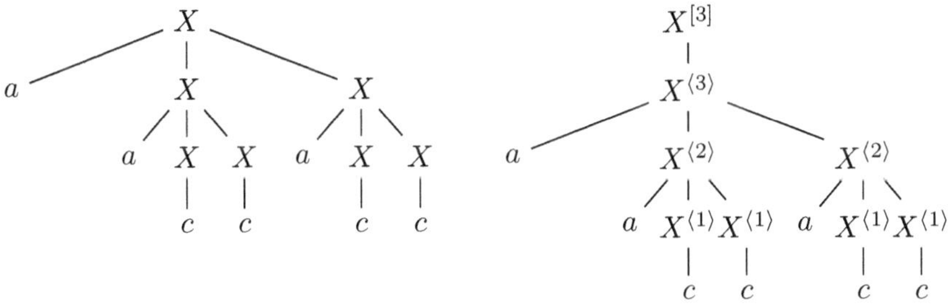

We can prove:

Theorem 3 ([EKL10]). *For every $i \geq 1$, there is a yield-preserving bijection between $T(H^{[i]})$ and the trees of $T(G)$ of dimension at most i.*

According to this theorem, the tree above must belong to $T(H^{[3]})$ and indeed this is the case, as shown by the derivation tree on the right. Note that along any path from the root to a leaf the sequence of numbers in the superscripts drops atmost by one in each step. One the other hand, moving from a leaf to the root, the superscript only increases from i to $i+1$ at a given node if this very node has at least a second child with superscript i. The superscripts in round (square) brackets happen just to be (an upper bound on) the dimension of the corresponding subtree. So we conclude:

Newton's approximation sequence is the result of evaluating the derivation trees of G by increasing dimension.

4.1 Convergence of Newton's Method in Commutative Semirings

The convergence speed of Newton's method over the reals is well-understood. In many cases – for example (7) – it converges *quadratically*, which in computer science terms means that the approximation error decreases exponentially in the number of iterations. In this section we analyze the convergence speed valid for *arbitrary* commutative semirings, i.e., semirings in which multiplication is commutative.

Recall that, by definition,

$$V(H^{[i]}) = \sum_{t \in T(H^{[i]})} V(t) \qquad \text{and} \qquad V(G) = \sum_{t \in T(G)} V(t)\,.$$

For every $s \in S$, let $\alpha^{[i]}(s)$ be the number of trees $t \in T(H^{[i]})$ such that $V(t) = s$, if the number is finite, and $\alpha^{[i]}(s) = \infty$ otherwise. Define $\alpha(s)$ similarly for $T(G)$. Then we have

$$V(H^{[i]}) = \sum_{s \in S} \sum_{i=1}^{\alpha^{[i]}(s)} s \qquad V(G) = \sum_{s \in S} \sum_{i=1}^{\alpha(s)} s$$

with the convention $\sum_{i=1}^{0} s = 0$. We estimate the convergence speed of Newton's method by analyzing how fast the sequence $(\alpha^{[i]}(s))_{i \in \mathbb{N}}$ converges to $\alpha(s)$. Our result shows that in a system of n equations after $(kn+1)$ iterations of Newton's method we have $\alpha^{[kn+1]}(s) \geq \min\{\alpha(s), k\}$.

Theorem 4 ([Lut]). *Let $X = f(X)$ be a formal polynomial system with n equations, and let V be a valuation over a commutative ω-continuous semiring S. We have $\alpha^{[k \cdot n+1]}(s) \geq \min\{\alpha(s), k\}$ for every $s \in S$ and every $k \in \mathbb{N}$.*

We sketch the proof of the theorem for the (very) special case $n = k = 1$. We have to show $\alpha^{[2]}(s) \geq \min\{\alpha(s), 1\}$, i.e., that $\alpha(s) > 0$ implies $\alpha^{[2]}(s) > 0$ or, equivalently, that for every $t \in T(G)$ some $t' \in T(H^{[2]})$ satisfies $V(t) = V(t')$. As $T(H^{[2]})$ is in bijection with the trees of $T(G)$ of dimension at most 2, it suffices to prove that for every $t \in T(G)$ there is $t' \in T(G)$ of dimension at most 2 such that $V(t') = V(t)$. If t has dimension 1 or 2, we take $t' = t$. Otherwise, we explain how to proceed using grammar (4) and the tree of dimension 3 deriving the word $aaccacc$:

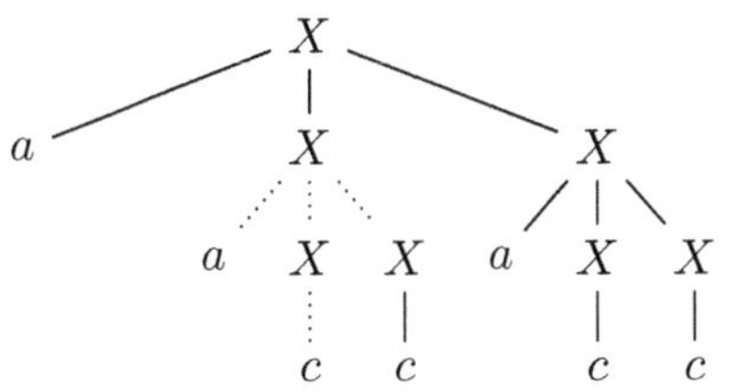

If we remove the dotted subtree (pump tree), the dimension of the second child of the root decreases by 1, and we are left with the tree of dimension 2 shown below, on the left. This tree only derives the word $acacc$, and so the idea is to reinsert the missing subtree so that the result (i) is again a derivation tree w.r.t. G, and (ii) we do not increase the dimension. If we achieve this, then the new tree derives a permutation w of $acacacc$ and, *since the semiring is commutative*, we have $V(w) = V(aaccacc)$. Condition (i) poses no problem in the univariate case, as as all inner nodes correspond to the same variable (nonterminal). In order to satisfy condition (ii), it suffices to pick any subtree derived from X of dimension 2 and replace the edge to its father by the missing dotted subtree as shown below, on the right.

It can be shown that this reallocation of subtrees is also possible in the multivariate case and allows to generate the required number of distinct derivations trees, although additional care is needed in order to satisfy the two conditions.

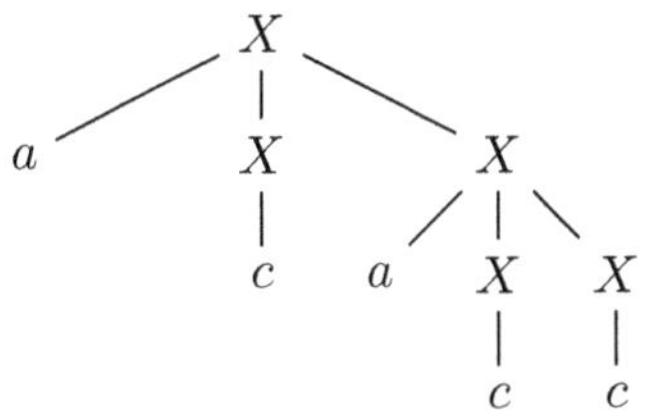

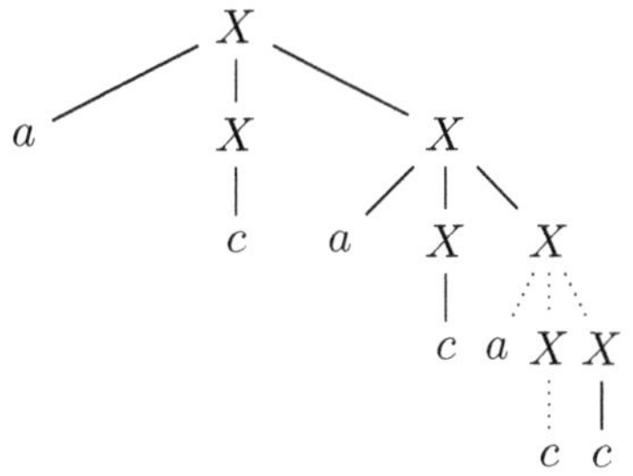

5 Derivation Tree Analysis for Idempotent Semirings

In the previous section, we have seen how to relocate subtrees of a derivation tree in order to reduce its dimension. In commutative semirings, relocating subtrees preserves the value of the tree, and we have used this fact to derive Theorem 4, a quantitative meassure of the speed at which the Newton approximations $V(H^{[i]})$ converge to $V(G)$. In particular, for $k = 1$ we obtain $\alpha^{[n+1]}(s) \geq \min\{\alpha(s), 1\}$ or, equivalently,

For every tree $t \in T(G)$ there is a tree $t' \in V(H^{[i]})$ such that $V(t) = V(t')$.

This has an important consequence for *idempotent* semirings, i.e., for semirings satisfying the identity $a + a = a$ for every $a \in S$. For any valuation V over an idempotent semiring, $V(t) = V(t')$ implies $V(t) + V(t') = V(t')$. So for idempotent and commutative ω-continuous semirings we get $V(G) + V(H^{[n+1]}) = V(H^{[n+1]})$, which together with $V(H^{[n+1]}) \sqsubseteq V(G)$ implies $V(H^{[n+1]}) = V(G)$. It follows:

Theorem 5 ([EKL10]). *Let $X = f(X)$ be a formal polynomial system with n equations. For every valuation V over an idempotent and commutative ω-continuous semiring*

$$\mu f_V = V(H^{[n+1]}) .$$

Intuitively, this result states that in order to compute μf_V we can safely "forget" the derivation trees of dimension greater than $n+1$, which implies that Newton's method terminates after at most $n+1$ iterations.

In the rest of the section we study two further classes of idempotent ω-continuous semirings for which a similar result can be proved: idempotence allows to "forget" derivation trees, and compute the least solution exactly after finitely many steps.

5.1 1-bounded Semirings

A semiring $\langle S, +, \cdot, 0, 1\rangle$ 1-*bounded* if it is idempotent and $a \sqsubseteq 1$ for all $a \in S$.

One-bounded semirings occur, for instance, in probabilistic settings when one is interested in the most likely path between two nodes of a Markov chain. The probability of a path is the product of the probabilities of its transitions, and we are interested on the maximum over all paths. This results in an equation system over the *Viterbi semiring* [DKV09] whose carrier is the interval $[0, 1]$, and has $\max$ and $\cdot$ as addition and multiplication operators, respectively.

We show that over 1-bounded semirings we may "forget" all derivation trees of height greater than n. Fix a formal polynomial system $X = f(X)$ with n equations and valuation V over a 1-bounded semiring. Let G be the associated context-free grammar. G then has also n nonterminals. A derivation tree $t \in T(G)$ is *pumpable* if it contains a path from its root to one of its leaves in which some variable occurs at least twice. Clearly, every tree of height at least $n+1$ is pumpable. It is well-known that a pumpable tree t induces a *pumpable* factorization $w = uvxyz$ of its yield w such that $uv^ixy^iz \in L(G)$ for every $i \geq 0$. In particular, for every $i \geq 0$ there is a derivation tree t^i that (i) yields uv^ixy^iz, and (ii) is derived from the same axiom as t. Now we have

$$\begin{aligned} V(t) + V(t^0) &= V(w) + V(uxz) \\ &= V(u)V(v)V(x)V(y)V(z) + V(u)V(x)V(z) \\ &\sqsubseteq V(u)\, 1\, V(x)\, 1\, V(z) + V(u)V(x)V(z) && \text{(1-boundedness)} \\ &= V(uxz) && \text{(idempotence)} \\ &= V(t^0) \end{aligned}$$

Repeating this procedere as long as possible, we eventually arrive from a pumpable tree t to another tree $\hat{t}$ of height at most n with $V(t) + V(\hat{t}) = V(\hat{t})$. So, denoting by $T^{[n]}(G)$ the trees of G of height at most n, we have

Theorem 6 ([EKL08a]). *For $X = f(X)$ a formal polynomial system in n variables, G its associated grammar, and V any valuation over a 1-bounded semiring, we have:*

$$\mu f_V = V(T(G)) = V(T^{[n]}(G)) = V(G^{[n]}) = f_V^h(0).$$

Since the Kleene sequence converges after at most n steps we can compute the least solution even if the semiring is not ω-continuous.

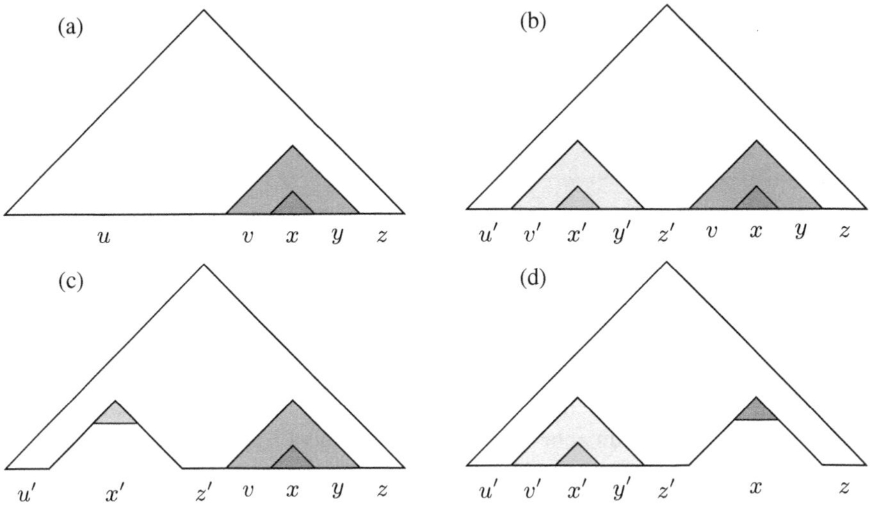

Fig. 1. "Unpumping" trees

5.2 Star-distributive Semirings

In an ω-continuous semiring we can define the *Kleene star* operation by $a^* = \sum_{i \geq 0} a^i$, where $a^0 = 1$. A semiring $\langle S, +, \cdot, 0, 1\rangle$ is *star-distributive* if it is ω-continuous, idempotent, commutative, and $(a+b)^* = a^* + b^*$ holds for every $a, b \in S$.

The *tropical semiring* $\langle \overline{\mathbb{N}}, \min, +, \infty, 0\rangle$ is a prominent example of star-distributive semiring. Actually, any ω-continuous commutative and idempotent semiring in which the natural order $\sqsubseteq$ is total is star-distributive. Indeed, for any two elements a, b, assuming w.l.o.g. $a \sqsubseteq b$, which implies $a^* \sqsubseteq b^*$, we get:

$$(a+b)^* = a^* = a^* + b^* \ .$$

Finally, for a last bit of motivation, a recent paper shows that the computation of several types of provenance of datalog queries can be reduced to the problem of (in our terminology) computing the least solution of a formal polynomial system over a commutative semiring S [GKT07]. Specifically, in the case of the why-provenance S is also idempotent and further augmented by the identity $a^2 = a$ for all $a \in \Sigma$. Clearly, such semirings are star-distributive.

We show that idempotent together with commutativity and star-distributivity allows us to forget most derivation trees of a grammar G associated with a formal polynomial system. In fact, we do not use star-distributivity directly, but the following two identities implied by it in conjunction with commutativity:

Proposition 2. *If S is star-distributive, then for every $a, b \in S$*

$$a^* + b^* = a^* b^* \qquad \textit{and} \qquad (ab^*)^* = a^* + ab^* .$$

Again, fix a formal polynomial system $X = f(X)$ with n equations, and let G be the grammar (without explicit axiom) associated to the system. Further, let V be a valuation over some star-distributive semiring S. We have:

Proposition 3. *Let $t \in T(G)$ be a pumpable tree deriving a word w with pumpable factorization $uvxyz$. Then there are pumpable trees $t_1, \ldots, t_r \in T(G)$ (derived from the same axiom as t) of height at most $n+1$ such that each t_i has a pumpable factorization $u_i v_i x_i y_i z_i$ satisfying*

$$V(w) \sqsubseteq \sum_{i=1}^{r} \sum_{j=0}^{\infty} V(u_i v_i^j x_i y_i^j z_i) . \tag{8}$$

In essence, this proposition tells us that we only need to evaluate derivation trees of G which are either of "unpumpable" (thus, of height at most n) or the result of pumping a fixed factorization in a pumpable derivation tree of height at most $n+1$, while we may "forget" the rest.

We sketch one case of the proof of the proposition. Fix a pumpable tree t with pumpable factorization $uvxyz$ as schematically described in Figure 1(a) where the middle (grey) and the lower (dark grey) part are derived from the same nonterminal, and the top part (white) may be empty. If t has height at most $n+1$, we set $t_1 := t$ and are done. Otherwise, one of the three parts of t (white, grey, or dark grey) contains a subtree of height at least $n+1$. Since G only has n variables, this subtree is also pumpable. We only consider the case that the pumpable tree is on the left part of the white zone (other cases are similar). Then there is a pumpable factorization of u, i.e. $u = u'v'x'y'z'$, as shown in Figure 1(b), and we have $u'(v')^i x'(y)^i z' u v^j x y^j z \in L(G)$ for every $i, j \geq 0$. Applying the properties of star-distributive semirings we get

$$\begin{aligned}
&\sum_{i\geq 0}\sum_{j\geq 0} u'(v')^i x'(y')^i z' v^j x y^j z \\
&= u'x'z'xz(v'y')^*(vy)^* && \text{(commutativity)} \\
&= u'x'z'xz((v'y')^* + (vy)^*) && (a^*b^* = a^* + b^*) \\
&= \sum_{i\geq 0} u'(v')^i x'(y')^i z'xz + \sum_{j\geq 0} u'x'z'v^j x y^j z
\end{aligned}$$

It is easy to see that G has derivation trees t_1 and t_2 (schematically shown in Figure 1(c) and (d)) with pumpable factorizations $w_1 = u_1 v_1 x_1 y_1 z_1$ and $w_2 = u_2 v_2 x_2 y_2 z_2$ given by

$$\begin{array}{lllll}
u_1 = u'x'z' & v_1 = v & x_1 = x & y_1 = y & z_1 = z \\
u_2 = u' & v_2 = v' & x_2 = x' & y_2 = y' & z_2 = z'xz
\end{array}$$

Therefore, we have

$$V(w) \sqsubseteq \sum_{j=0}^{\infty} V(u_1 v_1^j x_1 y_1^j z_1) + \sum_{j=0}^{\infty} V(u_2 v_2^j x_2 y_2^j z_2)$$

If t_1 and t_2 have height at most $n+1$, then we are done; otherwise, the step above is iterated. This concludes the proof sketch.

Let us now see how to apply the proposition. Let $L \subseteq L(G)$ be the language containing

- the words derived by the "unpumpable" trees of G, and
- the words of the form $uv^j xy^j z$, where $uvxyz$ is a pumpable factorization of a tree of $T(G)$ of height at most $n+1$.

Given $w \in L(G)$, there are two possible cases: if w is derived by some "umpumpable" tree, then $w \in L$, and so $V(w) \sqsubseteq V(L)$; if w is derived by some pumpable tree, then by (8) we also have $V(w) \sqsubseteq V(L)$. So $V(w) \sqsubseteq V(L)$ holds for every $w \in L(G)$. Since S is idempotent, we get

$$\begin{aligned} V(G) &= \textstyle\sum_{t \in T(G)} V(t) \\ &= \textstyle\sum_{w \in L(G)} V(w) && \text{(idempotence)} \\ &\sqsubseteq \textstyle\sum_{w \in L(G)} V(L) && (V(w) \sqsubseteq V(L)) \\ &= V(L) && \text{(idempotence and } \omega\text{-continuity)} \end{aligned}$$

Looking at the definition of L it is not difficult to show (see [EKL08a]) that it is subsumed by the words of $L(G)$ derived by the *bamboos* of $T(G)$, a set of derivation trees defined as follows:

Definition 5. *A derivation tree t is a* bamboo *if there is a path leading from the root of t to some leaf of t, the* stem, *such that the height of every subtree of t not containing a node of the stem is at most n.*

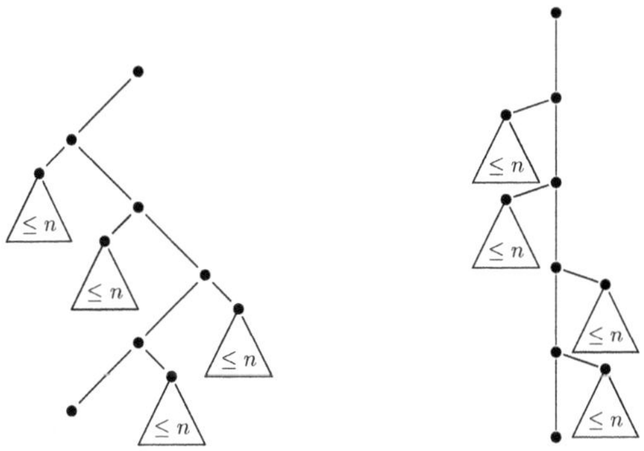

Fig. 2. An example of the structure of a bamboo: it consists of a stem of unbounded length from which subtrees of height at most n sprout; on the right it is shown with its stem straightened

Figure 2 illustrates the definition. The definition of "bamboo" directly leads to an unfolding rule for G: in every rule we limit the recursion depth of all but one terminal to n in the same way as we did in the case of the Kleene approximation. Notice that, since $V(G) = V(L)$ by idempotence, we do not need to ensure that each derivation tree of the unfolded grammars uniquely corresponds to a derivation tree of G. This very much simplifies the definition of the unfolding. For instance, if G has nonterminals $\{X, Y, Z, U, V\}$, then the productions

$$X \to aXY \mid bZ \mid c$$

are unfolded to

$$\begin{aligned} X &\to aXY^{[5]} \mid aX^{[5]}Y \mid bZ \mid c \\ X^{[5]} &\to aX^{[4]}Y^{[4]} \mid bZ^{[4]} \mid c \\ &\cdots \\ X^{[2]} &\to aX^{[1]}Y^{[1]} \mid bZ^{[1]} \mid c \\ X^{[1]} &\to c \end{aligned}$$

The structure of the grammar then again allows us to recursively compute the yield of the derivations trees derived from any nonterminal which gives us an algorithm for computing the least fixed point of any formal polynomial system w.r.t. any valuation over some star-distributive semiring:

Theorem 7 ([EKL08a]). *Let $X = f(X)$ be formal polynomial system consisting of n equations and let V be a valuation over a star-distributive semiring S.*

Then μf_V can be computed using n Kleene iteration steps and then solving a single linear system over S.

This result can be used to compute the provenance of datalog queries over the tropical semiring, a problem that was left open in [GKT07].

6 Conclusions

We have presented some old and some new links between computational algebra and language theory. We have shown how the formal similarity between fixed-point equations and context-free grammars goes very far, and leads to novel algorithms.

The unfolding of grammars leading to Newton's approximation has already found some applications in verification [GMM10, EG11] and Petri net theory [GA11]. Theorem 5 has lead to a simple algorithm for constructing an automaton whose language is Parikh-equivalent to the language of a given context-free grammar [EGKL11]. Theorem 7 was used in [EKL08a] to improve the complexity bound of [CCFR07] for computing the throughput of context-free grammars from $\mathcal{O}(n^4)$ to $\mathcal{O}(n^3)$.

An interesting question is whether the results we have obtained can be proved by purely algebraic means, e.g. without using "tree surgery". Further open questions concern data structures and efficient algorithms for the approximation schemes we have sketched.

Acknowledgments. Many thanks to Volker Diekert for his help with Theorem 4, to Rupak Majumdar for pointing us to applications of semirings to the provenance problem in databases [GKT07], and to Pierre Ganty for many discussions.

References

[Boz99] Bozapalidis, S.: Equational elements in additive algebras. Theory Comput. Syst. 32(1), 1–33 (1999)

[BR82] Berstel, J., Reutenauer, C.: Recognizable formal power series on trees. Theor. Comput. Sci. 18, 115–148 (1982)

[CCFR07] Caucal, D., Czyzowicz, J., Fraczak, W., Rytter, W.: Efficient computation of throughput values of context-free languages. In: Holub, J., Žd̆árek, J. (eds.) CIAA 2007. LNCS, vol. 4783, pp. 203–213. Springer, Heidelberg (2007)

[CS63] Chomsky, N., Schützenberger, M.P.: The Algebraic Theory of Context-Free Languages. In: Computer Programming and Formal Systems, pp. 118–161. North Holland, Amsterdam (1963)

[DKV09] Droste, M., Kuich, W., Vogler, H.: Handbook of Weighted Automata. Springer, Heidelberg (2009)

[EG11] Esparza, J., Ganty, P.: Complexity of pattern-based verification for multithreaded programs. In: POPL, pp. 499–510 (2011)

[EGKL11] Esparza, J., Ganty, P., Kiefer, S., Luttenberger, M.: Parikhs theorem: A simple and direct automaton construction. Inf. Process. Lett. 111(12), 614–619 (2011)

[EKL07a] Esparza, J., Kiefer, S., Luttenberger, M.: An extension of newton's method to ω-continuous semirings. In: Harju, T., Karhumäki, J., Lepistö, A. (eds.) DLT 2007. LNCS, vol. 4588, pp. 157–168. Springer, Heidelberg (2007)

[EKL07b] Esparza, J., Kiefer, S., Luttenberger, M.: On fixed point equations over commutative semirings. In: Thomas, W., Weil, P. (eds.) STACS 2007. LNCS, vol. 4393, pp. 296–307. Springer, Heidelberg (2007)

[EKL08a] Esparza, J., Kiefer, S., Luttenberger, M.: Derivation tree analysis for accelerated fixed-point computation. In: Ito, M., Toyama, M. (eds.) DLT 2008. LNCS, vol. 5257, pp. 301–313. Springer, Heidelberg (2008)

[EKL08b] Esparza, J., Kiefer, S., Luttenberger, M.: Newton's method for ω-continuous semirings. In: Aceto, L., Damgård, I., Goldberg, L.A., Halldórsson, M.M., Ingólfsdóttir, A., Walukiewicz, I. (eds.) ICALP 2008, Part II. LNCS, vol. 5126, pp. 14–26. Springer, Heidelberg (2008)

[EKL10] Esparza, J., Kiefer, S., Luttenberger, M.: Newtonian program analysis. J. ACM 57(6), 33 (2010)

[GA11] Ganty, P., Atig, M.: Approximating Petri net reachability along context-free traces. Technical report, arXiv:1105.1657v1 (2011)

[GKT07] Green, T.J., Karvounarakis, G., Tannen, V.: Provenance semirings. In: PODS, pp. 31–40 (2007)

[GMM10] Ganty, P., Majumdar, R., Monmege, B.: Bounded underapproximations. In: Touili, T., Cook, B., Jackson, P. (eds.) CAV 2010. LNCS, vol. 6174, pp. 600–614. Springer, Heidelberg (2010)

[Kle38] Kleene, S.C.: On notation for ordinal numbers. J. Symb. Log. 3(4), 150–155 (1938)

[Kui97] Kuich, W.: Semirings and Formal Power Series: Their Relevance to Formal Languages and Automata. In: Handbook of Formal Languages, ch. 9, vol. 1, pp. 609–677. Springer, Heidelberg (1997)

[Lut] Luttenberger, M.: An extension of Parikh's theorem. Technical report, Technische Universität München, Institut für Informatik (forthcoming)

[Lut10] Luttenberger, M.: Solving Systems of Polynomial Equations: A Generalization of Newton's Method. PhD thesis, Technische Universität München (2010)

[NNH99] Nielson, F., Nielson, H.R., Hankin, C.: Principles of Program Analysis. Springer, Heidelberg (1999)

[Tar55] Tarski, A.: A lattice-theoretical fixpoint theorem and its applications. Pacific J. Math. 5(2), 285–309 (1955)

[Tha67] Thatcher, J.W.: Characterizing derivation trees of context-free grammars through a generalization of finite automata theory. J. Comput. Syst. Sci. 1(4), 317–322 (1967)

Abstract Local Reasoning for Program Modules

Thomas Dinsdale-Young, Philippa Gardner, and Mark Wheelhouse

Imperial College London

Extended Abstract

Hoare logic ([7]) is an important tool for formally proving correctness properties of programs. It takes advantage of modularity by treating program fragments in terms of provable specifications. However, heap programs tend to break this type of modular reasoning by permitting pointer aliasing. For instance, the specification that a program reverses one list does not imply that it leaves a second list alone. To achieve this disjointness property, it is necessary to establish disjointness conditions throughout the proof.

O'Hearn, Reynolds, and Yang ([11]) introduced separation logic for reasoning *locally* about heap programs, in order to address this problem. The fundamental principle of local reasoning is that, if we know how a local computation behaves on some state, then we can infer the behaviour when the state is extended: it simply leaves the additional state unchanged. A program is specified in terms of its *footprint* — the resource necessary for it to operate — and a *frame rule* is used to infer that any additional resource is indeed unchanged. For example, given a proof that a program reverses a list, the frame rule can directly establish that the program leaves a second disjoint list alone. Consequently, separation logic enables modular reasoning about heap programs.

Abstraction (see *e.g.* Reynolds, [13]; Mitchell and Plotkin, [10]) and refinement (see *e.g.* Hoare, [8]; de Roever and Engelhardt, [3]) are also essential for modular reasoning. Abstraction takes a concrete program and produces an abstract specification; refinement takes an abstract specification and produces a correct implementation. Both approaches result in a program that correctly implements an abstract specification. Such a result is essential for modularity because it means that a program can be replaced by any other program that meets the same specification. Abstraction and refinement are well-established techniques in program verification, but have so far not been fully understood in the context of local reasoning.

Parkinson and Bierman ([12]) introduced abstract predicates in separation logic to provide abstract reasoning. An abstract predicate is, to the client, an opaque object that encapsulates the unknown representation of an abstract datatype. They inherit some of the benefits of locality from separation logic; an operation on one abstract predicate leaves others alone. However, the client cannot take advantage of local behaviour that is provided by the abstraction itself.

Consider a set module. The operation of removing, say, the value 3 from the set is local at the abstract level; it is independent of whether any other value is in the set. Yet, consider an implementation of the set as a sorted, singly-linked

A. Corradini, B. Klin, and C. Cîrstea (Eds.): CALCO 2011, LNCS 6859, pp. 36–39, 2011.

list in the heap, starting from address h. The operation of removing 3 from the set must traverse the list from h. The footprint, therefore, comprises the entire list segment from h up to the node with value 3. With abstract predicates, the abstract footprint corresponds to the concrete footprint and hence, in this case, includes all the elements of the set less than or equal to 3. Consequently, abstract predicates cannot be used to present a local abstract specification for removing 3.

Calcagno, Gardner, and Zarfaty ([1]) introduced context logic, a generalisation of separation logic, to provide *abstract local reasoning* about abstract data structures. Context logic has been used to reason about programs that manipulate data structures, such as sequences, multisets and trees (Calcagno et al., [2]). In particular, it has been successfully applied to reason about the W3C DOM tree update library (Gardner et al., [6]). Until recently, context logic reasoning has always been justified with respect to an operational semantics defined at the same level of abstraction as the reasoning. Recently, in Dinsdale-Young et al. ([4]), we combined abstract local reasoning with data refinement, to refine abstract module specifications into correct implementations.

Filipović, O'Hearn, Torp-Smith, and Yang ([5]); Mijajlović et al. ([9]) previously considered data refinement for local reasoning, studying modules built on the heap model. They observed that a client can violate a module's abstraction boundary by dereferencing pointers to the module's internal state, and thereby break the refinement between abstract modules and their concrete implementations. In their motivating example, a simple memory allocator, a client can violate the concrete allocator's free list through pointers to memory that has been deallocated; the abstract allocator, which maintains a free set, is unaffected by such an access, hence the refinement breaks. Their solution was to "blame the client" by introducing a modified operational semantics that treats such access violations as faulting executions. Using special simulation relations, they were able to recover soundness of data refinement. Their techniques adapt to different data models, however, both module and client use the same model.

We apply data refinement to local reasoning, demonstrating that abstract local reasoning is sound for module implementations. By contrast with Filipović et al. ([5]), we work with the axiomatic semantics, rather than operational semantics, of the language, defining proof transformations that establish that concrete implementations simulate abstract specifications. This avoids having to consider badly behaved clients, since the proof system only makes guarantees about well behaved clients. Furthermore, the abstract and concrete levels in our refinements typically have *different* data store models, meaning that the concept of locality itself is different at each level.

Our motivating example is the stepwise refinement of a tree module $\mathbb{T}$, illustrated in Fig. 1. We present two different refinements from the tree module $\mathbb{T}$ to the familiar heap module of separation logic $\mathbb{H}$. The first, labelled τ_1, uses a direct implementation of trees in the heap in which each tree node is represented by a contiguous block of heap cells.

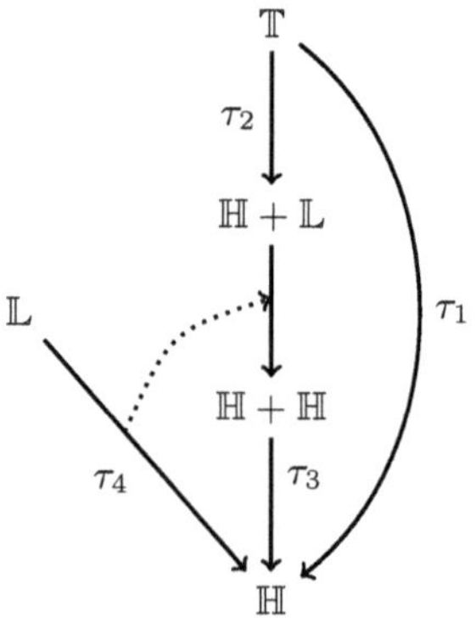

Fig. 1. Stepwise refinement of a tree module $\mathbb{T}$ to a heap module $\mathbb{H}$

The second refinement uses an abstract list module $\mathbb{L}$ as an intermediate step in the refinement. We show that the tree module $\mathbb{T}$ can be correctly implemented in the combination of the heap and list modules $\mathbb{H} + \mathbb{L}$, using translation τ_2. We also show that the list module $\mathbb{L}$ can be correctly implemented using the heap module $\mathbb{H}$, using translation τ_4. Since our approach is modular, this translation can be lifted to a translation from the combined heap and list module $\mathbb{H} + \mathbb{L}$ to the combination of two heap modules $\mathbb{H} + \mathbb{H}$. (This is illustrated by the dotted arrow in Fig. 1.) To complete the refinement, we show that the double-heap module $\mathbb{H} + \mathbb{H}$ can be trivially implemented by the heap module $\mathbb{H}$.

Our development introduces two general techniques for verifying module implementations with respect to their local specifications, using the data refinement technique known as *forward simulation* (*L-simulation* in de Roever and Engelhardt, [3]). We introduce *locality-preserving* and *locality-breaking* translations. Locality-preserving translations, broadly speaking, relate locality at the abstract level with locality of the implementation. However, implementations typically operate on a larger state than the abstract footprint, for instance, by performing pointer surgery on the surrounding state. We introduce the notion of *crust* to capture this additional state. This crust intrudes on the context, and breaks the disjointness that exists at the abstract level. We therefore relate abstract locality with implementation-level locality via a *fiction of disjointness.*

With locality-breaking translations, locality at the abstract level does not correspond to locality of the implementation. Even in this case, we can think about a locality-preserving translation using possibly the whole data structure as the crust. Instead, we prove soundness by establishing that the specifications of the module commands are preserved under translation in any abstract context, showing the soundness of the abstract frame rule. We thus establish a *fiction of locality* at the abstract level.

Acknowledgements. We acknowledge the support of EPSRC Programme Grant "Resource Reasoning". Dinsdale-Young and Wheelhouse acknowledge the support of EPSRC DTA awards. We thank Mohammad Raza and Uri Zarfaty for detailed discussions of this work. In particular, some of the technical details in our locality-preserving translations come from an unpublished technical report *Reasoning about High-level Tree Update and its Low-level Implementation*, written by Gardner and Zarfaty in 2008.

References

1. Calcagno, C., Gardner, P., Zarfaty, U.: Context Logic and tree update. In: POPL 2005, pp. 271–282 (2005)
2. Calcagno, C., Gardner, P., Zarfaty, U.: Local reasoning about data update. ENTCS 172, 133–175 (2007)
3. de Roever, W.-P., Engelhardt, K.: Data Refinement: Model-Oriented Proof Methods and Their Comparison. Cambridge University Press, Cambridge (1999)
4. Dinsdale-Young, T., Gardner, P., Wheelhouse, M.: Abstraction and refinement for local reasoning. In: Leavens, G.T., O'Hearn, P., Rajamani, S.K. (eds.) VSTTE 2010. LNCS, vol. 6217, pp. 199–215. Springer, Heidelberg (2010)
5. Filipović, I., O'Hearn, P., Torp-Smith, N., Yang, H.: Blaming the client: on data refinement in the presence of pointers. Formal Aspects of Computing (2009)
6. Gardner, P.A., Smith, G.D., Wheelhouse, M.J., Zarfaty, U.D.: Local Hoare reasoning about DOM. In: PODS, pp. 261–270 (2008)
7. Hoare, C.A.R.: An axiomatic basis for computer programming. Commun. ACM 12(10), 576–580 (1969)
8. Hoare, C.A.R.: Proof of correctness of data representations. Acta Inf. 1(4), 271–281 (1972)
9. Mijajlović, I., Torp-Smith, N., O'Hearn, P.W.: Refinement and separation contexts. In: Lodaya, K., Mahajan, M. (eds.) FSTTCS 2004. LNCS, vol. 3328, pp. 421–433. Springer, Heidelberg (2004)
10. Mitchell, J.C., Plotkin, G.D.: Abstract types have existential type. In: POPL, pp. 37–51 (1985)
11. O'Hearn, P.W., Reynolds, J., Yang, H.: Local reasoning about programs that alter data structures. In: Fribourg, L. (ed.) CSL 2001 and EACSL 2001. LNCS, vol. 2142, p. 1. Springer, Heidelberg (2001)
12. Parkinson, M.J., Bierman, G.M.: Separation logic and abstraction. In: POPL, pp. 247–258 (2005)
13. Reynolds, J.C.: Types, abstraction and parametric polymorphism. In: IFIP Congress, pp. 513–523 (1983)

Infinite Computation, Co-induction and Computational Logic

Gopal Gupta, Neda Saeedloei, Brian DeVries, Richard Min, Kyle Marple, and Feliks Kluźniak

Department of Computer Science,
University of Texas at Dallas,
Richardson, TX 75080

Abstract. We give an overview of the coinductive logic programming paradigm. We discuss its applications to modeling ω-automata, model checking, verification, non-monotonic reasoning, developing SAT solvers, etc. We also discuss future research directions.

1 Introduction

Coinduction is a technique for reasoning about unfounded sets [12], behavioral properties of programs [2], and proving liveness properties in model checking [16]. Coinduction also provides the foundation for lazy evaluation [9] and type inference [21] in functional programming as well as for interactive computing [33].

Coinduction is the dual of induction. Induction corresponds to well-founded structures that start from a basis which serves as the foundation: e.g., natural numbers are inductively defined via the base element zero and the successor function. Inductive definitions have 3 components: initiality, iteration and minimality. For example, the inductive definition of lists of numbers is as follows: (i) `[]` (empty list) is a list (initiality); (ii) `[H|T]` is a list if `T` is a list and `H` is some number (iteration); and, (iii) the set of lists is the smallest set satisfying (i) and (ii) (minimality). Minimality implies that infinite-length lists of numbers are not members of the inductively defined set of lists of numbers. Inductive definitions correspond to least fixed point (*LFP*) interpretations of recursive definitions.

Coinduction eliminates the initiality condition and replaces the minimality condition with maximality. The coinductive definition of *infinite* lists of numbers is: (i) `[H|T]` is a list if `T` is a list and `H` is some number (iteration); and, (ii) the set of lists is the largest set of lists satisfying (i) (maximality). There is no base case in a coinductive definition, and while it may appear circular, the definition is well formed since coinduction corresponds to the greatest fixed point (*GFP*) interpretation of recursive definitions: namely, the set of of all infinite lists of numbers. (Note, however, that if we had a recursive definition with a base case, then under the coinductive interpretation, the set would contain both finite and infinite-sized lists.) A coinductive proof is essentially an infinite-length proof.

A. Corradini, B. Klin, and C. Cîrstea (Eds.): CALCO 2011, LNCS 6859, pp. 40–54, 2011.

2 Coinduction and Logic Programming

Coinduction has been incorporated in logic programming in a systematic way only recently [30,10,29]. An operational semantics—similar to SLD resolution—was given for computing those answers to a query that are in the greatest fixed point of a logic program (the semantics is discussed below).

Consider the list example discussed in Sec. 1. The normal logic programming definition of a stream (list) of numbers is given as program P1 below:

```
stream([]).
stream([H|T]) :- number(H), stream(T).
```

Under SLD resolution, the query `?- stream(X).` will systematically produce all finite streams one by one, starting from the `[]` stream. Suppose now we remove the base case and obtain the program P2:

```
stream([H|T]) :- number(H), stream(T).
```

In standard logic programming the query `?- stream(X).` fails, since the model of P2 does not contain any instances of `stream/1`. The problems are two-fold: (i) the Herbrand universe does not contain infinite terms; (ii) the least Herbrand model does not allow for infinite proofs, such as the proof of `stream(X)`; yet these concepts are commonplace in computer science, and a sound mathematical foundation exists for them in the field of hyperset theory [2]. Coinductive LP extends the traditional declarative and operational semantics of LP to allow reasoning over infinite and cyclic structures and properties. In the coinductive LP paradigm the declarative semantics of the predicate `stream/1` above is given in terms of *infinitary Herbrand (or co-Herbrand) universe*, *infinitary Herbrand (or co-Herbrand) base* [15], and *maximal models* (computed using greatest fixed-points).

Under the coinductive interpretation of P2, the query `?- stream(X).` should produce all infinite sized streams as answers, e.g., `X = [1, 1, 1, ... ]`, `X = [1, 2, 1, 2, ... ]`, etc. The model of P2 does contain instances of `stream/1` (but proofs may be of infinite length).

If we take a coinductive interpretation of program P1, then we get all finite and infinite streams as answers to the query `?- stream(X).` Coinductive logic programming allows programmers to manipulate infinite structures. As a result, unification must be extended and the "occurs check" removed: unification equations such as `X = [1 | X]` are allowed in coinductive logic programming; in fact, such equations will be used to represent infinite (rational) structures in a finite manner.

The operational semantics of coinductive logic programming is given in terms of the *coinductive hypothesis rule*: during execution, if the current resolvent R contains a call C' that unifies with an ancestor call C, then the call C' succeeds; the new resolvent is $R'\theta$ where $\theta = mgu(C, C')$ and R' is obtained by deleting C' from R. With this extension, a clause such as

```
p([1|T]) :- p(T).
```

and the query `?- p(Y).` will produce an infinite answer `Y = [1|Y]`.

In coinductive logic programming the alternative computations started by a call are not only those that begin with unifying the call and the head of a clause, but also those that begin with unifying the call and one of its ancestors in a proof tree.

Regular logic programming execution extended with the coinductive hypothesis rule is termed *co-logic programming* [29]. The coinductive hypothesis rule will work for only those infinite proofs that are *regular* (or *rational*), i.e., infinite behavior is obtained by a finite number of finite behaviors interleaved an infinite number of times. More general implementations of coinduction are possible [29]. More complex examples of coinductive LP program can be found elsewhere [10].

Even with the restriction to rational proofs, there are many applications of coinductive logic programming, some of which are discussed next. These include model checking, modeling ω-automata, non-monotonic reasoning, etc. Implementations of co-LP have been realized: a meta-interpreter that includes both tabled and coinductive logic programming is available from the authors [14]. Recently, SWI Prolog [32] and Logtalk have also added support for coinduction.

The traditional model of declarative computing is based on recursion: a problem is solved by breaking it down into smaller subproblems, which are broken down further, until base cases are reached that are trivially solved. Solutions to the subproblems are then used to synthesize a solution to the problem. This model of computation is based on the theory of well-founded sets, and is not appropriate for computations that are cyclical in nature. Such cyclical computations arise when the solutions to the subproblems of a given problem are mutually interdependent. Solving the problem involves establishing the consistency of the interdependent solutions to the subproblems, and coinduction/corecursion provides the necessary framework [2].

Intuitively, a computational model in which both *LFP*-based and *GFP*-based computations can be conveniently expressed will allow one to elegantly express any computable function. As a result, logic programming extended with coinduction provides a basis for many powerful applications, in areas of verification, non-monotonic reasoning, modal logics, etc. A number of challenges remain in co-LP. These relate to:

- **Nesting of inductive and coinductive computations.** In such cases giving the proper semantics may not be simple. This is illustrated by the following program
```
p :- q.
q :- p.
```
where `q` has standard inductive semantics while `p` has coinductive semantics. A call to `q` should fail, and so should a call to `p` (since it calls `q`), but a call to `p` may succeed coinductively. (We will return to this issue in Sec. 6.)
- **Reporting isomorphic solutions only once.** Consider the coinductive predicate `p/1`.
```
p([1|T]) :- p(T).
```
The query `p(A)` will produce a solution `A = [1|A]`. However, if normal call expansion is also considered as an alternative along with the coinductive

hypothesis rule, then an infinite number of solutions will be produced, all of them isomorphic to the solution `A = [1 | A]`:

```
A = [1, 1 | A]
A = [1, 1, 1 | A]
A = [1, 1, 1, 1 | A]
....
```

One could avoid this problem by memoing the cyclical solutions, however, an operational semantics that incorporates this memoing has yet to be investigated [6].

- **Efficient implementation of co-LP.** The current operational semantics of co-LP require that every coinductive call be unified with every ancestor call to the same predicate: a naive implementation may be quite inefficient.
- **Reporting all solutions.** Consider the following program where both `p` and `q` are coinductive.

```
p([a|X]) :- q(X).
q([b|X]) :- p(X).
```

The query `?- p(A).` has an infinite number of solutions, both rational and irrational. In general, systematic enumeration of even the rational solutions requires a fair execution strategy (e.g., breadth-first search). Such an execution strategy should be coupled with a mechanism that eliminates redundant isomorphic solutions.

As discussed earlier, there are many extensions and applications of coinduction and co-LP. In the rest of the paper we outline some of them. We discuss the extension of co-SLD resolution with negation (termed co-SLDNF resolution) and its use in realizing goal-directed non-monotonic reasoners and SAT solvers. We also discuss how co-LP can be extended with constraints and used for modeling complex real-time systems, for example, how timed ω-automata, timed pushdown ω-automata, and timed grammars can be elegantly modeled. Co-LP can also be used for developing executable operational semantics of Π-calculus and linear-time temporal logic (LTL). Relationship between co-LP and various ω-automata (Büchi, Rabin, Streett) is also discussed.

3 Model Checking with Co-LP

Model checking is a popular technique used for verifying hardware and software systems. It is done by constructing a model of the system as a finite-state Kripke structure and then determining whether the model satisfies various properties specified as temporal logic formulae [3]. The verification is performed by means of systematically searching the state space of the Kripke structure for a counter-example that falsifies the given formula. The vast majority of properties that are to be verified can be classified into *safety* properties and *liveness* properties. Intuitively, a safety property asserts that "nothing bad will happen", while a liveness property asserts that "something good will eventually happen".

It is well known that safety properties can be verified by reachability analysis, i.e., if a counter-example to the postulated property exists, it can be finitely determined by enumerating all the reachable states of the Kripke structure. In the context of Logic Programming, verification of safety properties amounts to computing elements of the *LFP* of a program, and is thus elegantly handled by standard LP systems extended with tabling [24]. Verification of liveness properties is less straightforward, because counterexamples take the form of infinite traces, which are semantically equivalent to elements of the the *GFP* of a logic program: co-LP is more suitable for directly computing such counterexamples without the expensive transformations required by some other approaches suggested in the literature [24].

Intuitively, a state is live if it can be reached via an infinite loop (cycle). Liveness counterexamples can be found by (coinductively) enumerating all possible states that can be reached via infinite loops and then determining if any of these states constitute valid counterexamples.

To demonstrate the power of coinductive logic programming, we show how an interpreter for linear temporal logic can be written very elegantly. In LTL, one checks if a temporal logic formula is true along a path. Temporal operators whose meaning is given in terms of *LFP*s are realized via tabled logic programming, while those whose meaning is given in terms of *GFP*s are realized using coinductive logic programming.

The `verify/3` predicate in the interpreter of Fig. 1 takes as input a state and an LTL formula, and produces as an answer a path for which this formula is true. To verify that an LTL formula `F` holds in a given state `S`, it has to be negated. The negated formula is then converted to negation normal form (i.e., a negation symbol can appear only next to a proposition), and then given as input to the `verify/3` predicate along with the state `S`. If the call to `verify/3` fails, then there is no path on which the negated formula holds, implying that the formula `F` holds in state `S`. In contrast, if `verify/3` returns a path as an answer, then that specific path is a counterexample for which the original formula `F` does not hold. Note that the Kripke structure is represented as a transition table using the `trans/2` predicate, while information about which proposition(s) holds in which state(s) is given by the `holds/2` predicate. The temporal operators `X, F, G, U` and `R` are represented with corresponding lower case letters, while `^`, `v`, and `~` represent $\wedge$, $\vee$, and negation respectively. The program in Fig. 1 should be self-explanatory.

Note that while this program is elegant, its soundness and completeness depend on the execution strategy used for realizing coinduction, and on how the interleaving of coinduction and tabling (induction) is handled. The issue of interleaving of coinduction and induction is discussed in section 6.2.

4 Negation in Co-LP

As mentioned earlier, SLD resolution extended with the coinductive hypothesis rule is called co-SLD resolution. Co-SLDNF resolution further extends co-SLD resolution with negation as failure [15]. Essentially, it augments co-SLD with the negative coinductive hypothesis rule, which states that if a negated call `not(p)` is

```
verify( S, g A,   Path       ) :-  coverify( S, g A,   Path ).
verify( S, A r B, Path       ) :-  coverify( S, A r B, Path ).
verify( S, f A,   Path       ) :-  tverify( S, f A,   Path ).
verify( S, A u B, Path       ) :-  tverify( S, A u B, Path ).
verify( S, A,     [ S ]      ) :-  proposition( A ),    holds( S, A ).
verify( S, ~ A,   [ S ]      ) :-  proposition( A ), \+ holds( S, A ).
verify( S, A ^ B, Path       ) :-  verify( S, A, PathA ),
                                   verify( S, B, PathB ),
                                   prefix( PathA, PathB, Path ).
verify( S, A v B, Path       ) :-    verify( S, A, Path )
                                   ; verify( S, B, Path ).
verify( S, x A,   [ S | P ] ) :-  trans( S, S2 ), verify( S2, A, P ).

:- tabled tverify/3.
tverify( S, f A,   Path ) :-    verify( S, A, Path )
                              ; verify( S, x f A, Path ).
tverify( S, A u B, Path ) :-    verify( S, B, Path )
                              ; verify( S, A ^ x( A u B), Path ).

:- coinductive coverify/3.
coverify( S, g A,   Path ) :-  verify( S, A ^ x g A,       Path ).
coverify( S, A r B, Path ) :-  verify( S, A ^ B,           Path ).
coverify( S, A r B, Path ) :-  verify( S, B ^ x( A r B ),  Path ).

prefix( Prefix, Path, Path ) :-  append( Prefix, _, Path ),  !.
prefix( Path, Prefix, Path ) :-  append( Prefix, _, Path ),  !.
```

Fig. 1. An LTL model-checker

encountered during resolution, and another call to `not(p)` has been seen before in the same computation, then `not(p)` coinductively succeeds [20].

To implement co-SLDNF resolution, the set of positive and negative calls has to be maintained in the *positive hypothesis table* (PHT) and *negative hypothesis table* (NHT) respectively. An attempt to place the same call in both tables will induce failure of the computation. The framework based on maintaining a pair of sets (corresponding to a partial interpretation of success set and failure set, resulting in a partial model [8]) provides a good basis for the operational semantics of co-SLDNF resolution [20].

One of the most interesting applications of co-SLDNF resolution is in obtaining goal-directed strategies for executing answer set programs. Answer Set Programming (ASP) [1] is a powerful paradigm for performing non-monotonic reasoning within logic programming. Current ASP implementations are restricted to grounded range-restricted function-free normal programs [1] and use essentially an evaluation strategy that is bottom-up. Co-LP with co-SLDNF resolution has allowed the development of top-down goal evaluation strategies for ASP [10], which in turn allows ASP to be extended to predicates [18]. This co-LP based method eliminates the need for grounding, allows functions, and effectively handles a large class of predicate ASP programs including possibly infinite ASP programs. We have designed and implemented the techniques and algorithms to execute propositional and predicate ASP programs. Our techniques are applicable to the full range of propositional answer set programs, but for predicate ASP we are restricted to programs that are *call-consistent* or *order-consistent* [19].

We illustrate co-SLDNF resolution by showing how a boolean SAT solver can be readily obtained. A SAT solver attempts to find a truth assignment that satisfies a propositional formula. Our technique based on co-SLDNF resolution essentially makes assumptions about truth values of propositions, and checks that these assumptions bear out throughout the formula. A propositional formula X must satisfy the following rules:

```
(1)   t(X) :- not( neg(t(X)) ).
(2)   neg(t(X)) :- not( t(X) ).
```

The predicate `t/1` is a truth-assignment (or a valuation) where `X` is a propositional Boolean formula to be checked for satistifiability. Clause (1) asserts that `t(X)` is true if there is no counter-case for `neg(t(X))` (that is, `neg(t(X))` is false (coinductively), with the assumption that `t(X)` is true (coinductively)). Clause (2) asserts that `neg(t(X))` is true if there is no counter-case for `t(X)`. Next, any well-formed propositional Boolean formula constructed from a set of propositional symbols and logical connectives and in conjunctive normal form (CNF) can be translated to a query for the co-LP program as follows: first, each propositional symbol p is translated to t(p). Second, any negated proposition, that is ¬t(p), is translated to `neg(t(p))`. Third, the Boolean operators $\wedge$ and $\vee$ are translated to Prolog's conjunction (`,`) and disjunction (`;`) respectively.

The predicate `t(X)` determines the truth-assignment of formula `X` (if `X` is true, `t(X)` succeeds; else it fails). Note that each query is a Boolean expression whose satistifiability is to be coinductively determined. As an example, the formula $(p1 \vee p2 \vee p3) \wedge (p1 \vee \neg p3) \wedge (\neg p2 \vee \neg p4)$ will be translated into the query

`(t(p1); t(p2); t(p3)), (t(p1); not(t(p3))), (not(t(p2)); not(t(p4))).`

Propositions that must be assigned the value true will be found in the PHT, while those that must be assigned the value false will be found in NHT. The answer printed by our prototype SAT solver based on co-LP is shown below (the `print_ans` command prints the answer):

```
?- (t(p1);t(p2);t(p3)),(t(p1); neg(t(p3))),
          (neg(t(p2)); neg(t(p4))), print_ans.
     Answer:
       PHT == [t(p1)]
       NHT == [t(p2)]
     yes
```

The answer indicates that this propositional formula is satisfiable if `p1` is assigned true and `p2` is assigned false.

5 Modeling Complex Real-time Systems

Next, we consider the use of co-LP to model ω-automata and their various extensions. As is well known, finite state automata as well as grammars can be elegantly modeled as inductive logic programs, and recognizers for them are readily obtained (due to the relational nature of logic programming, these recognizers can also systematically generate all the words of the recognized language). Similar recognizers can be obtained for ω-automata and ω-grammars with co-LP. Moreover, co-LP can be augmented with constraint logic programming (e.g., over the real numbers [CLP(R)] [13]), allowing elegant modeling of real-time systems. We next give an overview of application of co-LP (combined with CLP(R)) to modeling complex real-time systems; in particular, we discuss how pushdown timed automata, timed grammars and timed π-calculus can be elegantly represented. The use of co-LP together with CLP(R) for modeling these system leads one naturally towards coinductive constraint programming, an area that we are beginning to investigate.

5.1 Pushdown ω-Automata and ω-Grammars

A timed automaton is an ω-automaton extended with clocks or stopwatches. A pushdown timed automaton (PTA) [4] extends a timed automaton with a stack, just as a pushdown automaton extends a finite automaton. Transitions from one state to another are made not only on the alphabet symbols of the language but also on constraints imposed on clocks (e.g., at least 2 units of time must have elapsed). Transitions may result in changing the stack by performing the push and pop operations. A PTA recognizes a language consisting of timed words, where a timed word is an infinite sequence of symbols from the alphabet, each of them paired with a time-stamp. The sequence of symbols in an infinite sequence accepted by a PTA must obey the rules of syntax laid down by the underlying untimed pushdown automaton, while the time-stamps must obey the timing constraints. Additionally, the stack must be empty whenever a final state is entered.

To model and reason about PTA (and timed grammars) we must account for the fact that: (i) the underlying language is context free, not regular; (ii) accepted strings are infinite; and (iii) clock constraints are posed over continuously flowing time. All three aspects can be elegantly handled—and PTAs naturally modeled—within co-LP. Additionally, the definite clause grammar (DCG) facility of Prolog allows one to easily obtain a parser for a grammar. Through co-LP, one can develop language processors that can act as recognizers for ω-pushdown automata and ω-grammars. Further, with incorporation of CLP(R) one can also model the timing aspects.

A PTA specification can be automatically transformed to a coinductive program extended with CLP(R) [28]. The method takes the description of a PTA and generates a coinductive constraint logic program over reals. The generated logic program models the PTA as a collection of transition rules (one rule per transition): each rule is extended with stack actions as well as clock constraints. Our coinductive constraint logic programming realization of pushdown timed automata and timed automata can be regarded as a general framework for modeling/verifying real-time systems.

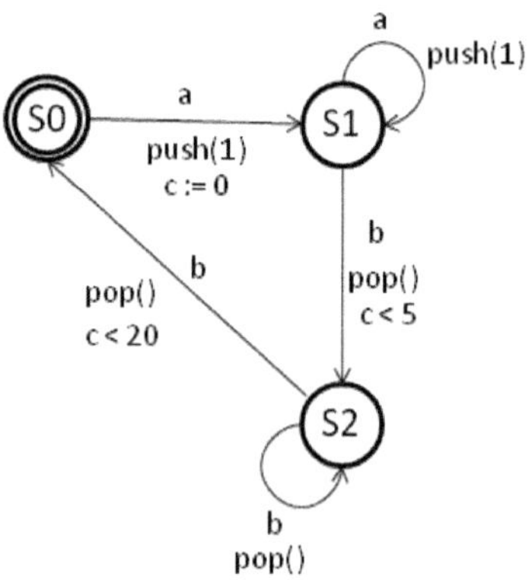

Fig. 2. A Pushdown Timed Automaton

Fig. 2 shows an example of a pushdown timed automaton. The following is an encoding of this PTA in the form of a coinductive logic program with constraints that was generated automatically by our system:

```
trans(s0, a, s1, T, Ti, To, [],     [1]) :- {To = T}.
trans(s1, a, s1, T, Ti, To, C, [1 | C]) :- {To = Ti}.
trans(s1, b, s2, T, Ti, To, [1 | C], C) :- {T - Ti < 5,  To = Ti}.
trans(s2, b, s2, T, Ti, To, [1 | C], C) :- {To = Ti}.
trans(s2, b, s0, T, Ti, To, [1 | C], C) :- {T - Ti < 20, To = Ti}.

:- coinductive(pta/6).
pta([ X | R], Si, T, Ti, C1, [ (X, T) | S]) :-
            trans(Si, X, So, T, Ti, To, C1, C2),
            {Ta > T}, pta(R, So, Ta, To, C2, S).
```

Once a timed system is modeled as a coinductive CLP(R) program, the program can be used to (i) check whether a particular timed string will be accepted or not; (ii) systematically generate all the possible timed strings that can be accepted; or, (iii) verify safety and liveness properties by posing appropriate queries.

The co-LP and CLP(R) based framework has been used to study the *generalized railroad crossing (GRC) problem* [11] and to verify its safety and utility properties by posing simple queries [28]. This approach based on coinductive CLP(R) for the GRC is considerably more elegant and simpler than other approaches (e.g., [22]).

Context Free Grammars can be extended to define languages consisting of (infinite) timed words. These extended grammars are called timed context-free ω-grammars (ω-TCFG) [27]. Such languages are useful for modeling complex real-time systems [25] that run forever.

As an example of an ω-TCFG, consider a language in which each sequence of a's is followed by a sequence of an equal number of b's, with each accepted string having at least two a's and two b's. For each pair of equinumerous sequences of a's and b's, the *first* symbol b must appear within 5 units of time from the *first* symbol a and the *final* symbol b must appear within 20 units of time from the *first* symbol a. The grammar annotated with clock expressions is shown below: c is a clock which is reset when the first symbol a is seen.

$S \to R\ S$
$R \to a\ \{c := 0\}\ T\ b\ \{c < 20\}$
$T \to a\ T\ b$
$T \to a\ b\ \{c < 5\}$

Definite Clause Grammars (DCG) [31] together with CLP(R) and coinduction can be used to develop an effective and elegant method for parsing the languages generated by ω-TCFGs. Specification of an ω-TCFG can be automatically transformed to a DCG augmented with coinduction and CLP(R) [27]. The resulting coinductive constraint logic program acts as a practical parser. Given this program, one can pose queries to check whether a timed word satisfies the timing constraints imposed by the timed grammar. Alternatively, one can generate possible legal timed words (note that in this case a CLP(R) system will output timed words in which time-stamps will be represented as variables; constraints that these time-stamps must satisfy will be output as well). Finally, one can verify properties of this timed language (e.g., checking the simple property that all the a's are generated within 5 units of time, in any timed string that is accepted).

5.2 Timed π-calculus

The π-calculus was introduced by Milner et al. [17] with the aim of modeling concurrent/mobile processes. The π-calculus provides a conceptual framework for describing systems whose components interact with each other. It contains an algebraic language for descriptions of processes in terms of the communication actions they can perform. Theoretically, the π-calculus can model mobility, concurrency and message exchange between processes as well as infinite computation (through the '!' operation). Operational semantics of π-calculus can also be elegantly modeled with the help of coinduction. Specifically, to model the replication operator faithfully, one needs coinduction. The π-calculus can also be extended to real time, and executable operational semantics obtained for it using coinduction and constraints. The extension to time is helpful in modeling controller processes that control physical devices where a notion of real-time is important [25]. Several extensions of π-calculus with time have been proposed to overcome this problem (e.g., [5]); all these approaches discretize time rather than represent it faithfully as a continuous quantity.

For a complete encoding of the operational semantics of timed π-calculus, we must model three aspects of timed π-calculus processes: concurrency, infinite computation and time constraints/clock expressions. An executable operational semantics of π-calculus in logic programming (LP) has been developed [26,25]. Channels are modeled as streams, *rational* infinite computations are handled by *coinduction* [29,10] and concurrency is handled by allowing *coroutining* within logic programming computations. This operational semantics is extended with continuous real time, which we have modeled with *constraint logic programming over reals* [13]. The executable operational semantics, thus realized, automatically leads to an implementation of the timed π-calculus in the form of a coinductive coroutined constraint logic program that can be regarded as an interpreter for timed π-calculus expressions, and can be used for modeling and verification of real-time systems.

Note that there is past work on developing LP-based executable operational semantics of the π-calculus (but not timed π-calculus) [34] but it is unable to model infinite processes and infinite replication, since coinductive logic programming has been developed relatively recently [29,10].

Note also that giving proper semantics to coinductive constraint logic programs is not straightforward, and a formal study of coinductive constraint programming must be undertaken. In our investigations thus far, to ensure soundness, we assume that all the clocks involved in a cycle are reset every time we go around the cycle [25].

6 Integrating Coinduction and Induction

We now turn our attention to the problems resulting from integrating coinductive logic programming with traditional (i.e., inductive) logic programming. As discussed earlier, the semantics of arbitrary interleaving of induction and coinduction are unclear. We begin this section with co-logic programming in its current form, along with some example applications and a discussion of its limitations, followed by a brief description of our ongoing efforts to relax these limitations and extend the expressiveness of the formalism.

6.1 Stratified Co-LP: Applications and Limitations

The current (in particular, operational) semantics of inductive and coinductive logic programming permit either a least or a greatest fixed point interpretation of a predicate: no reasonable semantics have yet been given to programs that involve mutual recursion between inductive and coinductive predicates. In practice, programs lacking this mutual recursion—known as *stratified* co-logic programs—are still capable of representing solutions to a wide range of problems.

We begin with the near-canonical example of filters over streams. Let S be an inductively defined set, such as the set of natural numbers. A *stream* over S is an infinite sequence of elements of S. A *filter* takes a stream as input and removes elements that are not in a designated subset of S.

Coinductive logic programs cannot construct finite proofs over arbitrary streams, but they can be applied to streams of the form uv^ω, where a finite suffix v is repeated ad infinitum. The following co-logic program filters a stream over the natural numbers so that only even numbers remain:

```
:- coinductive filter/2.
filter( [ H | T ], [ H | T2 ] ) :- even( H ), filter( T, T2 ).
filter( [ H | T ], T2 ) :- \+ even( H ), filter( T, T2 ).

even(0).
even( s( s( N ) ) ) :- even( N ).
```

Here, the `even` predicate is the usual inductive definition of even natural numbers, while `filter` is a coinductive definition of our desired filter. We can see that this program is stratified, as `filter` and `even` are not mutually recursive. We can make use of this program with queries such as the following:

```
?- L = [ 0, s( 0 ), s( s ( 0 ) ) | L ], filter( L, L2 ).
```

`L2` will be bound to the list `L2 = [ 0, s( s( 0 ) ) | L2 ]`.

Let us also consider the case of an inductive list containing coinductive elements. Let $\mathbb{N}^+ = \mathbb{N} \cup \{\infty\}$. The following program filters finite (inductive) lists over $\mathbb{N}^+$ so that only even natural numbers and infinity remain (the latter is accomplished through a coinductive definition of `even`):

```
filter( [], [] ).
filter( [ H | T ], [ H | T2 ] ) :- even( H ), filter( T, T2 ).
filter( [ H | T ], T2 ) :- \+ even( H ), filter( T, T2 ).

:- coinductive even/1.
even( 0 ).
even( s( s( N ) ) ) :- even( N ).
```

In general, inductive and coinductive predicates can be mixed in co-logic programs, as long as there exists a clear, acyclic (i.e., stratified) hierarchy between them. Co-LP provides a natural paradigm for constructing definitions and proofs over such predicates.

Co-LP can also be used to construct infinite, rational proofs in a finite manner. In such cases, it is typical to augment SLD resolution with *tabling* (called SLG resolution), which memoizes calls and solutions to inductive predicates and by extension prevents construction of (rational) infinite-depth inductive proofs. The combination of coinductive and tabled logic programming is particularly useful for implementing solvers for fixed point and modal logics; for example, tabling by itself has been used to implement model checkers for the alternation-free μ-calculus [34,24]. While the addition of coinduction permits consideration of more expressive forms of such logics, the stratification restriction must still be kept in mind: for example, co-logic programming cannot be used as a solver for the full μ-calculus, as it permits unrestricted nesting of least and greatest fixed point operators.

Part of our ongoing work is precisely describing restrictions of such logics that make them stratified. We have already discovered that the coinductive and tabled execution strategies, when considered as decision processes over proof trees (which is their use in constructing solvers for such fixed point logics), are equivalent to *stratified Büchi tree automata*, or SBTAs [6]. In SBTAs, cycles (via automaton transitions) between accepting and non-accepting states are forbidden. The class of SBTA-recognizable languages is distinguishable from the class of deterministic BTA-recognizable languages, though the two classes are not disjoint. In our equivalence proof, the non-accepting states of an SBTA correspond to calls to tabled predicates, while the coinductive calls correspond to the accepting states of the SBTA. The primary practical benefit of this equivalence is that by proving that solutions to a particular problem can or cannot be computed by an SBTA, one can determine whether the solutions can be computed directly by the co-logic programming execution strategy. If they can, then a declarative co-logic program can be constructed from the SBTA: the automaton states and transitions are translated to predicates and clauses (respectively) in the program, with the predicates for accepting and nonaccepting states respectively declared as coinductive and tabled.

6.2 Towards Non-stratified Co-LP

One significant hurdle to the use of co-logic programming is that existing co-logic programming systems perform no analysis to check whether a program is stratified, making the programmer responsible for ensuring the stratification of his program. According to the operational semantics of tabling and coinduction, non-stratified programs exhibit inconsistent behavior and produce non-intuitive answers, and the resulting errors are often difficult to diagnose. To ease the burden on both the programmer and the implementor of a co-logic programming system, we want to provide a consistent semantics to non-stratified programs. Because we consider execution strategies as being tree automata decision processes, we seek an acceptance condition with sufficient power to represent non-stratified programs.

To this end, we are currently focused on Rabin tree automata (RTAs) [23], again drawing a correspondence between categories of predicates and categories of states in an RTA. RTA states can be split into three categories: accepting, non-accepting, and rejecting. A tree is accepted by an RTA if, along each branch of the tree, an accepting state is encountered infinitely often without any rejecting state also being encountered infinitely often. As before, accepting states correspond to calls to coinductive predicates. In order to accommodate the other two categories, we use two categories of inductive predicates: *strongly inductive*, whose calls cannot occur infinitely often along any path in a proof and thereby forbid coinductive success when it would violate this requirement; and *weakly inductive*, whose calls cannot occur infinitely often along any proof path by themselves but still permit coinductive success [7].

By way of a brief example, let us consider the smallest non-stratified program:

```
:- coinductive p/0.
:- inductive q/0.
p :- q.
q :- p.
```

Under the existing co-logic programming semantics (using tabling for induction), the query `?- p, q.` succeeds, while `?- q, p.` fails. To resolve this inconsistency, we require that inductive predicates be declared to be either `weak_inductive` or `strong_inductive`, as shown in the following:

```
:- coinductive p/0.
:- strong_inductive q/0.
p :- q.
q :- p.
```

Both queries to this program will fail, as both proof trees contain infinite numbers of occurrences of the call to `p` and to `q`: the latter is forbidden by the `strong_inductive` declaration on `q`. A `weak_inductive` declaration would allow both queries to succeed. Finally, consider the following program:

```
:- coinductive p/0.
:- weak_inductive q/0.
p :- q.
q :- q.
```

Here, only calls to `q` occur infinitely often in both proofs, causing both queries to fail (as no coinductive call occurs infinitely often).

Armed with this additional execution strategy and its correspondence to RTAs, we will use co-logic programming to declaratively implement solvers for several pertinent domains, in particular, solvers and model checkers for temporal and modal logics. While it is worth noting that we are still in the process of deriving the declarative fixed-point semantics for non-stratified co-logic programming we already regard co-logic programming as a powerful, declarative technique for solving a wide range of logical and algebraic problems.

7 Conclusions

In this paper we gave an overview of the coinductive logic programming paradigm and illustrated many of its applications. Co-LP gives an operational semantics to declarative semantics that is based on the greatest fixpoint. We believe that a combination of inductive and coinductive LP allows one to implement any desired logic programming semantics (well-founded, stable model semantics, etc.). Many problems still remain open. These include arbitrary interleaving of inductive and coinductive computations and combining constraints with coinductive LP.

References

1. Baral, C.: Knowledge Representation, Reasoning and Declarative Problem Solving. Cambridge University Press, Cambridge (2003)
2. Barwise, J., Moss, L.: Vicious Circles: On the Mathematics of Non-Wellfounded Phenomena. CSLI Publications, Stanford (1996)
3. Clarke Jr., E.M., Grumberg, O., Peled, D.A.: Model Checking. The MIT Press, Cambridge (1999)
4. Dang, Z.: Binary reachability analysis of pushdown timed automata with dense clocks. In: Berry, G., Comon, H., Finkel, A. (eds.) CAV 2001. LNCS, vol. 2102, pp. 506–518. Springer, Heidelberg (2001)
5. Degano, P., Loddo, J., Priami, C.: Mobile processes with local clocks. In: LOMAPS, pp. 296–319. Springer, Heidelberg (1996)
6. DeVries, B., et al.: Semantics and Implementation of Co-Logic Programming (forthcoming)
7. DeVries, B., et al.: A Co-LP Execution Strategy Derived from Rabin Tree Automata (in preparation)
8. Fages, F.: Consistency of Clark's completion and existence of stable models. Journal of Methods of Logic in Computer Science 1, 51–60 (1994)
9. Gordon, A.: A Tutorial on Co-induction and Functional Programming. In: Glasgow Functional Programming Workshop, pp. 78–95. Springer, Heidelberg (1994)
10. Gupta, G., Bansal, A., Min, R., Simon, L., Mallya, A.: Coinductive logic programming and its applications. In: Dahl, V., Niemelä, I. (eds.) ICLP 2007. LNCS, vol. 4670, pp. 27–44. Springer, Heidelberg (2007)

11. Heitmeyer, C.L., Lynch, N.A.: The generalized railroad crossing: A case study in formal verification of real-time systems. In: IEEE RTSS, pp. 120–131 (1994)
12. Jacobs, B.: Introduction to Coalgebra: Towards Mathematics of States and Observation. Draft manuscript
13. Jaffar, J., Maher, M.J.: Constraint logic programming: A survey. J. Log. Program. 19/20, 503–581 (1994)
14. Kluźniak, F.: A logic programming meta-interpreter that combines tabling and coinduction, http://www.utdallas.edu/~gupta/meta.tar.gz
15. Lloyd, J.W.: Foundations of Logic Programming, 2nd edn. Springer, Heidelberg (1987)
16. Mallya, A.: Deductive Multi-valued Model Checking. Ph.D. thesis. University of Texas at Dallas (2006)
17. Milner, R., Parrow, J., Walker, D.: A calculus of mobile processes, parts i and ii. Inf. Comput. 100(1), 1–77 (1992)
18. Min, R., Bansal, A., Gupta, G.: Towards Predicate Answer Set Programming via Coinductive Logic Programming. In: AIAI 2009 (2009)
19. Min, R.: Predicate Answer Set Programming with Coinduction. Ph.D. Thesis. University of Texas at Dallas (2009)
20. Min, R., Gupta, G.: Coinductive Logic Programming with Negation. In: De Schreye, D. (ed.) LOPSTR 2009. LNCS, vol. 6037, pp. 97–112. Springer, Heidelberg (2010)
21. Pierce, B.: Types and Programming Languages. The MIT Press, Cambridge (2002)
22. Puchol, C.: A solution to the generalized railroad crossing problem in Esterel. Technical report, Dep. of Comp. Science, The University of Texas at Austin (1995)
23. Rabin, M.O.: Decidability of Second-Order Theories and Automata on Infinite Trees. Transactions of the American Mathematical Society 141, 1 (1969)
24. Ramakrishna, Y.S., Ramakrishnan, C.R., Ramakrishnan, I.V., Smolka, S.A., Swift, T., Warren, D.: Efficient Model Checking Using Tabled Resolution. In: Proc. CAV 1997, pp. 143–154. Springer, Heidelberg (1997)
25. Saeedloei, N.: Extending Infinite Systems with Real-time. Ph.D. Thesis. University of Texas at Dallas (forthcoming)
26. Saeedloei, N., Gupta, G.: Timed pi-calculus. University of Texas at Dallas technical report
27. Saeedloei, N., Gupta, G.: Timed definite clause omega-grammars. In: ICLP (Technical Communications), pp. 212–221 (2010)
28. Saeedloei, N., Gupta, G.: Verifying complex continuous real-time systems with coinductive CLP(R). In: Dediu, A.-H., Fernau, H., Martín-Vide, C. (eds.) LATA 2010. LNCS, vol. 6031, pp. 536–548. Springer, Heidelberg (2010)
29. Simon, L.: Coinductive Logic Programming. Ph,D thesis, University of Texas at Dallas (2006)
30. Simon, L., Mallya, A., Bansal, A., Gupta, G.: Coinductive Logic Programming. In: Etalle, S., Truszczyński, M. (eds.) ICLP 2006. LNCS, vol. 4079, pp. 330–345. Springer, Heidelberg (2006)
31. Sterling, L., Shapiro, E.: The Art of Prolog: Advanced Programming Techniques, 2nd edn. The MIT Press, Cambridge (1994)
32. Wielemaker, J.: SWI-Prolog, http://www.swi-prolog.org
33. Wegner, P., Goldin, D.: Mathematical models of interactive computing. Brown University Technical Report CS 99-13 (1999)
34. Yang, P., Ramakrishnan, C.R., Smolka, S.A.: A logical encoding of the π-calculus: Model checking mobile processes using tabled resolution. In: Zuck, L.D., Attie, P.C., Cortesi, A., Mukhopadhyay, S. (eds.) VMCAI 2003. LNCS, vol. 2575, pp. 116–131. Springer, Heidelberg (2002)

From Corecursive Algebras to Corecursive Monads

Jiří Adámek, Mahdieh Haddadi, and Stefan Milius

Institut für Theoretische Informatik, Technische Universität Braunschweig, Germany
adamek@iti.cs.tu-bs.de, mail@stefan-milius.eu,
haddadi_1360@yahoo.com

To the memory of Stephen Bloom

Abstract. An algebra is called corecursive if from every coalgebra a unique coalgebra-to-algebra homomorphism exists into it. We prove that free corecursive algebras are obtained as a coproduct of the final coalgebra (considered as an algebra) and with free algebras. The monad of free corecursive algebras is proved to be the free corecursive monad, where the concept of corecursive monad is a generalization of Elgot's iterative monads, analogous to corecursive algebras generalizing completely iterative algebras. We also characterize the Eilenberg-Moore algebras for the free corecursive monad and call them Bloom algebras.

1 Introduction

The study of structured recursive definitions is fundamental in many areas of computer science. It can use algebraic methods extended by suitable recursion concepts. One such example are completely iterative algebras: these are algebras in which recursive equations with parameters have unique solutions, see [14]. In the present paper we study corecursive algebras, which are algebras for a given endofunctor H in which recursive equations without parameters have unique solutions or, equivalently, which for every coalgebra have a unique coalgebra-to-algebra morphism. V. Capretta, T. Uustalu and V. Vene [10] present applications of corecursive algebras for the semantics of structural corecursion in languages of total functional programming [18], such as R. Cockett's charity [11], and other settings, where unrestricted general recursion is unavailable. The dual concept, recursive coalgebra, was introduced by G. Osius in [15] to categorically capture well-founded induction. For endofunctors weakly preserving pullbacks P. Taylor proved that recursive coalgebras are equivalent to parametrically recursive ones, see [16]. Recursive coalgebras were also studied by V. Capretta et al. [9]. In the dual situation, since weak preservation of pushouts is rare, the concepts of corecursive algebra and completely iterative one usually do not coincide. The former was studied by V. Capretta et al. [10], and various counter-examples demonstrating e.g. the difference of the two concepts for algebras can be found there.

In the present paper we contribute to the development of the mathematical theory of corecursive algebras. The goal is to eventually arrive at a useful body of results and constructions for these algebras. A major ingredient of any theory of algebraic structures is the study of how to freely endow an object with the structure of interest. So the main focus of the present paper are corecursive H-algebras freely generated by an object Y. Let FY denote the free H-algebra on Y and T the final H-coalgebra

A. Corradini, B. Klin, and C. Cîrstea (Eds.): CALCO 2011, LNCS 6859, pp. 55–69, 2011.

(which, due to Lambek's Lemma, can be regarded as an algebra). We prove that the coproduct of these two algebras

$$MY = T \oplus FY$$

is the free corecursive algebra on Y. Here $\oplus$ is the coproduct in the category of H-algebras. For example for the endofunctor $HX = X \times X$ the algebra MY consists of all (finite and infinite) binary trees with finitely many leaves labelled in Y.

We also introduce the concept of a corecursive monad. This is a weakening of completely iterative monads of C. Elgot, S. Bloom and R. Tindell [13] analogous to corecursive algebras as a weakening of completely iterative ones. The monad $Y \mapsto MY$ of free corecursive algebras is proved to be corecursive, actually, this is the free corecursive monad generated by H. For endofunctors of **Set** we also prove the converse: whenever H generates a free corecursive monad, then it has free corecursive algebras (and the free monad is then given by the corresponding adjunction).

Finally, we study the equational properties of the solution operation in corecursive algebras. In category-theoretic terms, we characterize the Eilenberg-Moore algebras for the free corecursive monad: they are H-algebras equipped with an operation $\dagger$ that assigns to every recursive equation without parameters a solution, where $\dagger$ is subject to one axiom stating that the assignment of solutions is functorial (or uniform). We call these algebras Bloom algebras; they are analogous to the complete Elgot algebras of [4] where the corresponding monad was the free completely iterative monad on H. The characterization of the Eilenberg-Moore algebras for the free corecursive monad can be understood as a kind of completeness result: all equational properties of $\dagger$ that hold in every corecursive algebra follow from the properties of $\dagger$ given in the definition of a Bloom algebra.

2 Corecursive Algebras

The following definition is the dual of the concept introduced by G. Osius in [15] and studied by P. Taylor [16]. We assume throughout the paper that a category $\mathcal{A}$ and an endofunctor $H : \mathcal{A} \to \mathcal{A}$ are given. We denote by $\mathsf{Alg}\, H$ the category of algebras $a : HA \to A$ and homomorphisms, and by $\mathsf{Coalg}\, H$ the category of coalgebras $e : X \to HX$ and homomorphisms. A coalgebra-to-algebra morphism from the latter to the former is a morphism $f : X \to A$ such that $f = a.Hf.e$.

Definition 2.1. *An algebra $a : HA \to A$ is called* corecursive *if for every coalgebra $e : X \to HX$ there exists a unique coalgebra-to-algebra homomorphism $e^\dagger : X \to A$. That is, the square*

$$\begin{array}{ccc} X & \xrightarrow{e^\dagger} & A \\ {\scriptstyle e}\downarrow & & \uparrow{\scriptstyle a} \\ HX & \xrightarrow[He^\dagger]{} & HA \end{array}$$

commutes. We call e an equation morphism *and $e^\dagger$ its* solution.

Example 2.2. (1) V. Capretta et al. [10] studied this concept of corecursive algebras and compared it with a number of related concepts. A concrete example of corecursive algebra from [10], for the endofunctor $HX = E \times X \times X$ of **Set**, is the set E^ω of all streams. The operation $a : E \times E^\omega \times E^\omega \to E^\omega$ is given by $a(e, u, v)$ having head e and continuing by the merge of u and v.

(2) If H has a final coalgebra $\tau : T \to HT$, then by Lambek's Lemma τ is invertible and the resulting algebra $\tau^{-1} : HT \to T$ is corecursive. In fact, this is the initial corecursive algebra, that is, for every corecursive algebra (A, a) a unique algebra homomorphism from (T, τ^{-1}) exists, see (the dual of) Proposition 2 in [9]. There also the converse is proved (dual of Proposition 7), that is, if the initial corecursive algebra exists, then it is a final coalgebra (via the inverse of the algebra structure).

(3) The trivial final algebra $H1 \to 1$, where 1 is the final object in $\mathcal{A}$, is clearly corecursive.

(4) If $a : HA \to A$ is a corecursive algebra, then so is $Ha : HHA \to HA$, see the dual of Proposition 6 in [9].

(5) Combining (3) and (4) we conclude that the final ω^{op}-chain

$$1 \xleftarrow{a} H1 \xleftarrow{Ha} HH1 \xleftarrow{HHa} \cdots$$

consists of corecursive algebras. Indeed, the continuation to H^i1 for all ordinals (with $H^i1 = \lim_{k \le i} H^k1$ for all limit ordinals) also yields corecursive algebras. This follows from Proposition 2.6 below.

Remark 2.3. For an endofunctor of **Set**, we can view $e : X \to HX$ as a system of recursive equations using variables from the set X and $e^\dagger : X \to A$ as the solution of the system. We illustrate this on classical algebras for a signature Σ. Equivalently, these are the algebras for the polynomial set functor

$$H_\Sigma X = \coprod_{\sigma \in \Sigma} X^n$$

where n is the arity of σ. For every set X (of recursion variables) and every system of mutually recursive equations

$$x = \sigma(x_1, \ldots, x_n),$$

one for every $x \in X$, where $\sigma \in \Sigma$ has arity n and $x_i \in X$, we get the corresponding coalgebra $e : X \to H_\Sigma X$with $x \mapsto (x_1, \ldots, x_n)$ in the σ-summand X^n. The square in Definition 2.1 tells us that the substitution of $e^\dagger(x)$ for $x \in X$ makes the formal equations $x = \sigma(x_1, \ldots, x_n)$ identities in A:

$$e^\dagger(x) = \sigma^A(e^\dagger(x_1), \ldots, e^\dagger(x_n)).$$

Example 2.4. Binary algebras: For $HX = X \times X$, every algebra (given by the binary operation "$*$" on a set A) which is corecursive has a unique *idempotent* $i = i * i$. This is the solution of the recursive equation

$$x = x * x$$

expressed by the isomorphism $e : 1 \xrightarrow{\sim} 1 \times 1$. Moreover the idempotent is *completely factorizable*, where the set of all completely factorizable elements is the largest subset of A such that every element a in it can be factorized as $a = b * c$, with b, c completely factorizable. The corecursiveness of A implies that no other element but i is completely factorizable: consider the system of recursive equations

$$x_\epsilon = x_0 * x_1, \qquad x_0 = x_{00} * x_{01}, \qquad \dots \qquad x_w = x_{w0} * x_{w1}, \qquad \dots$$

for all binary words w. Every completely factorizable element a provides a solution $e^\dagger$ with $e^\dagger(x_\epsilon) = a$. Since solutions are unique, $a = i$.

Conversely, every binary algebra A with an idempotent i which is the only completely factorizable element is corecursive. Indeed, given a morphism $e : X \to X \times X$, the constant map $e^\dagger : X \to A$ with value i is a coalgebra-to-algebra morphism. Conversely, if $e^\dagger$ is a coalgebra-to-algebra morphism, then for every $x \in X$, the element $e^\dagger(x)$ is clearly completely factorizable. Therefore, $e^\dagger(x) = i$.

Remark 2.5. Recall the concept of *completely iterative algebra* (*cia* for short) from [14]: it is an algebra $a : HA \to A$ such that for every "flat equation" morphism $e : X \to HX + A$ there exists a unique solution, i.e. a unique morphism $e^\dagger$ such that

$$e^\dagger = (X \xrightarrow{e} HX + A \xrightarrow{He^\dagger + A} HA + A \xrightarrow{[a,A]} A).$$

This is obviously stronger than corecursiveness because every coalgebra $e : X \to HX$ yields a flat equation morphism $\mathsf{inl}.e : X \to HX + A$, where $\mathsf{inl} : X \to X + A$ denotes the left-hand coproduct injection.[1] Then solutions of e are in bijective correspondence with coalgebra-to-algebra homomorphisms from e to a. Since the former exists uniquely, so does the latter. Thus, for example, in the category of complete metric spaces with distance less than one and nonexpanding functions, all algebras for contracting endofunctors (in the sense of P. America and J. Rutten [7]) are corecursive, because, as proved in [14], they are cia's. Here is a concrete example: $HX = X \times X$ equipped with the metric taking $1/2$ of the maximum of the two distances is contracting. Thus every binary algebra whose operation is contracting is corecursive.

Proposition 2.6. *Let $\mathcal{A}$ be a complete category. Then corecursive algebras are closed under limits in $\mathsf{Alg}\,H$. Thus, limits of corecursive algebras are formed on the level of $\mathcal{A}$.*

Lemma 2.7. *Every homomorphism $h : (A,a) \to (B,b)$ in $\mathsf{Alg}\,H$ with (A,a) and (B,b) corecursive, preserves solutions. That is, given a coalgebra $e : X \to HX$ with a solution $e^\dagger : X \to A$ in the domain algebra, then $h.e^\dagger : X \to B$ is the solution in the codomain one.*

We thus consider corecursive algebras as a full subcategory $\mathsf{Alg}_C\,H$ of $\mathsf{Alg}\,H$. We obtain a forgetful functor $\mathsf{Alg}_C\,H \to \mathcal{A}$ with $(A,a) \mapsto A$. In Section 4 we prove that this forgetful functor has a left adjoint, that is, free corecursive algebras exist, if and only

[1] Similarly, $\mathsf{inr} : A \to X + A$ denotes the right-hand coproduct injection.

if a terminal coalgebra T exists and every object Y generates a free algebra FY (i.e., the forgetful functor $\mathsf{Alg}\,H \to \mathcal{A}$ has a left adjoint). This equivalence holds for all set functors, and for them the formula for the free corecursive algebra is $T \oplus FY$, where $\oplus$ is the coproduct in $\mathsf{Alg}\,H$.

3 Bloom Algebras

For iterative algebras, it was proved in [5] that every finitary functor H of $\mathcal{A}$ has a free iterative algebra, and the resulting monad $\mathbb{R}$ of $\mathcal{A}$ is a free iterative monad. The next step was a characterization of the Eilenberg-Moore algebras for $\mathbb{R}$ that were called Elgot algebras, see [4]. An Elgot algebra has for every finitary flat equation e a solution $e^\dagger$, but not necessarily unique. Instead, Elgot algebras are equipped with a solution operation $e \mapsto e^\dagger$ satisfying two "natural" axioms.

In the present section we take the corresponding step for corecursive algebras. We introduce Bloom algebras as algebras equipped with an operation assigning to every coalgebra e a solution $e^\dagger$, and the operation $\dagger$ forms a functor. Later we prove that Bloom algebras are (analogously to Elgot algebras) precisely the Eilenberg-Moore algebras for the free corecursive monad, see Theorem 4.13.

Definition 3.1. *A* Bloom algebra *is a triple* $(A, a, \dagger)$ *where* $a : HA \to A$ *is an* H*-algebra and* $\dagger$ *is an operation assigning to every coalgebra* $e : X \to HX$ *a coalgebra-to-algebra homomorphism* $e^\dagger : X \to A$ *so that* $\dagger$ *is* functorial. *This means that we can define a functor* $\dagger : \mathsf{Coalg}\,H \to \mathcal{A}/A$. *More explicitly, for every* H*-coalgebra homomorphism* $h : (X, e) \to (Y, f)$ *we have* $f^\dagger.h = e^\dagger : X \to A$.

Example 3.2. (1) Every corecursive algebra can be considered as a Bloom algebra. Indeed, functoriality easily follows from the uniqueness of solutions due to the diagram

$$\begin{array}{ccccc} X & \xrightarrow{h} & X' & \xrightarrow{f^\dagger} & A \\ {\scriptstyle e}\downarrow & & \downarrow{\scriptstyle Hf} & & \uparrow{\scriptstyle a} \\ HX & \xrightarrow[Hh]{} & HX' & \xrightarrow[Hf^\dagger]{} & HA \end{array}$$

(2) Let $\mathcal{A}$ have finite products. An algebra $a : A \times A \to A$ for $HX = X \times X$ can be equipped with a Bloom algebra structure if and only if it has an idempotent global element, that is $i : 1 \to A$ satisfying $a.(i \times i) = i$ (recall that $1 \times 1 = 1$). More precisely:

(a) Given an idempotent i, we have a Bloom algebra $(A, a, \dagger)$, where $\dagger$ is given by

$$e^\dagger = (X \xrightarrow{!} 1 \xrightarrow{i} A).$$

(b) Given a Bloom algebra $(A, a, \dagger)$, there exists an idempotent i such that $\dagger$ is the constant function with value $e^\dagger = i.!$.

(3) Every group, considered as a binary algebra in **Set**, is thus a Bloom algebra in a unique sense.

(4) Every continuous algebra is a Bloom algebra if we define $e^\dagger$ to be the least solution of e. More detailed, let H be a locally continuous endofunctor of the category CPO of complete ordered sets (i.e., partially ordered sets with a least element and with joins of ω-chains) and continuous functions. For every continuous algebra $a : HA \to A$ and every equation morphism $e : X \to HA$, we can define in $\mathsf{CPO}(X, A)$ a function $e^\dagger : X \to A$ as the join of the sequence $e_n^\dagger$ defined by $e_0^\dagger = \mathsf{const}_\perp$ and $e_{n+1}^\dagger = a.He_n^\dagger.e$. Thus, the least solution is $e^\dagger = \bigvee e_n^\dagger$ and $(A, a, \dagger)$ is a Bloom algebra.

(5) Every limit of Bloom algebras is a Bloom algebra. Indeed, this is proved precisely as Proposition 2.6.

(6) Every complete Elgot algebra in the sense of [4] is a Bloom algebra.

Definition 3.3. *By a* homomorphism *of Bloom algebras from* $(A, a, \dagger)$ *to* $(B, b, \ddagger)$ *is meant an algebra homomorphism* $h : (A, a) \to (B, b)$ *preserving solutions, that is, for every coalgebra* $e : X \to HX$ *we have*

$$e^\ddagger = (X \xrightarrow{e^\dagger} A \xrightarrow{h} B).$$

We denote by $\mathsf{Alg}_B\, H$ *the corresponding category of Bloom algebras.*

Proposition 3.4. *An initial Bloom algebra is precisely a final coalgebra.*

More precisely, the statement in Example 2.2(2) generalizes from corecursive algebras to Bloom algebras. In fact, the proof in [9] can be used again.

Lemma 3.5. *If* $(A, a, \dagger)$ *is a Bloom algebra and* $h : (A, a) \to (B, b)$ *is a homomorphism of algebras, then there is a unique structure of a Bloom algebra on* (B, b) *such that* h *is a solution preserving morphism. We call it, the Bloom algebra* induced *by* h.

Remark 3.6. We are going to characterize the left adjoint of the forgetful functor

$$U : \mathsf{Alg}_B\, H \to \mathcal{A} \ \text{ with } \ (A, a, \dagger) \mapsto A.$$

In other words, free Bloom algebras exists. Moreover, these are coproducts $T \oplus FY$ of free algebras and the final coalgebra. For that we first attend to the existence of those ingredients.

Lemma 3.7. *Let* $\mathcal{A}$ *be a complete category. If* H *has a free Bloom algebra on an object* Y *with* $\mathcal{A}(Y, HY) \neq \emptyset$, *then* H *has a final coalgebra.*

Proof. The free Bloom algebra $(A, a, \dagger)$ on Y is weakly initial in $\mathsf{Alg}_B\, H$. To see this, choose a morphism $e : Y \to HY$. For every Bloom algebra $(B, b, \ddagger)$ the solution $e^\ddagger : Y \to B$ extends to a homomorphism $h : (A, a, \dagger) \to (B, b, \ddagger)$ of Bloom algebras.

Since $\mathsf{Alg}_B\, H$ is complete by Example 3.2(5), we can use Freyd's Adjoint Functor Theorem. The existence of a weakly initial object implies that $\mathsf{Alg}_B\, H$ has an initial object. Now apply Proposition 3.4. □

Proposition 3.8. *Let* (T, τ) *be a final coalgebra for* H. *The category of Bloom algebras for* H *is isomorphic to the slice category* $(T, \tau^{-1})/\mathsf{Alg}\, H$.

Construction 3.9. Free-algebra chain. Recall from [2] that if $\mathcal{A}$ is cocomplete, we can define a chain constructing the free algebra on Y as follows:

$$Y \xrightarrow{\mathsf{inr}} HY + Y \xrightarrow{H\mathsf{inr}+Y} H(HY+Y)+Y \longrightarrow \cdots$$

We mean the essentially unique chain $V : \mathsf{Ord} \to \mathcal{A}$ with

$$V_0 = Y, \quad V_{i+1} = HV_i + Y, \quad \text{and} \quad V_j = \mathrm{colim}_{k<j} V_k \text{ for limit ordinals } j,$$

whose connecting morphisms $v_{ij} : V_i \to V_j$ are defined by

$$v_{0,1} \equiv \mathsf{inr} : Y \to HY + Y \qquad \text{and} \qquad v_{i+1,j+1} \equiv Hv_{i,j} + Y,$$

and for limit ordinals j, $(v_{k,j})_{k<j}$ is the colimit cocone.

This chain is called the *free-algebra chain*. If it *converges* at some ordinal λ, which means that $v_{\lambda,\lambda+1}$ is an isomorphism, then V_λ is a free algebra on Y. More detailed: this isomorphism turns V_λ into a coproduct $V_\lambda = HV_\lambda + Y$ and thus V_λ is an H-algebra via the right-hand injection, and the left-hand one $Y \to V_\lambda$ is the universal arrow.

Definition 3.10. *(See [17]) We say that monomorphisms are* constructive *provided that*

(a) *if $m_i : A_i \to B_i$ are monomorphisms for $i = 1, 2$ then $m_1 + m_2 : A_1 + A_2 \to B_1 + B_2$ is a monomorphism,*
(b) *coproduct injections are monomorphisms, and*
(c) *if $a_i : A_i \to A$, $(i < \alpha)$, is a colimit of an α-chain and $m : A \to B$ has all composites $m.a_i$ monic, then m is monic.*

Example 3.11. Sets, posets, graphs and abelian groups have constructive monomorphisms. If $\mathcal{A}$ has constructive monomorphisms, then all functor categories $\mathcal{A}^{\mathcal{C}}$ do. In all locally finitely presentable categories (c) holds (see [6]) but (a) and (b) can fail. For example, in the category of rings (b) fails because the initial ring $\mathbb{Z}$ has morphisms $f : \mathbb{Z} \to A$ that fail to be monomorphisms. And f is a coproduct injection of $A = A + \mathbb{Z}$.

Proposition 3.12. *Let $\mathcal{A}$ be a cocomplete, wellpowered category with constructive monomorphisms. If H has a free Bloom algebra on Y and preserves monomorphisms, then it also has a free algebra on Y.*

Sketch of proof. Given a free Bloom algebra MY the coproduct $HMY + Y$ also carries the structure of a Bloom algebra. From that we derive $MY \simeq HMY + Y$ and obtain a cone of the free-algebra chain formed by monomorphisms $m_i : V_i \to MY$. Since $\mathcal{A}$ is wellpowered, the chain converges. □

Notation 3.13. The coproduct in $\mathsf{Alg}\, H$ is denoted by $(A, a) \oplus (B, b)$.

Theorem 3.14. *Suppose that H has a terminal coalgebra T, a free algebra FY on Y, and their coproduct $T \oplus FY$. Then the last algebra is the free Bloom algebra induced by* $\mathsf{inl} : T \to T \oplus FY$ *(cf. Lemma 3.5) with the universal arrow* $\mathsf{inr}.\eta : Y \to T \oplus FY$.

Proof. Given a Bloom algebra $(B, b, \ddagger)$ and morphism $g : Y \to B$, we obtain a unique homomorphism $\overline{g} : (FY, \varphi_Y) \to (B, b)$ with $g = \overline{g}.\eta$. We also have a unique solution preserving homomorphism $f : (T, \tau^{-1}, \dagger) \to (B, b, \ddagger)$, see Proposition 3.4. This yields a homomorphism $[f, \overline{g}] : T \oplus FY \to B$ which is solution preserving since solutions in $T \oplus FY$ have the form $\mathsf{inl}.e^\dagger$. Thus, $[f, \overline{g}].(\mathsf{inl}.e^\dagger) = f.e^\dagger = e^\ddagger$. And this is the desired morphism since $[f, g].\mathsf{inr}.\eta = \overline{g}.\eta = g$.

Conversely, given a solution preserving homomorphism $h : T \oplus FY \to B$ with $h.\mathsf{inr}.\eta = g$, then $h = [f, \overline{g}]$, because $h.\mathsf{inl} : T \to B$ is clearly solution preserving, hence $h.\mathsf{inl} = f$. Also $h.\mathsf{inr}$ is a homomorphism from FY with $h.\mathsf{inr}.\eta = g$, thus $h.\mathsf{inr} = \overline{g}$. □

Recall from [6] that given an infinite cardinal number λ, a functor is called *λ-accessible* if it preserves λ-filtered colimits. An object X whose hom-functor $\mathcal{A}(X, -)$ is λ-accessible is called *λ-presentable*. A category $\mathcal{A}$ is *locally λ-presentable* if it has (a) colimits, and (b) a set of λ-presentable objects whose closure under λ-filtered colimits is all of $\mathcal{A}$. For a λ-accessible endofunctor H, the category $\mathsf{Alg}\, H$ is also locally λ-presentable (see [6]).

Corollary 3.15. *Every accessible endofunctor of a locally presentable category has free Bloom algebras. They have the form $T \oplus FY$.*

4 Free Corecursive Algebras

For accessible functors H we now prove that free corecursive algebras MY exist and, if H preserves monomorphisms, they coincide with the free Bloom algebras $MY = T \oplus FY$. Moreover an iterative construction of these free algebras (closely related to the free algebra chain in Construction 3.9) is presented.

We first prove that the category of corecursive algebras is strongly epireflective in the category of Bloom algebras. That is, the full embedding is a right adjoint, and the components of the unit of the adjunction are strong epimorphisms.

Proposition 4.1. *For every accessible endofunctor of a locally presentable category, corecursive algebras form a strongly epireflective subcategory of the category of Bloom algebras. In particular, every Bloom subalgebra of a corecursive algebra is corecursive.*

Corollary 4.2. *Every accessible endofunctor of a locally presentable category has free corecursive algebras.*

Indeed, since the functors $\mathsf{Alg}_C\, H \hookrightarrow \mathsf{Alg}_B\, H$ and $\mathsf{Alg}_B\, H \to \mathcal{A}$ have left adjoints by Proposition 4.1 and Corollary 3.15, this composite has a left adjoint, too.

Remark 4.3. We conjecture that in the generality of the above corollary, the free corecursive algebras are $T \oplus FY$ (as in Corollary 3.15). But we can only prove this in case H preserves monomorphisms and monomorphisms are constructive. We are going to apply the following transfinite construction of free corecursive algebras closely related to the free-algebra construction of 3.9

Construction 4.4. Free Corecursive Chain. Let $\mathcal{A}$ be cocomplete and H have a final coalgebra (T, τ). For every object Y we define an essentially unique chain $U : \mathsf{Ord} \to \mathcal{A}$ by

$$U_0 = T, \quad U_{i+1} = HU_i + Y, \quad \text{and} \quad U_j = \mathrm{colim}_{k<j} U_k \text{ for limit ordinals } j.$$

The connecting morphisms $u_{i,j} : U_i \to U_j$ are defined by

$$u_{0,1} = (T \xrightarrow{\tau} HT \xrightarrow{\mathsf{inl}} HT + Y), \quad u_{i+1,j+1} = Hu_{i,j} + id_Y$$

and for limit ordinals j, $(u_{k,j})_{k<j}$ is the colimit cocone.
We say that the chain *converges* at λ if the connecting morphism $u_{\lambda,\lambda+1}$ is an isomorphism, thus $U_\lambda = HU_\lambda + Y$ so that U_λ is an H-algebra (via inl) connected to Y via $\mathsf{inr} : Y \to U_\lambda$.

Proposition 4.5. *Let $\mathcal{A}$ be a cocomplete and wellpowered category with constructive monomorphisms, and let H preserve monomorphisms and have a final coalgebra. If the corecursive chain for Y converges in λ steps, then $U_\lambda = T \oplus FY$.*

Sketch of proof. The free algebra FY exists because we have a natural transformation $m_i : V_i \to U_{i+1}$ from the free-algebra chain to the corecursive chain: put $m_0 = inr : Y \to HT + Y$ and $m_{i+1} = Hm_i + id_Y$. Since the m_i's are monomorphisms and U_i converges, so does V_i. Thus $FY = V_\rho$ for some $\rho \geq \lambda$. One readily proves that U_λ is a coproduct (in $\mathsf{Alg}\, H$) of V_ρ and T w.r.t. the injections $m_\rho : FY \to MY$ and $u_{0,\lambda} : T \to MY$. □

Theorem 4.6. *Let $\mathcal{A}$ be a locally presentable category with constructive monomorphisms. Every accessible endofunctor preserving monomorphisms has free corecursive algebras $MY = T \oplus FY$.*

Sketch of proof. Let X be an accessible endofunctor of $\mathcal{A}$. From Theorem 3.14 we know that $T \oplus FY$ is a free Bloom algebra, thus, it is sufficient to prove that this algebra is corecursive. For that, we use Proposition 4.1 and find a corecursive algebra such that $T \oplus FY$ is its subalgebra; this will finish the proof.

The endofunctor $H(-) + Y$ is also accessible. Thus, it also has a final coalgebra. We denote it by TY. The components of the inverse of its algebra structure $TY \xrightarrow{\sim} H(TY) + Y$ are denoted by $\tau_Y : H(TY) \to TY$ and $\eta_Y : Y \to TY$. As proved in [14], the algebra TY is a cia for H, cf. Remark 2.5. Next one proves that $T \oplus FY$ is a subalgebra of this H-algebra TY.

Then we use the final opchain of $H(-) + Y$ which converges and yields TY. This yields a canonical monomorphism from T to TY in $\mathsf{Alg}\, H$, and this is used to start a cocone of monomorphisms in $\mathsf{Alg}\, H$ on the corecursive chain U_i for H with vertex TY. Since $\mathcal{A}$ is wellpowered, this implies that U_i converges. □

Example 4.7. Free corecursive algebras MY obtained as U_ω.

(1) For $H = Id$ we have

$$MY = U_\omega = 1 + Y + Y + Y + \cdots$$

Indeed, the final H-coalgebra is $T = 1$, and

$$U_1 = T + Y, \quad U_2 = T + Y + Y, \quad \cdots$$

with colimit $U_\omega = 1 + Y + Y + Y + \cdots$.

(2) More generally, let $H : \mathcal{A} \to \mathcal{A}$ preserve countable coproducts and have a final coalgebra T. Then $MY = U_\omega = T + Y + HY + H^2Y + \cdots$.

(3) For the endofunctor $HX = X \times X$ of **Set** (of binary algebras) we have that the free corecursive algebra MY consists of

$$MY = \text{all binary trees with finitely many leaves which are labelled in } Y.$$

(4) More generally, let $\Sigma = (\Sigma_k)_{k<\omega}$ be a signature. Then Σ-algebras are precisely the algebras for the polynomial endofunctor H_Σ as explained in Remark 2.3.

Recall that the final H_Σ-coalgebra is the coalgebra of all Σ-*trees*, that is, (possibly infinite, rooted and ordered) trees labelled in Σ so that every node with a label of arity n has precisely n children. And FY is the algebra of all finite $(\Sigma + Y)$-trees, where members of Y are considered to have arity 0. Then U_n is the set of all $(\Sigma + Y)$-trees with no leaf of depth greater than n having a label from Y. (That means that all leaves with level n or more are labelled by a nullary symbol in Σ_0.) Consequently, the free corecursive algebra is

$$MY = T \oplus FY = \text{ all } (\Sigma + Y)\text{-trees with only finitely many } Y\text{-labelled leaves.}$$

Remark 4.8. (1) A *pre-fixed point* of a functor H is an object A such that HA is a subobject of A.

(2) A fixed point, i.e. an object $A \simeq HA$, can be considered as an algebra or a coalgebra for H. When we speak about *corecursive fixed points*, we mean fixed points $HA \xrightarrow{\sim} A$ that are corecursive algebras.

Theorem 4.9. *For every set endofunctor, the following statements are equivalent:*

(i) *all free corecursive algebras exist,*
(ii) *all free algebras and a final coalgebra exist, and*
(iii) *arbitrarily large pre-fixed points and a corecursive fixed point exist.*

They imply that the free corecursive algebra on Y is $T \oplus FY$.

Sketch of proof. Without loss of generality, H preserves monomorphisms. Then (i)⇒(ii) by Propositions 3.8 and 3.12 (which are true for corecursive algebras). The proof of (ii)⇒(i) is analogous to that of Theorem 4.6, just in lieu of the final coalgebra TY for $H(-) + Y$ we use members of the final opchain of that functor. The equivalence of (ii) and (iii) follows from the fact that (a) arbitrarily large pre-fixed points are necessary and sufficient for the existence of free algebras, see Theorem II.4 in [17] and (b) every corecursive fixed point is an initial corecursive algebra, thus, a final coalgebra, see Proposition 7 in [9]. □

Notation 4.10. Let H have free corecursive algebras MY. Denote by $\delta_Y : HMY \to MY$ the algebra structure and by $\eta_Y : Y \to MY$ the universal map. Then we obtain a unique homomorphism $\mu_Y : (MMY, \delta_{MY}) \to (MY, \delta_Y)$ with $\mu_Y.\eta_{MY} = id$:

$$\begin{array}{ccccc} HMMY & \xrightarrow{\delta_{MY}} & MMY & \xleftarrow{\eta_{MY}} & MY \\ \downarrow{\scriptstyle H\mu_Y} & & \downarrow{\scriptstyle \mu_Y} & \swarrow{\scriptstyle id_{MY}} & \\ HMY & \xrightarrow[\delta_Y]{} & MY & & \end{array} \tag{4.1}$$

The triple $\mathbb{M} = (M, \mu, \eta)$ is the monad generated by the adjoint situation $\mathsf{Alg}_C\, H \leftrightarrows \mathcal{A}$.

Example 4.11. We have $MY = \mathbb{N} \times Y + 1$ for the identity functor on **Set**, see Example 4.7(1). And for $HX = X \times X$ we have

$$MY = \text{binary trees with finitely many leaves labelled in } Y;$$

see Example 4.7(3). The functor $HX = \coprod_{n<\omega} X^n$ generates the monad

$$MY = \text{finitely branching trees with finitely many leaves labelled in } Y,$$

cf. Example 4.7(4).

Remark 4.12. Since μ_Y is, by definition, a homomorphism we have $\mu_Y.\delta_{MY} = \delta_Y.H\mu_Y$, and the unit law $\mu_Y.\eta_{MY} = id$ yields $\delta_Y = \mu_Y.\delta_{MY}.H\eta_{MY}$. It easy to prove that the δ_Y's are the components of a natural transformation $\delta : HM \to M$, and so are η and μ being part of the monad $\mathbb{M}$.

Theorem 4.13. *Let $\mathcal{A}$ be a locally presentable category with constructive monomorphisms. Let H be an accessible endofunctor preserving monomorphisms. Then Bloom algebras are precisely the Eilenberg-Moore algebras for $\mathbb{M}$, i.e., the category* $\mathsf{Alg}_B\, H$ *is isomorphic to $\mathcal{A}^{\mathbb{M}}$.*

Sketch of proof. From Theorem 3.14 and Proposition 4.5 we know that the monad generated by free Bloom algebras is $\mathbb{M}$. By Beck's Theorem we only need to verify that the forgetful functor U of $\mathsf{Alg}_B\, H$ creates coequalizers of U-split pairs. □

5 Corecursive Monads

The iterative theories (or iterative monads) of C. Elgot [12] were introduced as a formalization of iteration in an algebraic setting, and in [13] completely iterative theories are studied. We first recall the concept of a completely iterative monad, and then introduce the weaker concept of a corecursive monad. The relationship between these two concepts is analogous to the relationship between cia's (see Remark 2.5) and corecursive algebras. The following definition is, for the base category **Set**, equivalent to completely iterative theories, as shown in [1].

Definition 5.1. *(See [1]) (1) An* ideal monad *is a six-tuple* $\mathbb{S} = (S, \eta, \mu, S', \sigma, \mu')$ *consisting of a monad* (S, η, μ)*, a subfunctor* $\sigma : S' \to S$ *(called the* ideal of $\mathbb{S}$ *) such that* $S = S' + Id$ *with injections* σ *and* η*, and a natural transformation* $\mu' : S'S \to S'$ *restricting* μ*, i.e., with* $\sigma.\mu' = \mu.\sigma S$*.*

(2) An equation morphism with parameters *for* $\mathbb{S}$ *is a morphism* $e : X \to S(X+Y)$*, we call* X *the* variables *and* Y *the* parameters *of* e*. It is called* ideal *if it factorizes through* σ_{X+Y}*. A solution of* e *is a morphism* $e^\dagger : X \to SY$ *such that*

$$e^\dagger = (X \xrightarrow{e} S(X+Y) \xrightarrow{S[e^\dagger, \eta_Y]} SSY \xrightarrow{\mu_Y} SY) \tag{5.1}$$

(3) An ideal monad is called completely iterative *provided that every ideal equation morphism has a unique solution.*

Example 5.2. (See [1]) Let H be an endofunctor of $\mathcal{A}$ such that for every object Y a final coalgebra TY of $H(-)+Y$ exists. Then the assignment $Y \mapsto TY$ yields a monad (T, η, μ), which is the monad of free cia's for H. This is an ideal monad w.r.t. $T' = HT$ and $\mu' = H\mu$. Moreover, this monad is completely iterative, indeed, the free completely iterative monad on H.

For example the set functor $HX = X \times X$ generates the free completely iterative monad $\mathbb{T}$ where TY consists of all binary trees with leaves labelled in Y.

Definition 5.3. *Let* $\mathbb{S}$ *be an ideal monad. An* equation morphism *(without parameters) is a morphism* $e : X \to SX$*, and* e *is called* ideal *if it factorizes through* σ_X*, i.e., there exist* $e_0 : X \to S'X$ *such that* $e = \sigma_X \cdot e_0$*. The monad* $\mathbb{S}$ *is called* corecursive *if every ideal equation morphism* e *has a unique solution* $e^\dagger$*, i.e.,* $e^\dagger : X \to SY$ *such that* $e^\dagger = \mu_Y \cdot Se^\dagger \cdot e$*.*

Example 5.4. Examples of corecursive monads on **Set**.

(1) All the monads of Example 4.11 are corecursive, as we will see in Theorem 6.4 below.

(2) All completely iterative monads are corecursive, e.g, $\mathbb{S}$ where SY consists of all finitely branching trees with leaves labelled in Y. This is the free completely iterative monad on the functor $HX = \coprod_{n<\omega} X^n$.

(3) The monad

$$RY = \text{all rational, finitely branching trees with leaves labelled in } Y,$$

where *rational* means that the tree has up to isomorphism only finitely many subtrees. This is a corecursive monad that is neither free on any endofunctor, nor completely iterative.

(4) More generally, every submonad of $\mathbb{S}$ in item (2) containing the complete binary tree is corecursive.

Proposition 5.5. *The monad* $\mathbb{M} = (M, \eta, \mu)$ *of free corecursive algebras (of Notation 4.10) is ideal w.r.t. the ideal* $M' = HM$ *where* $\sigma = \delta : HM \to M$ *and* $\mu' = H\mu : HMM \to HM$*.*

Theorem 5.6. *The monad $\mathbb{M}$ of free corecursive algebras is corecursive.*

Sketch of proof. For every ideal equation morphism $e : X \to MX$ we form an equation morphism $\overline{e} : MX \to HMX$ by composing the isomorphism $MX \simeq HMX + X$ with $[HMX, e_0] : HMX + X \to MX$. Every algebra MY has a unique solution of $\overline{e}$, then one proves that $\overline{e}^{\dagger}.\eta_X$ is the unique solution of e. □

6 Free Corecursive Monad

In this section we prove that the corecursive monad $\mathbb{M}$ given by the free corecursive algebras is a free corecursive monad. For that we need the appropriate concept of morphism:

Definition 6.1. *(1) An* ideal monad morphism *from an ideal monad $(S, \eta^S, \mu^S, S', \sigma, \mu'^S)$ to an ideal monad $(U, \eta^U, \mu^U, U', \omega, \mu'^U)$ is a monad morphism $(S, \eta^S, \mu^S) \to (U, \eta^U, \mu^U)$ which has a domain-codomain restriction to the ideal, that is, there exists a natural transformation $\lambda' : S' \to U'$ with $\lambda.\sigma = \omega.\lambda'$.*

(2) Given a functor H, a natural transformation $\lambda : H \to S$ is called ideal *if it factorizes through $\sigma : S' \to S$.*

(3) By a free corecursive monad *on an endofunctor H is meant a corecursive monad $\mathbb{S} = (S, \mu, \eta, S', \sigma, \mu')$ together with an ideal natural transformation $\kappa : H \to S$ having the following universal property: For every ideal natural transformation $\lambda : H \to \overline{\mathbb{S}}$, where $\overline{\mathbb{S}}$ is a corecursive monad, there exists a unique ideal monad morphism $\hat{\lambda} : \mathbb{S} \to \overline{\mathbb{S}}$ such that $\lambda = \hat{\lambda} \cdot \kappa$.*

Remark 6.2. Let $\mathsf{CMon}(\mathcal{A})$ denote the category of corecursive monads and ideal monad morphisms. We have a forgetful functor to $\mathsf{Fun}(\mathcal{A})$, the category of all endofunctors of $\mathcal{A}$, assigning to every corecursive monad $\mathbb{S}$ its ideal S'. A free corecursive monad on $H \in \mathsf{Fun}(\mathcal{A})$ is precisely a universal arrow from H to the above forgetful functor.

Example 6.3. If H has free corecursive algebras, then we have the corecursive monad $\mathbb{M}$ of Proposition 5.5. And the natural transformation

$$\kappa = (H \xrightarrow{H\eta} HM \xrightarrow{\delta} M)$$

is obviously ideal. We prove that κ has the universal property:

Theorem 6.4. *If an endofunctor H has free corecursive algebras, then the corresponding monad $\mathbb{M}$ is the free corecursive monad on H.*

Remark 6.5. The proof is analogous to the corresponding theorem for free completely iterative monads, scc [14], Theorem 4.3.

Are there any other free corecursive monads than the monads $\mathbb{M}$ of free corecursive algebras? Not for endofunctors of **Set**:

Proposition 6.6. *If a set functor generates a free corecursive monad, then it has free corecursive algebras.*

Proof. Let $H : \mathbf{Set} \to \mathbf{Set}$ generate a free corecursive monad $\mathbb{S} = (S, \mu^S, \eta^S, S', \sigma, \mu')$, and let $\kappa : H \to S$ be the universal arrow. Following Theorem 4.9 we need to prove the existence of (a) arbitrary large pre-fixed points and (b) a corecursive fixed point.

The main technical statement is the fact that the ideal S' is naturally isomorphic to HS. This proof is analogous to the same proof concerning free completely iterative monads, see Sections 5 and 6 in [14]. We therefore omit it.

Ad (a). Arbitrarily large pre-fixed points. Since $SY = S'Y + Y = HSY + Y$ for every set Y, we see that SY is a pre-fixed point of cardinality at least card Y.

Ad (b). A corecursive fixed point. The isomorphism $\sigma_\emptyset : HS\emptyset \to S\emptyset$ defines a corecursive algebra for H. To prove this, consider an arbitrary equation morphism $e : X \to HX$ and form the equation morphism $\overline{e} = \kappa_X.e : X \to SX$. Then solutions of $\overline{e}$ w.r.t $\mathbb{S}$ are precisely the solutions of e (in $S\emptyset$). This is easy to prove, the details are as in the proof of Theorem 6.1 of [14]. □

7 Conclusions and Further Work

Whereas for coalgebras the concept of recursivity has several equivalent formulations (assuming the given endofunctor weakly preserves pullbacks), as proved by P. Taylor [16], in the dual situation we need to study non-equivalent variations. The present paper is dedicated to corecursive algebras A, where corecursity means that every recursive system of equations represented by a coalgebra has a unique solution in A. This is strictly weaker than the concept of a completely iterative algebra, where every *parameterized* recursive system of equations has a unique solution. For example, if we consider the endofunctor $X \mapsto X \times X$ of one binary operation in **Set**, the algebra of all binary trees with finitely many leaves is corecursive, but not completely iterative.

The main result of our paper is the description of free corecursive algebras on Y as coproducts $MY = T \oplus FY$ of the final coalgebra T and a free algebra FY (in the category of all algebras). The above example of binary trees is the free corecursive algebra $M1$ on one generator. Our description is true for all accessible (= bounded) endofunctors of **Set** and, moreover, for all endofunctors of **Set** having free corecursive algebras. For accessible functors preserving monomorphisms, more general base categories (posets, groups, monoids etc.) also allow for the above description of the free corecursive algebras.

We introduced the concept of a corecursive monad, a weakening of completely iterative monad. We proved that the assignment $Y \mapsto MY = T \oplus FY$ is the free corecursive monad on the given accessible endofunctor. And we characterized the Eilenberg-Moore algebras for this monad. We called them Bloom algebras in honor of Stephen Bloom. They play the analogous role that Elgot algebras, studied in [4], play for iterative monads: solutions of recursive equations are not required to be unique, but have to satisfy some "basic" properties. In the case of Bloom algebras, the only property needed is functoriality.

One feature that has not been treated in our paper is that of finitariness of equations: If we consider systems of recursive equations as coalgebras $e : X \to HX$, then finite systems of recursive equation are represented by coalgebras in which X is a finite set (or more generally, a finitely presentable object). We can speak about finitely corecursive algebras as those in which these finite systems have unique solutions.

Another question is: what is the analogy of the notion of an iteration monad of S. Bloom and Z. Ésik [8] in the realm of corecursive algebras? We do not know the answer but at least we can formulate the question precisely. The idea of iteration monads is to collect all "equational" properties that the operation $e \mapsto e^\dagger$ of solving recursive systems e has in trees for a signature. This can be understood as forming the monad of free iterative theories (or monads) on the category $\mathsf{Fin}(\mathcal{A})$ of finitary endofunctors, and characterizing its Eilenberg-Moore algebras: these are, as recently proved in [3], precisely the iteration theories of S. Bloom and Z. Ésik that are functorial. So we state the following problem for future work: form the monad of free finitely corecursive theories on $\mathsf{Fin}(\mathcal{A})$, what are its Eilenberg-Moore algebras?

Acknowledgment. We are grateful to Paul Levy for interesting discussions and his formulation of Proposition 3.8.

References

[1] Aczel, P., Adámek, J., Milius, S., Velebil, J.: Infinite trees and completely iterative theories: A coalgebraic view. Theoret. Comput. Sci. 300, 1–45 (2003)
[2] Adámek, J.: Free algebras and automata realizations in the language of categories. Comment. Math. Univ. Carolinæ 15, 589–602 (1974)
[3] Adámek, J., Milius, S., Velebil, J.: Elgot theories: a new perspective of the equational properties of iteration. Math. Structures Comput. Sci. (to appear)
[4] Adámek, J., Milius, S., Velebil, J.: Elgot algebras. Log. Methods Comput. Sci. 2(5:4), 31 (2006)
[5] Adámek, J., Milius, S., Velebil, J.: Iterative algebras at work. Math. Structures Comput. Sci. 16(6), 1085–1131 (2006)
[6] Adámek, J., Rosický, J.: Locally presentable and accessible categories. Cambridge University Press, Cambridge (1994)
[7] America, P., Rutten, J.: Solving reflexive domain equations in a category of complete metric. J. Comput. System Sci. 39, 343–375 (1989)
[8] Bloom, S.L., Ésik, Z.: Iteration Theories: the equational logic of iterative processes. EATCS Monographs on Theoretical Computer Science. Springer, Heidelberg (1993)
[9] Capretta, V., Uustalu, T., Vene, V.: Recursive coalgebras from comonads. Inform. and Comput. 204, 437–468 (2006)
[10] Capretta, V., Uustalu, T., Vene, V.: Corecursive algebras: A study of general structured corecursion. In: Oliveira, M.V.M., Woodcock, J. (eds.) SBMF 2009. LNCS, vol. 5902, pp. 84–100. Springer, Heidelberg (2009)
[11] Cockett, R., Fukushima, T.: About charity. Yellow Series Report 92/480/18, Dept. of Computer Science University of Calgary (1992)
[12] Elgot, C.C.: Monadic computation and iterative algebraic theories. In: Rose, H.E., Shepherdson, J.C. (eds.) Logic Colloquium '73. North-Holland Publishers, Amsterdam (1975)
[13] Elgot, C.C., Bloom, S.L., Tindell, R.: On the algebraic structure of rooted trees. J. Comput. System Sci. 16, 361–399 (1978)
[14] Milius, S.: Completely iterative algebras and completely iterative monads. Inform. and Comput. 196, 1–41 (2005)
[15] Osius, G.: Categorical set theory: a characterization of the category of set. J. Pure Appl. Algebra 4, 79–119 (1974)
[16] Taylor, P.: Practical Foundations of Mathematics. Cambridge University Press, Cambridge (1999)
[17] Trnková, V., Adámek, J., Koubek, V., Reiterman, J.: Free algebras, input processes and free monads. Comment. Math. Univ. Carolinæ 16, 339–351 (1975)
[18] Turner, D.A.: Total functional programing. J. Univ. Comput. Sci. 10(7), 751–768 (2004)

A Categorical Semantics for Inductive-Inductive Definitions

Thorsten Altenkirch[1,⋆,⋆⋆], Peter Morris[1,⋆⋆], Fredrik Nordvall Forsberg[2,⋆], and Anton Setzer[2,⋆]

[1] School of Computer Science, University of Nottingham, UK
[2] Department of Computer Science, Swansea University, UK

Abstract. Induction-induction is a principle for defining data types in Martin-Löf Type Theory. An inductive-inductive definition consists of a set A, together with an A-indexed family $B : A \to \mathrm{Set}$, where both A and B are inductively defined in such a way that the constructors for A can refer to B and vice versa. In addition, the constructors for B can refer to the constructors for A. We extend the usual initial algebra semantics for ordinary inductive data types to the inductive-inductive setting by considering dialgebras instead of ordinary algebras. This gives a new and compact formalisation of inductive-inductive definitions, which we prove is equivalent to the usual formulation with elimination rules.

1 Introduction

Induction is an important principle of definition and reasoning, especially so in constructive mathematics and computer science, where the concept of inductively defined set and data type coincide. There are two well-established approaches to model the semantics of such data types: in Martin-Löf Type Theory [14], each set A comes equipped with an eliminator which at the same time represents reasoning by induction over A and the definition of recursive functions out of A. A more categorical approach [10] models data types as initial T-algebras for a suitable endofunctor T.

At first, it would seem that the eliminator approach is stronger, as it allows us to define dependent functions $(x : A) \to P(x)$, in contrast with the non-dependent arrows $A \to B$ given by the initiality of the algebra. However, Hermida and Jacobs [12] showed that an eliminator can be defined for every initial T-algebra, where T is a polynomial functor. Ghani et. al. [9] then extended this to arbitrary endofunctors. This covers many forms of induction and data type definitions such as indexed inductive definitions [5] and induction-recursion [7] (Dybjer and Setzer [8] also give a direct proof for induction-recursion).

There are, however, other meaningful forms of data types which are not covered by these results. One such example are inductive-inductive definitions [16],

⋆ Supported by EPSRC grant EP/G033374/1.
⋆⋆ Supported by EPSRC grant EP/G034109/1.

A. Corradini, B. Klin, and C. Cîrstea (Eds.): CALCO 2011, LNCS 6859, pp. 70–84, 2011.

where a set A and a function $B : A \to \mathrm{Set}$ are simultaneously inductively defined (compare with induction-recursion, where A is defined inductively and B recursively). In addition, the constructors for B can refer to the constructors for A.

In earlier work [16], a subset of the authors gave an eliminator-based axiomatisation of a type theory with inductive-inductive definitions and showed it to be consistent. In this article, we describe a generalised initial algebra semantics for induction-induction, and prove that it is equivalent to the original axiomatisation.

One could imagine that that inductive-inductive definitions could be described by functors mapping families of sets to families of sets (similar to the situation for induction-recursion [8]), but this fails to take into account that the constructors for B should be able to refer to the constructors for A. Thus, we will see that the constructor for B can be described by an operation

$$\mathrm{Arg_B} : (A : \mathrm{Set})(B : A \to \mathrm{Set})(c : \mathrm{Arg_A}(A, B) \to A) \to \mathrm{Arg_A}(A, B) \to \mathrm{Set}$$

where $c : \mathrm{Arg_A}(A, B) \to A$ refers to the already defined constructor for A. However, $(\mathrm{Arg_A}, \mathrm{Arg_B})$ is then no longer an endofunctor and we move to the more general setting of dialgebras [11,18] to describe algebras of such functors. The equivalence between initiality and having an eliminator still carries over to this new setting.

1.1 Examples of Inductive-Inductive Definitions

Danielsson [4] and Chapman [3] define the syntax of dependent type theory in the theory itself by inductively defining contexts, types in a given context and terms of a given type. Let us concentrate on contexts and types for simplicity. There should be an empty context ε, and if we have any context Γ and a valid type σ in that context, then we should be able to extend the context with a fresh variable of that type. We end up with the following inductive definition of the set of contexts:

$$\frac{}{\varepsilon : \mathrm{Ctxt}} \qquad \frac{\Gamma : \mathrm{Ctxt} \quad \sigma : \mathrm{Type}(\Gamma)}{\Gamma \rhd \sigma : \mathrm{Ctxt}}$$

For types, let us have a base type ι (valid in any context) and a dependent function type: if σ is a type in context Γ, and τ is a type in Γ extended with a fresh variable of type σ (the variable from the domain), then $\Pi(\sigma, \tau)$ is a type in the original context. This leads us to the following inductive definition of $\mathrm{Type} : \mathrm{Ctxt} \to \mathrm{Set}$:

$$\frac{\Gamma : \mathrm{Ctxt}}{\iota_\Gamma : \mathrm{Type}(\Gamma)} \qquad \frac{\Gamma : \mathrm{Ctxt} \quad \sigma : \mathrm{Type}(\Gamma) \quad \tau : \mathrm{Type}(\Gamma \rhd \sigma)}{\Pi_\Gamma(\sigma, \tau) : \mathrm{Type}(\Gamma)}$$

Note that the definition of Ctxt refers to Type, so both sets have to be defined simultaneously. Another peculiarity is how the introduction rule for Π explicitly focuses on a specific constructor in the index of the type of τ.

For an example with more of a programming flavour, consider defining a data type consisting of sorted lists (of natural numbers, say). With induction-induction, we can simultaneously define the set SortedList of sorted lists and the predicate $\leq_{\mathrm{L}}$: $(\mathbb{N} \times \mathrm{SortedList}) \to \mathrm{Set}$ with $n \leq_{\mathrm{L}} \ell$ true if n is less than or equal to every element of ℓ.

The empty list is certainly sorted, and if we have a proof p that n is less than or equal to every element of the list ℓ, we can put n in front of ℓ to get a new sorted list $\mathrm{cons}(n, \ell, p)$. Translated into introduction rules, this becomes:

$$\frac{}{\mathrm{nil} : \mathrm{SortedList}} \qquad \frac{n : \mathbb{N} \qquad \ell : \mathrm{SortedList} \qquad p : n \leq_{\mathrm{L}} \ell}{\mathrm{cons}(n, \ell, p) : \mathrm{SortedList}}$$

For $\leq_{\mathrm{L}}$, we have that every $m : \mathbb{N}$ is trivially smaller than every element of the empty list, and if $m \leq n$ and inductively $m \leq_{\mathrm{L}} \ell$, then $m \leq_{\mathrm{L}} \mathrm{cons}(n, \ell, p)$:

$$\frac{}{\mathrm{triv}_m : m \leq_{\mathrm{L}} \mathrm{nil}} \qquad \frac{q : m \leq n \qquad p_{m,\ell} : m \leq_{\mathrm{L}} \ell}{\ll q, p_{m,\ell} \gg_{m,n,\ell,p} : m \leq_{\mathrm{L}} \mathrm{cons}(n, \ell, p)}$$

Of course, there are many alternative ways to define such a data type using ordinary induction, but the inductive-inductive one seems natural and might be more convenient for some purposes. It is certainly more pleasant to work with in the proof assistant/ programming language Agda [17] which allows inductive-inductive definitions using the `mutual` keyword. One aim of our investigation into inductive-inductive definitions is to justify their existence in Agda.

It might be worth pointing out that inductive-inductive and inductive-recursive definitions are different. Not every inductive-inductive definition can be directly translated into an inductive-recursive definition, since the inductive definition of the second type B may not proceed according to the recursive ordering. The contexts and types example above is an example of this. On the other hand, inductive-recursive definitions can use negative occurrences of B, which is not possible for inductive-inductive definitions. For instance, a universe closed under Π-types can be defined using induction-recursion but not induction-induction.

1.2 Preliminaries and Notation

We work in an extensional type theory [15] with the following ingredients:

Set We use Set to denote our universe of small types, and we write $B : A \to \mathrm{Set}$ for an A-indexed family of sets.

Π-types Given $A : \mathrm{Set}$ and $B : A \to \mathrm{Set}$, then $\big((x : A) \to B(x)\big) : \mathrm{Set}$. Elements of $(x : A) \to B(x)$ are functions f that map $a : A$ to $f(a) : B(a)$.

Σ-types Given $A : \mathrm{Set}$ and $B : A \to \mathrm{Set}$, then $\Sigma x : A.\, B(x) : \mathrm{Set}$. Elements of $\Sigma x : A.\, B(x)$ are dependent pairs $\langle a, b \rangle$ where $a : A$ and $b : B(a)$. We write $\pi_0 : \Sigma x : A.\, B(x) \to A$ and $\pi_1 : (y : \Sigma x : A.\, B(x)) \to B(\pi_0(y))$ for the projections. We write $\{\, a : A \mid B(a)\}$ for $\Sigma x : A.\, B(x)$ if $B : A \to \mathrm{Set}$ is propositional, i.e. there is at most one inhabitant in $B(a)$ for every $a : A$.

$+$ Given $A, B : \mathrm{Set}$, we denote their coproduct $A + B$ with coprojections $\mathrm{inl} : A \to A + B$ and $\mathrm{inr} : B \to A + B$. We use $[f, g]$ for cotupling.

Equality and unit types Given $a, b : A$ we write $a = b : \mathrm{Set}$ for the equality type, inhabited by refl if and only if $a = b$. In contrast, the unit type $\mathbf{1}$ always has a unique element $\star : \mathbf{1}$.

We call a type expression *strictly positive* in X if X never appears in the domain of a Π-type. It is a requirement for inductive definitions in predicative Type Theory that the inductively defined types appear only strictly positive in the domain of the constructors.

2 Inductive-Inductive Definitions as Dialgebras

In this section, our goal is to describe each inductive-inductively defined set as the initial object in a category constructed from a description of the set. Just as for ordinary induction and initial algebras, this description will be a functor of sorts, but because of the more complicated structure involved, this will no longer be an endofunctor. The interesting complication is the fact that the constructor for the second set B can refer to the constructor for the first set A (as for example the argument $\tau : \mathrm{Type}(\Gamma \rhd \sigma)$ referring to $\cdot \rhd \cdot$ in the introduction rule for the Π-type). Thus we will model the constructor for B as (the second component of) a morphism $(c, d) : \mathrm{Arg}(A, B, c) \to (A, B)$ where $c : \mathrm{Arg}_{\mathrm{A}}(A, B) \to A$ is the constructor for A. Here, (c, d) is a morphism in the category of families of sets:

Definition 2.1. *The category Fam(Set) of families of sets has as objects pairs (A, B), where A is a set and $B : A \to$ Set is an A-indexed family of sets. A morphism from (A, B) to (A', B') is a pair (f, g) where $f : A \to A'$ and $g : (x : A) \to B(x) \to B'(f(x))$.*

Note that there is a forgetful functor $U : \mathrm{Fam}(\mathrm{Set}) \to \mathrm{Set}$ sending (A, B) to A and (f, g) to f. Now, $c : \mathrm{Arg}_{\mathrm{A}}(A, B) \to A$ is not an $\mathrm{Arg}_{\mathrm{A}}$-algebra, since $\mathrm{Arg}_{\mathrm{A}} : \mathrm{Fam}(\mathrm{Set}) \to \mathrm{Set}$ is not an endofunctor. However, we have $c : \mathrm{Arg}_{\mathrm{A}}(A, B) \to U(A, B)$. This means that c is a $(\mathrm{Arg}_{\mathrm{A}}, U)$-dialgebra, as introduced by Hagino [11]:

Definition 2.2. *Let $F, G : \mathbb{C} \to \mathbb{D}$ be functors. The category Dialg(F, G) has as objects pairs (A, f) where $A \in \mathbb{C}$ and $f : F(A) \to G(A)$. A morphism from (A, f) to (A', f') is a morphism $h : A \to A'$ in $\mathbb{C}$ such that $G(h) \circ f = f' \circ F(h)$.*

There is a forgetful functor $V : \mathrm{Dialg}(F, G) \to \mathbb{C}$ defined by $V(A, f) = A$.

Putting things together, we will model the constructor for A as a morphism $c : \mathrm{Arg}_{\mathrm{A}}(A, B) \to A$ in Set and the constructor for B as the second component of a morphism $(c, d) : \mathrm{Arg}(A, B, c) \to (A, B)$ in Fam(Set). Thus, we see that the data needed to describe (A, B) as inductively generated with constructors c, d are the functors $\mathrm{Arg}_{\mathrm{A}}$ and Arg. However, we must also make sure that the first component of Arg coincides with $\mathrm{Arg}_{\mathrm{A}}$, i.e. that $U \circ \mathrm{Arg} = \mathrm{Arg}_{\mathrm{A}} \circ V$.

Definition 2.3. *An inductive-inductive definition is given by two functors*

$$\mathit{Arg}_A : \mathit{Fam}(\mathit{Set}) \to \mathit{Set} \qquad \mathit{Arg} : \mathit{Dialg}(\mathit{Arg}_A, U) \to \mathit{Fam}(\mathit{Set})$$

such that $U \circ \mathit{Arg} = \mathit{Arg}_A \circ V$.

Since the first functor is determined by the second, we often write such a pair as $\mathrm{Arg} = (\mathrm{Arg}_{\mathrm{A}}, \mathrm{Arg}_{\mathrm{B}})$ where

$$\mathrm{Arg}_{\mathrm{B}} : (A : \mathrm{Set})(B : A \to \mathrm{Set})(c : \mathrm{Arg}_{\mathrm{A}}(A, B) \to A) \to \mathrm{Arg}_{\mathrm{A}}(A, B) \to \mathrm{Set} \; .$$

Example 2.4 (Contexts and types). The inductive-inductive definition of Ctxt : Set and Type : Ctxt → Set from the introduction is given by

$$\mathrm{Arg}_{\mathrm{Ctxt}}(A,B) = \mathbf{1} + \Sigma\,\Gamma{:}A.\,B(\Gamma)$$
$$\mathrm{Arg}_{\mathrm{Type}}(A,B,c,x) = \mathbf{1} + \Sigma\,\sigma{:}B(c(x)).\,B(c(\mathrm{inr}(c(x),\sigma)))\ .$$

For $\mathrm{Arg}_{\mathrm{Ctxt}}$, the left summand $\mathbf{1}$ corresponds to the constructor ε taking no arguments, and the right summand $\Sigma\,\Gamma{:}A.\,B(\Gamma)$ corresponds to $\rhd$'s two arguments Γ : Ctxt and σ : Type(Γ). Similar considerations apply to $\mathrm{Arg}_{\mathrm{Type}}$.

Example 2.5 (Sorted lists). The sorted list example does not fit into our framework, since $\leq_{\mathrm{L}}$: ($\mathbb{N}$ × SortedList) → Set is indexed by $\mathbb{N}$ × SortedList and not simply SortedList. It is however straightforward to generalise the construction to include this example as well: instead of considering ordinary families, consider "$\mathbb{N}\times A$-indexed" families (A, B) where A is a set and $B : (\mathbb{N} \times A) \to$ Set. The inductive-inductive definition of SortedList : Set and $\leq_{\mathrm{L}}$: ($\mathbb{N}$×SortedList) → Set is then given by

$$\mathrm{Arg}_{\mathrm{SList}}(A,B) = \mathbf{1} + (\Sigma\,n{:}\mathbb{N}.\,\Sigma\,\ell{:}A.\,B(n,\ell))$$
$$\mathrm{Arg}_{\leq_{\mathrm{L}}}(A,B,c,m,\mathrm{inl}(\star)) = \mathbf{1}$$
$$\mathrm{Arg}_{\leq_{\mathrm{L}}}(A,B,c,m,\mathrm{inr}(\langle n,\ell,p\rangle)) = \Sigma\,m \leq n.\,B(m,\ell)\ .$$

For ease of presentation, we will only consider ordinary families of sets.

2.1 A Category for Inductive-Inductive Definitions

Given Arg = ($\mathrm{Arg}_{\mathrm{A}}$, $\mathrm{Arg}_{\mathrm{B}}$) representing an inductive-inductive definition, we will now construct a category $\mathbb{E}_{\mathrm{Arg}}$ whose initial object (if it exists) is the intended interpretation of the inductive-inductive definition. Figure 1 summarises the functors and categories involved (U, V and W are all forgetful functors).

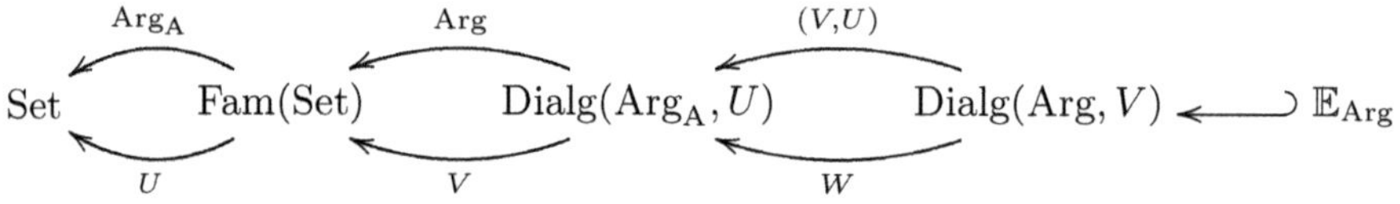

Fig. 1. The functors and categories involved

One might think that the category we are looking for is Dialg(Arg, V), where V : Dialg($\mathrm{Arg}_{\mathrm{A}}$, U) → Fam(Set) is the forgetful functor. Dialg(Arg, V) has objects $(A, B, c, (d_0, d_1))$, where A : Set, $B : A \to$ Set, $c : \mathrm{Arg}_{\mathrm{A}}(A, B) \to A$ and $(d_0, d_1) : \mathrm{Arg}(A, B, c) \to (A, B)$. The function $d_0 : \mathrm{Arg}_{\mathrm{A}}(A, B) \to A$ looks like the constructor for A that we want, but

$$d_1 : (x : \mathrm{Arg}_{\mathrm{A}}(A,B)) \to \mathrm{Arg}_{\mathrm{B}}(A,B,c,x) \to B(d_0(x))$$

does not look quite right – we need c and d_0 to be the same!

To this end, we will consider the equalizer of the forgetful functor W : $\text{Dialg}(\text{Arg}, V) \to \text{Dialg}(\text{Arg}_{\text{A}}, U)$, $W(A, B, c, (d_0, d_1)) = (A, B, c)$, and the functor (V, U) defined by

$$(V,U)(A,B,c,(d_0,d_1)) := (V(A,B,c), U(d_0,d_1)) = (A,B,d_0)$$
$$(V,U)(f,g) := (f,g)$$

Note that $U(d_0, d_1) : U(\text{Arg}(A, B, c)) \to U(V(A, B, c))$ but $U \circ \text{Arg} = \text{Arg}_{\text{A}} \circ V$, so that $U(d_0, d_1) : \text{Arg}_{\text{A}}(V(A, B, c)) \to U(V(A, B, c))$. In other words, $(V(A, B, c), U(d_0, d_1))$ is an object in $\text{Dialg}(\text{Arg}_{\text{A}}, U)$, so (V, U) really is a functor from $\text{Dialg}(\text{Arg}, V)$ to $\text{Dialg}(\text{Arg}_{\text{A}}, U)$.

Definition 2.6. *For $Arg = (Arg_A, Arg_B)$ representing an inductive-inductive definition, let $\mathbb{E}_{Arg}$ be the underlying category of the equaliser of (V, U) and the forgetful functor $W : Dialg(Arg, V) \to Dialg(Arg_A, U)$.*

Explicitly, the category $\mathbb{E}_{\text{Arg}}$ has

- Objects (A, B, c, d), where A : Set, $B : A \to \text{Set}$, $c : \text{Arg}_{\text{A}}(A, B) \to A$, $d : (x : \text{Arg}_{\text{A}}(A, B)) \to \text{Arg}_{\text{B}}(A, B, c, x) \to B(c(x))$.
- Morphisms from (A, B, c, d) to (A', B', c', d') are morphisms $(f, g) : (A, B, c) \Rightarrow_{\text{Dialg}(\text{Arg}_{\text{A}}, U)} (A', B', c')$ such that in addition

$$g(c(x), d(x, y)) = d'(\text{Arg}_{\text{A}}(f, g)(x), \text{Arg}_{\text{B}}(f, g)(x, y)) \ .$$

Example 2.7. Consider the functors Arg_{Ctxt}, Arg_{Type} from Example 2.4:

$$\text{Arg}_{\text{Ctxt}}(A, B) = \mathbf{1} + \Sigma\, \Gamma : A.\, B(\Gamma)$$
$$\text{Arg}_{\text{Type}}(A, B, c, x) = \mathbf{1} + \Sigma\, \sigma : B(c(x)).\, B(c(\text{inr}(c(x), \sigma))) \ .$$

An object in $\mathbb{E}_{(\text{Arg}_{\text{Ctxt}}, \text{Arg}_{\text{Type}})}$ consists of A : Set, $B : A \to \text{Set}$ and morphisms $c = [\varepsilon_{A,B}, \triangleright_{A,B}]$ and $d = \lambda \Gamma.[\iota_{A,B}(\Gamma), \Pi_{A,B}(\Gamma)]$ which can be split up into[1]

$$\varepsilon_{A,B} : \mathbf{1} \to A \ , \qquad \triangleright_{A,B} : ((\Gamma : A) \times B(\Gamma)) \to A \ ,$$
$$\iota_{A,B} : (\Gamma : \text{Arg}_{\text{Ctxt}}(A, B)) \to \mathbf{1} \to B(c(\Gamma)) \ ,$$
$$\Pi_{A,B} : (\Gamma : \text{Arg}_{\text{Ctxt}}(A, B)) \to ((\sigma : B(c(\Gamma))) \times (\tau : B(\triangleright_{A,B}(c(\Gamma), \sigma)))) \to B(c(\Gamma)) \ .$$

Remark 2.8. The intended interpretation of the inductive-inductive definition given by $\text{Arg} = (\text{Arg}_{\text{A}}, \text{Arg}_{\text{B}})$ is the initial object in $\mathbb{E}_{\text{Arg}}$. Depending on the meta-theory, this might of course not exist. However, we will show that it does if and only if an eliminator for the inductive-inductive definition exists.

Remark 2.9. From Figure 1, it should be clear how to generalise the current construction to the simultaneous definition of A : Set, $B : A \to \text{Set}$, $C : (x : A) \to B(x) \to \text{Set}$, etc.: for a definition of n sets, replace Fam(Set) with the category FAM_n of families $(A_1, A_2, A_3, \ldots, A_n)$ and consider $\text{Arg}_{\text{A}} : \text{FAM}_n \to \text{Set}$, $\text{Arg} : \text{Dialg}(\text{Arg}_{\text{A}}, U) \to \text{Fam}(\text{Set})$, $\text{Arg}_{\text{C}} : \mathbb{E}_{\text{Arg}} \to \text{FAM}_3, \ldots$ taking an equalizer where necessary to make the constructors in different positions equal.

[1] Notice that $\iota_{A,B} : (\Gamma : \text{Arg}_{\text{Ctxt}}(A, B)) \to \ldots$ and not $\iota_{A,B} : (\Gamma : A) \to \ldots$ as one would maybe expect. There is no difference for initial A, as we have $\text{Arg}_{\text{Ctxt}}(A, B) \cong A$ by (a variant of) Lambek's Lemma.

2.2 How to Exploit Initiality: An Example

Let us consider an example of how to use initiality to derive a program dealing with the contexts and types from the introduction. Suppose that we want to define a concatenation $+\!\!+ : \mathrm{Ctxt} \to \mathrm{Ctxt} \to \mathrm{Ctxt}$ of contexts – such an operation could be useful to formulate more general formation rules, such as:

$$\frac{\sigma : \mathrm{Type}(\Gamma) \qquad \tau : \mathrm{Type}(\Delta)}{\sigma \times \tau : \mathrm{Type}(\Gamma +\!\!+ \Delta)}$$

Such an operation should satisfy the equations

$$\begin{aligned} \Delta +\!\!+ \quad \varepsilon \quad &= \Delta \\ \Delta +\!\!+ (\Gamma \triangleright \sigma) &= (\Delta +\!\!+ \Gamma) \triangleright (\mathrm{wk}_\Gamma(\sigma, \Delta)) \quad , \end{aligned}$$

where $\mathrm{wk} : (\Gamma : \mathrm{Ctxt}) \to (\sigma : \mathrm{Type}(\Gamma)) \to (\Delta : \mathrm{Ctxt}) \to \mathrm{Type}(\Delta +\!\!+ \Gamma)$ is a weakening operation, i.e. if $\sigma : \mathrm{Type}(\Gamma)$, then $\mathrm{wk}_\Gamma(\sigma, \Delta) : \mathrm{Type}(\Delta +\!\!+ \Gamma)$. A moment's thought should convince us that we want wk to satisfy

$$\begin{aligned} \mathrm{wk}_\Gamma(\iota_\Gamma, \Delta) &= \iota_{\Delta +\!\!+ \Gamma} \\ \mathrm{wk}_\Gamma(\Pi_\Gamma(\sigma, \tau), \Delta) &= \Pi_{\Delta +\!\!+ \Gamma}(\mathrm{wk}_\Gamma(\sigma, \Delta), \mathrm{wk}_{\Gamma \triangleright \sigma}(\tau, \Delta)) \ . \end{aligned}$$

Our hope is now to exploit the initiality of $(\mathrm{Ctxt}, \mathrm{Type})$ to get such operations. Recall from Example 2.4 that Ctxt, Type are the underlying sets for the inductive-inductive definition given by the functors

$$\begin{aligned} \mathrm{Arg}_{\mathrm{Ctxt}}(A, B) &= \mathbf{1} + \Sigma\, \Gamma : A.\, B(\Gamma) \\ \mathrm{Arg}_{\mathrm{Type}}(A, B, c, x) &= \mathbf{1} + (\Sigma\, \sigma : B(c(x)).\, \tau : B(c(\mathrm{inr}(c(x), \sigma)))) \ . \end{aligned}$$

From the types of $+\!\!+ : \mathrm{Ctxt} \to \mathrm{Ctxt} \to \mathrm{Ctxt}$ and $\mathrm{wk} : (\Gamma : \mathrm{Ctxt}) \to (A : \mathrm{Type}(\Gamma)) \to (\Delta : \mathrm{Ctxt}) \to \mathrm{Type}(\Delta +\!\!+ \Gamma)$, we see that if we can equip (A, B) where $A = \mathrm{Ctxt} \to \mathrm{Ctxt}$ and $B(f) = (\Delta : \mathrm{Ctxt}) \to \mathrm{Type}(f(\Delta))$ with an $(\mathrm{Arg}_{\mathrm{Ctxt}}, \mathrm{Arg}_{\mathrm{Type}})$ structure, initiality will give us functions of the right type. Of course, we must choose the right structure so that our equations will be satisfied:

$$\begin{aligned} &\mathrm{in}_A : \mathrm{Arg}_{\mathrm{Ctxt}}(A, B) \to A \\ &\mathrm{in}_A(\mathrm{inl}(\star)) \qquad\quad = \lambda\Delta.\, \Delta \\ &\mathrm{in}_A(\mathrm{inr}(\langle f, g \rangle)) \quad = \lambda\Delta.\, (f(\Delta) \triangleright g(\Delta)) \quad , \end{aligned}$$

$$\begin{aligned} &\mathrm{in}_B : (x : \mathrm{Arg}_{\mathrm{Ctxt}}(A, B)) \to \mathrm{Arg}_{\mathrm{Type}}(A, B, \mathrm{in}_A, x) \to B(\mathrm{in}_A(x)) \\ &\mathrm{in}_B(\Delta, \mathrm{inl}(\star)) \qquad = \lambda\Gamma.\, \iota_{\mathrm{in}_A(\Delta)(\Gamma)} \\ &\mathrm{in}_B(\Delta, \mathrm{inr}(\langle g, h \rangle)) = \lambda\Gamma.\, \Pi_{\mathrm{in}_A(\Delta)(\Gamma)}(g(\Gamma), h(\Gamma)) \ . \end{aligned}$$

Since $(A, B, \mathrm{in}_A, \mathrm{in}_B)$ is an object in $\mathbb{E}_{\mathrm{Arg}}$, initiality gives us a morphism $(+\!\!+, \mathrm{wk}) : (\mathrm{Ctxt}, \mathrm{Type}) \to (A, B)$ such that $(+\!\!+, \mathrm{wk}) \circ ([\varepsilon, \triangleright], [\iota, \Pi]) = (\mathrm{in}_A, \mathrm{in}_B) \circ (\mathrm{Arg}_{\mathrm{Ctxt}}, \mathrm{Arg}_{\mathrm{Type}})(+\!\!+, \mathrm{wk})$. In particular, this means that

$$\begin{aligned} +\!\!+(\varepsilon) &= \mathrm{in}_A(\mathrm{Arg}_{\mathrm{Ctxt}}(+\!\!+, \mathrm{wk})(\mathrm{inl}(\star))) = \mathrm{in}_A(\mathrm{inl}(\star)) = \lambda\Delta.\, \Delta \\ +\!\!+(\Gamma \triangleright \sigma) &= \mathrm{in}_A(\mathrm{Arg}_{\mathrm{Ctxt}}(+\!\!+, \mathrm{wk})(\mathrm{inr}(\langle \Gamma, \sigma \rangle))) = \mathrm{in}_A(\mathrm{inr}(\langle +\!\!+(\Gamma), \mathrm{wk}(\Gamma, \sigma) \rangle)) \\ &= \lambda\Delta.\, +\!\!+ (\Gamma, \Delta) \triangleright \mathrm{wk}(\Gamma, \sigma, \Delta) \ . \end{aligned}$$

Thus, we see that $\Delta +\!\!+\, \varepsilon = \Delta$ and $\Delta +\!\!+\, (\Gamma \triangleright \sigma) = (\Delta +\!\!+\, \Gamma) \triangleright \mathrm{wk}_\Gamma(\sigma, \Delta)$ as required.[2] In the same way, the equations for the weakening operation hold.

2.3 Relationship to Induction-Induction as Axiomatised in [16]

In short, the earlier axiomatisation [16] postulated the existence of a universes SP'_A, SP'_B of codes for inductive-inductive sets, together with decoding functions Arg'_A, Arg'_B and $\mathrm{Index}'_\mathrm{B}$. Intuitively, Arg'_A gives the domain of the constructor $\mathrm{intro}_\mathrm{A}$ for A, Arg'_B the domain for the constructor $\mathrm{intro}_\mathrm{B}$ for B and $\mathrm{Index}'_\mathrm{B}(x)$ the index of the type of $\mathrm{intro}_\mathrm{B}(x)$. More formally, they have types

$$\begin{aligned}
\mathrm{Arg}'_\mathrm{A} &: (\gamma_A : \mathrm{SP}'_\mathrm{A})(A : \mathrm{Set})(B : A \to \mathrm{Set}) \to \mathrm{Set} \ , \\
\mathrm{Arg}'_\mathrm{B} &: (\gamma_A : \mathrm{SP}'_\mathrm{A})(\gamma_B : \mathrm{SP}'_\mathrm{B}(\gamma_A)) \\
&\qquad \to (A : \mathrm{Set})(B_0 : A \to \mathrm{Set})(B_1 : \mathrm{Arg}'_\mathrm{A}(\gamma_A, A, B_0) \to \mathrm{Set}) \\
&\qquad \to \ldots \to (B_n : \mathrm{Arg}'^{n}_\mathrm{A}(\gamma_A, A, \vec{B}_{(n)}) \to \mathrm{Set}) \to \mathrm{Set} \ , \\
\mathrm{Index}'_\mathrm{B}&(\gamma_A, \gamma_B, A, B_0, \ldots, B_n) : \\
&\qquad \mathrm{Arg}'_\mathrm{B}(\gamma_A, \gamma_B, A, B_0, \ldots, B_n) \to \coprod_{i=0}^{i} \mathrm{Arg}'^{n}_\mathrm{A}(\gamma_A, A, \vec{B}_{(i)}) \ ,
\end{aligned}$$

where $\vec{B}_{(i)} = (B_0, \ldots, B_{i-1})$ and $\mathrm{Arg}'^{i}_\mathrm{A}(\gamma_A, A, B_{(i)})$ is defined by

$$\begin{aligned}
\mathrm{Arg}'^{0}_\mathrm{A}(\gamma_A, A, B_{(0)}) &:= A \\
\mathrm{Arg}'^{n+1}_\mathrm{A}(\gamma_A, A, \vec{B}_{(n), B_{n+1}}) &:= \mathrm{Arg}'_\mathrm{A}(\gamma_A, \coprod_{i=0}^{n} \mathrm{Arg}'^{i}_\mathrm{A}(\gamma_A, A, \vec{B}_{(i)}), [B_0, \ldots, B_n]) \ .
\end{aligned}$$

The axiomatisation then states that we have introduction and elimination rules, i.e. that for each code $\gamma = (\gamma_A, \gamma_B)$ there exists is a family $A_\gamma : \mathrm{Set}$, $B_\gamma : A_\gamma \to \mathrm{Set}$ with constructors $\mathrm{intro}_\mathrm{A} : \mathrm{Arg}'_\mathrm{A}(\gamma_A, A_\gamma, B_\gamma) \to A_\gamma$ and $\mathrm{intro}_\mathrm{B} : (x : \mathrm{Arg}'_\mathrm{B}(\gamma, A_\gamma, B_\gamma, B_1, \ldots, B_n)) \to B_\gamma(\mathrm{index}(x))$, and a suitable eliminator (see Section 3). Here, $B_i = B \circ k_i$ and $\mathrm{index}(x) = [k_0, \ldots, k_n](\mathrm{Index}'_\mathrm{B}(\gamma, A, B_0, \ldots, B_n, x))$ where $k_0 = \mathrm{id}$ and $k_{i+1} = \mathrm{intro}_\mathrm{A} \circ \mathrm{Arg}'^{i}_\mathrm{A}([k_0, \ldots, k_i], [\mathrm{id}', \ldots, \mathrm{id}'])$. The codes are chosen so that all occurrences of A and B in the domains of $\mathrm{intro}_\mathrm{A}$ and $\mathrm{intro}_\mathrm{B}$ are strictly positive.

The relationship between the codes from this axiomatisation and the formalisation in this article can now be summed up in the following proposition:

Proposition 2.10. *For each code $\gamma = (\gamma_A, \gamma_B)$, the operations $Arg_{\gamma_A} : Fam(Set) \to Set$ and $Arg_\gamma = (Arg_{\gamma_A}, Arg_{\gamma_B}) : Dialg(Arg_{\gamma_A}, U) \to Fam(Set)$ given by*

$$\begin{aligned}
Arg_{\gamma_A}(A, B) &:= Arg'_A(\gamma_A, A, B) \ , \\
Arg_{\gamma_B}(A, B, c, x) &:= \{\, y : Arg'_B(\gamma_A, \gamma_B, A, B_0, \ldots, B_n) \mid c(x) = index(y)\}
\end{aligned}$$

are functorial. □

[2] Actually, the order of the arguments is reversed, so we would have to define $\Delta +\!\!+' \Gamma := +\!\!+ (\Gamma, \Delta)$.

We will call a functor F strictly positive if it arises as $F = \mathrm{Arg}_\gamma$ for some code γ. In Section 3.3 , we show that that the original introduction and elimination rules hold if and only if $\mathbb{E}_{\mathrm{Arg}_\gamma}$ has an initial object.

3 The Elimination Principle

In this section, we introduce the elimination principle for inductive-inductive definitions. We show that every initial object has an eliminator (Proposition 3.8), and that every object with an eliminator is weakly initial (Proposition 3.9). Under the added assumption of strict positivity, we can also show uniqueness. Hence the two notions are equivalent for strictly positive functors (Theorem 3.10).

3.1 Warm-Up: A Generic Eliminator for an Inductive Definition

The traditional type-theoretical way of defining recursive functions like the context concatenation $+\!\!+$ in Section 2.2 is to define them in terms of eliminators. Roughly, the eliminator for an F-algebra (A, c) is a term

$$\frac{P : A \to \mathrm{Set} \qquad \mathrm{step}_c : (x : F(A)) \to \Box_F(P, x) \to P(c(x))}{\mathrm{elim}_F(P, \mathrm{step}_c) : (x : A) \to P(x)}$$

with computation rule $\mathrm{elim}_F(P, \mathrm{step}_c, c(x)) = \mathrm{step}_c(x, \mathrm{dmap}_F(P, \mathrm{elim}(P, \mathrm{step}_c), x))$. Here, $\Box_F(P) : F(A) \to \mathrm{Set}$ is the type of inductive hypothesis for P; it consists of proofs that P holds at all F-substructures of x, and $\mathrm{dmap}_F(P) : (x : F(A) \to P(x)) \to (x : F(A)) \to \Box_F(P, x)$ takes care of recursive calls.

Example 3.1. Let $F(X) = \mathbf{1} + X$, i.e. F is the functor whose initial algebra is $(\mathbb{N}, [0, \mathrm{suc}])$. We then have

$$\Box_{\mathbf{1}+X}(P, \mathrm{inl}(\star)) \cong \mathbf{1} \qquad \Box_{\mathbf{1}+X}(P, \mathrm{inr}(n)) \cong P(n)$$

so that the eliminator for $(\mathbb{N}, [0, \mathrm{suc}])$ becomes

$$\frac{P : \mathbb{N} \to \mathrm{Set} \qquad \begin{array}{c}\mathrm{step}_0 : \mathbf{1} \to P(0) \\ \mathrm{step}_{\mathrm{suc}} : (n : \mathbb{N}) \to P(n) \to P(\mathrm{suc}(n))\end{array}}{\mathrm{elim}_{\mathbf{1}+X}(P, \mathrm{step}_0, \mathrm{step}_{\mathrm{suc}}) : (x : \mathbb{N}) \to P(x)}$$

For polynomial functors F, $\Box_F$ can be defined inductively over the structure of F as is given in e.g. Dybjer and Setzer [8]. However, $\Box_F$ and dmap_F can be defined for any functor $F : \mathrm{Set} \to \mathrm{Set}$ by defining

$$\begin{aligned} \Box_F(P, x) &:= \{y : F(\Sigma\, z \!:\! A.\, P(z)) | F(\pi_0)(y) = x\} \\ \mathrm{dmap}_F(P, \mathrm{step}_c, x) &:= F(\lambda y. \langle y, \mathrm{step}_c(y)\rangle)(x) \ . \end{aligned}$$

We see that indeed $\Box_{\mathbf{1}+X}(P, \mathrm{inl}(\star)) \cong \mathbf{1}$ and $\Box_{\mathbf{1}+X}(P, \mathrm{inr}(n)) \cong P(n)$ as in Example 3.1.

3.2 The Generic Eliminator for an Inductive-Inductive Definition

Let us now generalise the preceding discussion from inductive definitions (i.e. endofunctors on Set) to inductive-inductive definitions (i.e. functors $\mathrm{Arg} = (\mathrm{Arg_A}, \mathrm{Arg_B})$ as in Definition 2.3). Since we replace the carrier set A with a carrier family (A, B), we should also replace the predicate $P : A \to \mathrm{Set}$ with a "predicate family" (P, Q) where $P : A \to \mathrm{Set}$ and $Q : (x : A) \to B(x) \to P(x) \to \mathrm{Set}$. This forces us to refine the step function $\mathrm{step}_c : (x : F(A)) \to \Box_F(P, x) \to P(c(x))$ into two functions

$$\begin{aligned}
&\mathrm{step}_c : (x : \mathrm{Arg_A}(A, B)) \to \Box_{\mathrm{Arg_A}}(P, Q, x) \to P(c(x)) \ , \\
&\mathrm{step}_d : (x : \mathrm{Arg_A}(A, B)) \to (y : \mathrm{Arg_B}(A, B, c, x)) \to (\widetilde{x} : \Box_{\mathrm{Arg_A}}(P, Q, x)) \\
&\qquad\qquad \to \Box_{\mathrm{Arg_B}}(P, Q, c, \mathrm{step}_c, x, y, \widetilde{x}) \to Q(c(x), d(x, y), \mathrm{step}_c(x, \widetilde{x})) \ .
\end{aligned}$$

As can already be seen in the types of step_c and step_d above, we replace $\Box_F$ with $\Box_{\mathrm{Arg_A}}$ and $\Box_{\mathrm{Arg_B}}$ of type

$$\begin{aligned}
&\Box_{\mathrm{Arg_A}}(P, Q) : \mathrm{Arg_A}(A, B) \to \mathrm{Set} \ , \\
&\Box_{\mathrm{Arg_B}}(P, Q) : \big(\mathrm{step}_c : (x : \mathrm{Arg_A}(A, B)) \to \Box_{\mathrm{Arg_A}}(P, Q, x) \to P(c(x))\big) \to \\
&\qquad (x : \mathrm{Arg_A}(A, B)) \to (y : \mathrm{Arg_B}(A, B, c, x)) \to \\
&\qquad (\widetilde{x} : \Box_{\mathrm{Arg_A}}(P, Q, x)) \to \mathrm{Set}
\end{aligned}$$

and we replace dmap_F with $\mathrm{dmap}_{\mathrm{Arg_A}}$, $\mathrm{dmap}_{\mathrm{Arg_B}}$ of type

$$\begin{aligned}
\mathrm{dmap}_{\mathrm{Arg_A}}(P, Q) : \ & \big(f : (x : A) \to P(x)\big) \to \\
& \big(g : (x : A) \to (y : B(x)) \to Q(x, y, f(x))\big) \to \\
& (x : \mathrm{Arg_A}(A, B)) \to \Box_{\mathrm{Arg_A}}(P, Q, x) \\
\mathrm{dmap}_{\mathrm{Arg_B}}(P, Q) : \ & \big(\mathrm{step}_c : (x : \mathrm{Arg_A}(A, B)) \to \Box_{\mathrm{Arg_A}}(P, Q, x) \to P(c(x))\big) \to \\
& \big(f : (x : A) \to P(x)\big) \to \\
& \big(g : (x : A) \to (y : B(x)) \to Q(x, y, f(x))\big) \to \\
& (x : \mathrm{Arg_A}(A, B)) \to (y : \mathrm{Arg_B}(A, B, c, x)) \\
& \quad \to \Box_{\mathrm{Arg_B}}(P, Q, \mathrm{step}_c, x, y, \mathrm{dmap}_{\mathrm{Arg_A}}(P, Q, f, g, x)) \ .
\end{aligned}$$

We can define $\Box_{\mathrm{Arg_A}}$, $\Box_{\mathrm{Arg_B}}$, $\mathrm{dmap}_{\mathrm{Arg_A}}$ and $\mathrm{dmap}_{\mathrm{Arg_B}}$ for arbitrary functors representing inductive-inductive definitions. First, define:

Definition 3.2. *Let $(A, B) \in \mathit{Fam(Set)}$, $P : A \to \mathit{Set}$, $Q : (x : A) \to B(x) \to P(x) \to \mathit{Set}$.*

(i) Define $\Sigma_{\mathit{Fam(Set)}}(A, B)\ (P, Q) \in \mathit{Fam(Set)}$ by

$$\Sigma_{\mathit{Fam(Set)}}(A, B)\ (P, Q) := (\Sigma\ A\ P, \lambda\langle a, p\rangle.\Sigma b{:}B(a).\ Q(a, b, p))$$

(ii) In addition, for $(f, g) : (A, B) \to (A', B')$ and

$$h : (x : A) \to P(f(x)) \qquad k : (x : A) \to (y : B(x)) \to Q(f(x), g(x, y), h(x)) \ ,$$

define $\langle (f,g),(h,k)\rangle : (A,B) \to \Sigma_{Fam(Set)}(A',B')\ (P,Q)$ *by*

$$\langle (f,g),(h,k)\rangle := (\lambda x.\, \langle f(x), h(x)\rangle, \lambda x\, y.\, \langle g(x,y), k(x,y)\rangle) \ .$$

(iii) *For* $h : (x : A) \to P(x)$ *and* $k : (x : A) \to (y : B(x)) \to Q(x,y,h(x))$, *define* $\overline{(h,k)} : (A,B) \to \Sigma_{Fam(Set)}(A,B)\ (P,Q)$ *by* $\overline{(h,k)} := \langle id, (h,k)\rangle$.

We have $(\pi_0, \pi_0') := (\pi_0, \lambda x.\, \pi_0) : \Sigma_{\mathrm{Fam(Set)}}(A,B)\ (P,Q) \to (A,B)$ with $(\pi_0,\pi_0')\circ \overline{(h,k)} = \mathrm{id}$. Note also that we can extend $\Sigma_{\mathrm{Fam(Set)}}$ to morphisms by defining $[(f,g),(h,k)] : \Sigma_{\mathrm{Fam(Set)}}(A,B)\ (P,Q) \to \Sigma_{\mathrm{Fam(Set)}}(A',B')\ (P',Q')$ for appropriate f,g,h,k by $[(f,g),(h,k)] = \langle (f,g)\circ(\pi_0,\pi_0'), (h,k)\rangle$. We can now define $\Box_{\mathrm{Arg_A}}$ and $\mathrm{dmap}_{\mathrm{Arg_A}}$:

Definition 3.3. *Define* $\Box_{Arg_A}$ *and* $dmap_{Arg_A}$ *with types as above by*

$$\Box_{Arg_A}(P,Q,x) := \{y : Arg_A(\Sigma_{Fam(Set)}(A,B)\ (P,Q))\ | Arg_A(\pi_0,\pi_0')(y) = x\} \ ,$$

$$dmap_{Arg_A}(P,Q,f,g) := Arg_A(\overline{(f,g)}) \ .$$

Note that we have an isomorphism

$$\varphi_{\mathrm{Arg_A}} : \mathrm{Arg_A}(\Sigma_{\mathrm{Fam(Set)}}(A,B)\ (P,Q)) \to \Sigma\, x{:}\mathrm{Arg_A}(A,B).\, \Box_{\mathrm{Arg_A}}(P,Q,x)$$

defined by $\varphi_{\mathrm{Arg_A}}(x) = \langle \mathrm{Arg_A}(\pi_0,\pi_0')(x), x\rangle$.

Definition 3.4. *Given* P, Q, $step_c$, x, y, $\widetilde{x}$ *as above, define*

(i) $\Sigma_{Dialg}(A,B,c)\ (P,Q,step_c) := (\Sigma_{Fam(Set)}\ (A,B)\ (P,Q), [c, step_c]\circ\varphi_{Arg_A})$,
(ii) $\Box_{Arg_B}(P,Q,step_c,x,y,\widetilde{x}) :=$
$\{z : Arg_B((\Sigma_{Dialg}(A,B,c)\ (P,Q,step_c)),\widetilde{x}) \mid Arg_B(\pi_0,\pi_0',\widetilde{x},z) = y\}$,
(iii) $dmap_{Arg_B}(P,Q,step_c,f,g) := Arg_B(\overline{(f,g)})$.

We can now define what the eliminators for inductive-inductive definitions are:

Definition 3.5. *We say that* (A,B,c,d) *in* $\mathbb{E}_{Arg}$ *has an eliminator, if there exist two terms*

$$\frac{\begin{array}{c} P : A \to Set \\ Q : (x : A) \to B(x) \to P(x) \to Set \\ step_c : (x : Arg_A(A,B)) \to \Box_{Arg_A}(P,Q,x) \to P(c(x)) \\ step_d : (x : Arg_A(A,B)) \to (y : Arg_B(A,B,c,x)) \to (\widetilde{x} : \Box_{Arg_A}(P,Q,x)) \\ \to \Box_{Arg_B}(P,Q,c,step_c,x,y,\widetilde{x}) \to Q(c(x),d(x,y),step_c(x,\widetilde{x})) \end{array}}{\begin{array}{c} elim_{Arg_A}(P,Q,step_c,step_d) : (x : A) \to P(x) \\ elim_{Arg_B}(P,Q,step_c,step_d) : (x : A) \to (y : B(x)) \to Q(x,y,elim_{Arg_A}(P,Q,step_c,step_d,x)) \end{array}}$$

with

$$elim_{Arg_A}(P,Q,step_c,step_d,c(x)) = step_c(x, dmap'_{Arg_A})$$

$$elim_{Arg_B}(P,Q,step_c,step_d,c(x),d(x,y)) = step_d(x,y,dmap'_{Arg_A}, dmap'_{Arg_B})$$

where

$$dmap'_{Arg_A} = dmap_{Arg_A}(elim_{Arg_A}(P,Q,step_c,step_d), elim_{Arg_B}(P,Q,step_c,step_d),x)$$

$$dmap'_{Arg_B} = dmap_{Arg_B}(step_c, elim_{Arg_A}(P,Q,step_c,step_d), elim_{Arg_B}(P,Q,step_c,step_d),x,y) \ .$$

Example 3.6 (The eliminator for sorted lists). Recall from Example 2.5 that sorted lists were given by the functors $\mathrm{Arg}_{\mathrm{SList}}$, $\mathrm{Arg}_{\leq_{\mathrm{L}}}$, where

$$\mathrm{Arg}_{\mathrm{SList}}(A, B) = \mathbf{1} + (\Sigma\, n{:}\mathbb{N}.\, \Sigma\, \ell{:}A.\, B(n, \ell))$$

Thus, we see that e.g.

$$\Box_{\mathrm{Arg}_{\mathrm{SList}}}(P, Q, \mathrm{inl}(\star)) = \{y : \mathbf{1} + \ldots \mid (\mathrm{id} + \ldots)(y) = \mathrm{inl}(\star)\} \cong \mathbf{1}$$

$$\begin{aligned} &\Box_{\mathrm{Arg}_{\mathrm{SList}}}(P, Q, \mathrm{inr}(\langle n, \ell, p\rangle)) \cong \\ &\{y : \Sigma\, n'{:}\mathbb{N}.\, \Sigma\, \langle \ell', \widetilde{\ell}\rangle{:}(\Sigma A P).\, \Sigma p'{:}B(n, \ell).\, Q(n', \ell', p', \widetilde{\ell}) \mid \Sigma(\mathrm{id}, \Sigma(\pi_0, \pi_0'))(y) \\ &\qquad = \langle n, \ell, p\rangle\} \\ &\cong \Sigma \widetilde{\ell}{:}P(\ell).\, Q(n, \ell, p, \widetilde{\ell}) \end{aligned}$$

and similarly for $\Box_{\mathrm{Arg}_{\leq_{\mathrm{L}}}}$, so that the eliminators are equivalent to

$$\begin{aligned} \mathrm{elim}_{\mathrm{SortedList}} : {} & (P : \mathrm{SortedList} \to \mathrm{Set}) \to \\ & (Q : (n : \mathbb{N}) \to (\ell : \mathrm{SortedList}) \to n \leq_{\mathrm{L}} \ell \to P(\ell) \to \mathrm{Set}) \to \\ & (\mathrm{step}_{\mathrm{nil}} : P(\mathrm{nil})) \to \\ & \big(\mathrm{step}_{\mathrm{cons}} : (n : \mathbb{N}) \to (\ell : \mathrm{SortedList}) \to (p : n \leq_{\mathrm{L}} \ell) \to (\widetilde{\ell} : P(\ell)) \\ & \qquad \to Q(n, \ell, p, \widetilde{\ell}) \to P(\mathrm{cons}(n, \ell, p))\big) \to \\ & \big(\mathrm{step}_{\mathrm{triv}} : (n : \mathbb{N}) \to Q(n, \mathrm{nil}, \mathrm{triv}_n, \mathrm{step}_{\mathrm{nil}})\big) \to \\ & \big(\mathrm{step}_{\ll\cdot\gg} : (m : \mathbb{N}) \to (n : \mathbb{N}) \to (\ell : \mathrm{SortedList}) \to (p : n \leq_{\mathrm{L}} \ell) \\ & \qquad \to (q : m \leq n) \to (p' : m \leq_{\mathrm{L}} \ell) \to (\widetilde{\ell} : P(\ell)) \\ & \qquad \to (\widetilde{p} : Q(n, \ell, p, \widetilde{\ell})) \to (\widetilde{p'} : Q(m, \ell, p', \widetilde{\ell})) \\ & \qquad \to Q(m, \mathrm{cons}(n, \ell, p), \ll q, p' \gg_{p,m,n,\ell}, \mathrm{step}_{\mathrm{cons}}(n, \ell, p, \widetilde{\ell}, \widetilde{p}))\big) \to \\ & (\ell : \mathrm{SortedList}) \to P(\ell)\ , \\ \mathrm{elim}_{\leq_{\mathrm{L}}} : {} & \ldots \to \\ & (n : \mathbb{N}) \to (\ell : \mathrm{SortedList}) \to (p : n \leq_{\mathrm{L}} \ell) \\ & \quad \to Q(n, \ell, p, \mathrm{elim}_{\mathrm{SortedList}}(P, Q, \mathrm{step}_{\mathrm{nil}}, \mathrm{step}_{\mathrm{cons}}, \mathrm{step}_{\mathrm{triv}}, \mathrm{step}_{\ll\cdot\gg}, \ell))\,. \end{aligned}$$

3.3 The Equivalence between Having an Eliminator and Being Initial

We now prove the promised equivalence. In what follows, let $\mathrm{Arg} = (\mathrm{Arg}_{\mathrm{A}}, \mathrm{Arg}_{\mathrm{B}})$ be functors for an inductive-inductive definition.

Lemma 3.7. *There is an isomorphism*

$$\begin{aligned} \varphi_{Arg} = (\varphi_{Arg_A}, \varphi_{Arg_B}) : {} & Arg(\Sigma_{Dialg}(A, B, c)\ (P, Q, step_c)) \\ & \to \Sigma_{Fam(Set)} Arg(A, B, c)\ (\Box_{Arg}(P, Q, step_c)) \end{aligned}$$

such that $(\pi_0, \pi_0') \circ \varphi_{Arg} = Arg(\pi_0, \pi_0')$ *and*

$$\varphi_{Arg} \circ Arg(\overline{(f, g)}) = \overline{(dmap_{ArgA}(P, Q, f, g), dmap_{Arg_B}(P, Q, step_c, f, g))}\ . \qquad \Box$$

Proposition 3.8. *Every initial object (A, B, c, d) in $\mathbb{E}_{Arg}$ has an eliminator.*

Proof. Let P, Q, step_c, step_d as in the type signature for $\text{elim}_{\text{Arg}_\text{A}}$ and elim_{Arg} be given. Define $\text{in}_\Sigma : \text{Arg}(\Sigma_{\text{Dialg}}(A, B, c)\ (P, Q, \text{step}_c)) \to V(\Sigma_{\text{Dialg}}(A, B, c)\ (P, Q, \text{step}_c))$ by $\text{in}_\Sigma = [(c, d), (\text{step}_c, \text{step}_d)] \circ \varphi_{\text{Arg}}$. This makes $\Sigma_{\text{Dialg}}(A, B, c)\ (P, Q, \text{step}_c)$ an object of $\mathbb{E}_{\text{Arg}}$.

Since (A, B, c, d) is initial in $\mathbb{E}_{\text{Arg}}$, we get a morphism $(h, h') : (A, B) \to \Sigma_{\text{Fam(Set)}}(A, B)\ (P, Q)$ which makes the top part of the following diagram commute:

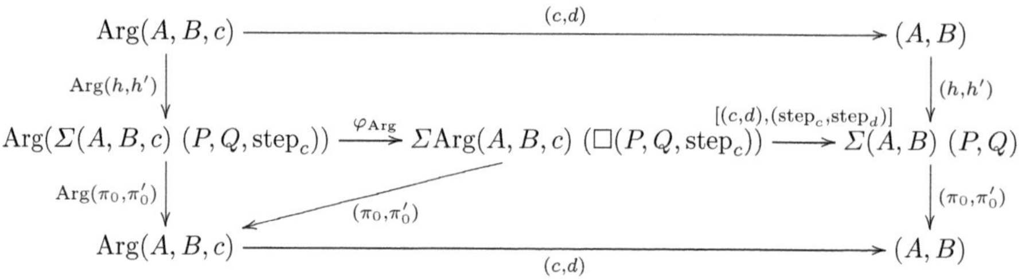

The bottom part commutes by Lemma 3.7 and calculation. Hence $(\pi_0, \pi_0') \circ (h, h')$ is a morphism in $\mathbb{E}_{\text{Arg}}$ and we must have $(\pi_0, \pi_0') \circ (h, h') = \text{id}$ by initiality. Thus $\pi_1 \circ h : (x : A) \to P(x)$ and $\pi_1(h'(x, y)) : Q(x, y, \pi_1(h(x)))$ for $x : A$, $y : B(x)$, so we can define $\text{elim}_{\text{Arg}_\text{A}}(P, Q, \text{step}_c, \text{step}_d) = \pi_1 \circ h$ and $\text{elim}_{\text{Arg}_\text{B}}(P, Q, \text{step}_c, \text{step}_d, x, y) = \pi_1(h'(x, y))$.

To verify the computation rules, note that since $(\pi_0, \pi_0') \circ (h, h') = \text{id}$, we have $(h, h') = \overline{(\pi_1, \pi_1') \circ (h, h')}$. We only show the calculation for Arg_A:

$$
\begin{aligned}
\text{elim}_{\text{Arg}_\text{A}}(P, Q, \text{step}_c, \text{step}_d, c(x))) &= \pi_1(h(c(x))) \\
&= \text{step}_c(\varphi_{\text{Arg}_\text{A}}(\text{Arg}_\text{A}(h, h')(x))) \\
&= \text{step}_c(\varphi_{\text{Arg}_\text{A}}(\text{Arg}_\text{A}(\overline{(\pi_1, \pi_1') \circ (h, h')})(x))) \\
&= \text{step}_c(x, \text{dmap}_{\text{Arg}_\text{A}}((\pi_1, \pi_1') \circ (h, h'))(x)) \\
&= \text{step}_c(x, \text{dmap}'_{\text{Arg}_\text{A}})
\end{aligned}
$$

□

Proposition 3.9. *Every (A, B, c, d) with an eliminator is weakly initial in $\mathbb{E}_{Arg}$.*

Proof. Let (A', B', c', d') be another object in $\mathbb{E}_{\text{Arg}}$. Notice that for $P(x) = A'$, $Q(x, y, \widetilde{x}) = B'(\widetilde{x})$, the usually dependent second projections π_1, π_1' become non-dependent and make up a morphism $(\pi_1, \pi_1') : \Sigma_{\text{Fam(Set)}}(A, B)\ (P, Q) \to (A', B')$. Since

$$\pi_1 \circ [c, c' \circ \text{Arg}_\text{A}(\pi_1, \pi_1') \circ \varphi^{-1}_{\text{Arg}_\text{A}}] \circ \varphi_{\text{Arg}_\text{A}} = c' \circ \text{Arg}_\text{A}(\pi_1, \pi_1') \ ,$$

this lifts to $(\pi_1, \pi_1') : \Sigma_{\mathrm{Dialg}}(A, B, c)\ (P, Q, c' \circ \mathrm{Arg}_{\mathrm{A}}(\pi_1, \pi_1') \circ \varphi^{-1}_{\mathrm{Arg}_{\mathrm{A}}}) \to (A', B', c')$. By currying $(f, g) := (c', d') \circ \mathrm{Arg}(\pi_1, \pi_1') \circ \varphi^{-1}_{\mathrm{Arg}}$, we get

$$\begin{aligned}
\widehat{f} &: (x : \mathrm{Arg}_{\mathrm{A}}(A, B)) \to \Box_{\mathrm{Arg}_{\mathrm{A}}}(P, Q, x) \to A' \\
\widehat{g} &: (x : \mathrm{Arg}_{\mathrm{A}}(A, B)) \to (y : \mathrm{Arg}_{\mathrm{B}}(A, B, c, x)) \to (\widetilde{x} : \Box_{\mathrm{Arg}_{\mathrm{A}}}(P, Q, x)) \\
&\qquad \to \Box_{\mathrm{Arg}_{\mathrm{B}}}(P, Q, c, \widehat{f}, x, y, \widetilde{x}) \to B'(\widehat{f}(x, \widetilde{x}))
\end{aligned}$$

so that $(h, h') := (\mathrm{elim}_{\mathrm{Arg}_{\mathrm{A}}}(P, Q, \widehat{f}, \widehat{g}), \mathrm{elim}_{\mathrm{Arg}_{\mathrm{B}}}(P, Q, \widehat{f}, \widehat{g})) : (A, B) \to (A', B')$.

We have to check that $(h, h') \circ (c, d) = (c', d') \circ \mathrm{Arg}(h, h')$.

$$\begin{aligned}
(h, h') \circ (c, d) &= (\mathrm{elim}_{\mathrm{Arg}_{\mathrm{A}}}(P, Q, \widehat{f}, \widehat{g}), \mathrm{elim}_{\mathrm{Arg}_{\mathrm{B}}}(P, Q, \widehat{f}, \widehat{g})) \circ (c, d) \\
&= (\widehat{f}, \widehat{g}) \circ \overline{(\mathrm{dmap}_{\mathrm{Arg}_{\mathrm{A}}}(h, h'), \mathrm{dmap}_{\mathrm{Arg}_{\mathrm{B}}}(h, h'))} \\
&= (\widehat{f}, \widehat{g}) \circ \varphi_{\mathrm{Arg}} \circ \mathrm{Arg}(\overline{h, h'}) \\
&= (c', d') \circ \mathrm{Arg}(\pi_1, \pi_1') \circ \mathrm{Arg}(\overline{h, h'}) \\
&= (c', d') \circ \mathrm{Arg}(h, h')
\end{aligned}$$

□

For strictly positive functors, we can say more, since we can argue by induction over their construction:

Theorem 3.10. *The functors $Arg_\gamma = (Arg_{\gamma_A}, Arg_{\gamma_B})$ from the original axiomatisation as described in Section 2.3 have eliminators if and only if $\mathbb{E}_{Arg_\gamma}$ has an initial object.*

Proof. Putting Proposition 3.8 and Proposition 3.9 together, all that is left to prove is that given an eliminator, the arrow (h, h') we construct is actually unique. Assume that (k, k') is another arrow with $(k, k') \circ (c, d) = (c', d') \circ \mathrm{Arg}_\gamma(k, k')$.

We use the eliminator (and extensional equality) to prove that $(h, h') = (k, k')$; let $P(x) = (h(x) = k(x))$ and $Q(x, y, \widetilde{x}) = (h'(x, y) = k'(x, y))$. It is enough to prove $P(c(x))$ and $Q(c(x), d(x, y), \_)$ for arbitrary $x : \mathrm{Arg}_{\gamma_{\mathrm{A}}}(A, B)$, $y : \mathrm{Arg}_{\gamma_{\mathrm{B}}}(A, B, c, x)$, given the induction hypothesis $\Box_{\mathrm{Arg}_{\mathrm{A}}}(P, Q)$ and $\Box_{\mathrm{Arg}_{\mathrm{B}}}(P, Q)$. By induction on the buildup of $\mathrm{Arg}_{\gamma_{\mathrm{A}}}$ and $\mathrm{Arg}_{\gamma_{\mathrm{B}}}$, we can prove that $\Box_{\mathrm{Arg}_{\mathrm{A}}}(P, Q)$ and $\Box_{\mathrm{Arg}_{\mathrm{B}}}(P, Q)$ give that $\mathrm{Arg}(h, h') = \mathrm{Arg}(k, k')$, and hence

$$(h, h') \circ (c, d) = (c', d') \circ \mathrm{Arg}(h, h') = (c', d') \circ \mathrm{Arg}(k, k') = (k, k') \circ (c, d) \ .$$

Using the elimination principle, we conclude that $(h, h') = (k, k')$. □

4 Conclusions and Future Work

We have shown how to give a categorical semantics for inductive-inductive definitions, a principle for defining data types in Martin-Löf Type Theory. In order to do this, we generalised the usual initial algebra semantics to a dialgebra setting and showed that there is still an equivalence between this semantics and the more traditional formulation in terms of elimination and computation rules.

Future work includes extending the notion of containers [1] to inductive-inductive definitions. We also conjecture that W-types are enough to ensure the existence of inductive-inductive definitions in an extensional theory. More precisely, it should be possible to interpret inductive-inductive definitions as indexed inductive definitions, for which W-types are enough [2].

It could also be worthwhile to generalise this work to a unified setting including other forms of inductive definitions: let $F, G : \mathbb{C} \to \mathbb{D}$ be functors between categories having all finite limits. One can then extend $\mathbb{C}$ and $\mathbb{D}$ to Categories with Families [6,13] and use that structure to define the concept of an eliminator for F and G. If G is left exact, one can show that having an eliminator and being initial in (a subcategory of) $\mathrm{Dialg}(F, G)$ is equivalent.

References

1. Abbott, M., Altenkirch, T., Ghani, N.: Containers: Constructing strictly positive types. Theoretical Computer Science 342(1), 3–27 (2005)
2. Altenkirch, T., Morris, P.: Indexed containers. In: 24th Annual IEEE Symposium on Logic In Computer Science, LICS 2009, pp. 277–285 (2009)
3. Chapman, J.: Type theory should eat itself. Electronic Notes in Theoretical Computer Science 228, 21–36 (2009)
4. Danielsson, N.A.: A formalisation of a dependently typed language as an inductive-recursive family. In: Altenkirch, T., McBride, C. (eds.) TYPES 2006. LNCS, vol. 4502, pp. 93–109. Springer, Heidelberg (2007)
5. Dybjer, P.: Inductive families. Formal Aspects of Computing 6(4), 440–465 (1994)
6. Dybjer, P.: Internal type theory. In: Berardi, S., Coppo, M. (eds.) TYPES 1995. LNCS, vol. 1158, pp. 120–134. Springer, Heidelberg (1996)
7. Dybjer, P., Setzer, A.: A finite axiomatization of inductive-recursive definitions. In: Girard, J.-Y. (ed.) TLCA 1999. LNCS, vol. 1581, pp. 129–146. Springer, Heidelberg (1999)
8. Dybjer, P., Setzer, A.: Induction–recursion and initial algebras. Annals of Pure and Applied Logic 124(1-3), 1–47 (2003)
9. Ghani, N., Johann, P., Fumex, C.: Fibrational induction rules for initial algebras. In: Dawar, A., Veith, H. (eds.) CSL 2010. LNCS, vol. 6247, pp. 336–350. Springer, Heidelberg (2010)
10. Goguen, J., Thatcher, J., Wagner, E., Wright, J.: Initial algebra semantics and continuous algebras. Journal of the ACM 24(1), 68–95 (1977)
11. Hagino, T.: A Categorical Programming Language. Ph.D. thesis, University of Edinburgh (1987)
12. Hermida, C., Jacobs, B.: Structural induction and coinduction in a fibrational setting. Information and Computation 145(2), 107–152 (1998)
13. Hofmann, M.: Syntax and semantics of dependent types. In: Semantics and Logics of Computation, pp. 79–130. Cambridge University Press, Cambridge (1997)
14. Martin-Löf, P.: Intuitionistic type theory. Bibliopolis Naples (1984)
15. Nordström, B., Petersson, K., Smith, J.: Programming in Martin-Löf's type theory: an introduction. Oxford University Press, Oxford (1990)
16. Nordvall Forsberg, F., Setzer, A.: Inductive-inductive definitions. In: Dawar, A., Veith, H. (eds.) CSL 2010. LNCS, vol. 6247, pp. 454–468. Springer, Heidelberg (2010)
17. Norell, U.: Towards a practical programming language based on dependent type theory. Ph.D. thesis, Chalmers University of Technology (2007)
18. Poll, E., Zwanenburg, J.: From algebras and coalgebras to dialgebras. Electronic Notes in Theoretical Computer Science 44(1), 289–307 (2001)

Finitary Functors: From Set to Preord and Poset

Adriana Balan[1,*] and Alexander Kurz[2]

[1] University Politehnica of Bucharest, Romania
[2] University of Leicester, UK

Abstract. We investigate how finitary functors on Set can be extended or lifted to finitary functors on Preord and Poset and discuss applications to coalgebra.

Keywords: extension, lifting, relator, simulation, (final) coalgebra, exact square, embedding.

1 Introduction

Endofunctors T : Set $\to$ Set play a crucial role in the theory of coalgebras and the rich body of results on them [4] has been exploited over the years to prove results about the category Coalg(T), and about logics for T-coalgebras, uniformly in the functor T.

Not as dominant as Set-functors, functors on preorders and on posets have made their appearance, for example, if one is interested in simulation rather than only bisimulation [8,12]. Moreover, we think of the categories Preord and Poset as the natural link between universal coalgebra [14] and domain theory [1], as domains are special posets.

A general plan of work would be the comprehensive study of Preord- and Poset-functors and their relationship to Set-functors and to coalgebras. In this paper, we restrict ourselves to the modest approach of transforming Set-functors into Preord and Poset-functors and study how some properties important from the coalgebraic point of view are transfered. Two notions arise here: extension and lifting of a Set-functor T, where extension means a functor which coincides with T on discrete set and lifting means that underlying sets are kept but some order is added.

For extensions, the final coalgebra is discrete, but nevertheless the associated notion of (bi)simulation on posets can be interesting. For liftings, the order on the final coalgebra is similarity, an insight going back to [13, Thm 4.1] and [17, Thm 5.9].

We start from the observation that every finitary Set-functor T has a (canonical) presentation as a coequalizer of two polynomial functors. Since sets are discrete preorders, this coequalizer can be computed in preorders (or posets) to yield a functor $\tilde{T}$: Preord $\to$ Preord, which simultaneously lifts and extends T. As shown in [16] this leads to interesting examples: if T is the finite powerset

* Supported by the CNCSIS project PD-56 no. 19/03.08.10.

A. Corradini, B. Klin, and C. Cîrstea (Eds.): CALCO 2011, LNCS 6859, pp. 85–99, 2011.

functor, then $\tilde{T}$ on Poset yields the finite(ly generated) convex powerset functor. On the other hand, the final $\tilde{T}$-coalgebra is always discrete and, therefore, does not capture a notion of simulation.

So we study quotients of polynomial functors, but now with *ordered coefficients* and show that this is equivalent to the notion of an order $\bar{T}$: Set $\to$ Preord on a functor T from [8]. The latter investigate conditions under which $\bar{T}$ can be lifted to a functor $\hat{T}$: Preord $\to$ Preord. On the other hand, again interpreting the coequalizer of polynomial functors in Preord, we obtain another lifting $\check{T}$: Preord $\to$ Preord, which always exists. We show $\hat{T} = \check{T}$ under the conditions which ensure the existence of $\hat{T}$. The table below summarizes the notation for the various extensions and liftings met in the paper.

T : Set $\to$ Set
$\bar{T}$: Set $\to$ Preord (Def. 9)
$\tilde{T}$: Preord $\to$ Preord Extension (3.2)
$\hat{T}, \check{T}$: Preord $\to$ Preord Liftings (Def.-Prop. 16, resp. Def. 24)

The last section of the paper focuses on Poset-functors obtained from the previous constructions by taking quotients, with similar results obtained.

Finally, further topics pursued in the paper are the preservation of exact squares (the ordered analogue of weak pullbacks) and of embeddings. The latter is motivated by the result on the expressiveness of modal logic over posets [9], while the former comes from the fact that it replaces preservation of weak pullbacks as a condition for the existence of the relation lifting on preorders or posets [5].

2 Preliminaries

We denote by Preord and Poset the categories of preordered sets and of posets, respectively, and monotone maps. We write $D \dashv U$: Preord $\to$ Set for the adjunction between the discrete and the forgetful functor. As U has also a right adjoint (which endows a set with the indiscrete preorder), it preserves all limits and colimits. In particular, coequalizers in Preord are computed as in Set, namely for any pair of monotone maps

$$(X, \leq) \rightrightarrows (Y, \leq) \xrightarrow{\pi} Z \ ,$$

their Set-coequalizer Z, with the smallest preorder such that π is monotone, becomes the coequalizer in Preord. We denote by $Q \dashv J$: Poset $\to$ Preord the adjunction between the quotient functor (sending every preordered set to the quotient poset obtained by identifying all x, y with $x \leq y$ and $y \leq x$) and the inclusion functor. For later use, recall that coequalizers in Poset are computed in two steps: first, take the coequalizer in Preord, then quotient it to obtain a poset.

An *embedding* in Preord or Poset is an injective, monotone and order reflecting map. In both categories, the embeddings are precisely the strong monomorphisms. An *exact square* [7] in Preord or in Poset is a diagram

$$\begin{array}{ccc} P & \xrightarrow{\alpha} & X \\ {\scriptstyle\beta}\downarrow & & \downarrow{\scriptstyle f} \\ Y & \xrightarrow{g} & Z \end{array} \tag{2.1}$$

with $f\alpha \leq g\beta$, such that

$$\forall x \in X, y \in Y.\ f(x) \leq g(y) \ \Rightarrow \exists p \in P.\ x \leq \alpha(p) \wedge \beta(p) \leq y\ . \tag{2.2}$$

If P is $\{(x,y) \in X \times Y \mid f(x) \leq g(y)\}$ with the product order and α and β are the usual projections, then (2.2) is obviously satisfied. (2.1) is then called a comma square.

The terminology is borrowed from [7], where exact squares where introduced in the framework of 2-categories. See also [11], where equalizers were similarly replaced by subequalizers. In Set, an exact square is precisely a weak pullback. In [12], a commutative square having the property (2.2) is called a preorder quasi-pullback.

Let T be a Set-functor. It is well-known that T is finitary (commutes with ω-filtered colimits) if and only if it admits a coend representation

$$TX \cong \int^{n<\omega} \mathsf{Set}(\underline{n}, X) \bullet T\underline{n}\ ,$$

that is, TX has a presentation given by the coequalizer

$$\coprod_{m,n<\omega} \mathsf{Set}(\underline{m}, \underline{n}) \times T\underline{m} \times X^n \underset{\lambda_X}{\overset{\rho_X}{\rightrightarrows}} \coprod_{n<\omega} T\underline{n} \times X^n \xrightarrow{\pi_X} TX\ , \tag{2.3}$$

where $\underline{n}$ refers to $\{0, 1, \ldots, n-1\}$, and the pair (λ_X, ρ_X) is given by $\lambda_X(f, \sigma, x) = (Tf(\sigma), x)$ and $\rho_X(f, \sigma, x) = (\sigma, x \circ f)$, for $f : \underline{m} \to \underline{n}$, $x : \underline{n} \to X$ and $\sigma \in T\underline{m}$.[1] Also, $\pi_X(\sigma, x) = Tx(\sigma)$. Intuitively, $T\underline{n}$ can be seen as the set of operations of arity n applied to the variables in X. In the sequel, we shall omit the subscript X when referring to the maps λ_X, ρ_X and π_X if the context is clear.

3 From Set to Preord

Section 3.1 considers the notions of extensions and liftings. Section 3.2 introduces the preordification $\tilde{T}$ of a Set-functor T based on the presentation of T. Section 3.3 shows that putting an order on the coefficients of the presentation

[1] We shall identify functions $x : \underline{n} \to X$ with tuples $x = (x_0, \ldots, x_{n-1})$, where $x_i = x(i), \forall i \in \{0, \ldots, n-1\}$.

agrees with the notion of an order $\bar{T}$ on a Set-functor T from [8]. Section 3.4 recalls how [8] uses relation lifting to extend $\bar{T}$ to an endofunctor $\hat{T}$ on Preord and shows that if T preserves weak pullbacks then $\tilde{T}$ preserves exact squares and, therefore, embeddings. Section 3.5 emphasizes that the order on the final $\hat{T}$-coalgebra coincides with the similarity given by relation lifting. Whereas [8] use relation lifting to extend $\bar{T}$ to a functor $\hat{T}$ on preorder, we can now also use the presentation to extend $\bar{T}$ to a functor $\check{T}$ on preorders. Section 3.6 shows that $\check{T} = \hat{T}$ under the conditions given in [8] for the existence of $\hat{T}$.

3.1 Extension and Lifting

Definition 1. *Let T be a* Set-*functor. An extension of T to* Preord *is a locally monotone functor Γ :* Preord $\to$ Preord *such that $\Gamma D = DT$. A lifting of T to* Preord *is a locally monotone functor Γ :* Preord $\to$ Preord *such that $U\Gamma = TU$.*

In the following, if T is finitary, we also require Γ (extension or lifting) to be so.[2] Extensions and liftings of a Set-functor to Poset are defined similarly.

It follows that both a lifting and an extension will satisfy the relation $T = U\Gamma D$. Intuitively, an extension will coincide with T on discrete sets, while a lifting means that we put a preorder (respectively a partial order) on TX. Also, there is an immediate test to decide whether a (finitary) locally monotone Preord (or Poset)-functor Γ is a lifting or an extension of a Set-functor: namely, compute $T = U\Gamma D$ and check if $\Gamma D = DT$ or $U\Gamma = TU$.

Remark 2. An extension is not necessarily unique. Let Γ be the functor sending a preordered set (a poset) to the (discrete) set of its connected components. Then $U\Gamma D = Id^3$. But also the identity on Preord (respectively on Poset) is an extension of Id, showing that a Set-functor can have different extensions.

The local monotonicity requirement is natural, as the categories Preord and Poset are enriched over themselves (in the sense that the hom-sets are ordered) and enriched functors coincide with locally monotone ones. In all constructions that we shall perform, the local monotonicity of the lifted/extended functor will come for free.

Note that although Set is (discretely) enriched over Preord (and over Poset), the adjunction $D \dashv U$ is not, since U is not locally monotone. In particular, $\Gamma = DTU$ will not in general be an extension/lifting and we shall replace it by (3.2) instead.

---

[2] An extension Γ of a finitary functor T need not be finitary: consider the finitary functor $TX = \{l : \mathbb{N} \to X \mid l(n) = l(n+1)$ for all but a finite number of $n\}$. It admits the Preord-extension $\Gamma(X, \leq_X) = \{l : (\mathbb{N}, \leq_{\mathbb{N}}) \to (X, \leq_X) \mid l(n) \leq l(n+1)$ for all but a finite number of $n\}$. But this Γ is not finitary: take the family of finite sets $(\underline{n}, \leq)$ ordered as usual, with inclusion maps, whose colimit is $(\mathbb{N}, \leq_{\mathbb{N}})$. Then one can check that $\mathsf{colim}\,\Gamma(\underline{n}, \leq) \ncong \Gamma(\mathsf{colim}(\underline{n}, \leq))$. We would like to thank the anonymous referee for pointing us this example. We didn't succeed in constructing a similar example for liftings.

[3] Here Id stands for the identity functor.

If T is a (finitary) Set-functor and Γ is an extension of T, then T-coalgebras and Γ-coalgebras are related by an adjunction $\tilde{C} \dashv \tilde{D} : \mathsf{Coalg}(T) \to \mathsf{Coalg}(\Gamma)$ which can be easily derived from the adjunction $C \dashv D : \mathsf{Set} \to \mathsf{Preord}$ between the connected components functor and the discrete functor, using that $T = C\Gamma D$ if Γ is an extension of T.[4] Consequently, $\tilde{D}$ will preserve limits, in particular, the final coalgebra (if it exists).

Proposition 3. *For any (finitary) extension Γ of a finitary functor T, the final Γ-coalgebra does exist and is discrete.*

The situation slightly changes when we consider a lifting instead of an extension. In this case there is an obvious forgetful functor $\tilde{U} : \mathsf{Coalg}(\Gamma) \to \mathsf{Coalg}(T)$, which has a left adjoint still denoted $\tilde{D} : \mathsf{Coalg}(T) \to \mathsf{Coalg}(\Gamma)$ (this is not hard to check). Thus $\tilde{U}$ preserves limits; in particular the underlying set of the final Γ-coalgebra (if it exists) will be the final T-coalgebra.

Proposition 4. *For any (finitary) lifting Γ of a finitary functor T, the final Γ-coalgebra exists [10] and is built on the same set as the final T-coalgebra.*

3.2 First Construction: Order on the Variables

Going back to (2.3), and following [16], we are now interested in this coequalizer if we replace the set X by a preorder $(X, \leq)$. The other sets involved in (2.3) remain discretely ordered, except for X^n, which carries the product order from $(X, \leq)$. Then λ, ρ are monotone. The coequalizer in Preord of this monotone pair of maps (λ, ρ) has the same underlying set TX, but now with a preorder $\trianglelefteq$:

$$\coprod_{m,n<\omega} \mathsf{Set}(\underline{m}, \underline{n}) \times T\underline{m} \times (X^n, \leq) \underset{\lambda}{\overset{\rho}{\rightrightarrows}} \coprod_{n<\omega} T\underline{n} \times (X^n, \leq) \xrightarrow{\pi} (TX, \trianglelefteq). \quad (3.1)$$

If $f : (X, \leq) \to (Y, \leq)$ is a monotone map, it follows that $Tf : (TX, \trianglelefteq) \to (TY, \trianglelefteq)$ is monotone. Thus we obtain a functor which is also locally monotone

$$\tilde{T} : \mathsf{Preord} \to \mathsf{Preord}, \; \tilde{T}(X, \leq) = (TX, \trianglelefteq). \quad (3.2)$$

It simultaneously defines an extension and a lifting of T. In fact, it is an enriched coend $\tilde{T}(X, \leq) \cong \int^{n<\omega} [D\underline{n}, (X, \leq)] \bullet DT\underline{n}$, where $[D\underline{n}, (X, \leq)]$ refers to the preordered set (internal hom) of all monotone maps from $D\underline{n}$ to X.

A functor may have different presentations, [5] but we have

Proposition 5. *$\tilde{T}$ is independent of the chosen presentation of T.*

Example 6. 1. Let $T = \mathcal{P}_f$, the finite powerset functor. For $(X, \leq)$ a preordered set, the above construction leads to the Egli-Milner preorder on $\mathcal{P}_f X$: $u \trianglelefteq v$ for $u, v \subseteq X$ finite iff $\forall a \in u \; \exists b \in v. \; a \leq b$ and $\forall b \in v \; \exists a \in u. \; a \leq b$.

---

[4] We leave the details to the reader. Notice there is a similar adjunction for Poset-extensions.

[5] For example, the finite powerset functor $\mathcal{P}_f$ can be presented as in (2.3), but also as a quotient of the list-functor $\coprod_{n<\omega} X^n$.

2. Take $TX = 1 + X$ the lift functor. For $(X, \leq)$ a preordered set, the corresponding order $\trianglelefteq$ on $1 + X$ will be the coproduct order.
3. For the list functor $TX = X^*$, a preorder on lists is obtained as follows:

$$[x_0 \ldots x_{n-1}] \trianglelefteq [y_0 \ldots y_{m-1}] \Leftrightarrow m = n \wedge x_i \leq y_i, \forall i < n .$$

In Sect. 3.4 we will see another description of $\trianglelefteq$ based on relation liftings.

3.3 Second Construction: Order on the Operations

We now equip a functor T with an order on the coefficients of its presentation.

Definition 7. *Let T be a finitary Set-functor. We say that T has a presentation with a preorder, if for each finite arity n, there is a preorder $\leq$ on $T\underline{n}$ such that $Tf : (T\underline{m}, \leq) \to (T\underline{n}, \leq)$ is monotone for all $f : \underline{m} \to \underline{n}$.*

There are many functors who carry a natural order, as eg the powerset functor (with the inclusion order), or the lift functor $TX = \{\bot\} + X$, with the flat order $\bot \leq x, \forall x \in X$ (see Example 11). The latter is a special case of the following:

Example 8. Let T be a (finitary) Set-functor, but not the constant functor mapping everything to the empty set. Then $T1 \neq \emptyset$. Specify a preorder on $T1$. This will induce a preorder on $T\underline{n}$ for all $n < \omega$ via the image of the map $\underline{n} \to 1$ through T, namely the preorder on $T\underline{n}$ is the inverse image of the order on $T1$. Then we obtain an order on T.

Definition 9. *Let T be a finitary functor with preorder $\leq$. Consider on TX the preorder $\sqsubseteq$ obtained from the coequalizer*

$$\coprod_{m,n<\omega} \mathsf{Set}(\underline{m}, \underline{n}) \times (T\underline{m}, \leq) \times X^n \underset{\lambda}{\overset{\rho}{\rightrightarrows}} \coprod_{n<\omega} (T\underline{n}, \leq) \times X^n \xrightarrow{\pi} (TX, \sqsubseteq). \quad (3.3)$$

This defines a functor $\bar{T} : \mathsf{Set} \to \mathsf{Preord}$, $\bar{T}X = (TX, \sqsubseteq)$.

Notice that ρ is always monotone, while λ is so by Def. 7. Therefore it makes sense to compute the above coequalizer in Preord. The functor $\bar{T}$ is finitary and satisfies $U\bar{T} = T$. In [8], such a functor is called an order on T. We keep the same terminology. This means that on each TX there is a preorder $\sqsubseteq$, and these preorders must be preserved by renaming: for each map $f : X \to Y$, its image $Tf : (TX, \sqsubseteq) \to (TY, \sqsubseteq)$ is monotone. Choosing $\underline{n}$ for X, we find

Proposition 10. $(T\underline{n}, \sqsubseteq) = (T\underline{n}, \leq)$.

Example 11. 1. Take all $T\underline{n}$ to be discretely ordered. Then T automatically satisfies Def. 7. The preorder obtained on TX will be the discrete one, as any coequalizer of discrete preordered sets is again discrete (D preserves all colimits being left adjoint, in particular coequalizers).

2. Take all $T\underline{n}$ to be indiscretely ordered. If T is finitary then all $(TX, \sqsubseteq) = TX \times TX$ are again indiscrete. Indeed, take any $u, v \in TX$. As T is finitary, we can find a finite set $\underline{n}$ and an injection $x : \underline{n} \to X$ such that u, v lie in the image of the map $Tx : T\underline{n} \to TX$. So $u = Tx(\sigma)$ and $v = Tx(\tau)$ with $\sigma, \tau \in T\underline{n}$. As Tx is monotone and σ, τ are comparable, it follows that $u \sqsubseteq v$.
3. Let $T = \mathcal{P}_f$ be the finite powerset functor, with the inclusion order on $\mathcal{P}_f(\underline{n})$. Then the resulting order on any $\mathcal{P}_f X$ is again the inclusion: take any finite subsets $u, v \subseteq X$. Then $u \sqsubseteq v$ if we can find $\sigma \subseteq \underline{n}$, $x : \underline{n} \to X$, $\tau \subseteq \underline{m}$, $y : \underline{m} \to X$ such that $\pi(\sigma, x) = u$, $\pi(\tau, y) = v$, and (σ, x) and (τ, y) are comparable in $\coprod_{n<\omega} \mathcal{P}_f(\underline{n}) \times X^n$. But this can be possible only if both lie in the same component, so $m = n$, and share same variables, $x = y$. It follows $\sigma \subseteq \tau$, hence $u = \mathcal{P}_f(x)(\sigma) \subseteq \mathcal{P}_f(x)(\tau) = v$. Similarly, if on $\mathcal{P}_f(\underline{n})$ we consider the converse inclusion, the resulting preorder $\sqsubseteq$ is $\subseteq^{op}$.
4. Take now $TX = \{\perp\} + X$. On the associated signature $T\underline{n} = \{\perp\} + \underline{n}$ consider the flat order $\perp < i, \forall i < n$.

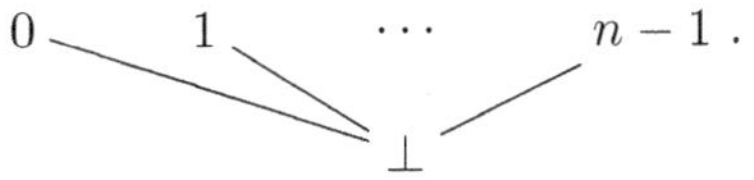

The quotient function $\pi : \coprod_{n<\omega} T\underline{n} \times X^n \to TX$ maps $(\perp, x : n \to X)$ to $\perp$ and $(i, x : n \to X)$ to x_i. It follows now easily that $\perp$ will be the least element in $(\{\perp\} + X, \sqsubseteq)$ and that different elements of X are not comparable, hence on TX we get the same flat order.
5. Consider the polynomial functor $TX = \mathbb{N} \times X$. On each $\mathbb{N} \times \underline{n}$, take the preorder given by $(N, i) \leq (M, j) \Leftrightarrow N < M$ or $(N = M$ and $i = j)$. This is precisely the lexicographic order with respect to the usual ordering of $\mathbb{N}$, when $\underline{n}$ is considered discrete. Then the induced preorder on the quotient TX is similar: two pairs (N, x) and (M, y) in $\mathbb{N} \times X$ are comparable if either are equal or the first components are comparable. In the future, we shall denote this functor by $TX = \mathbb{N} \ltimes X$ to emphasize the special preorder.
6. Let $T = (-)^3_2$ be the functor introduced by Aczel and Mendler in [2], given on objects by $X^3_2 = \{x = (x_1, x_2, x_3) \in X^3 \mid |(x_1, x_2, x_3)| \leq 2\}$. There is a natural (pre)ordering $\sqsubseteq$ on X^3_2 as follows: all triples $(x_1, x_2, x_3) \in X^3_2$ with equal components are minimal elements, all the others are maximals, and a minimal element is comparable with a maximal one only if they share a common component, as in the picture below:

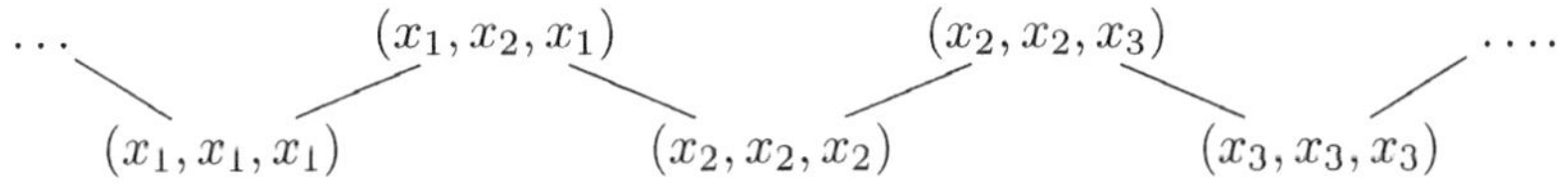

We shall call this the zig-zag preorder. If we restrict it to $T\underline{n}$ and compute $(TX, \sqsubseteq)$ as in (3.3), we obtain again precisely the zig-zag order.
7. Take the list functor $TX = X^*$. Put on each TX the following order

$$[x_0 \dots x_{n-1}] \leq [y_0 \dots y_{m-1}] \Leftrightarrow \exists \varphi : \underline{n} \to \underline{m} \; . \; x_i = y_{\varphi(i)}, \forall i < n \; ,$$

see [8, Example 2.2.(3)], where the function φ is required to be strictly monotone. It means that two lists are comparable if one can be obtained from the other by removing some elements. In particular, two lists of same length are comparable only if they are equal. By a similar reasoning as in the previous examples, restricting this preorder to all $T\underline{n}$ and computing $\sqsubseteq$ gives the same order.

The previous examples suggest a correspondence between orders $\bar{T}$ on T and preorders on the associated signature as in Def. 7. This is indeed the case.

Proposition 12. *Let T be a finitary* Set*-functor. Then there is a bijective correspondence between orders on T and presentations with preorders.*

3.4 Lifting T to $\hat{T}$ Using Relators

Starting from an order $\bar{T}$ on T, we will see in this section how a weak pullback preserving functor T lifts to a Preord-endofunctor $\hat{T}$ using relators. We present below a very brief overview on relators, for more details we refer to [15] or [8].

Let T be a Set-functor. For two sets X, Y and a relation $R \subseteq X \times Y$, the T-relation lifting of R is

$$\mathsf{Rel}_T(R) = \{(u, v) \in TX \times TY \mid \exists w \in TR \,.\, T\pi_1(w) = u \ \wedge \ T\pi_2(w) = v\},$$

where π_1, π_2 are the projections $X \xleftarrow{\pi_1} R \xrightarrow{\pi_2} Y$. The relation lifting satisfies the following properties ([15]):

1. Equality: $=_{TX} \ = \ \mathsf{Rel}_T(=_X)$.
2. Inclusion: if $R \subseteq S$ then $\mathsf{Rel}_T(R) \subseteq \mathsf{Rel}_T(S)$.
3. Composition: if $R \subseteq X \times Y$ and $S \subseteq Y \times Z$, then $\mathsf{Rel}_T(S \circ R) \subseteq \mathsf{Rel}_T(S) \circ \mathsf{Rel}_T(R)$, with equality if and only if T preserves weak pullbacks.
4. Inverse images (substitution): given functions $f : X \to X'$, $g : Y \to Y'$ and relation $R' \subseteq X' \times Y'$, then

$$\mathsf{Rel}_T((f \times g)^{-1}(R')) \subseteq (Tf \times Tg)^{-1}(\mathsf{Rel}_T(R')) \tag{3.4}$$

   with equality if T preserves weak pullbacks.

An immediate consequence is the following: if T preserves weak pullbacks and $\leq$ is a preorder on a set X, then $\mathsf{Rel}_T(\leq)$ is a preorder on TX.

Proposition 13. *Let T be a finitary* Set*-functor which preserves weak pullbacks. Then for each preordered set $(X, \leq)$, $\mathsf{Rel}_T(\leq)$ coincides with the preorder $\trianglelefteq$ on TX constructed in Sect. 3.2.*[6]

[6] Notice that the only thing needed in Prop.13 is that $\mathsf{Rel}_T(\leq)$ is a preorder. This is of course implicit when T preserves weak pullbacks, as mentioned earlier. We do not know if there are examples when $\mathsf{Rel}_T(\leq)$ is a preorder, for any preordered set $(X, \leq)$, without requesting T to preserve weak pullbacks.

Assume now that there is an order on T, given by $\bar{T}X = (TX, \sqsubseteq)$. For any relation $R \subseteq X \times Y$, the associated relation on $TX \times TY$ given by

$$\mathsf{Rel}^{\sqsubseteq}_{T}(R) = \sqsubseteq \circ \mathsf{Rel}_T(R) \circ \sqsubseteq,$$

is usually called a T-relator, or lax T-relation lifting ([15], [8]). Equivalently,

$$(u, v) \in \mathsf{Rel}^{\sqsubseteq}_{T}(R) \iff \exists w \in T(R)\ .\ u \sqsubseteq T\pi_1(w) \text{ and } T\pi_2(w) \sqsubseteq v.$$

The T-relator satisfies the following properties ([8], Lemma 4.2):

1. $\sqsubseteq_{TX} = \mathsf{Rel}^{\sqsubseteq}_{T}(=_X)$.
2. If $R \subseteq S$ then $\mathsf{Rel}^{\sqsubseteq}_{T}(R) \subseteq \mathsf{Rel}^{\sqsubseteq}_{T}(S)$.
3. If $R \subseteq X \times Y$ and $S \subseteq Y \times Z$, then $\mathsf{Rel}^{\sqsubseteq}_{T}(S \circ R) \subseteq \mathsf{Rel}^{\sqsubseteq}_{T}(S) \circ \mathsf{Rel}^{\sqsubseteq}_{T}(R)$.
4. For any functions $f : X \to X'$, $g : Y \to Y'$ and any relation $R' \subseteq X' \times Y'$, $\mathsf{Rel}^{\sqsubseteq}_{T}((f \times g)^{-1}(R')) \subseteq (Tf \times Tg)^{-1}(\mathsf{Rel}^{\sqsubseteq}_{T}(R'))$.

Definition 14. *Let T be a* Set*-endofunctor. We say that an order on T, given by $\bar{T}$:* Set $\to$ Preord*, $\bar{T}X = (TX, \sqsubseteq)$,*

1. *is stable (preserves inverse images) if for any two functions $f : X \to X'$, $g : Y \to Y'$ and any relation $R' \subseteq X' \times Y'$,*

$$\mathsf{Rel}^{\sqsubseteq}_{T}((f \times g)^{-1}(R')) = (Tf \times Tg)^{-1}(\mathsf{Rel}^{\sqsubseteq}_{T}(R')). \tag{3.5}$$

2. *preserves composition of relations if for any $R \subseteq X \times Y$ and $S \subseteq Y \times Z$,*

$$\mathsf{Rel}^{\sqsubseteq}_{T}(S \circ R) = \mathsf{Rel}^{\sqsubseteq}_{T}(S) \circ \mathsf{Rel}^{\sqsubseteq}_{T}(R). \tag{3.6}$$

3. *preserves composition of preorders if for any preordered set $(X, \leq)$,*

$$\mathsf{Rel}^{\sqsubseteq}_{T}(\leq) \circ \mathsf{Rel}^{\sqsubseteq}_{T}(\leq) \subseteq \mathsf{Rel}^{\sqsubseteq}_{T}(\leq). \tag{3.7}$$

Proposition 15. *For any order on T, we have (3.5)$\Rightarrow$(3.6)$\Rightarrow$(3.7).*

Now the purpose of all these preparations is the following

Definition-Proposition 16. *Let T be a finitary functor having an order $\bar{T}$ which preserves composition of preorders. Then it lifts to a* Preord*-endofunctor $\hat{T}$ ([8], Lemma 5.5), given by $\hat{T}(X, \leq) = (TX, \mathsf{Rel}^{\sqsubseteq}_{T}(\leq))$.*

The functor $\hat{T}$ is locally monotone (defining thus a lifting in the sense of Def. 1): assume $f, g : (X, \leq) \to (Y, \leq)$ are monotone maps such that $f \leq g$ pointwise. Then for any $u \in TX$, we have $u \sqsubseteq u$ and

$$\sqsubseteq\ =\ \mathsf{Rel}^{\sqsubseteq}_{T}(=_X)\ \subseteq\ \mathsf{Rel}^{\sqsubseteq}_{T}((f \times g)^{-1}(\leq))\ \subseteq\ (Tf \times Tg)^{-1}(\mathsf{Rel}^{\sqsubseteq}_{T}(\leq)),$$

therefore $(Tf(u), Tg(u)) \in \mathsf{Rel}^{\sqsubseteq}_{T}(\leq)$.

It follows that any finitary functor having an order which preserves composition of preorders has a lifting[7] to Preord. We also point out that having an order which preserves composition of relations (preorders) is not equivalent nor implied by the preservation of weak pullbacks by the functor T itself, unless discrete preorder is involved (see Ex. 17.1 and also Ex. 17.5 and Ex. 17.6). Any polynomial functor has an order which preserves composition of relations (see [8], Def. 4.4 and the following paragraph there), but this property is not necessarily preserved by their quotients (see below Ex. 17.5).

In all examples below, T is a finitary Set-functor.

Example 17. 1. Assume the ordering on the operations of T is discrete. Then $\sqsubseteq$ is equality and $\mathsf{Rel}_T^{\sqsubseteq}(R) = \mathsf{Rel}_T(R)$. Therefore the order on T is stable iff T preserves weak pullbacks. In this case, the lifting of T to Preord will be $\hat{T}(X, \leq) = (TX, \mathsf{Rel}_T(\leq))$. In view of Prop. 13, we obtain that $\hat{T} = \tilde{T}$.

2. Assume that the order on the operations of T is indiscrete. We have seen that $\sqsubseteq\ = TX \times TX$, hence $\mathsf{Rel}_T^{\sqsubseteq}(R) = TX \times TY$, for any $R \subseteq X \times Y$, provided $\mathsf{Rel}_T(R)$ is not empty. Actually, what we only need is that $T(\leq)$ to be non empty, which happens for all (finitary) functors T except for the constant one mapping everything to the empty set. Then property 4. of T-relators holds with equality, hence again we get a lifting by $\hat{T}(X, \leq) = (TX, TX \times TX)$.
3. Let T be now the finite power-set functor $\mathcal{P}_f$, with inclusion as (pre)order on each $\mathcal{P}_f X$. Then for $R \subseteq X \times Y$, $\mathsf{Rel}_{\mathcal{P}_f X}(R)$ can be described as follows (see for example [15], Thm. 2.3.2):

$$(u, v) \in \mathsf{Rel}_{\mathcal{P}_f X}(R) \Leftrightarrow \forall a \in u\ \exists b \in v\ .\ (a, b) \in R \wedge \forall b \in v\ \exists a \in u\ .\ (a, b) \in R.$$

An easy computation shows now that the order $X \mapsto (\mathcal{P}_f X, \subseteq)$ preserves composition (is even stable), hence $\mathcal{P}_f$ lifts to a functor $\hat{\mathcal{P}}_f(X, \leq) = (\mathcal{P}_f X, \mathsf{Rel}_{\mathcal{P}_f X}^{\subseteq}(\leq))$ on Preord, with ordering

$$(u, v) \in \mathsf{Rel}_{\mathcal{P}_f}^{\subseteq}(\leq) \ \Leftrightarrow\ \forall a \in u\ \exists b \in v\ .\ a \leq b.$$

4. For $TX = \{\bot\} + X$, The order from Ex. 11.4 preserves at least composition of preorders, as the relation extension is $\mathsf{Rel}_T(R) = R \cup \{(\bot, \bot)\}$. The resulting functor $\hat{T}$ will then add a bottom element to any preordered set $(X, \leq)$.
5. Take now the finitary functor $TX = X_2^3$. The relation lifting associated to this functor is

$$((x_1, x_2, x_3), (y_1, y_2, y_3)) \in \mathsf{Rel}_{(-)_2^3}(R) \Leftrightarrow ((x_1, y_1), (x_2, y_2), (x_3, y_3)) \in R_2^3$$

for $R \subseteq X \times Y$. It is well-known that this functor does not preserve weak pullbacks. Recall that we have introduced the zig-zag preorder on TX (Ex. 11). This order is not stable nor preserves composition: for stability, take $X = \{0\}$, $Y = \{1\}$ and $X' = Y' = \{0, 1\}$, with f, g the inclusion

[7] This is no longer an extension: for discrete sets, $\hat{T}(X, =) = (TX, \sqsubseteq)$ is not necessarily discrete.

maps, and relation $R' = =_{X'}$. Then $(Tf \times Tg)^{-1}(\mathsf{Rel}^{\sqsubseteq}_{T}(=)) = TX \times TY$, while $\mathsf{Rel}^{\sqsubseteq}_{T}((f \times g)^{-1}(=)) = \emptyset$. For preservation of composition, take again $X = \{0, 1\}$ and the (preorder) relation $R = \{(0,0), (0,1), (1,1)\}$. Then for example $((0,0,1),(0,1,0)) \in \mathsf{Rel}^{\sqsubseteq}_{(-)^3_2}(R) \circ \mathsf{Rel}^{\sqsubseteq}_{(-)^3_2}(R)$, but $((0,0,1),(0,1,0)) \notin \mathsf{Rel}^{\sqsubseteq}_{(-)^3_2}(R \circ R) = \mathsf{Rel}^{\sqsubseteq}_{(-)^3_2}(R)$.

6. For the polynomial functor $TX = \mathbb{N} \ltimes X$ lexicographically ordered, a similar argument to the one in [8] shows that it is not stable. But it preserves composition with respect to preorders: if $(X, \leq)$ is a preordered set, then $\mathsf{Rel}^{\sqsubseteq}_{T}(\leq)$ is again a preorder, namely the usual lexicographic one on $\mathbb{N} \ltimes X$:

$$((n,x),(m,y)) \in \mathsf{Rel}^{\sqsubseteq}_{T}(\leq) \Leftrightarrow n < m \text{ or } (n = m \text{ and } x \leq y)$$

Proposition 18. *Let T be a finitary* Set*-functor having an order $\bar{T}(X, \leq) = (TX, \sqsubseteq)$. Then the following are equivalent:*

1. *The order is stable.*
2. *$\bar{T}$ maps weak pullbacks to exact squares.*
3. *The lifted functor $\hat{T}$ preserves exact squares.*

Corollary 19. *Let T be a finitary* Set*-functor which preserves weak pullbacks. Then the* Preord*-lifting $\tilde{T}$ from (3.2) preserves exact squares.*

Intuitively, we could simply say that if a Set-functor T preserves exact squares, then its lifting $\tilde{T}$ does so.

It is well known that all Set-functors preserve injective maps with non-empty domain. In Preord we are more interested in embeddings, and they are preserved if the functor preserves exact squares, see [7]. Hence we have the following.

Proposition 20. *If the order on T is stable, then $\hat{T}$ preserves embeddings.*

Corollary 21. *If T preserves weak pullbacks, then $\tilde{T}$ preserves embeddings.*

3.5 Preorder on the Final Coalgebra

There are several papers in the literature describing order relations on the final T-coalgebra (see for example [3] or [8]); as it is expected, there is a connection with the order on the final coalgebra of the lifted functor, first emphasized in [12]. This section intends to present a direct approach of that.

We shall assume that T has a stable order $\bar{T}X = (TX, \sqsubseteq)$ and we shall work with the associated lifting $\hat{T}(X, \leq) = (TX, \mathsf{Rel}^{\sqsubseteq}_{T}(\leq))$. Recall from [8] that a T-simulation with respect to the order $\sqsubseteq$ between two coalgebras $X \xrightarrow{c} TX$ and $Y \xrightarrow{d} TY$ is a relation $R \subseteq X \times Y$ such that $(x, y) \in R \Rightarrow (c(x), d(y)) \in \mathsf{Rel}^{\sqsubseteq}_{T}(R)$. In particular, for any $\hat{T}$-coalgebra $(X, \leq) \xrightarrow{c} (TX, \mathsf{Rel}^{\sqsubseteq}_{T}(\leq))$, the monotonicity of c implies the preorder $\leq$ is a simulation on X. The greatest simulation between two coalgebras is called similarity and denoted by $\lesssim$. It satisfies the following ([8], Prop. 5.4):

1. For any T-coalgebra homomorphisms $X \xrightarrow{f} Z$, $Y \xrightarrow{g} W$ and $x \in X, y \in Y$, we have $x \lesssim y \Leftrightarrow f(x) \lesssim g(y)$.
2. Similarity on a coalgebra $X \xrightarrow{c} TX$ is a preorder.

Let now $Z \xrightarrow{z} TZ$ be the final T-coalgebra (which exists by the finitarity assumption on T). As $\hat{T}$ is also finitary and $\tilde{U}$ preserves limits, the final $\hat{T}$-coalgebra is also Z, but now with some preorder $\leq_Z$ such that $(Z, \leq_Z) \xrightarrow{z} (TZ, \mathsf{Rel}_T^{\sqsubseteq}(\leq_Z))$ is an isomorphism in Preord. In particular, $\leq_Z$ is a simulation, hence $\leq_Z \subseteq \lesssim_Z$. Take $(X, \leq) \xrightarrow{c} (TX, \mathsf{Rel}_T^{\sqsubseteq}(\leq))$ a $\hat{T}$-coalgebra, with (monotone) anamorphism $(X, \leq) \xrightarrow{!} (Z, \leq_Z)$. By property (1) above, $(X, \leq) \xrightarrow{!} (Z, \lesssim_Z)$ is monotone and a $\hat{T}$- coalgebra map, hence in the following diagram the identity on Z is monotone[8], implying $\lesssim_Z \subseteq \leq_Z$.

$$\begin{array}{ccc} (X, \leq) & \xrightarrow{\;!\;} & (Z, \lesssim_Z) \\ & \searrow^{!} & \downarrow{\scriptstyle id} \\ & & (Z, \leq_Z). \end{array}$$

We have thus the following:

Proposition 22. *The preorder on the final $\hat{T}$-coalgebra is the similarity.*[9]

Remark 23. By [8], Thm. 6.2, if the order satisfies the condition

$$\mathsf{Rel}_T^{\sqsubseteq}(R_1) \cap \mathsf{Rel}_T^{\sqsubseteq^{op}}(R_2) \subseteq \mathsf{Rel}_T(R_1 \cap R_2) \tag{3.8}$$

for any two relations $R_1, R_2 \subseteq X \times Y$, then the two-way similarity $\lesssim \cap \lesssim^{op}$ is the same as bisimilarity. This holds for all coalgebras, in particular for the final coalgebra. But bisimilarity on final coalgebra is equality, hence the above condition implies that similarity on the final coalgebra Z is a partial order.

3.6 Third Construction: Order the Variables and Operations

Here we lift an order $\bar{T}$ on T to a Preord-endofunctor $\check{T}$ even in the case that T does not preserve weak pullbacks. The idea is to subsume the constructions in Sect 3.2 and 3.3 in a single one: putting order on the signature (as in Def. 7) and building the coequalizer of (2.3) in Preord.

Definition 24. *Let T be a functor with preorder $\sqsubseteq$. Denote by $\check{T}$ the functor given by the coequalizer $(TX, \preceq)$ below*

$$\coprod_{m,n<\omega} \mathsf{Set}(\underline{m}, \underline{n}) \times (T\underline{m}, \sqsubseteq) \times (X^n, \leq) \overset{\rho}{\underset{\lambda}{\rightrightarrows}} \coprod_{n<\omega} (T\underline{n}, \sqsubseteq) \times (X^n, \leq) \xrightarrow[\pi]{} (TX, \preceq)$$

[8] As $(Z, \lesssim_Z)$ is a $\hat{T}$-coalgebra, there is unique monotone map from it to the final $\hat{T}$-coalgebra. Via the forgetful functor, this is mapped to the unique T-coalgebra map $Z \to Z$ which is obviously the identity.

[9] See also [13], Thm. 4.1 and [17], Thm. 5.6.

*where the domain and the codomain of the coequalizer pair carry the coproduct preorder and each component of the coproduct has the product preorder (*Set$(\underline{m}, \underline{n})$ *is discrete, while* X^n *has the product preorder obtained from the one on* X*).*

Observe that $\check{T}$ is locally monotone and that $\sqsubseteq \subseteq \preceq$. Moreover, we have

Theorem 25. *Let* T *be a finitary* Set*-endofunctor which preserves weak pullbacks and has an order which preserves composition of preorders. Then the liftings* $\check{T}$ *and* $\hat{T}$ *coincide.*

Example 26. The construction presented at the beginning of this section says that we can still get a lifting, independently of T preserving weak pullbacks or the order preserving composition of relations; for example, consider the functor $TX = X_2^3$ with the zigzag order. The corresponding preorder $\preceq$ can then be described as follows: for $(x_1, x_2, x_3), (y_1, y_2, y_3) \in X_2^3$, $(x_1, x_2, x_3) \preceq (y_1, y_2, y_3)$ if $x_i \leq y_i, \forall 1 \leq i \leq 3$, or $x_1 = x_2 = x_3$ and there is some $1 \leq i \leq 3$ with $x_i \leq y_i$.

4 From Preord to Poset

Given a finitary Set-functor T, assume that we have an extension (or a lifting) Γ to Preord. We can move further to Poset by taking the locally monotone functor $T' = Q\Gamma J$: Poset $\to$ Preord, where $Q \dashv J$ is the (monadic) adjunction between the quotient and the inclusion functor mentioned in the preliminaries. If Γ is finitary, then T' is also, since both J and Q preserve filtered colimits.[10] Regarding coalgebras, notice that each Γ-coalgebra can be quotiented to a T'-coalgebra, thus there is a functor Q' : Coalg(Γ) $\to$ Coalg(T'), which sends a Γ-coalgebra $X \xrightarrow{c} \Gamma X$ to $QX \xrightarrow{Qc} Q\Gamma X \to T'QX$.

Now the discussion bifurcates according to Γ being an extension or a lifting.

If Γ is an extension, a simple computation shows that T' maps discrete sets to discrete sets, thus it is a Poset-extension of T. Moreover, a similar discussion to the one in Section 3.1 shows that the final T'-coalgebra exists and is discrete, with same carrier as the final T-coalgebra, once we assume T (and Γ) finitary.

For the particular extension $\tilde{T}(X, \leq) = (TX, \trianglelefteq)$ of a finitary functor T from Section 3.2, relations (3.1) and (3.2), let $T'(X, \leq)$ be the quotient Poset-functor, whose ordering will be denoted for convenience with same symbol $\trianglelefteq$. In case T preserves weak pullbacks, $\trianglelefteq$ on TX has been expressed in terms of the relation lifting (Prop. 13). An immediate result is

Proposition 27. *Let* T *be a finitary weak pullbacks preserving* Set*-functor and* $T' = Q\tilde{T}J$ *its* Poset*-extension, with* $\tilde{T}$ *as in (3.1). Then* T' *preserves exact squares and embeddings.*

Remark 28. If T does not preserve weak pullbacks, then T' may fail to preserve embeddings, as we can see from the following example: take T to be the functor part of the Boolean algebra monad. On finite sets, we can identify T with the

[10] Q is left adjoint, while for J it follows from [6], vol. 2, Prop. 5.5.6.

composition of the contravariant power-set functor with itself. Then T does not preserve weak pullbacks [14]. We shall show that the corresponding extension to Poset does not preserve embeddings. For this, take the embedding of the discrete two-elements poset $\{a, b\}$ into the poset $\{a, b, c\}$ ordered by $a < c$, $b < c$. Then $T'(\{a, b\}, =)$ is the (discrete) free Boolean algebra on two generators with 16 elements. For the poset (Boolean algebra) $(T'(\{a, b, c\}, \leq), \trianglelefteq)$, notice that monotonicity of operations implies $\perp \trianglelefteq a \wedge \neg a \trianglelefteq a \wedge \neg c \trianglelefteq c \wedge \neg c = \perp$, thus $a \trianglelefteq c$. Similarly, $\perp \trianglelefteq a \wedge \neg a \trianglelefteq c \wedge \neg a \trianglelefteq c \wedge \neg c = \perp$ implies $c \trianglelefteq a$. Thus the images of a and c in $T'(\{a, b, c\}, \leq)$ coincide; similarly for the images of b and c, which makes us conclude that $T'(\{a, b, c\}, \leq)$ has only 4 elements (the free Boolean algebra on only one generator). Hence T' cannot preserve the embedding $(\{a, b\}, =) \hookrightarrow (\{a, b, c\}, \leq)$.

In case Γ is a lifting, there is a preorder on TX for each $(X, \leq)$ and $T'(X, \leq)$ is the quotient of TX with respect to the equivalence relation induced by that preorder. The resulting functor T' is, in general, no longer a lifting of T to Poset nor an extension.

However, if we consider a particular lifting of T, namely $\hat{T}$, with respect to an order $\sqsubseteq$ which is already a partial order, then by restricting to posets we obtain that $\mathsf{Rel}_T^{\sqsubseteq}(\leq)$ is a partial order on TX (for each poset $(X, \leq)$), once we assume that $\sqsubseteq$ preserve compositions of preorders and satisfies (3.8). In this case, $Q\hat{T}J$ can be identified with $\hat{T}J$ and defines a lifting of T to Poset. In the general case, however, the best that we can say is that the analogue of Prop. 10 holds, namely that T' will coincide with T on finite sets $\underline{n}$, with partial order $\sqsubseteq$.

Example 29. 1. Let $\mathcal{D}$ be the finite subdistribution functor, $\mathcal{D}X = \{d : X \to [0, 1] \mid \sum_{x \in X} d(x) \leq 1, |\mathsf{supp}(d)| < \infty\}$, with the pointwise order $d \sqsubseteq d' \Leftrightarrow d(x) \leq d'(x), \forall x \in X$. The corresponding $\hat{\mathcal{D}}$ maps posets to posets (see comments after Def. 11 in [12]).
2. Take now $\mathcal{P}_f$, the finite powerset functor, with the inclusion order. Then this time the Poset-functor is indeed a quotient, namely the finite convex power set.

Although T' is not a lifting nor an extension, it still behaves well with respect to exact squares (and embeddings):

Proposition 30. *Let T be a finitary* Set*-functor having an order $\bar{T}(X, \leq) = (TX, \sqsubseteq)$ which is stable and $T' = Q\hat{T}J$ as above. Then T' preserves exact squares, thus also embeddings.*

5 Conclusion

Considering the rich body of results on Set-functors, see eg [4], our investigations suggest that analogous results for functors on preorders or on posets would be of interest to coalgebra. For example, characterizations of functors that preserve exact squares or embeddings would be of interest. This also links the current paper with investigations on coalgebraic logic over preorders or posets, where first steps have been taken on logics given by predicate liftings in [9] and on Moss's ∇ in [5]. Another direction is to follow the connection with coalgebraic (bi)simulations like in [8,12].

Acknowledgements. We would like to thank J. Velebil for pointing out the importance of exact squares and the reference [7], and the referees for their valuable suggestions.

References

1. Abramsky, S., Jung, A.: Domain Theory. In: Abramsky, S., Gabbay, D.M., Maibaum, T.S.E. (eds.) Handbook of Logic in Computer Science, vol. 3, pp. 1–168. Oxford Univ. Press, New York (1994)
2. Aczel, P., Mendler, N.: A Final Coalgebra Theorem. In: Dybjer, P., Pitts, A.M., Pitt, D.H., Poigné, A., Rydeheard, D.E. (eds.) Category Theory and Computer Science. LNCS, vol. 389, pp. 357–365. Springer, Heidelberg (1989)
3. Adámek, J.: Final Coalgebras Are Ideal Completions of Initial Algebras. J. Logic Computat. 12(2), 217–242 (2002)
4. Adámek, J., Trnková, V.: Automata and Algebras in Categories. Mathematics and Its Applications: East European Series, vol. 37. Kluwer Academic Publishers, Dordrecht (1990)
5. Bilková, M., Kurz, A., Petrişan, D., Velebil, J.: Relation Liftings on Preorders and Posets. In: Corradini, A., Klin, B., Cârstea, C. (eds.) CALCO 2011. LNCS, vol. 6859, pp. 115–129. Springer, Heidelberg (2011)
6. Borceux, F.: Handbook of Categorical Algebra. In: Encycl. Mathem. Appl., vol. 50-52. Cambridge Univ. Press, Cambridge (1994)
7. Guitart, R.: Relations et Carrés Exacts. Ann. Sci. Math. Québec 4(2), 103–125 (1980)
8. Hughes, J., Jacobs, B.: Simulations in Coalgebra. Theor. Comput. Sci. 327, 71–108 (2004)
9. Kapulkin, K., Kurz, A., Velebil, J.: Expressivity of Coalgebraic Logic over Posets. In: Jacobs, B.P.F., Niqui, M., Rutten, J.J.M.M., Silva, A.M. (eds.) CMCS 2010 Short contributions, CWI Technical report SEN-1004, pp. 16–17 (2010)
10. Karazeris, P., Matzaris, A., Velebil, J.: Final Coalgebra in Accessible Categories, accepted for publication in Math. Struct. Comput. Sci., `http://xxx.lanl.gov/abs/0905.4883`
11. Lambek, J.: Subequalizers. Canad. Math. Bull. 13, 337–349 (1970)
12. Levy, P.: Similarity Quotients as Final Coalgebras. In: Hofmann, M. (ed.) FOSSACS 2011. LNCS, vol. 6604, pp. 27–41. Springer, Heidelberg (2011)
13. Rutten, J.: Relators and Metric Bisimulations (Extended Abstract). In: Jacobs, B., Moss, L., Reichel, H., Rutten, J. (eds.) First Workshop on Coalgebraic Methods in Computer Science, CMCS 1998. Electr. Notes Theor. Comput. Sci., vol. 11, pp. 252–258 (1998)
14. Rutten, J.: Universal Coalgebra: A Theory of Systems. Theor. Comput. Sci. 249, 3–80 (2000)
15. Thijs, A.: Simulation and Fixed Point Sematics, Ph. D. Thesis, University of Groningen (1996)
16. Velebil, J., Kurz, A.: Equational Presentations of Functors and Monads. Math. Struct. Comput. Sci. 21(2), 363–381 (2011)
17. Worrell, J.: Coinduction for Recursive Data Types: Partial Orders, Metric Spaces and Ω-Categories. In: Reichel, H. (ed.) Coalgebraic Methods in Computer Science, CMCS 2000. Electr. Notes Theor. Comput. Sci., vol. 33, pp. 337–356 (2000)

Model Constructions for Moss' Coalgebraic Logic

Jort Bergfeld[1,*] and Yde Venema[2,**]

[1] Artificial Intelligence and Cognitive Engineering, Rijksuniversiteit Groningen, Nijenborgh 9, 9747 AG Groningen, The Netherlands.
[2] Institute for Logic, Language and Computation, Universiteit van Amsterdam, Science Park 904, 1098XH Amsterdam, The Netherlands

Abstract. We discuss two model constructions related to the coalgebraic logic introduced by Moss. Our starting point is the derivation system $\mathbf{M}_T$ for this logic, given by Kupke, Kurz and Venema. Based on the one-step completeness of this system, we first construct a finite coalgebraic model for an arbitrary $\mathbf{M}_T$-consistent formula. This construction yields a simplified completeness proof for the logic $\mathbf{M}_T$ with respect to the intended, coalgebraic semantics. Our second main result concerns a strong completeness result for $\mathbf{M}_T$, provided that the functor T satisfies some additional constraints. Our proof for this result is based on the construction, for an $\mathbf{M}_T$-consistent set of formulas A, of a coalgebraic model in which A is satisfiable.

Keywords: coalgebra, modal logic, completeness, finite model property, strong completeness.

1 Introduction

Universal Coalgebra [16] provides the notion of a *coalgebra* as the natural mathematical generalization of state-based evolving systems such as streams, (infinite) trees, Kripke models, (probabilistic) transition systems, and many others. This approach combines simplicity with generality and wide applicability: many features, including input, output, nondeterminism, probability, and interaction, can easily be encoded in the coalgebra type, which in this paper we will take to be an endofunctor T on the category Set of sets as objects with functions as arrows. Logic enters the picture if one wants to specify and reason about *behavior*, one of the most fundamental notions admitting a coalgebraic formalization. With Kripke structures constituting key examples of coalgebras, it should come as no surprise that most coalgebraic logics are some kind of modification or generalization of *modal logic* [5].

* This work is part of the VIDI research programme with number 639.072.904, which is financed by the Netherlands Organisation for Scientific Research.
** The research of this author has been made possible by VICI grant 639.073.501 of the Netherlands Organization for Scientific Research (NWO).

A. Corradini, B. Klin, and C. Cîrstea (Eds.): CALCO 2011, LNCS 6859, pp. 100–114, 2011.

This approach was initiated by Moss [13], who generalized the so-called 'cover modality' ∇_P from Kripke structures to coalgebras of arbitrary type T. The fascinating novelty of Moss' language is that his modality has a rather non-standard arity: Moss' syntax specifies that $\nabla_T\alpha$ is a formula for all $\alpha \in T\mathcal{L}$ (where $\mathcal{L}$ is the collection of formulas), while its semantics is given by a categorical notion of *relation lifting* $\overline{T}$. This approach is completely uniform in the functor T, but as a drawback, for $\overline{T}$ to behave well T must satisfy the category-theoretic property of preserving weak pullbacks. In order to overcome the shortcomings of Moss' logic, Kurz [11], Pattinson [15] and others considered coalgebraic modal formalisms, that use standard syntax and work for coalgebras of arbitrary type. The success of this approach, in which the semantics of each modality is determined by a so-called *predicate lifting*, directed attention away from Moss' logic.

Interest in Moss' logic revived when it became clear that an approach based on his modality could have some advantages. In particular, some key results on the modal μ-calculus were obtained by Janin & Walukiewicz [8], on the basis of proofs that crucially involve a reconstruction of the classical modal language on the basis of the nabla modality (which they introduced, independently of Moss, as a primitive connective). Kupke & Venema [10] showed that many fundamental results in the area of (fixpoint) logic and automata theory could be lifted to the abstraction level of coalgebra.

Given the nonstandard syntax of Moss' language it was not a priori clear whether the collection of coalgebraic validities would allow nice derivation systems. As a first result, Palmigiano & Venema [14] gave a complete axiomatization for the cover modality ∇_P (i.e. in the case of Kripke frames). This calculus was streamlined into a formulation that admits a straightforward generalization to a calculus $\mathbf{M}_T$ for an arbitrary set functor T, by Bílková, Palmigiano & Venema [3], who also provided suitable Gentzen systems for the logic based on ∇_P. Kupke, Kurz & Venema [10] solved the outstanding problem by proving the soundness and completeness of the calculus $\mathbf{M}_T$ with respect to the coalgebraic semantics.

In this paper, which originated in the first author's MSc thesis [2] supervised by the second author, we continue the line of investigations of [10], taking their result on one-step soundness and completeness as our starting point. (As a minor difference with [10], we add explicit proposition letters to the language.) Our main contribution is two-fold. First, based on adapting ideas from Schröder [17] to the setting of Moss' logic, we provide a coalgebraic construction that, given an $\mathbf{M}_T$-consistent formula a, yields a finite model in which a is satisfied. As a corollary, we considerably simplify the second part of the completeness proof of [10] for the logic $\mathbf{M}_T$ with respect to its intended, coalgebraic semantics. Our second main result concerns a *strong* completeness result for $\mathbf{M}_T$, provided that the functor T restricts to finite sets and weakly preserves limits of surjective ω-cochains of finite sets. Our proof for this result is based on the quasi-canonical model method of Pattinson & Schröder [18].

2 Preliminaries

Categories and Coalgebras. We assume familiarity with basic notions from category theory (such as categories, functors, and natural transformations), and from coalgebra. Here we fix some notation and terminology. We restrict attention to Set-based coalgebras, where Set denotes the category with sets as objects and functions as arrows.

Convention 1. *Throughout the paper we fix a functor $T : \mathsf{Set} \to \mathsf{Set}$, which we assume to preserve inclusions and weak pullbacks.*

The restriction that T preserves inclusions is for reasons of presentation only; we motivate the other restriction in Remark 1. Many (but not all) examples of coalgebraically interesting set functors fall in the scope of our work. We mention the inductively defined class *EKPF* of *extended Kripke polynomial functors* given as follows

$$T := Id \mid C \mid P \mid B_\omega \mid D_\omega \mid T_0 \circ T_1 \mid T_0 + T_1 \mid T_0 \times T_1 \mid T^D,$$

where C is an abitrary constant functor, P is power set, B_ω is finitary multiset, D_ω is finitary probability distribution and T^D is exponentiation with respect to an arbitrary set. An example of a functor that does not preserve weak pullbacks is the (monotone) neighborhood functor.

The *finitary version* $T_\omega : \mathsf{Set} \to \mathsf{Set}$ of T is given, on objects, by $T_\omega X := \bigcup\{TY \mid Y \subseteq X, Y \text{ finite }\}$, and on arrows by $T_\omega f := Tf$. It can be proved that T_ω also preserves inclusions and weak pullbacks. Given an object $\alpha \in T_\omega A$, we let $Base_A(\alpha)$ denote the smallest finite subset of A such that $\alpha \in TBase_A(\alpha)$; in fact, the family of operations $Base_A : T_\omega A \to P_\omega A$ constitutes a natural transformation $Base : T_\omega \dot{\to} P_\omega$ [7].

Definition 1. *A* T-coalgebra *is a pair (S, σ) where S is a set and $\sigma : S \to TS$ is a function. A* morphism of T-coalgebras *from (S, σ) to (S', σ'), written $f : (S, \sigma) \to (S, \sigma')$, is a function $f : S \to S'$ such that $Tf \circ \sigma = \sigma \circ f$.*

Relation Lifting. The coalgebraic semantics of Moss' coalgebraic language is based on the notion of *relation lifting* that we now briefly discuss (see [10] for more information). First we introduce some notation for relations and functions. The *graph* of a function $f : X \to X'$ is the relation $Grf := \{(x, f(x)) \in X \times X' \mid x \in X\}$. The *diagonal* relation on a set X is denoted as Id_X, and the *converse* of a relation R as $R^\smile$. Given subsets $Y \subseteq X$, $Y' \subseteq X'$, the *restriction* of R to Y and Y' is given as $R\restriction_{Y \times Y'} := R \cap (Y \times Y')$. The composition of two relations $R \subseteq X \times X'$ and $R' \subseteq X' \times X''$ is denoted by $R \,;\, R'$, whereas the composition of two functions $f : X \to X'$ and $f' : X' \to X''$ is denoted by $f' \circ f$. Thus, we have $Gr(f' \circ f) = Grf \,;\, Grf'$.

Definition 2. *[1] Given a binary relation $R \subseteq X_1 \times X_2$ with projection functions $\pi_i : R \to X_i$, we define its* T-lifting *$\overline{T}R \subseteq TX_1 \times TX_2$ as follows:*

$$\overline{T}R := \{((T\pi_1^R)\rho, (T\pi_2^R)\rho) \mid \rho \in TR\}.$$

Throughout the paper, we will use properties of the relation $\overline{T}R$; unless explicitly stated otherwise, these can always be derived by elementary means from the following fact.

Fact 2 (Properties of Relation Lifting). *The relation lifting $\overline{T}$ satisfies the following properties, for all functions $f : X \to X'$, all relations $R, S \subseteq X \times X'$, $R' \subseteq X' \times X''$, and all subsets $Y \subseteq X$, $Y' \subseteq X'$:*

1. *$\overline{T}$ extends T: $\overline{T}(Grf) = Gr(Tf)$;*
2. *$\overline{T}$ preserves the diagonal: $\overline{T}(Id_X) = Id_{TX}$;*
3. *$\overline{T}$ commutes with relation converse: $\overline{T}(R^{\smile}) = (\overline{T}R)^{\smile}$;*
4. *$\overline{T}$ is monotone: if $R \subseteq S$ then $\overline{T}(R) \subseteq \overline{T}(S)$;*
5. *$\overline{T}$ distributes over composition: $\overline{T}(R \,;\, R') = (\overline{T}R) \,;\, (\overline{T}R')$;*
6. *$\overline{T}$ commutes with restriction: $\overline{T}(R\restriction_{Y \times Y'}) = \overline{T}R\restriction_{TY \times TY'}$.*
7. *$\overline{T}_\omega$ coincides with $\overline{T}$: $\overline{T}_\omega R = (\overline{T}R)\restriction_{T_\omega X \times T_\omega X'}$.*

Remark 1. The main reason why we restrict our attention to coalgebra types T that preserve weak pullbacks is that for these functors, $\overline{T}$ distributes over relation composition (Fact 2(5)) [1,19].

Applying relation lifting to the membership relation $\in$, we obtain an interesting operation. Given a set X, we let $\in_X \subseteq X \times PX$ denote the membership relation, restricted to X. We define the map $\lambda^T_X : TPX \to PTX$ by

$$\lambda^T_X(\Phi) := \{\alpha \in TX \mid \alpha \, \overline{T}{\in_X} \, \Phi\},$$

and call elements of $\lambda^T_X(\Phi)$ *lifted members* of Φ. Related to Fact 2(5) is that the family of maps $\lambda^T_X : TPX \to PTX$ constitutes a distributive laws of T over the monad P (see [10] for a discussion). Of more immediate importance is the following distributive law relative to the *contravariant power set functor* $\breve{P}$ [4].

Fact 3. *The family of maps λ^T provides a natural transformation $\lambda^T : T\breve{P} \dot{\to} \breve{P}T$.*

The following concept is needed in the axioms describing the interaction between ∇ and conjunctions.

Definition 3. *An object $\Phi \in TPX$ is a* redistribution *of $A \in PTX$ if $A \subseteq \lambda^T_X(\Phi)$. In case $A \in P_\omega T_\omega X$, we call a redistribution Φ* slim *if $\Phi \in T_\omega P_\omega(\bigcup_{\alpha \in A} Base(\alpha))$. The set of slim redistributions of A is denoted as $SRD(A)$.*

Fact 4. *[21] Given sets X, Y, a set $\Gamma \in PTX$ and a surjection $f : X \to Y$, we have*

$$\{TPf(\Phi) \mid \Phi \in SRD(\Gamma)\} - SRD(PTf(\Gamma)).$$

Propositional Logic. Given a set X, we define the set $\mathcal{L}_0(X)$ of propositional formulas over X by the following grammar:

$$a \ ::= \ x \mid \neg a \mid \bigwedge A \mid \bigvee A,$$

where $x \in X$, and $A \in P_\omega \mathcal{L}_0(X)$. That is, as the primitive connectives of our propositional language we take the unary symbol $\neg$ and the finitary meet and join symbols, $\bigwedge$ and $\bigvee$. We abbreviate $\bot := \bigvee \varnothing$ and $\top := \bigwedge \varnothing$.

Given sets X and S, an *X-valuation on S* is a map $V : X \to PS$; such a map can be naturally extended to a homomorphism $\widehat{V} : \mathcal{L}_0(X) \to PS$ by putting $\widehat{V}(\bigwedge A) := \bigcap\{V_0(a) \mid a \in A\}$, etc.

3 Moss' Logic and Its Axiomatization

In this section we briefly recall the syntax and semantics of Venema's finitary version of Moss' coalgebraic logic [13,20], and the axiomatization of its coalgebraic valid formulas, given by Kupke, Kurz and Venema [10].

3.1 Moss' Logic

The finitary version $\mathcal{L}$ of Moss' language is defined as follows.

Definition 4. *Given a set* Prop *of variables, the set* $\mathcal{L}(\mathsf{Prop})$ *of Moss formulas in* Prop *is given by the following grammar:*

$$a \ ::= \ p \mid \neg a \mid \bigwedge A \mid \bigvee A \mid \nabla \alpha$$

where $p \in \mathsf{Prop}$, $A \in P_\omega \mathcal{L}$ *and* $\alpha \in T_\omega \mathcal{L}$.

Despite its unconventional appearance, the language $\mathcal{L}$ admits fairly standard definitions of most syntactical notions. For example, we may define the (finite!) set $Sfor(a)$ of *subformulas* of a by a straightforward formula induction, of which the only nonstandard clause concerns the nabla operator:

$$Sfor(\nabla\alpha) := \{\nabla\alpha\} \cup \bigcup_{a \in Base(\alpha)} Sfor(a).$$

The elements of $Base(\alpha) \subseteq Sfor(\nabla\alpha)$ will be called the *immediate* subformulas of $\nabla\alpha$. Given a formula a, we define the *single negation* of a formula a as ${\sim}a := b$ if $a = \neg b$ for some formula b, and as ${\sim}a := \neg a$ otherwise.

Since in this paper we will not only be dealing with formulas and sets of formulas, but also with elements of the sets $T_\omega\mathcal{L}$, $P_\omega T_\omega \mathcal{L}$ and $T_\omega P_\omega \mathcal{L}$, it will be convenient to use the following *naming convention*:

Set	Prop	$\mathcal{L}$	$T_\omega\mathcal{L}$	$P_\omega\mathcal{L}$	$P_\omega T_\omega\mathcal{L}$	$T_\omega P_\omega\mathcal{L}$
Elements	$p, q, \ldots$	$a, b, \ldots$	$\alpha, \beta, \ldots$	$A, B, \ldots$	$\Gamma, \Delta, \ldots$	$\Phi, \Psi, \ldots$

We may see the boolean connectives $\bigvee$ and $\bigwedge$ as maps from finite sets of formulas to formulas, $\bigvee, \bigwedge : P_\omega\mathcal{L} \to \mathcal{L}$. Applying the functor to these maps, we obtain functions $T\bigvee, T\bigwedge : T_\omega P_\omega \mathcal{L} \to T\mathcal{L}$. In particular, for any object $\Phi \in T_\omega P_\omega \mathcal{L}$, we obtain well-formed formulas of the form $\nabla(T\bigvee)\Phi$ and $\nabla(T\bigwedge)\Phi$.

Since we consider a version of Moss' language with proposition letters, in order to interpret this language we have to introduce valuations and models.

Definition 5. *A valuation on a T-coalgebra (S, σ) is a valuation V : $\mathsf{Prop} \to PS$; the induced structure (S, σ, V) will be called a T-*model. *For such a model, the satisfaction relation $\Vdash_{\sigma,V} \subseteq S \times \mathcal{L}$ is defined by the following induction on the complexity of formulas:*

$$\begin{array}{ll} s \Vdash_{\sigma,V} p & \textit{if } s \in V(p), \\ s \Vdash_{\sigma,V} \neg a & \textit{if } s \not\Vdash_{\sigma,V} a, \\ s \Vdash_{\sigma,V} \bigwedge A & \textit{if } s \Vdash_{\sigma,V} a \textit{ for all } a \in A, \\ s \Vdash_{\sigma,V} \bigvee A & \textit{if } s \Vdash_{\sigma,V} a \textit{ for some } a \in A, \\ s \Vdash_{\sigma,V} \nabla\alpha & \textit{if } \sigma(s)\, \overline{T}\Vdash_{\sigma,V} \alpha. \end{array}$$

When no confusion is likely, we may write $\Vdash$ instead of $\Vdash_{\sigma,V}$. If $s \Vdash_{\sigma,V} a$ we say that a is true, *or* holds *at s in $\mathbb{S}$, and we may write $\mathbb{S}, s \Vdash a$, where $\mathbb{S}$ denotes the T-model (S, σ, V).*

Two important observations about Moss' logic are that it is *adequate* with respect to behavioral equivalence (or, equivalently, bisimilarity), and *expressive* when we confine attention to finitely branching coalgebras.

3.2 The Derivation System M

When it comes to axiomatics, following [10] we find it convenient to take an approach based on derivation systems that manipulate equations, or rather, inequalities. An *inequality* is a pair consisting of two formulas a and b, usually written $a \preccurlyeq b$. Readers may think of this as the formula $a \to b$, as is obvious from the semantics.

Definition 6. *An inequality $a \preccurlyeq b$ is* valid, *notation: $\models_T a \preccurlyeq b$, if for every coalgebraic model $\mathbb{S} = (S, \sigma, V)$, and any $s \in S$, if $\mathbb{S}, s \Vdash a$, then $\mathbb{S}, s \Vdash b$.*

The following axiomatization for this logic was proved to be sound and complete in [10].

Definition 7. *The derivation system* **M** *is given by the derivation rules of Table 1, together with any complete set of axioms and rules (in inequational format) for classical propositional logic.*

Observe that unless T restricts to finite sets, **M** is an infinitary derivation system, in that the rules (∇2) and (∇3) may have infinitely many premisses. To get some intuitive understanding of this derivation system, we first note that (∇1) functions as a combined congruence and monotonicity rule. It has a side condition expressing that it may only be applied when the set of premisses is indexed by a relation Z such that (α, β) belongs to the lifted relation $\overline{T}Z$. Each of the other two rules should be seen as a distributive law (in the logical sense of the word). To see this, first consider the case that T preserve finiteness. Then we may replace the rules (∇2) and (∇3) with the following *axioms*:

$$\bigwedge\left\{\nabla\alpha \mid \alpha \in \Gamma\right\} \preccurlyeq \bigvee\left\{\nabla(T\textstyle\bigwedge)\Phi \mid \Phi \in SRD(\Gamma)\right\} \qquad (\nabla 2_f)$$

$$\nabla(T\textstyle\bigvee)\Phi \preccurlyeq \bigvee\left\{\nabla\beta \mid \beta\, \overline{T}\!\in \Phi\right\} \qquad (\nabla 3_f)$$

Table 1. Modal derivation rules of the system **M**

$$(\nabla 1)\ \frac{\{a \preccurlyeq b \mid (a,b) \in Z\}}{\nabla\alpha \preccurlyeq \nabla\beta}\ (\alpha,\beta) \in \overline{T}Z$$

$$(\nabla 2)\ \frac{\{\nabla(T\bigwedge)(\Phi) \preccurlyeq a \mid \Phi \in SRD(\Gamma)\}}{\bigwedge\{\nabla\alpha \mid \alpha \in \Gamma\} \preccurlyeq a}$$

$$(\nabla 3)\ \frac{\{\nabla\alpha \preccurlyeq a \mid \alpha \overline{T}{\in}\ \Phi\}}{\nabla(T\bigvee)(\Phi) \preccurlyeq a}$$

Roughly speaking, $(\nabla 3_f)$ expresses how ∇ distributes over disjunctions, while $(\nabla 2_f)$ shows how a conjunction of nabla formulas can be rewritten as a disjunction of nabla formulas of conjunctions of the collection of immediate subformulas of the nabla formulas. If T does not restrict to finite sets, we may still think of $(\nabla 2)$ and $(\nabla 3)$ as these identities: the only problem is that the expressions on the right hand side of $(\nabla 2_f)$ and $(\nabla 3_f)$ may no longer denote properly defined formulas.

The notions of derivability with respect to this system is standard. A *derivation* is a well-founded tree, labelled with inequalities, such that the leaves of the tree are labelled with axioms of **M**, whereas with each parent node we may associate a derivation rule of which the conclusion labels the parent node itself, and the premisses label its children. If there is a derivation of the inequality $a \preccurlyeq b$, we write $\vdash_T a \preccurlyeq b$. A formula a is **M**-*consistent* if the inequality $a \preccurlyeq \bot$ is not derivable in **M**; a set A of formulas is consistent if the formula $\bigwedge A_0$ is consistent for each finite subset $A_0 \subseteq A$.

The following theorem is the main result of Kupke, Kurz & Venema [10]

Fact 5 (Soundness and Completeness of M). *[10] For each pair of formulas $a, b \in \mathcal{L}$:*

$$\models_T a \preccurlyeq b \textit{ iff } \vdash_T a \preccurlyeq b.$$

The completeness proof in [10] proceeds in two steps. First the authors prove a so-called one-step completeness result for their system; then they apply Pattinson's stratification method, involving the terminal sequence of the functor T, to prove Fact 5. The construction given in our paper will provide a much simpler alternative for the second part of their proof.

3.3 One-Step Soundness and Completeness

Given a set X, we define the set $\mathcal{L}_\nabla(X)$ of *rank-1 formulas* in X by putting

$$\mathcal{L}_\nabla(X) := \mathcal{L}_0\{\nabla\alpha \mid \alpha \in T_\omega\mathcal{L}_0X\}.$$

It will sometimes be convenient to think of $\mathcal{L}_\nabla(X)$ as a propositional language, generated from the set $T_\omega^\nabla(X) := \{\nabla\alpha \mid \alpha \in T_\omega\mathcal{L}_0X\}$ as proposition letters.

Any valuation $V : X \to PS$, interpreting elements of X as subsets of some set S, not only extends to a propositional meaning function $\widehat{V} : \mathcal{L}_0(X) \to PS$, it also induces an interpretation $\widetilde{V} : \mathcal{L}_\nabla(X) \to PTS$ of rank-1 formulas in X as subsets of TS. For the definition of $\widetilde{V}$, observe that the map $T\widehat{V} : T_\omega\mathcal{L}_0 X \to TPS$ naturally yields a $T_\omega^\nabla(X)$-valuation $\lambda_S^T \circ T\widehat{V}$ on TS given by, for $\alpha \in T_\omega\mathcal{L}_0 X$:

$$\nabla\alpha \mapsto \lambda_S^T(T\widehat{V}(\alpha)).$$

Then we define $\widetilde{V} := \widehat{\lambda^T \circ T\widehat{V}}$.

We may take the set PS itself as a collection of proposition letters; then the identity map on PS becomes a special PS-valuation on PS: the *identity valuation on* S, notation $i_S : PS \to PS$. We say that an $\mathcal{L}_0(PS)$-inequality $a \preccurlyeq b$ is a *true fact on* PS, notation: $\models_0^S a \preccurlyeq b$, if $\widehat{i_S}(a) \subseteq \widehat{i_S}(b)$; an $\mathcal{L}_\nabla(PS)$-inequality $a \preccurlyeq b$ is *one-step valid*, notation: $\models_1^S a \preccurlyeq b$, if $\widetilde{i_S}(a) \subseteq \widetilde{i_S}(b)$.

On the axiomatic side, we modify the derivation $\mathbf{M}$ into a *one-step derivation system* $\mathbf{M}^S$, which only uses $\mathcal{L}_0(PS)$ and $\mathcal{L}_\nabla(PS)$-formulas. More precisely, a $\mathbf{M}^S$-derivation is a well-founded tree, labelled with $\mathcal{L}_0(PS)$- and $\mathcal{L}_\nabla(PS)$-inequalities, such that (1) the leaves of the tree are labelled with true facts on PS, whereas (2) with each parent node we may associate a derivation rule of which (a) the conclusion is an $\mathcal{L}_\nabla(PS)$-inequality labelling the parent node itself, (b) the premisses label its children, and (c) these premisses are either all $\mathcal{L}_\nabla(PS)$-inequalities, or all $\mathcal{L}_0(PS)$-inequalities; in the latter case the children are all leaves and the derivation rule is (∇1). Hence if we do induction on the complexity of one-step derivations, we may assume that the base case is given by an application of rule (∇1). If there is such a one-step derivation of the inequality $a \preccurlyeq b$, we write $\vdash_1^S a \preccurlyeq b$.

Fact 6 (One-Step Soundness and Completeness). *[10] Given a set S, for each pair of rank-1 formulas $a, b \in \mathcal{L}_\nabla(PS)$:*

$$\models_1^S a \preccurlyeq b \text{ iff } \vdash_1^S a \preccurlyeq b.$$

4 A Finite Model Construction

In this section we will give the main construction of this paper, serving to prove Theorem 7 below. As a corollary we obtain Fact 5, the Soundness and Completeness Theorem of [10].

Theorem 7. *Every consistent formula is satisfied in a finite T-coalgebra.*

Our construction is based on ideas from Schröder [17]. To give a rough idea, we need to introduce some terminology. We call a set of formulas *closed* if it is closed under taking subformulas and single negations ($\sim$). Given a closed set R of formulas, we call a subset $A \subseteq R$ an *R-atom* if A is a maximal consistent subset of R. Any R-atom A has the properties, for every $a \in R$, that $a \in A$ iff ${\sim}a \notin A$, and for every $a \wedge b \in R$, that $a \wedge b \in A$ iff both $a \in A$ and $b \in A$, etc. As usual it is straightforward to prove a *Lindenbaum Lemma* stating that every consistent subset of R can be extended to an R-atom.

Definition 8. *Given a formula c, let $C(c)$ denote the smallest closed set containing c, and define the closed set $R(c)$ by*

$$R(c) := \{ \bigvee_{A \in \mathcal{A}} \bigwedge A, \neg \bigvee_{A \in \mathcal{A}} \bigwedge A \mid \mathcal{A} \subseteq PC(c)\}.$$

We let $S(c)$ denote the set of $R(c)$-atoms.

Clearly $S(c)$ is a finite set. Hence, by the Lindenbaum Lemma, in order to prove Theorem 7, it suffices to build a model $(S(c), \sigma, V)$ on the set $S(c)$ for which we can prove a *Truth Lemma* stating that for all atoms/states $A \in S(c)$ and all formulas $a \in C(a)$:

$$a \in A \text{ iff } A \Vdash_{\sigma,V} a. \tag{1}$$

The proof of this Truth Lemma will proceed by a formula induction. It should be obvious how to define a valuation $V : \mathsf{Prop} \to S(c)$ ensuring (1) for atomic formulas a; the earlier mentioned properties of atoms takes care of the boolean cases of the inductive step of the proof. In order to prove the ∇-case of the induction, we have to come up with a proper definition of the coalgebra map $\sigma : S(c) \to TS(c)$. This definition will be crucially based on the one-step soundness and completeness (Fact 6).

Turning to the technicalities, we fix a consistent formula c, and write C, R and S instead of $C(c)$, $R(c)$ and $S(c)$. For technical reasons, it will be convenient to see formulas in R as separate proposition letters; formally we define

$$\underline{R} := \{\underline{b} \mid b \in R\},$$

and we assume the existence of a bijection $q : \underline{R} \to R$ given by $q(\underline{b}) := b$. In order to apply the one-step soundness and completeness, we link this set with PS by defining the valuation $j : \underline{R} \to PS$ as follows:

$$j(\underline{b}) := \{A \in S \mid q(\underline{b}) = b \in A\}.$$

It is straightforward to verify that j is surjective: For each $A \in S$ we have $j(\underline{\bigwedge(A \cap C)}) = \{A\}$ and for each $Z \subseteq S$ we have $j(\underline{\bigvee_{A \in Z} \bigwedge(A \cap C)}) = Z$. We extend j to a function $j_0 : \mathcal{L}_0\underline{R} \to \mathcal{L}_0 PS$ by inductively defining $j_0(\neg b) = \neg j_0(b)$, $j_0(\bigwedge B) = \bigwedge\{j_0(b') \mid b' \in B\}$ and $j_0(\bigvee B) = \bigvee\{j_0(b') \mid b' \in B\}$ for all $b \in \mathcal{L}_0\underline{R}$ and $B \subseteq \mathcal{L}_0\underline{R}$. We now lift j_0 to a function $j_1 : \mathcal{L}_\nabla\underline{R} \to \mathcal{L}_\nabla PS$, by putting $j_1(\nabla\beta) = \nabla T j_0(\beta)$, $j_1(\neg b) = \neg j_1(b)$, $j_1(\bigwedge B) = \bigwedge\{j_1(b') \mid b' \in B\}$ and $j_1(\bigvee B) = \bigvee\{j_1(b') \mid b' \in B\}$ for all $\nabla\beta \in \mathcal{L}_\nabla\underline{R}$, $b \in \mathcal{L}_\nabla\underline{R}$ and $B \subseteq \mathcal{L}_\nabla\underline{R}$.

In the same way we obtain q_0 and q_1 from q. Note that all these functions are surjective, however, q_0 and q_1 are not necessarily injective. For example take any $a \wedge b \in R$, then $q_0(\underline{a \wedge b}) = q_0(\underline{a} \wedge \underline{b}) = a \wedge b$.

Lemma 1 below, our main technical lemma, links $\mathbf{M}$-derivations of formulas in $\mathcal{L}_0 R$ and $\mathcal{L}_\nabla R$ to, respectively, true facts on PS, and one-step derivations of formulas in $\mathcal{L}_\nabla PS$. The $\underline{R}$-formulas serve as a bridge between R-formulas and PS-formulas.

Lemma 1. *1. For $a, b \in \mathcal{L}_0\underline{R}$ we have $\vdash_{\mathbf{M}} q_0(a) \preccurlyeq q_0(b)$ iff $\models_0^S j_0(a) \preccurlyeq j_0(b)$.*
2. For $a, b \in \mathcal{L}_\nabla\underline{R}$ we have $\vdash_{\mathbf{M}} q_1(a) \preccurlyeq q_1(b)$ iff $\vdash_1^S j_1(a) \preccurlyeq j_1(b)$.

Proof. For part 1, first observe that by some routine propositional reasoning, we may reduce the problem to the case where $a = \bigwedge A$ and $b = \bigvee B$ for some $A, B \subseteq \underline{R}$. (To see this, note that $\bigvee A' \preccurlyeq b$ corresponds to the set $\{a' \preccurlyeq b \mid a' \in A'\}$, etc.)

First suppose that $\vdash_{\mathbf{M}} q_0(\bigwedge A) \preccurlyeq q_0(\bigvee B)$, then the set $\{q(a') \mid a' \in A\} \cup \{\neg q(b') \mid b' \in B\}$ is $\mathbf{M}$-inconsistent. We claim that if $\{q(a') \mid a' \in A\} \subseteq D$ for some $D \in S$, then $q(b') \in D$ for some $b' \in B$. (If not, then since D is an R-atom, we would obtain $\neg q(b') \in D$ for all $b' \in B$, contradicting the consistency of D.) Therefore we have $\widehat{i_S} j_0(\bigwedge A) = \bigcap_{a' \in A} j(a') \subseteq \bigcup_{b' \in B} j(b') = \widehat{i_S} j_0(\bigvee B)$, thus $j_0(\bigwedge A) \preceq j_0(\bigvee B)$ is a true fact on PS.

For the other direction, suppose that $\not\vdash_{\mathbf{M}} q_0(\bigwedge A) \preccurlyeq q_0(\bigvee B)$, then the set $\{q(a') \mid a' \in A\} \cup \{\neg q(b') \mid b' \in B\}$ is $\mathbf{M}$-consistent. By the Lindenbaum Lemma there exists a $D \in S$ extending this set. Thus $D \in \widehat{i_S} j_0(\bigwedge A) = \bigcap_{a' \in A} j(a')$, but $D \notin \widehat{i_S} j_0(\bigvee B) = \bigcup_{b' \in B} j(b)$. Therefore $\bigcap_{a' \in A} j(a') \nsubseteq \bigcup_{b' \in B} j(b')$ and thus $j_0(\bigwedge A) \preceq j_0(\bigvee B)$ is not a true fact on PS.

For part 2, we only consider the direction from right to left. (The other direction, which we do not need in the remainder, is proved similarly.) By induction on the complexity of one-step derivation trees we will show that any one-step derivation tree $\mathcal{D}$, of which the root is labelled with an inequality $j_1(a) \preccurlyeq j_1(b)$, can be transformed into an $\mathbf{M}$-derivation tree for the inequality $q_1(a) \preccurlyeq q_1(b)$.

Base case: $\nabla 1$ By definition of our one-step derivation tree, we may just as well assume that in the base case of our inductive proof we are dealing with an instance of the rule $\nabla 1$, of which the premisses are all true facts on PS. More precisely, in this case the conclusion $j_1(a) \preccurlyeq j_1(b)$ stems from some $a = \nabla\alpha$ and $b = \nabla\beta$ (where $\alpha, \beta \in T\mathcal{L}_0\underline{R}$) in the sense that $j_1(a) = \nabla T j_0(\alpha)$ and $j_1(b) = \nabla T j_0(\beta)$, and the last applied rule was

$$(\nabla 1)\ \frac{\{a' \preceq b' \mid (a', b') \in Z\}}{\nabla T j_0(\alpha) \preceq \nabla T j_0(\beta)}\ (T j_0(\alpha), T j_0(\beta)) \in \overline{T} Z.$$

for some relation $Z \subseteq \mathcal{L}_0 PS \times \mathcal{L}_0 PS$. In fact, given the properties of relation lifting, we may assume without loss of generality that

$$Z = \{(a', b') \in \mathit{Base}(T j_0(\alpha)) \times \mathit{Base}(T j_0(\beta)) \mid a' \preccurlyeq b' \text{ is a true fact }\}.$$

By the naturality of Base we have for all $\delta \in T\mathcal{L}_0\underline{R}$

$$\mathit{Base}(T j_0(\delta)) = \{j_0(d) \mid d \in \mathit{Base}(\delta)\}. \tag{2}$$

Now define

$$\hat{Z} := \{(a', b') \in \mathit{Base}(\alpha) \times \mathit{Base}(\beta) \mid q_0(a') \preccurlyeq q_0(b') \text{ is derivable}\},$$

then by equation (2) and by part 1, we have for all $a' \in Base(\alpha)$, $b' \in Base(\beta)$ that $(j_0(a'), j_0(b')) \in Z$ iff $(a', b') \in \hat{Z}$. From this it follows by the properties of relation lifting that for all $\alpha' \in TBase(\alpha)$, $\beta' \in TBase(\beta)$:

$$(Tj_0(\alpha'), Tj_0(\beta')) \in \overline{T}Z \text{ iff } (\alpha', \beta') \in \overline{T}\hat{Z}.$$

In particular, we obtain that $(\alpha, \beta) \in \overline{T}\hat{Z}$. Again using the properties of relation lifting we may conclude from this that

$$(Tq_0(\alpha), Tq_0(\beta)) \in \overline{T}\Big(\{(q_0(a'), q_0(b')) \mid (a', b') \in \hat{Z}\}\Big).$$

But then, since $q_1(a) = \nabla Tq_0(\alpha)$ and $q_1(b) = \nabla Tq_0(\beta)$, we can derive the inequality $q_1(a) \preccurlyeq q_1(b)$, as follows:

$$(\nabla 1)\ \frac{\{q_0(a') \preceq q_0(b') \mid (a', b') \in \hat{Z}\}}{\nabla Tq_0(\alpha) \preceq \nabla Tq_0(\beta)},$$

where all premisses are derivable by definition of $\hat{Z}$.

In the inductive step we make a case distinction; we only consider the cases where the last applied rule was $(\nabla 3)$.

Inductive case: $(\nabla 3)$ Suppose that $j_1(a)$ is of the form $\nabla(T\bigvee)(\Psi)$ for some $\Psi \in TP\mathcal{L}_0PS$, and that the last applied rule is:

$$(\nabla 3) \qquad \frac{\{\nabla\beta \preceq j_1(b) \mid \beta \,\overline{T}{\in}\, \Psi\}}{\nabla(T\bigvee)(\Psi) \preceq j_1(b)} \tag{3}$$

We claim that a is of the form $\nabla(T\bigvee)(\Phi)$ for some $\Phi \in T_\omega P_\omega \mathcal{L}_0\underline{R}$ such that $\Psi = TPj_0(\Phi)$. To see this, first observe that a must obviously be of the form $\nabla\alpha$ for some $\alpha \in T_\omega\mathcal{L}_0\underline{R}$; we will show that α is of the form $T\bigvee(\Phi)$ with Φ as above. For this purpose, note that by definition of j_0, if $a' \in \mathcal{L}_0\underline{R}$ is such that $j_0(a') = \bigvee A$ for some $A \in P_\omega\mathcal{L}_0PS$, then a' must be of the form $\bigvee B$ for some $B \in P_\omega\mathcal{L}_0\underline{R}$ with $A = \{j_0(b') \mid b' \in B\}$. This condition can be expressed as $Gr(j_0) \,;\, (Gr\bigvee)^{\smile} \subseteq (Gr\bigvee)^{\smile} \,;\, Gr(Pj_0)$. Then by the properties of relation lifting we find that $Gr(Tj_0) \,;\, (GrT\bigvee)^{\smile} \subseteq (GrT\bigvee)^{\smile} \,;\, Gr(TPj_0)$. From this the existence of the required object Ψ is immediate.

In order to find an **M**-derivation for the inequality $q_1(a) \preccurlyeq q_1(b)$, we calculate $q_1(a) = q_1(\nabla(T\bigvee)(\Phi)) = \nabla(T\bigvee)(TPq_0(\Phi))$. Aiming to derive $\nabla(T\bigvee)(TPq_0(\Phi)) \preccurlyeq q_1(b)$ via the rule $(\nabla 3)$, let β be an arbitrary lifted member of $TPq_0(\Phi)$. From $\beta \in \lambda^T(TPq_0(\Phi)) = PTq_0(\lambda^T(\Phi))$, we obtain that β is of the form $Tq_0(\alpha')$, for some lifted member α' of Φ. But from $\alpha' \in \lambda^T(\Phi)$ it follows that $Tj_0(\alpha') \in PTj_0(\lambda^T(\Phi)) = \lambda^T(TPj_0(\Phi)) = \lambda^T(\Psi)$. Now observe that $j_1(\nabla\alpha') = \nabla Tj_0(\alpha')$, and so the inequality $j_1(\nabla\alpha') \preccurlyeq j_1(b)$ is one of the premisses of (3). Thus by the inductive hypothesis, we have a **M**-derivation for the inequality $q_1(\nabla\alpha') \preccurlyeq q_1(b)$, which is nothing but $\nabla Tq_0(\alpha') \preccurlyeq q_1(b)$, that is, $\nabla\beta \preccurlyeq q_1(b)$.

It follows that we can derive $q_1(a) = \nabla(T\bigvee)(TPq_0(\Phi)) \preceq q_1(b)$ by

$$(\nabla 3) \qquad \frac{\{\nabla\beta \preceq q_1(b) \mid \beta \,\overline{T}{\in}\, TPq_0(\Phi)\}}{\nabla(T\bigvee)(TPq_0(\Phi)) \preceq q_1(b)}$$

On the basis of Lemma 1(2) we can prove the existence of a coalgebra map $\sigma : S \to TS$ with the right properties. Note that $jq^{-1}(b) = \{B \in S \mid b \in B\}$

Lemma 2 (Existence Lemma). *There is a map $\sigma : S \to TS$ such that for all atoms $A \in S$ and all formulas of the form $\nabla\alpha \in R$:*

$$\nabla\alpha \in A \textit{ iff } \sigma(A)\, \overline{T}{\in}\, T(jq^{-1})(\alpha). \tag{4}$$

Proof. Suppose towards contradiction that for some $A \in S$ there is no $\sigma(A)$ that satisfies equation (4). Define $b := \bigwedge_{\nabla\alpha\in A} \nabla Tq^{-1}(\alpha) \vee \bigwedge_{\neg\nabla\alpha\in A} \neg\nabla Tq^{-1}(\alpha)$. By assumption we have

$$\widetilde{i_S} j_1(b) = \bigcap_{\nabla\alpha\in A} \lambda^T(T(jq^{-1})(\alpha)) \cap \bigcap_{\neg\nabla\alpha\in A} TS \setminus \lambda^T(T(jq^{-1})(\alpha)) = \emptyset.$$

In other words, $\models_1^S j_1(b) \preceq \bot$, and so we have $\vdash_1^S j_1(b) \preceq \bot$ by one-step completeness. Then Lemma 1 provides an **M**-derivation of $q_0(b) \preceq \bot$, contradicting the consistency of A.

Lemma 3 (Truth Lemma). *Let (S, σ, v) be a model where S is the set of $R(c)$-atoms, σ is any map satisfying condition (4) of Lemma 2, and $V : \mathsf{Prop} \to PS$ is given by $V(p) := \{A \in S \mid p \in A\}$. Then (1) holds for all $a \in C$ and all $A \in S$.*

Proof. Via a straightforward induction on the complexity of a. We only discuss the case $a = \nabla\alpha$. By definition of the semantics we have $A \Vdash \nabla\beta$ iff $\sigma(A)\, \overline{T}{\Vdash}\, \beta$, and by Lemma 2 we have $\nabla\beta \in A$ iff $\sigma(A)\, \overline{T}{\in}\, T(jq^{-1})(\beta)$. So in order to finish the proof, it suffices to show that $\overline{T}{\Vdash} = \overline{T}{\in}\, ;\, Gr(T(jq^{-1}))^{\smile}$, or, equivalently, that (*) $\overline{T}{\Vdash} = \overline{T}{\in}\, ;\, Gr(Tj)^{\smile}\, ;\, Gr(Tq)$. But inductively, if we restrict to formulas $b \in R$ of smaller complexity than $\nabla\alpha$, we have that $jq^{-1}(b) = \{B \in S \mid b \in B\} = \{B \in S \mid B \Vdash b\}$. This means that $\Vdash\, = \,\in\, ;\, Gr(j)^{\smile}\, ;\, Gr(q)$ so that (*) directly follows by the properties of relation lifting.

The proof of Theorem 7 is now straightforward. If c is a consistent formula, it belongs to some R-atom A by the Lindenbaum Lemma. Then by the Existence Lemma and the Truth Lemma, we can endow the (finite!) set S of R-atoms with a coalgebra structure σ and a valuation V such that $A \Vdash_{\sigma,V} a$.

5 Strong Completeness

In this section we will prove strong completeness of the axiom system **M**. That is, we will prove that, given some restrictions on the functor T, every **M**-consistent set of formulas is satisfiable. It might be possible to see our strong completeness result as a special case of Theorem 8.1 in Kurz & Rosický [12] (see also [9]). Nevertheless, we believe our short, direct proof, which follows the ideas of Pattinson & Schröder [18], to be of value.

Our purpose is to endow the set S of maximal **M**-consistent sets of formulas with a coalgebra structure $\sigma : S \to TS$ and a valuation V such that for all $A \in S$ we can prove the following Truth Lemma, stating that for all formulas a:

$$a \in A \text{ iff } A \Vdash_{\sigma,V} a. \tag{5}$$

The idea underlying the construction of σ is that, for each $A \in S$, we may *approximate* $\sigma(A)$ by considering finite versions of S. For this purpose, enumerate $\mathsf{Prop} = \{p_i \mid i \in \omega\}$, and define $\mathsf{Prop}_n := \{p_i \mid 0 \leq i < n\}$. Let $\mathcal{L}_0 = \mathcal{L}_0(\varnothing)$ and let $\mathcal{L}_n$ denote the set $\mathcal{L}_0(\mathsf{Prop}_{n+1} \cup \{\nabla\alpha \mid \alpha \in T\mathcal{L}_n\})$. Then clearly $\mathcal{L}(\mathsf{Prop}) = \bigcup_{n\in\omega} \mathcal{L}_n$. Let S_n be the set of $\mathcal{L}_n$-atoms. It is not hard to show that if T restricts to finite sets, then each $\mathcal{L}_n$ is finite modulo equivalence, whence each S_n is finite. We let $h_n : S_{n+1} \to S_n$ and $\pi_n : S \to S_n$ be defined by $h_n(A) := A \cap \mathcal{L}_n$ and $\pi_n(A) := A \cap \mathcal{L}_n$. By the Lindenbaum Lemma all h_n and π_n are surjective.

On the basis of the results in the previous section we can prove the following lemma.

Lemma 4. *For each maximal consistent set $A \in S$ there is a family $(\tau_n)_{n\in\omega}$, with $\tau_n \in TS_n$, and such that for all n:*

$$(Th_n)\tau_{n+1} = \tau_n, \tag{6}$$

and for all $\alpha \in T\mathcal{L}_n$ it holds that

$$\nabla\alpha \in \pi_{n+1}(A) \text{ iff } \alpha \overline{T}{\in} \tau_n. \tag{7}$$

Note that (7) requires a relation between elements of S_{n+1} and objects, not in TS_{n+1}, but in TS_n.

Proof. (Sketch) For each n let $A_n \in S_n$ denote the atom $\pi_n(A)$. Using the methods of the previous section it is straightforward to show that for the atom $A_n \in S_n$ there is an object $\rho \in TS_n$ which *works for* A_n in the sense that for all $\alpha \in T\mathcal{L}_{n-1}$ it holds that $\nabla\alpha \in A_n$ iff $\alpha \overline{T}{\in} \rho$. It is not hard to show that if $\rho \in TS_{n+1}$ works for A_{n+1} then $(Th_n)\rho$ works for A_n. Consider the tree with nodes $N := \bigcup_{n\in\omega}\{\rho \in TS_n \mid \rho \text{ works for } A_n\}$, and edge relation E given by $\rho E \rho'$ iff $\rho = (Th_n)\rho'$ for some n. By König's Lemma this tree has an infinite path $(\tau_n)_{n\in\omega}$, and it is a routine exercise to verify that this family satisfies the required properties.

The point of considering the sequence $(\tau_n)_{n\in\omega}$ is that, under some condition on T, they *approximate* some object $\tau \in TS$, that we can take for our $\sigma(A)$. To formulate this condition, we define a *surjective ω-cochain of finite sets* to be a sequence $(X_n)_{n\in\omega}$ of finite sets, with surjections $h_n : X_{n+1} \to X_n$ that are called *projections*. In Set, such a diagram has a limit X with limit projections $T\pi_n : X \to X_n$. Clearly also in Set each endofunctor T transforms a surjective ω-cochain of sets into a surjective ω-cochain of sets. Now, to say that T *weakly* preserve limits of these diagrams means that whenever we have a diagram as above, with limit $(X, (\pi_n)_{n\in\omega})$, the set TX with the maps $T\pi_n : TX \to TX_n$ is a *weak* limit of the diagram $((TX_n)_{n\in\omega}, (p_n)_{n\in\omega})$. An equivalent requirement is that for each so-called coherent family $(\tau_n \in TX_n)_{n\in\omega}$ (that is, satisfying (6) for all n), there is a (not necessarily unique) element $\tau \in TX$ such that $T\pi_n(\tau) = \tau_n$ for all n.

On the basis of Lemma 4 we can now prove the following.

Lemma 5. *Let T restrict to finite sets and weakly preserve limits of surjective ω-cochains of finite sets. Then there is a coalgebra map $\sigma : S \to TS$ and a valuation $V : \mathsf{Prop} \to PS$ such that for all $a \in \mathcal{L}(\mathsf{Prop})$, and all $A \in S$, the Truth Lemma* (5) *holds.*

Proof. We define $V : \mathsf{Prop} \to PS$ by putting $V(p) := \{A \in S \mid p \in A\}$. For the definition of σ, take an arbitrary $A \in S$, and consider the coherent family $(\tau_n)_{n\in\omega}$ of Lemma 4. By the assumptions on T, we may fix an element $\sigma(A) \in TS$ such that $T\pi_n(\sigma(A)) = \tau_n$ for each n.

By induction on n we prove that for all $a \in \mathcal{L}_n$ we have

$$A \Vdash_{\sigma,V} a \text{ iff } a \in \pi_n A. \tag{8}$$

Confining our attention to the inductive case where $n = k + 1$, we prove (8) by formula induction, and we only cover the case where $a = \nabla\alpha$. Note that here we have $\alpha \in T\mathcal{L}_n$, and this enables us to apply the outer induction hypothesis.

Now we prove (8) by the following chain of equivalences (writing $\Vdash$ rather than $\Vdash_{\sigma,V}$):

$$\begin{aligned}
A \Vdash \nabla\alpha \text{ iff } & (\sigma(A),\alpha) \in \overline{T}{\Vdash} && \text{(definition of } \Vdash) \\
\text{iff } & (\sigma(A),\alpha) \in \overline{T}(\Vdash\restriction_{S\times\mathcal{L}_k}) && \text{(Fact 2)} \\
\text{iff } & (\sigma(A),\alpha) \in \overline{T}(Gr(\pi_k)\,;\,(\in)^{\smile}) && \text{(inductive hypothesis)} \\
\text{iff } & \alpha \,\overline{T}{\in}\, T\pi_k(\sigma(A)) && \text{(Fact 2)} \\
\text{iff } & \alpha \,\overline{T}{\in}\, \tau_k && \text{(definition of } \sigma) \\
\text{iff } & \nabla\alpha \in \pi_{k+1}(A) && \text{(equation (7))}
\end{aligned}$$

Finally, the Truth Lemma is immediate from (8) by the fact that $\mathcal{L}(\mathsf{Prop}) = \bigcup_{n\in\omega} \mathcal{L}_n$ and the definitions.

On the basis of this the following is immediate.

Theorem 8. *Let T restrict to finite sets and weakly preserve limits of surjective ω-cochains of finite sets. Then the logic $\mathbf{M}^T$ is strongly complete with respect to its coalgebraic semantics.*

References

1. Barr, M.: Relational algebras. In: Reports of the Midwest Category Seminar IV. Lecture Notes in Mathematics, vol. 137, pp. 39–55 (1970)
2. Bergfeld, J.: Moss's coalgebraic logic: Examples and completeness results. Master's thesis, ILLC, University of Amsterdam (2009)
3. Bílková, M., Palmigiano, A., Venema, Y.: Proof systems for the coalgebraic cover modality. In: Areces, C., Goldblatt, R. (eds.) Advances in Modal Logic 7, pp. 1–21. College Publications (2008)
4. Cîrstea, C.: A compositional approach to defining logics for coalgebras. Theoretical Computer Science 327, 45–69 (2004)

5. Cîrstea, C., Kurz, A., Pattinson, D., Schröder, L., Venema, Y.: Modal logics are coalgebraic. The Computer Journal 54, 524–538 (2011)
6. Fiadeiro, J.L., Harman, N.A., Roggenbach, M., Rutten, J. (eds.): CALCO 2005. LNCS, vol. 3629. Springer, Heidelberg (2005)
7. Gumm, H.P.: From T-coalgebras to filter structures and transition systems. In: Fiadeiro [6], pp. 194–212
8. Janin, D., Walukiewicz, I.: Automata for the modal μ-calculus and related results. In: Hájek, P., Wiedermann, J. (eds.) MFCS 1995. LNCS, vol. 969, pp. 552–562. Springer, Heidelberg (1995)
9. Kupke, C., Kurz, A., Pattinson, D.: Ultrafilter extensions for coalgebras. In: Fiadeiro [6], pp. 263–277
10. Kupke, C., Kurz, A., Venema, Y.: Completeness for the coalgebraic cover modality (2010) (submitted); An earlier version appeared in Advances in Modal Logic 7. College Publications (2008)
11. Kurz, A.: Specifying coalgebras with modal logic. Theoretical Computer Science 260, 119–138 (2001)
12. Kurz, A., Rosický, J.: Strongly complete logics for coalgebras (unpublished) (2006)
13. Moss, L.: Coalgebraic logic. Annals of Pure and Applied Logic 96, 277–317 (1999); Erratum published APAL 1999, 241–259 (1999)
14. Palmigiano, A., Venema, Y.: Nabla algebras and chu spaces. In: Mossakowski, T., Montanari, U., Haveraaen, M. (eds.) CALCO 2007. LNCS, vol. 4624, pp. 394–408. Springer, Heidelberg (2007)
15. Pattinson, D.: Coalgebraic modal logic: Soundness, completeness and decidability of local consequence. Theoretical Computer Science 309, 177–193 (2003)
16. Rutten, J.: Universal coalgebra: A theory of systems. Theoretical Computer Science 249, 3–80 (2000)
17. Schröder, L.: A finite model construction for coalgebraic modal logic. The Journal of Logic and Algebraic Programming 73, 97–110 (2007)
18. Schröder, L., Pattinson, D.: Strong completeness of coalgebraic modal logics. In: Albers, S., Marion, J.-Y. (eds.) 26th International Symposium on Theoretical Aspects of Computer Science (STACS 2009). Leibniz International Proceedings in Informatics (LIPIcs), Dagstuhl, Germany, vol. 3, pp. 673–684 (2009)
19. Trnková, V.: General theory of relational automata. Fundamenta Informaticae 3(2), 189–234 (1980)
20. Venema, Y.: Automata and fixed point logic: a coalgebraic perspective. Information and Computation 204, 637–678 (2006)
21. Venema, Y., Vickers, S., Vosmaer, J.: Powerlocales via relation lifting (2010) (submitted)

Relation Liftings on Preorders and Posets

Marta Bílková[1], Alexander Kurz[2,⋆], Daniela Petrişan[2], and Jiří Velebil[3,⋆⋆]

[1] Faculty of Philosophy, Charles University, Prague, Czech Republic
marta.bilkova@ff.cuni.cz
[2] Department of Computer Science, University of Leicester, United Kingdom
{kurz,petrisan}@mcs.le.ac.uk
[3] Faculty of Electrical Engineering, Czech Technical University in Prague, Czech Republic
velebil@math.feld.cvut.cz

Abstract. The category Rel(Set) of sets and relations can be described as a category of spans and as the Kleisli category for the powerset monad. A set-functor can be lifted to a functor on Rel(Set) iff it preserves weak pullbacks. We show that these results extend to the enriched setting, if we replace sets by posets or preorders. Preservation of weak pullbacks becomes preservation of exact lax squares. As an application we present Moss's coalgebraic over posets.

1 Introduction

Relation lifting [Ba, CKW, HeJ] plays a crucial role in coalgebraic logic, see eg [Mo, Bal, V].

On the one hand, it is used to explain bisimulation: If $T : \mathsf{Set} \longrightarrow \mathsf{Set}$ is a functor, then the largest bisimulation on a coalgebra $\xi : X \longrightarrow TX$ is the largest fixed point of the operator $(\xi \times \xi)^{-1} \circ \overline{T}$ on relations on X, where $\overline{T}$ is the lifting of T to $\mathsf{Rel}(\mathsf{Set}) \longrightarrow \mathsf{Rel}(\mathsf{Set})$. (The precise meaning of 'lifting' will be given in the Extension Theorem 5.3.)

On the other hand, Moss's coalgebraic logic [Mo] is given by adding to propositional logic a modal operator ∇, the semantics of which is given by applying $\overline{T}$ to the forcing relation $\Vdash\, \subseteq X \times \mathcal{L}$, where $\mathcal{L}$ is the set of formulas: If $\alpha \in T(\mathcal{L})$, then $x \Vdash \nabla\alpha \;\Leftrightarrow\; \xi(x)\ \overline{T}(\Vdash)\ \alpha$.

In much the same way as Set-coalgebras capture bisimulation, Pre-coalgebras and Pos-coalgebras capture simulation [R, Wo, HuJ, Kl, L, BK]. This suggests that, in analogy with the Set-based case, a coalgebraic understanding of logics for simulations should derive from the study of Pos-functors together with on the one hand their predicate liftings and on the other hand their ∇-operator. The study of predicate liftings of Pos-functors was begun in [KaKuV], whereas here we lay the foundations for the ∇-operator of a Pos-functor. In order to do this, we start with the notion of monotone relation for the following reason. Let

⋆ Alexander Kurz acknowledges the support of EPSRC, EP/G041296/1.
⋆⋆ Marta Bílková and Jiří Velebil acknowledge the support of the grant No. P202/11/1632 of the Czech Science Foundation.

A. Corradini, B. Klin, and C. Cîrstea (Eds.): CALCO 2011, LNCS 6859, pp. 115–129, 2011.

$(X, \leq)$ and $(X', \leq')$ be the carriers of two coalgebras, with the preorders $\leq, \leq'$ encoding the simulation relations on X and X', respectively. Then a simulation between the two systems will be a relation $R \subseteq X \times X'$ such that $\geq ; R ; \geq' \subseteq R$, that is, R is a monotone relation. Similarly, $\Vdash$ will be a monotone relation. To summarise, the relations we are interested in are monotone, which enables us to use techniques of enriched category theory (of which no prior knowledge is assumed of the reader).

For the reasons outlined above, the purpose of the paper is to develop the basic theory of relation liftings over preorders and posets. That is, we replace the category Set of sets and functions by the category Pre of preorders or Pos of posets, both with monotone (i.e. order-preserving) functions. Section 2 introduces notation and shows that (monotone) relations can be presented by spans and by arrows in an appropriate Kleisli-category. Section 3 recalls the notion of exact squares. Section 4 characterises the inclusion of functions into relations $(-)_\diamond : \mathsf{Pre} \longrightarrow \mathsf{Rel}(\mathsf{Pre})$ by a universal property and shows that the relation lifting $\overline{T}$ exists iff T satisfies the Beck-Chevalley-Condition (BCC), which says that T preserves exact squares. The BCC replaces the familiar condition known from Rel(Set), namely that T preserves weak pullbacks. Section 5 lists examples of functors (not) satisfying the BCC and Section 6 gives the application to Moss's coalgebraic logic over posets.

Related Work. The universal property of the embedding of a (regular) category to the category of relations is stated in Theorem 2.3 of [He]. Theorem 4.1 below generalizes this in passing from a category to a simple 2-category of (pre)orders.

Liftings of functors to categories of relations within the realm of regular categories have also been studied in [CKW].

2 Monotone Relations

In this section we summarize briefly the notion of monotone relations on preorders and we show that their resulting 2-category can be perceived in two ways:

1. Monotone relations are certain *spans*, called *two-sided discrete fibrations*.
2. Monotone relations form a *Kleisli category* for a certain *KZ doctrine* on the category of preorders.

Definition 2.1. *Given preorders $\mathscr{A}$ and $\mathscr{B}$, a* monotone relation *R from $\mathscr{A}$ to $\mathscr{B}$, denoted by*

$$\mathscr{A} \xrightarrow{R} \mathscr{B}$$

is a monotone map $R : \mathscr{B}^{op} \times \mathscr{A} \longrightarrow 2$ where by 2 we denote the two-element poset on $\{0,1\}$ with $0 \leq 1$.

Remark 2.2. Unravelling the definition: for a binary relation R, $R(b,a) = 1$ means that a and b are related by R. Monotonicity of R then means that if $R(b,a) = 1$ and $b_1 \leq b$ in $\mathscr{B}$ and $a \leq a_1$ in $\mathscr{A}$, then $R(b_1, a_1) = 1$.

Relations compose in the obvious way. Two relations as on the left below

$$\mathscr{A} \xrightarrow{R} \!\!\!\!\!\!\!\!\!+\;\; \mathscr{B} \qquad \mathscr{B} \xrightarrow{S} \!\!\!\!\!\!\!\!\!+\;\; \mathscr{C} \qquad\qquad \mathscr{A} \xrightarrow{S\cdot R} \!\!\!\!\!\!\!\!\!\!\!+\;\;\; \mathscr{C}$$

compose to the relation on the right above by the formula

$$S \cdot R(c,a) = \bigvee_b R(b,a) \wedge S(c,b) \tag{2.1}$$

hence the validity of $S \cdot R(c,a)$ is witnessed by at least one b such that both $R(b,a)$ and $S(c,b)$ hold.

Remark 2.3. The supremum in formula (2.1) is, in fact, exactly a coend in the sense of enriched category theory, see [Ke].

The above composition of relations is associative and it has monotone relations $\mathscr{A} \xrightarrow{\mathscr{A}} \!\!\!\!\!\!\!\!\!+\;\; \mathscr{A}$ as units, where $\mathscr{A}(a,a')$ holds, if $a \leq a'$. Moreover, the relations can be ordered pointwise: $R \longrightarrow S$ means that $R(b,a)$ entails $S(b,a)$, for every a and b. Hence we have a 2-category of monotone relations $\mathsf{Rel}(\mathsf{Pre})$.

Remark 2.4. Observe that one can form analogously the 2-category $\mathsf{Rel}(\mathsf{Pos})$ of monotone relations on *posets*. In all what follows one can work either with preorders or posets. We will focus on preorders in the rest of the paper, the modifications for posets always being straightforward. Observe that both $\mathsf{Rel}(\mathsf{Pre})$ and $\mathsf{Rel}(\mathsf{Pos})$ have the crucial property: The only isomorphism 2-cells are identities.

2.A The Functor $(-)_\diamond : \mathsf{Pre} \longrightarrow \mathsf{Rel}(\mathsf{Pre})$

We describe now the functor $(-)_\diamond : \mathsf{Pre} \longrightarrow \mathsf{Rel}(\mathsf{Pre})$ and show its main properties. The case of posets is completely analogous. For a monotone map $f : \mathscr{A} \longrightarrow \mathscr{B}$ define two relations

$$\mathscr{A} \xrightarrow{f_\diamond} \!\!\!\!\!\!\!\!\!+\;\; \mathscr{B} \qquad\qquad \mathscr{B} \xrightarrow{f^\diamond} \!\!\!\!\!\!\!\!\!+\;\; \mathscr{A}$$

by the formulas $f_\diamond(b,a) = \mathscr{B}(b,fa)$ and $f^\diamond(a,b) = \mathscr{B}(fa,b)$.

Lemma 2.5. *For every* $f : \mathscr{A} \longrightarrow \mathscr{B}$ *in* Pre *there is an adjunction in* $\mathsf{Rel}(\mathsf{Pre})$

$$f_\diamond \dashv f^\diamond : \mathscr{B} \longrightarrow \!\!\!\!\!\!\!\!+\;\; \mathscr{A}\ .$$

Remark 2.6. Left adjoint morphisms in $\mathsf{Rel}(\mathsf{Pre})$ can be characterized as *exactly* those of the form $f_\diamond$ for some monotone map f having a *poset* as its codomain. Therefore, if $L \dashv R : \mathscr{B} \longrightarrow \!\!\!\!\!\!\!\!+\;\; \mathscr{A}$ in $\mathsf{Rel}(\mathsf{Pre})$ and $\mathscr{B}$ is a poset, then there exists a monotone map $f : \mathscr{A} \longrightarrow \mathscr{B}$ such that $f_\diamond = L$ and $f^\diamond = R$.

Observe that if $f \longrightarrow g$, then $f_\diamond \longrightarrow g_\diamond$ holds. For if $\mathscr{B}(b,fa) = 1$ then $\mathscr{B}(b,ga) = 1$ holds by transitivity, since $fa \leq ga$ holds. Moreover, taking the lower diamond clearly maps an identity monotone map $id_{\mathscr{A}} : \mathscr{A} \longrightarrow \mathscr{A}$ to the identity monotone relation $\mathscr{A} \xrightarrow[\quad]{\mathscr{A}=(id_{\mathscr{A}})_\diamond} \mathscr{A}$ Further, taking the lower diamond preserves composition:

$$(g \cdot f)_\diamond(c,a) = \mathscr{C}(c, gfa) = \bigvee_b \mathscr{C}(c, gb) \wedge \mathscr{B}(b, fa) = g_\diamond \cdot f_\diamond(c,a)$$

Hence we have a functor $(-)_\diamond : \mathsf{Pre} \longrightarrow \mathsf{Rel}(\mathsf{Pre})$ enriched in preorders. Moreover, $(-)_\diamond$ is *locally fully faithful*, i.e., $f_\diamond \longrightarrow g_\diamond$ holds iff $f \longrightarrow g$ holds.

2.B Rel(Pre) as a Kleisli Category

The 2-functor $(-)_\diamond : \mathsf{Pre} \longrightarrow \mathsf{Rel}(\mathsf{Pre})$ is a *proarrow equipment with power objects* in the sense of Section 2.5 [MRW]. This means that $(-)_\diamond$ has a right adjoint $(-)^\dagger$ such that the resulting 2-monad on Pre is a KZ doctrine and Rel(Pre) is (up to equivalence) the corresponding Kleisli 2-category. All of the following results are proved in the paper [MRW], we summarize it here for further reference.

The 2-functor $(-)^\dagger$ works as follows:

1. On objects, $\mathscr{A}^\dagger = [\mathscr{A}^{op}, 2]$, the lowersets on $\mathscr{A}$, ordered by inclusion.
2. For a relation R from $\mathscr{A}$ to $\mathscr{B}$, the functor $R^\dagger : [\mathscr{A}^{op}, 2] \longrightarrow [\mathscr{B}^{op}, 2]$ is defined as the left Kan extension of $a \mapsto R(-,a)$ along the Yoneda embedding $\mathsf{y}_{\mathscr{A}} : \mathscr{A} \longrightarrow [\mathscr{A}^{op}, 2]$. This can be expressed by the formula:

$$R^\dagger(W) = b \mapsto \bigvee_a Wa \wedge R(b,a)$$

i.e., b is in the lowerset $R^\dagger(W)$ iff there exists a in W such that $R(b,a)$ holds.

It is easy to prove that $(-)^\dagger$ is a 2-functor and that $(-)^\dagger \dashv (-)_\diamond$ is a 2-adjunction of a KZ type. The latter means that if we denote by

$$(\mathbb{L}, \mathsf{y}, \mathsf{m}) \tag{2.2}$$

the resulting 2-monad on Pre, then we obtain the string of adjunctions $\mathbb{L}(\mathsf{y}_{\mathscr{A}}) \dashv \mathsf{m}_{\mathscr{A}} \dashv \mathsf{y}_{\mathbb{L}\mathscr{A}}$, see [M$_1$], [M$_2$], for more details.

The unit of the above KZ doctrine is the Yoneda embedding $\mathsf{y}_{\mathscr{A}} : \mathscr{A} \longrightarrow [\mathscr{A}^{op}, 2]$ and the multiplication $\mathsf{m}_A : [[\mathscr{A}^{op}, 2]^{op}, 2] \longrightarrow [\mathscr{A}^{op}, 2]$ is the left Kan extension of identity on $[\mathscr{A}^{op}, 2]$ along $\mathsf{y}_{[\mathscr{A}^{op}, 2]}$. In more detail:

$$\mathsf{m}_{\mathscr{A}}(\mathscr{W}) = a \mapsto \bigvee_W \mathscr{W}(W) \wedge W(a)$$

where $\mathscr{W}$ is in $[[\mathscr{A}^{op}, 2]^{op}, 2]$ and W is in $[\mathscr{A}^{op}, 2]$. Hence a is in the lowerset $\mathsf{m}_{\mathscr{A}}(\mathscr{W})$ iff there exists a lowerset W in $\mathscr{W}$ such that a is in W. The following result is proved in Section 2.5 of [MRW]:

Proposition 2.7. *The 2-functor* $(-)_\diamond : \mathsf{Pre} \longrightarrow \mathsf{Rel}(\mathsf{Pre})$ *exhibits* Rel(Pre) *as a Kleisli category for the KZ doctrine* $(\mathbb{L}, \mathsf{y}, \mathsf{m})$.

2.C Relations as Spans

Monotone relations are going to be exactly certain spans, called *two-sided discrete fibrations*. For more information see [S$_4$].

Definition 2.8. *A* span $(d_0, \mathscr{E}, d_1) : \mathscr{B} \longrightarrow \mathscr{A}$ *from* $\mathscr{B}$ *to* $\mathscr{A}$ *is a diagram*

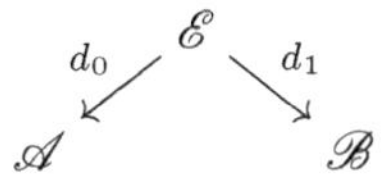

of monotone maps. The preorder $\mathscr{E}$ *is called the* vertex *of the span* $(d_0, \mathscr{E}, d_1)$.

Remark 2.9. Given a span $(d_0, \mathscr{E}, d_1) : \mathscr{B} \longrightarrow \mathscr{A}$, the following intuitive notation might prove useful: a typical element of $\mathscr{E}$ will be denoted by a wiggly arrow

$$d_0(e) \overset{e}{\rightsquigarrow} d_1(e)$$

and $d_0(e)$ will be the *domain* of e and $d_1(e)$ the *codomain* of e.

Definition 2.10. *A span* $(d_0, \mathscr{E}, d_1) : \mathscr{B} \longrightarrow \mathscr{A}$ *in* Pre *is a* two-sided discrete fibration *(we will say just* fibration *in what follows), if the following three conditions are satisfied. For every situation below on the left, there is a unique fill in on the right, denoted by* $(d_0)_*(e')$*, respectively* $(d_1)_*(e)$*:*

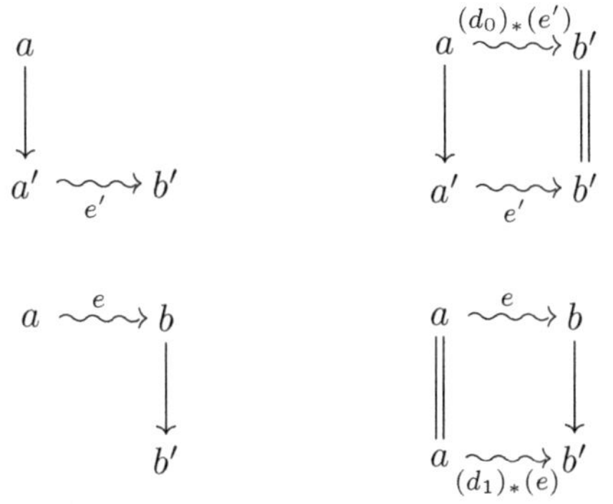

Every situation on the left can be written as depicted on the right:

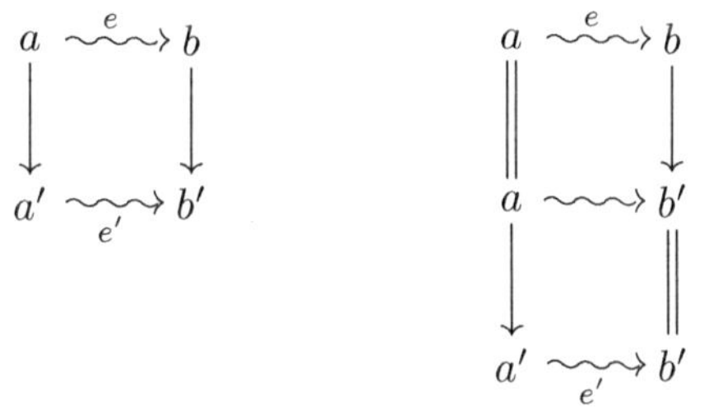

Definition 2.11. *A* comma object *of monotone maps* $f : \mathscr{A} \longrightarrow \mathscr{C}$, $g : \mathscr{B} \longrightarrow \mathscr{C}$ *is a diagram*

$$\begin{array}{ccc} f/g & \xrightarrow{p_1} & \mathscr{B} \\ {\scriptstyle p_0}\downarrow & \nearrow & \downarrow{\scriptstyle g} \\ \mathscr{A} & \xrightarrow[f]{} & \mathscr{C} \end{array}$$

where elements of the preorder f/g are pairs (a, b) with $f(a) \leq g(b)$ in $\mathscr{C}$, the preorder on f/g is defined pointwise and p_0 and p_1 are the projections. The whole "lax commutative square" as above will be called a comma square.

Example 2.12. Every span $(p_0, f/g, p_1) : \mathscr{A} \longrightarrow \mathscr{B}$ arising from a comma object of $f : \mathscr{A} \longrightarrow \mathscr{C}$, $g : \mathscr{B} \longrightarrow \mathscr{C}$ is a fibration.

A monotone relation $\mathscr{B} \overset{R}{\longrightarrow\!\!\!\!\!\!\!\!+\;\;} \mathscr{A}$ induces a fibration $(d_0, \mathscr{E}, d_1) : \mathscr{B} \longrightarrow \mathscr{A}$ with $\mathscr{E} = \{(a, b) \mid R(a, b) = 1\}$ ordered by $(a, b) \leq (a', b')$, if $a \leq a'$ and $b \leq b'$; and $(d_0, \mathscr{E}, d_1)$ induces the relation $R(a, b) = 1 \Leftrightarrow \exists e \in \mathscr{E} \,.\, d_0(e) = a, d_1(e) = b$.

Proposition 2.13. *Fibrations in* Pre *correspond exactly to monotone relations. Moreover, if $(d_0, \mathscr{E}, d_1) : \mathscr{B} \longrightarrow \mathscr{A}$ is the fibration corresponding to a relation $R : \mathscr{B} \longrightarrow\!\!\!\!\!\!\!\!+\;\; \mathscr{A}$, then $R = (d_0)_\diamond \cdot (d_1)^\diamond$.*

Remark 2.14. The proposition can be extended to any category enriched in Pre.

Example 2.15. Suppose that $f : \mathscr{A} \longrightarrow \mathscr{B}$ is monotone. Recall the relations $f_\diamond : \mathscr{A} \longrightarrow\!\!\!\!\!\!\!+\;\; \mathscr{B}$ and $f^\diamond : \mathscr{B} \longrightarrow\!\!\!\!\!\!\!+\;\; \mathscr{A}$. Their corresponding fibrations are the spans

$$\begin{array}{ccccc} & & id_{\mathscr{B}}/f & & \\ & {\scriptstyle p_0}\swarrow & & \searrow{\scriptstyle p_1} & \\ \mathscr{B} & & & & \mathscr{A} \end{array} \qquad \begin{array}{ccccc} & & f/id_{\mathscr{B}} & & \\ & {\scriptstyle p_0}\swarrow & & \searrow{\scriptstyle p_1} & \\ \mathscr{A} & & & & \mathscr{B} \end{array}$$

arising from the respective comma squares.

Example 2.16. The relation $(\mathsf{y}_{\mathscr{A}})^\diamond$ from $\mathbb{L}\mathscr{A}$ to $\mathscr{A}$ will be called the *elementhood* relation and denoted by $\in_{\mathscr{A}}$, since $(\mathsf{y}_{\mathscr{A}})^\diamond(a, A) = \mathbb{L}\mathscr{A}(\mathsf{y}_{\mathscr{A}} a, A) = A(a)$ holds by the Yoneda Lemma.

2.D Composition of Fibrations

Suppose that we have two fibrations as on the left below. We want to form their composite $\mathscr{E} \otimes \mathscr{F}$ *as a fibration.*

$$\begin{array}{ccccc} & & \mathscr{E} & & \\ & {\scriptstyle d_0^{\mathscr{E}}}\swarrow & & \searrow{\scriptstyle d_1^{\mathscr{E}}} & \\ \mathscr{C} & & & & \mathscr{B} \end{array} \begin{array}{ccccc} & & \mathscr{F} & & \\ & {\scriptstyle d_0^{\mathscr{F}}}\swarrow & & \searrow{\scriptstyle d_1^{\mathscr{F}}} & \\ \mathscr{B} & & & & \mathscr{A} \end{array} \qquad \begin{array}{ccccc} & & \mathscr{E} \otimes \mathscr{F} & & \\ & {\scriptstyle d_0^{\mathscr{E}\otimes\mathscr{F}}}\swarrow & & \searrow{\scriptstyle d_1^{\mathscr{E}\otimes\mathscr{F}}} & \\ \mathscr{C} & & & & \mathscr{A} \end{array}$$

The idea is similar to the ordinary relations: the composite is going to be a quotient of a pullback of spans, this time the quotient will be taken by a map that is surjective on objects, hence *absolutely dense*.

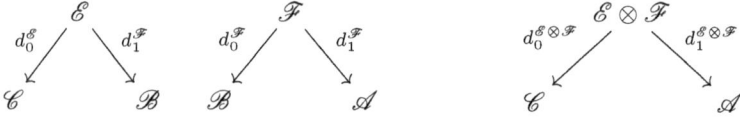

Remark 2.17. A monotone map $e : \mathscr{A} \longrightarrow \mathscr{B}$ is called *absolutely dense* (see [ABSV] and [BV]) iff there is an isomorphism

$$\mathscr{B}(b, b') \cong \bigvee_{a} \mathscr{B}(b, ea) \wedge \mathscr{B}(ea, b')$$

natural in b and b'. Clearly, every monotone map surjective on objects has this property. The converse is true if $\mathscr{B}$ is a poset. If $\mathscr{B}$ is a preorder, then e is absolutely dense when each strongly connected component of $\mathscr{B}$ contains at least an element in the image of e.

In defining the composition of fibrations we proceed as follows: construct the pullback

$$\begin{array}{ccc} \mathscr{E} \circ \mathscr{F} & \xrightarrow{q_1} & \mathscr{F} \\ {\scriptstyle q_0}\downarrow & & \downarrow{\scriptstyle d_0^{\mathscr{F}}} \\ \mathscr{E} & \xrightarrow[d_1^{\mathscr{E}}]{} & \mathscr{B} \end{array}$$

and define $\mathscr{E} \otimes \mathscr{F}$ to be the following preorder:

1. Objects are wiggly arrows of the form $c \rightsquigarrow a$ such that there exists $b \in \mathscr{B}$ with $(c \rightsquigarrow b, b \rightsquigarrow a) \in \mathscr{E} \circ \mathscr{F}$.
2. Put $c \rightsquigarrow a$ to be less or equal to $c' \rightsquigarrow a'$ iff $c \leq c'$ and $a \leq a'$.

Define a monotone map $w : \mathscr{E} \circ \mathscr{F} \longrightarrow \mathscr{E} \otimes \mathscr{F}$ in the obvious way and observe that it is surjective on objects.

We equip now $\mathscr{E} \otimes \mathscr{F}$ with the obvious projections $d_0^{\mathscr{E} \otimes \mathscr{F}} : \mathscr{E} \otimes \mathscr{F} \longrightarrow \mathscr{C}$ and $d_1^{\mathscr{E} \otimes \mathscr{F}} : \mathscr{E} \otimes \mathscr{F} \longrightarrow \mathscr{A}$. Then the following result is obvious.

Lemma 2.18. *The span* $(d_0^{\mathscr{E} \otimes \mathscr{F}}, \mathscr{E} \otimes \mathscr{F}, d_1^{\mathscr{E} \otimes \mathscr{F}}) : \mathscr{A} \longrightarrow \mathscr{C}$ *is a fibration.*

3 Exact Squares

The notion of *exact squares* replaces the notion of weak pullbacks in the preorder setting and exact squares will play a central rôle in our extension theorem. Exact squares were introduced and studied by René Guitart in [Gu].

Definition 3.1. *A lax square in* Pre

$$\begin{array}{ccc} \mathscr{P} & \xrightarrow{p_1} & \mathscr{B} \\ {\scriptstyle p_0}\downarrow & \nearrow & \downarrow{\scriptstyle g} \\ \mathscr{A} & \xrightarrow[f]{} & \mathscr{C} \end{array} \tag{3.3}$$

is exact *iff the canonical comparison in* Rel(Pre) *below is an iso (identity).*

$$\begin{array}{ccc} \mathscr{P} & \overset{(p_1)^{\diamond}}{\longleftarrow\!\!\!\!\!\!\!+\;\;} & \mathscr{B} \\ {\scriptstyle (p_0)_{\diamond}}\big\downarrow & \searrow & \big\downarrow{\scriptstyle g_{\diamond}} \\ \mathscr{A} & \underset{f^{\diamond}}{\longleftarrow\!\!\!\!\!\!\!+\;\;} & \mathscr{C} \end{array} \tag{3.4}$$

Remark 3.2. In defining the canonical comparison, we use the adjunctions $(p_1)_{\diamond} \dashv (p_1)^{\diamond}$ and $f_{\diamond} \dashv f^{\diamond}$ guaranteed by Lemma 2.5.

Using the formula (2.1) we obtain an equivalent criterion for exactness: there is an isomorphism, natural in a and b,

$$\mathscr{C}(fa, gb) \cong \bigvee_{w} \mathscr{A}(a, p_0 w) \wedge \mathscr{B}(p_1 w, b) \tag{3.5}$$

Remark 3.3 ([Gu], Example 1.14). Exact squares can be used to characterise order embeddings, absolutely dense morphisms, (relative) adjoints, and absolute Kan extensions. Further, (op-)comma squares are exact.

Example 3.4. Every square (3.3) where f and p_1 are *left* adjoints, is exact iff $p_0 \cdot p_1^r \cong f^r \cdot g$, where we denote by f^r and p_1^r the respective right adjoints.

Example 3.5. If the square on the left is exact, then so is the square on the right:

$$\begin{array}{ccc} \mathscr{P} & \xrightarrow{p_1} & \mathscr{B} \\ {\scriptstyle p_0}\big\downarrow & \nearrow & \big\downarrow{\scriptstyle g} \\ \mathscr{A} & \xrightarrow[f]{} & \mathscr{C} \end{array} \qquad\qquad \begin{array}{ccc} \mathscr{P}^{op} & \xrightarrow{p_0^{op}} & \mathscr{A}^{op} \\ {\scriptstyle p_1^{op}}\big\downarrow & \nearrow & \big\downarrow{\scriptstyle f^{op}} \\ \mathscr{B}^{op} & \xrightarrow[g^{op}]{} & \mathscr{C}^{op} \end{array}$$

Lemma 3.6. *Suppose that* $(d_0^S, \mathscr{E}^S, d_1^S)$ *and* $(d_0^R, \mathscr{E}^R, d_1^R)$ *are two-sided discrete fibrations. Then the pullback*

$$\begin{array}{ccc} \mathscr{E}^S \circ \mathscr{E}^R & \xrightarrow{q_1} & \mathscr{E}^R \\ {\scriptstyle q_0}\big\downarrow & & \big\downarrow{\scriptstyle d_0^R} \\ \mathscr{E}^S & \xrightarrow[d_1^S]{} & \mathscr{B} \end{array}$$

considered as a lax commutative square where the comparison is identity, is exact.

Given monotone relations $\mathscr{A} \overset{R}{\longrightarrow\!\!\!\!\!\!\!+\;\;} \mathscr{B}$ and $\mathscr{B} \overset{S}{\longrightarrow\!\!\!\!\!\!\!+\;\;} \mathscr{C}$, the two-sided fibration corresponding to the composition $S \cdot R$ is the composition of the fibrations corresponding to S and R as described in Section 2.D. The properties described in the next Corollary are essential for the proof of Theorem 4.1.

Corollary 3.7. *Form, for a pair R, S, of monotone relations the following commutative diagram*

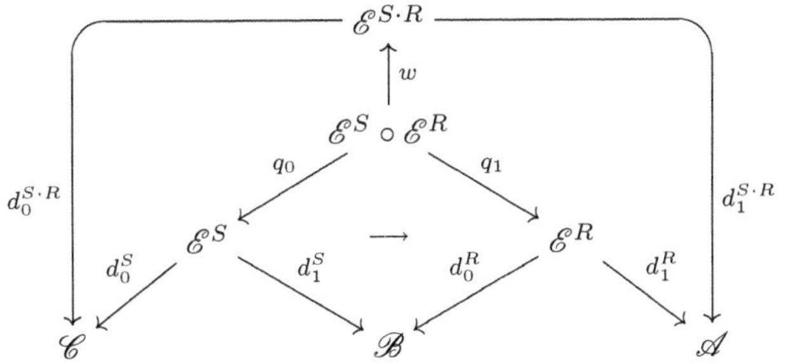

where the lax commutative square in the middle is a pullback square (hence the comparison is the identity), and w is a map, surjective on objects, coming from composing $\mathscr{E}^S$ and $\mathscr{E}^R$ as fibrations. Then the square is exact and w is an absolutely dense monotone map.

4 The Universal Property of $(-)_\diamond : \mathsf{Pre} \longrightarrow \mathsf{Rel(Pre)}$

We prove now that the 2-functor $(-)_\diamond : \mathsf{Pre} \longrightarrow \mathsf{Rel(Pre)}$ has an analogous universal property to the case of sets. From that, the result on a unique lifting of T to $\overline{T}$ will immediately follow, see Theorem 5.3 below.

Theorem 4.1. *The 2-functor $(-)_\diamond : \mathsf{Pre} \longrightarrow \mathsf{Rel(Pre)}$ has the following three properties:*

1. *Every $f_\diamond$ is a left adjoint.*
2. *For every exact square (3.3) the equality $f^\diamond \cdot g_\diamond = (p_0)_\diamond \cdot (p_1)^\diamond$ holds.*
3. *For every absolutely dense monotone map e, the relation $e_\diamond$ is a split epimorphism with the splitting given by $e^\diamond$.*

Moreover, the functor $(-)_\diamond$ is universal w.r.t. these three properties in the following sense: if K is any 2-category where the isomorphism 2-cells are identities, to give a 2-functor $H : \mathsf{Rel(Pre)} \longrightarrow \mathsf{K}$ is the same thing as to give a 2-functor $F : \mathsf{Pre} \longrightarrow \mathsf{K}$ with the following three properties:

1. *Every Ff has a right adjoint, denoted by $(Ff)^r$.*
2. *For every exact square (3.3) the equality $Ff^r \cdot Fg = Fp_0 \cdot (Fp_1)^r$ holds.*
3. *For every absolutely dense monotone map e, Fe is a split epimorphism, with the splitting given by $(Fe)^r$.*

Proof (Sketch.). It is trivial to see that $(-)_\diamond$ has the above three properties.

Given a 2-functor $H : \mathsf{Rel(Pre)} \longrightarrow \mathsf{K}$, define F to be the composite $H \cdot (-)_\diamond$. Such F clearly has the above three properties, since 2-functors preserve adjunctions.

Conversely, given $F : \mathsf{Pre} \longrightarrow \mathsf{K}$, define $H\mathscr{A} = F\mathscr{A}$ on objects, and on a relation $R = (d_0^R)_\diamond \cdot (d_1^R)^\diamond$ define $H(R) = Fd_0^R \cdot (Fd_1^R)^r$, where $(Fd_1^R)^r$ is the right adjoint of Fd_1^R in K.

That H is a well-defined functor follows using Corollary 3.7 and our assumption on F. □

5 The Extension Theorem

Definition 5.1. *We say that a locally monotone functor* T : $\mathsf{Pre} \longrightarrow \mathsf{Pre}$ *satisfies the* Beck-Chevalley Condition *(BCC) if it preserves exact squares.*

Remark 5.2. A functor satisfying the BCC has to preserve order-embeddings, absolutely dense monotone maps and absolute left Kan extensions. This follows from Example 1.14 of [Gu], see also Remark 3.3. Examples of functors (not) satisfying the BCC can be found in Section 6.

Theorem 5.3. *For a 2-functor* $T : \mathsf{Pre} \longrightarrow \mathsf{Pre}$ *the following are equivalent:*

1. *There is a 2-functor* $\overline{T} : \mathsf{Rel(Pre)} \longrightarrow \mathsf{Rel(Pre)}$ *such that*

$$\begin{array}{ccc} \mathsf{Rel(Pre)} & \xrightarrow{\overline{T}} & \mathsf{Rel(Pre)} \\ {\scriptstyle (-)_\diamond}\uparrow & & \uparrow{\scriptstyle (-)_\diamond} \\ \mathsf{Pre} & \xrightarrow[T]{} & \mathsf{Pre} \end{array} \tag{5.6}$$

2. *The functor* T *satisfies the BCC.*
3. *There is a distributive law* $T \cdot \mathbb{L} \longrightarrow \mathbb{L} \cdot T$ *of* T *over the KZ doctrine* $(\mathbb{L}, \mathsf{y}, \mathsf{m})$ *described in (2.2) above.*

Proof. The equivalence of 1. and 3. follows from general facts about distributive laws, using Proposition 2.7 above. See, e.g., [S_1]. For the equivalence of 1. and 2., observe that T satisfies the BCC iff

$$\mathsf{Pre} \xrightarrow{T} \mathsf{Pre} \xrightarrow{(-)_\diamond} \mathsf{Rel(Pre)}$$

satisfies the three properties of Theorem 4.1 above. □

Corollary 5.4. *If* T *is a locally monotone functor satisfying the BCC, the lifting* $\overline{T}$ *is computed as follows:* $\overline{T}(R) = (Td_0)_\diamond \cdot (Td_1)^\diamond$ *where* $(d_0, \mathscr{E}, d_1)$ *is the two-sided discrete fibration corresponding to* R.

6 Examples

Example 6.1. All the "Kripke-polynomial" functors satisfy the Beck-Chevalley Condition. This means the functors defined by the following grammar:

$$T ::= const_{\mathscr{X}} \mid Id \mid T^{\partial} \mid T + T \mid T \times T \mid \mathbb{L}T$$

where $const_{\mathscr{X}}$ is the constant-at-$\mathscr{X}$, T^{∂} is the *dual* of T, defined by putting

$$T^{\partial}\mathscr{A} = (T\mathscr{A}^{op})^{op}$$

and $\mathbb{L}\mathscr{X} = [\mathscr{X}^{op}, 2]$ (the lowersets on $\mathscr{X}$, ordered by inclusion). Observe that $\mathbb{L}^{\partial}\mathscr{X} = [\mathscr{X}, 2]^{op}$, hence $\mathbb{L}^{\partial}\mathscr{X} = \mathbb{U}\mathscr{X}$ (the uppersets on $\mathscr{X}$, ordered by reversed inclusion).

Example 6.2. Recall the adjunction $Q \dashv I : \mathsf{Pos} \longrightarrow \mathsf{Pre}$, where I is the inclusion functor and $Q(\mathscr{A})$ is the quotient of $\mathscr{A}$ obtained by identifying a and b whenever $a \leq b$ and $b \leq a$. The functors Q and I are locally monotone and map exact squares to exact squares. Hence, if $T : \mathsf{Pre} \longrightarrow \mathsf{Pre}$ satisfies the BCC, so does $QTI : \mathsf{Pos} \longrightarrow \mathsf{Pos}$.

Example 6.3. The *powerset functor* $\mathbb{P} : \mathsf{Pre} \longrightarrow \mathsf{Pre}$ is defined as follows. The order on $\mathbb{P}\mathscr{A}$ is the Egli-Milner preorder, that is, $\mathbb{P}(A, B) = 1$ if and only if

$$\forall a \in A\ \exists b \in B\ a \leq b \text{ and } \forall b \in B\ \exists a \in A\ a \leq b \tag{6.7}$$

$\mathbb{P}f(A)$ is the direct image of A. The functor $\mathbb{P}$ is locally monotone and satisfies the BCC.

The *finitary powerset functor* $\mathbb{P}_{\omega}$ is defined similarly: $\mathbb{P}_{\omega}\mathscr{A}$ consists of the finite subsets of $\mathscr{A}$ equipped with the Egli-Milner preorder. $\mathbb{P}_{\omega}$ is locally monotone and satisfies the BCC.

Example 6.4. Given a preorder $\mathscr{A}$, a subset $A \subseteq \mathscr{A}$ is called *convex* if $x \leq y \leq z$ and $x, z \in A$ imply $y \in A$.

The *convex powerset functor* $\mathbb{P}^c : \mathsf{Pos} \longrightarrow \mathsf{Pos}$ is defined as follows. $\mathbb{P}^c\mathscr{A}$ is the set of convex subsets of $\mathscr{A}$ endowed with the Egli-Milner order. $\mathbb{P}^c f(A)$ is the direct image of A. This is a well defined locally monotone functor. Notice that $\mathbb{P}^c \simeq Q\mathbb{P}I$, so by Example 6.2, $\mathbb{P}^c$ satisfies the BCC.

The *finitely-generated convex powerset* $\mathbb{P}^c_{\omega}$ is defined similarly to $\mathbb{P}^c$. The only difference is that the convex sets appearing in $\mathbb{P}^c_{\omega}\mathscr{A}$ are convex hulls of finitely many elements of $\mathscr{A}$. Then $\mathbb{P}^c_{\omega}$ is locally monotone and is isomorphic to $Q\mathbb{P}_{\omega}I$, thus it also satisfies the BCC.

Observe that both functors are self-dual: $(\mathbb{P}^c)^{\partial} = \mathbb{P}^c$ and $(\mathbb{P}^c_{\omega})^{\partial} = \mathbb{P}^c_{\omega}$.

Example 6.5. Since the lowerset functor $\mathbb{L} : \mathsf{Pre} \longrightarrow \mathsf{Pre}$ satisfies the Beck-Chevalley Condition by Example 6.1, we can compute its lifting $\overline{\mathbb{L}} : \mathsf{Rel}(\mathsf{Pre}) \longrightarrow \mathsf{Rel}(\mathsf{Pre})$. We show how $\overline{\mathbb{L}}$ works on the relation $\mathscr{A} \overset{R}{\longrightarrow\!\!\!\!\!\!\!+\,} \mathscr{B}$. The value $\overline{\mathbb{L}}(R)$ is, by Theorems 4.1 and 5.3, given by $(\mathbb{L}d_0)_{\diamond} \cdot (\mathbb{L}d_1)^{\diamond}$ where $(d_0, \mathscr{E}^R, d_1) : \mathscr{A} \longrightarrow \mathscr{B}$ is the two-sided discrete fibration corresponding to R. Using the formula (2.1) for relation composition, we can write

$$\overline{\mathbb{L}}(R)(B, A) = \bigvee_{W} \mathbb{L}\mathscr{B}(B, \mathbb{L}d_0(W)) \wedge \mathbb{L}\mathscr{A}(\mathbb{L}d_1(W), A) \tag{6.8}$$

where $B : \mathscr{B}^{op} \longrightarrow 2$ and $A : \mathscr{A}^{op} \longrightarrow 2$ are arbitrary lowersets. Since $\mathbb{L}d_1$ is a left adjoint to restriction along $d_1^{op} : (\mathscr{E}^R)^{op} \longrightarrow \mathscr{A}^{op}$, we can rewrite (6.8) to

$$\overline{\mathbb{L}}(R)(B, A) = \bigvee_{W} \mathbb{L}\mathscr{B}(B, \mathbb{L}d_0(W)) \wedge \mathbb{L}\mathscr{E}^R(W, A \cdot d_1^{op})$$

and, by the Yoneda Lemma, to

$$\overline{\mathbb{L}}(R)(B,A) = \mathbb{L}\mathscr{B}(B, \mathbb{L}d_0(A \cdot d_1^{op}))$$

Hence the lowersets B and A are related by $\overline{\mathbb{L}}(R)$ if and only if the inclusion

$$B \subseteq \mathbb{L}d_0(A \cdot d_1^{op})$$

holds in $[\mathscr{B}^{op}, 2]$. Recall that

$$\mathbb{L}d_0(A \cdot d_1^{op}) = b \mapsto \bigvee_w \mathscr{B}(b, d_0 w) \wedge (A \cdot d_1^{op})(w)$$

Therefore the inclusion $B \subseteq \mathbb{L}d_0(A \cdot d_1^{op})$ is equivalent to the statement: For all b in B there is (b_1, a_1) such that $R(b_1, a_1)$ and $b \leq b_1$ and a_1 in A.

Observe that the above condition is reminiscent of one half of the Egli-Milner-style of the relation lifting of a powerset functor. This is because $\mathbb{L}$ is the "lower half" of two possible "powerpreorder functors". The "upper half" is given by $\mathbb{U} : \mathsf{Pre} \longrightarrow \mathsf{Pre}$ where $\mathbb{U} = \mathbb{L}^{\partial}$.

Example 6.6. The relation liftings $\overline{\mathbb{P}}$, $\overline{\mathbb{P}^c}$, $\overline{\mathbb{P}_\omega}$, $\overline{\mathbb{P}^c_\omega}$ of the (convex) powerset functor and their finitary versions yield the "Egli-Milner" style of the relation lifting. More precisely, for a relation $\mathscr{B} \xrightarrow{R} \mathscr{A}$ we have $\overline{\mathbb{P}}(R)(B,A)$ (respectively $\overline{\mathbb{P}_\omega}(R)(B,A)$, $\overline{\mathbb{P}^c}(R)(B,A)$, $\overline{\mathbb{P}^c_\omega}(R)(B,A)$) if and only if

$$\forall a \in A\ \exists b \in B\ R(b,a) \text{ and } \forall b \in B\ \exists a \in A\ R(b,a).$$

Example 6.7. To find a functor that does not satisfy the BCC, it suffices, by Remark 5.2, to find a locally monotone functor $T : \mathsf{Pre} \longrightarrow \mathsf{Pre}$ that does not preserve order-embeddings. For this, let T be the *connected components functor*, i.e., T takes a preorder $\mathscr{A}$ to the discretely ordered poset of connected components of $\mathscr{A}$. T does not preserve embedding $f : \mathscr{A} \longrightarrow \mathscr{B}$ indicated below.

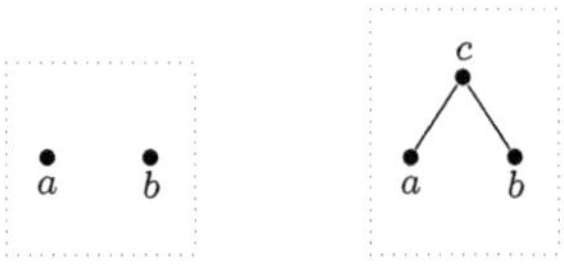

$\mathscr{A}$ $\mathscr{B}$

7 An Application: Moss's Coalgebraic Logic over Posets

We show how to develop the basics of Moss's coalgebraic logic over posets. For reasons of space, this development will be terse and assume some familiarity with, e.g., Sections 2.2 and 3.1 of [KuL].

Since the logics will have propositional connectives but no negation (to capture the semantic order on the logical side) we will use the category DL of bounded distributive lattices. We write $F \dashv U : \mathsf{DL} \longrightarrow \mathsf{Pos}$ for the obvious adjunction; and $P : \mathsf{Pos}^{op} \longrightarrow \mathsf{DL}$ where $UP\mathscr{X} = [\mathscr{X}, 2]$ and $S : \mathsf{DL} \longrightarrow \mathsf{Pos}^{op}$ where

$SA = \mathsf{DL}(A, 2)$. Note that $UP = [-, 2]$ and recall $\mathbb{L} = [(-)^{op}, 2]$. Further, let $T : \mathsf{Pos} \longrightarrow \mathsf{Pos}$ be a locally monotone finitary functor that satisfies the BCC.

We define coalgebraic logic abstractly by a functor $L : \mathsf{DL} \longrightarrow \mathsf{DL}$ given as

$$L = FT^{\partial}U$$

where the functor $T^{\partial} : \mathsf{Pos} \longrightarrow \mathsf{Pos}$ is given by $T^{\partial}\mathscr{X} = (T(\mathscr{X}^{op}))^{op}$. By Example 6.1, T^{∂} satisfies the BCC. The formulas of the logic are the elements of the initial L-algebra $FT^{\partial}U(\mathcal{L}) \longrightarrow \mathcal{L}$. The formula given by some $\alpha \in T^{\partial}U(\mathcal{L})$ is written as $\nabla\alpha$. The semantics is given by a natural transformation

$$\delta : LP \longrightarrow PT^{op}$$

Before we define δ, we need for every preorder $\mathscr{A}$, the relation[1]

$$[\mathscr{A}, 2] \xrightarrow{\ \ni_{\mathscr{A}}\ } \mathscr{A}^{op}$$

given by the evaluation map $\mathsf{ev}_{\mathscr{A}} : \mathscr{A} \times [\mathscr{A}, 2] \longrightarrow 2$. Observe that

$$\ni_{\mathscr{A}} = (\mathsf{y}_{\mathscr{A}^{op}})^{\diamond} \tag{7.9}$$

since $(\mathsf{y}_{\mathscr{A}^{op}})^{\diamond}(a, V) = [\mathscr{A}, 2](\mathsf{y}_{\mathscr{A}^{op}}a, V) = Va$ holds by the Yoneda Lemma.

Lemma 7.1. *For every monotone map $f : \mathscr{A} \longrightarrow \mathscr{B}$ we have*

$$\begin{array}{ccc} [\mathscr{A}, 2] & \xrightarrow{\ \ni_{\mathscr{A}}\ } & \mathscr{A}^{op} \\ {\scriptstyle [f,2]^{\diamond}} \uparrow & & \uparrow {\scriptstyle (f^{op})^{\diamond}} \\ [\mathscr{B}, 2] & \xrightarrow[\ \ni_{\mathscr{B}}\ ]{} & \mathscr{B}^{op} \end{array}$$

Corollary 7.2. *For every locally monotone functor T that satisfies the Beck-Chevalley Condition and for every monotone map $f : \mathscr{A} \longrightarrow \mathscr{B}$, we have*

$$\begin{array}{ccc} \overline{T}[\mathscr{A}, 2] & \xrightarrow{\ \overline{T}\ni_{\mathscr{A}}\ } & \overline{T}\mathscr{A}^{op} \\ {\scriptstyle \overline{T}[f,2]^{\diamond}} \uparrow & & \uparrow {\scriptstyle \overline{T}(f^{op})^{\diamond}} \\ \overline{T}[\mathscr{B}, 2] & \xrightarrow[\ \overline{T}\ni_{\mathscr{B}}\ ]{} & \overline{T}\mathscr{B}^{op} \end{array}$$

Coming back to $\delta : LP \longrightarrow PT^{op}$. It suffices, due to $F \dashv U$, to give

$$\tau : T^{\partial}UP \longrightarrow UPT^{op}$$

Observe that, for every preorder $\mathscr{X}$, we have

$$UPT^{op}(\mathscr{X}) = [T^{op}\mathscr{X}, 2] = \mathbb{L}((T^{op}\mathscr{X})^{op})$$

By Proposition 2.7, to define $\tau_{\mathscr{X}}$ it suffices to give a relation from $T^{\partial}UP\mathscr{X}$ to $(T^{op}\mathscr{X})^{op}$, and we obtain it from Theorem 5.3 by applying $\overline{T^{\partial}}$ to the relation $\ni_{\mathscr{X}}$. That $\tau_{\mathscr{X}}$ so defined is natural, follows from Corollary 7.2. This follows [KKuV] with the exception that here now we need to use T^{∂}.

[1] The type of $\ni_{\mathscr{X}}$ conforms with the logical reading of $\ni$ as $\Vdash$. Indeed, $\ni(x, \varphi)$ & $\varphi \subseteq \psi \Rightarrow \ni(x, \psi)$ and $\ni(x, \varphi)$ & $x \leq y \Rightarrow \ni(y, \varphi)$, where φ, ψ are uppersets of $\mathscr{X}$.

Example 7.3. Recall the functor $\mathbb{P}^c_\omega$ of Example 6.4 and consider a coalgebra $c : \mathscr{X} \longrightarrow \mathbb{P}^c_\omega \mathscr{X}$. On the logical side we allow ourselves to write $\nabla\alpha$ for any finite subset α of $U(\mathcal{L})$. Of course, we then have to be careful that the semantics of α agrees with the semantics of the convex closure of α. Interestingly, this is done automatically by the machinery set up in the previous section, since $\mathbb{P}^c_\omega = Q\mathbb{P}_\omega I$ and all these functors are self-dual. By Example 6.6, the semantics of $\nabla\alpha$ is given by

$$x \Vdash \nabla\alpha \quad \Leftrightarrow \quad \forall y \in c(x) \exists \varphi \in \alpha . y \Vdash \varphi \text{ and } \forall \varphi \in \alpha \exists y \in c(x) . y \Vdash \varphi .$$

8 Conclusions

We hope to have illustrated in the previous two sections that, after getting used to handle the $(-)_\diamond, (-)^\diamond$ and $(-)^{op}$, the techniques developed here work surprisingly smoothly and will be useful in many future developments. For example, an observation crucial for both [KKuV, KuL] is that composing the singleton map $X \longrightarrow \mathscr{P}X$, $x \mapsto \{x\}$, with the relation $\ni_X : \mathscr{P}X \longrightarrow\!\!\!\!\!\!\!\!+\;\; X$ is id_X. Referring back to (7.9), we find here the same relationship

$$\ni_{\mathscr{A}} \circ (\mathbb{y}_{\mathscr{A}^{op}})_\diamond = (\mathbb{y}_{\mathscr{A}^{op}})^\diamond \circ (\mathbb{y}_{\mathscr{A}^{op}})_\diamond = id_{\mathscr{A}^{op}}$$

The question whether the completeness proof of [KKuV] and the relationship between ∇ and predicate liftings of [KuL] can be carried over to our setting are a direction of future research.

Another direction is the generalisation to categories which are enriched over more general structures than $\mathbb{2}$, such as commutative quantales. Simulation, relation lifting and final coalgebras in this setting have been studied in [Wo].

References

[ABSV] Adámek, J., El Bashir, R., Sobral, M., Velebil, J.: On functors that are lax epimorphisms. Theory Appl. Categ. 8.20, 509–521 (2001)

[BK] Balan, A., Kurz, A.: Finitary Functors: from Set to Preord and Poset. In: Corradini, A., Klin, B., Crstea, C. (eds.) CALCO 2011. LNCS, vol. 6859, pp. 85–99. Springer, Heidelberg (2011)

[Bal] Baltag, A.: A logic for coalgebraic simulation. Electron. Notes Theor. Comput. Sci. 33, 41–60 (2000)

[Ba] Barr, M.: Relational algebras. In: Reports of the Midwest Category Seminar IV. Lecture Notes in Mathematics, vol. 137, pp. 39–55. Springer, Heidelberg (1970)

[BV] El Bashir, R., Velebil, J.: Reflective and coreflective subcategories of presheaves. Theory Appl. Categ. 10.16, 410–423 (2002)

[CKW] Carboni, A., Kelly, G.M., Wood, R.J.: A 2-categorical approach to change of base and geometric morphisms I. Cahiers de Top. et Géom. Diff. XXXII.1, 47–95 (1991)

[Gu] Guitart, R.: Relations et carrés exacts. Ann. Sci. Math. Québec IV.2, 103–125 (1980)

[He] Hermida, C.: A categorical outlook on relational modalities and simulations, preprint,
`http://maggie.cs.queensu.ca/chermida/papers/sat-sim-IandC.pdf`

[HeJ] Hermida, C., Jacobs, B.: Structural induction and coinduction in the fibrational setting. Inform. and Comput. 145, 107–152 (1998)

[HuJ] Hughes, J., Jacobs, B.: Simulations in coalgebra. Theor. Comput. Sci. 327, 71–108 (2004)

[KaKuV] Kapulkin, K., Kurz, A., Velebil, J.: Expressivity of Coalgebraic Logic over Posets. In: CMCS 2010 Short contributions, CWI Technical report SEN-1004, pp. 16–17 (2010)

[Ke] Kelly, G.M.: Basic concepts of enriched category theory. London Math. Soc. Lecture Notes Series, vol. 64. Cambridge Univ. Press, New York (1982)

[Kl] Klin, B.: An Abstract Coalgebraic Approach to Process Equivalence for Well-Behaved Operational Semantics. University of Aarhus (2004)

[KKuV] Kupke, C., Kurz, A., Venema, Y.: Completeness of the finitary Moss logic. In: Advances in Modal Logic, pp. 193–217. College Publications (2008)

[KuL] Kurz, A., Leal, R.: Equational coalgebraic logic. Electron. Notes Theor. Comput. Sci. 249, 333–356 (2009)

[L] Levy, P.B.: Similarity quotients as final coalgebras. In: Hofmann, M. (ed.) FOSSACS 2011. LNCS, vol. 6604, pp. 27–41. Springer, Heidelberg (2011)

[M_1] Marmolejo, F.: Doctrines whose structure forms a fully faithful adjoint string. Theor. Appl. Categ. 3(2), 24–44 (1997)

[M_2] Marmolejo, F.: Distributive laws for pseudomonads. Theor. Appl. Categ. 5(5), 91–147 (1999)

[MRW] Marmolejo, F., Rosebrugh, R., Wood, R.J.: Duality for CCD lattices. Theor. Appl. Categ. 22(1), 1–23 (2009)

[Mo] Moss, L.: Coalgebraic logic. Ann. Pure Appl. Logic 96, 277–317 (1999)

[R] Rutten, J.: Relators and Metric Bisimulations (Extended Abstract). Electr. Notes Theor. Comput. Sci. 11, 252–258 (1998)

[S_1] Street, R.: The formal theory of monads. J. Pure Appl. Algebra 2, 149–168 (1972)

[S_2] Street, R.: Fibrations and Yoneda's lemma in a 2-category. In: Category Seminar, Sydney 1974. Lecture Notes in Mathematics, vol. 420, pp. 104–133 (1974)

[S_3] Street, R.: Elementary cosmoi I. In: Category Seminar, Sydney 1974. Lecture Notes in Mathematics, vol. 420, pp. 134–180. Springer, Heidelberg (1974)

[S_4] Street, R.: Fibrations in bicategories. Cahiers de Top. et Géom. Diff. XXI.2, 111–159 (1980)

[V] Venema, Y.: Automata and fixed point logic: a coalgebraic perspective. Inform. and Comput. 204.4, 637–678 (2006)

[Wo] Worrell, J.: Coinduction for recursive data types: partial orders, metric spaces and Ω-categories. Electron. Notes Theor. Comput. Sci. 33, 337–356 (2000)

Model Checking Linear Coalgebraic Temporal Logics: An Automata-Theoretic Approach

Corina Cîrstea

University of Southampton
cc2@ecs.soton.ac.uk

Abstract. We extend the theory of maximal traces of pointed non-deterministic coalgebras by providing an automata-based characterisation of the set of maximal traces for finite such coalgebras. We then consider *linear* coalgebraic temporal logics interpreted over non-deterministic coalgebras, and show how to reduce the model checking problem for such logics to the problem of deciding the winner in a regular two-player game. Our approach is inspired by the automata-theoretic approach to model checking Linear Temporal Logic over transition systems.

Keywords: coalgebra, trace, temporal logic, automata, model checking.

1 Introduction

A coalgebraic version of the well-established notion of trace of a state in a transition system was defined in [4] for coalgebras of type $\mathcal{P} \circ F$ or $\mathcal{P}^+ \circ F$, with $\mathcal{P} : \mathsf{Set} \to \mathsf{Set}$ ($\mathcal{P}^+ : \mathsf{Set} \to \mathsf{Set}$) the (non-empty) powerset functor and $F : \mathsf{Set} \to \mathsf{Set}$ a polynomial functor. This was extended in [2] to coalgebras of type $T \circ F$, with $F : \mathsf{C} \to \mathsf{C}$ an endofunctor and $T : \mathsf{C} \to \mathsf{C}$ a monad that distributes suitably over F. Here we return to the setting of [4] and provide an automata-based characterisation of the traces of a pointed $\mathcal{P}^+ \circ F$-coalgebra.

Coalgebra automata were introduced in [8] as devices that accept pointed coalgebras. For a weak pullback preserving functor F, any pointed F-coalgebra $\mathbb{X} = (X, \gamma, x_0)$ with finite carrier defines a deterministic F-automaton accepting precisely the pointed F-coalgebras that are bisimilar to $\mathbb{X}$. The key idea underlying this paper is to view a finite, pointed $\mathcal{P}^+ \circ F$-coalgebra $\mathbb{X}$ not as a deterministic $\mathcal{P}^+ \circ F$-automaton, but as a *non-deterministic* F-automaton. Our first result provides a characterisation of the set of traces of $\mathbb{X}$ as those elements of the final F-coalgebra that are accepted by this automaton.

The second part of the paper exploits this insight in order to obtain a decision procedure for model-checking the coalgebraic counterparts of linear temporal logic. Such logics arise as certain fragments of the path-based coalgebraic temporal logics defined in [2], and are interpreted over coalgebras of type $\mathcal{P}^+ \circ F$. Specifically, we consider logics whose formulas are of the form $\Box\varphi$, with φ a formula of a coalgebraic fixed point logic determined by a set Λ of predicate liftings

A. Corradini, B. Klin, and C. Cîrstea (Eds.): CALCO 2011, LNCS 6859, pp. 130–144, 2011.

for the functor F. The formula $\Box\varphi$ requires *all* traces of a state in a $\mathcal{P}^+ \circ F$-coalgebra to satisfy the property expressed by φ. We define a regular two-player graph game for deciding whether a formula φ as above holds in *some* trace of a finite pointed $\mathcal{P}^+ \circ F$-coalgebra $\mathbb{X}$, by combining the Λ-automaton induced by φ (as defined in [3]) with the non-deterministic F-automaton induced by $\mathbb{X}$.

Our approach is inspired by the automata-theoretic approach to model-checking Linear Temporal Logic [7]. This exploits the observation that both the models and the formulas to be checked can be represented as *Büchi automata*. Then, checking a formula against a model amounts to verifying that the product of the model automaton (whose runs correspond to possible traces of the model) with the automaton induced by the negation of the formula to be verified (which accepts precisely those traces that violate the original formula) does not accept any trace. In our setting, the automata-based characterisation of the traces of pointed $\mathcal{P}^+ \circ F$-coalgebras is not directly useful, as our model automaton is an F-automaton whereas our formula automaton is a Λ-automaton. However, our approach is similar in nature: we define a regular game $\mathcal{G}_{\neq\emptyset}(\mathbb{X}, \mathbb{A})$ for every finite, pointed $\mathcal{P}^+ \circ F$-coalgebra $\mathbb{X}$ and every Λ-automaton $\mathbb{A}$, with the crucial property that $\mathcal{G}_{\neq\emptyset}(\mathbb{X}, \mathbb{A})$ admits a winning strategy for $\exists$ precisely when $\mathbb{X}$ contains a trace that is accepted by $\mathbb{A}$. This constitutes our second result.

2 Preliminaries

Graph Games. A *graph game* played by two players $\exists$ and $\forall$ is a tuple $\mathcal{G} = (B_\exists, B_\forall, E, Win)$ where $B = B_\exists \cup B_\forall$ is the disjoint union of *positions* owned by $\exists$ and respectively $\forall$, $E \subseteq B \times B$ indicates the allowed moves, and $Win \subseteq B^\omega$. A *play* in $\mathcal{G}$ is a finite or infinite sequence of positions $(b_0, b_1, \ldots)$ such that $(b_i, b_{i+1}) \in E$ for all i. A finite play is lost by the player who cannot move, whereas an infinite play $(b_0, b_1, \ldots)$ is won by $\exists$ if and only if $(b_0, b_1, \ldots) \in Win$.

A *strategy* in $\mathcal{G}$ for a player $P \in \{\exists, \forall\}$ is a function that maps plays that end in a position $b \in B_P$ to a position $b' \in B$ such that $(b, b') \in E$. Intuitively, a strategy determines a player's next move, depending on the history of the play. A strategy for player P is said to be (i) *history-free* when next moves only depend on the current position, and (ii) *winning from position* $b \in B$ if P wins all plays $(b_0, b_1, \ldots)$ with $b_0 = b$ when playing according to that strategy.

A graph game is called *regular* if there exists an ω-regular language L over a finite alphabet C and a map $col : B \to C$ such that $Win = \{(b_0, b_1, \ldots) \in B^\omega \mid col(b_0)col(b_1)\ldots \in L\}$. A *parity game* is a graph game whose winning condition is defined using a *parity map* $\Omega : B \to \omega$ with finite range, by $Win = \{ (b_0, b_1, \ldots) \mid max\{ k \mid k = \Omega(b_i) \text{ for infinitely many } i \in \omega \} \text{ is even} \}$. Any regular game can be transformed into an equivalent parity game [6]. Regular games are determined, that is, either $\exists$ or $\forall$ has a winning strategy from any given position. In addition, parity games admit history-free winning strategies.

Maximal Traces and Executions. A definition of infinite traces for coalgebras of type $\mathcal{P} \circ F$ or $\mathcal{P}^+ \circ F$ with F a *polynomial endofunctor* (i.e. constructed from identity and constant functors using products, coproducts and exponents) was given in [4]. This was generalised in [2], where it was shown how to define, for a coalgebra (X, γ) of type $T \circ F$ with $F : \mathsf{C} \to \mathsf{C}$ an endofunctor and $T : \mathsf{C} \to \mathsf{C}$ an affine monad that distributes over F via a distributive law $\lambda : FT \Longrightarrow TF$, a *maximal trace map* $\mathsf{tr}_\gamma : X \to T(Z)$ as well as a *maximal execution map* $\mathsf{exec}_\gamma : X \to T(Z_X)$[1]. Here, Z is the carrier of the final F-coalgebra (Z, ζ), and Z_X is the carrier of the final $X \times F$-coalgebra (Z_X, ζ_X). The states of Z and Z_X represent potential maximal traces, respectively executions. In addition to the information provided by a trace, an execution also records the states visited, including the initial state. The maximal trace (execution) map then assigns to each state in X a suitably-structured collection of traces (respectively executions). For $T = \mathcal{P}^+$, this collection is structured as a set.

Example 1. 1. Let $T = \mathcal{P}^+$ and $F = A \times \mathsf{Id}$. That is, $T \circ F$-coalgebras are labelled transition systems with the additional requirement that each state has at least one successor[2]. In this case, the potential maximal traces (elements of the final $A \times \mathsf{Id}$-coalgebra) are given by infinite sequences of elements of A. Also, for a $\mathcal{P}^+ \circ F$-coalgebra (X, γ), the potential maximal executions are given by infinite sequences of the form $x_0 a_0 x_1 a_1 \ldots$ with $x_i \in X$ and $a_i \in A$ for $i \in \omega$. The maximal execution map $\mathsf{tr}_\gamma : X \to \mathcal{P}^+(Z)$ assigns to each state $x_0 \in X$, the set of *computation paths* (infinite sequences $x_0 a_0 x_1 a_1 \ldots$ with $(a_i, x_{i+1}) \in \gamma(x_i)$ for $i \in \omega$) originating in x_0. Also, the maximal trace map assigns to each state x_0, the set of infinite sequences of elements of A that appear in computation paths starting in x_0.
2. By changing F to $1 + A \times \mathsf{Id}$ with $1 = \{*\}$, one moves to labelled transition systems with explicit termination (with the same restriction as above). This time, the potential maximal traces and executions also include *finite* ones of the form $a_0 a_1 \ldots a_n *$, and respectively $x_0 a_0 x_1 \ldots a_n x_{n+1} *$.
3. Taking $F = A + \mathsf{Id} \times \mathsf{Id}$ results in maximal executions for a $\mathcal{P}^+ \circ F$-coalgebra (X, γ) being given by (possibly infinite) binary trees whose nodes (including any leaves) are labelled by states of X, and whose leaves (if any) are labelled by elements of A. As before, maximal traces can be obtained from maximal executions by removing the labelings with elements of X from the nodes.

For $T = \mathcal{P}^+$, the trace map of [2] is given by:

$$\mathsf{tr}_\gamma(x) = \{z \in Z \mid \pi_i(z) \in \gamma_i(x) \text{ for all } i \in \omega\} \tag{1}$$

where $(Z, (\pi_i : Z \to F^i 1)_{i \in \omega})$ defines the limit of the final sequence of F, and where the maps $\gamma_i : X \to \mathcal{P}^+(F^i 1)$ are defined by induction on i:

- $\gamma_0 = \eta_1 \circ !_X : X \to \mathcal{P}^+ 1$, where $\eta : \mathsf{Id} \Longrightarrow \mathcal{P}^+$ denotes the unit of the monad $\mathcal{P}^+$, and $!_X : X \to 1$ is the unique map from X to the one-element set 1,
- $\gamma_{i+1} = \mu_{F^{i+1}1} \circ \mathcal{P}^+ \lambda_{F^i 1} \circ \mathcal{P}^+ F \gamma_i \circ \gamma : X \to \mathcal{P}^+(F^{i+1} 1)$ for $i \in \omega$, where $\mu : \mathcal{P}^+ \circ \mathcal{P}^+ \Longrightarrow \mathcal{P}^+$ denotes the multiplication of the monad $\mathcal{P}^+$.

[1] Some additional assumptions on F and T were needed in loc. cit., including preservation of certain ω^{op}-limits by F and $X \times F$.

[2] This restriction is typical in logics such as CTL*, where it allows the definition of a notion of computation path that only accounts for infinite computations.

Thus, the elements of $\mathsf{tr}_\gamma(x)$ are given by infinite sequences $(z_i)_{i\in\omega}$ with $z_i \in \gamma_i(x)$ for $i \in \omega$. Here, the elements of F^i1 are to be thought of as finite approximations of maximal traces, and the maps γ_i can be regarded as finite approximations of the trace map.

An equivalent definition of the trace map when $T = \mathcal{P}^+$ and when the distributive law $\lambda : F\mathcal{P}^+ \Longrightarrow \mathcal{P}^+F$ arises via *relation lifting* is given in [4]. For a relation $\langle r_1, r_2\rangle : R \rightarrowtail X \times Y$, the relation $\mathsf{Rel}(F)(R) \subseteq FX \times FY$ is obtained using the epi-mono factorisation of $\langle F(r_1), F(r_2)\rangle$ as follows:

$$\begin{array}{ccc} F(R) & \twoheadrightarrow & \mathsf{Rel}(F)(R) \\ & \searrow^{\langle F(r_1),F(r_2)\rangle} & \downarrow \\ & & F(X) \times F(Y) \end{array}$$

Then, when fixing $\lambda : F\mathcal{P}^+ \Longrightarrow \mathcal{P}^+F$ to be given by:

$$a \in \lambda_X(u) \ \text{ iff } \ (a, u) \in \mathsf{Rel}(F)(\in_X) \tag{2}$$

with $\in_X \subseteq X \times \mathcal{P}^+X$ the membership relation, the trace map $\mathsf{tr}_\gamma : X \to \mathcal{P}^+Z$ associated to a $\mathcal{P}^+ \circ F$-coalgebra (X, γ) can be defined by exploiting the observation that the following set carries F-coalgebra structure:

$$U := \{(u_i)_{i\in\omega} \in \prod_{i\in\omega} F^iX \mid (u_{i+1}, u_i) \in B_i \text{ for } i \in \omega\}$$

where the relations $B_i \subseteq F^{i+1}X \times F^iX$ are defined inductively by:

- $B_0 := \{(y, x) \in FX \times X \mid y \in \gamma(x)\}$,
- $B_{i+1} := \mathsf{Rel}(F)(B_i)$ for $i \in \omega$.

Specifically, [4] shows the existence of an isomorphism $\varphi : U_- \to FU$, with $U_- := \{(u_{i+1})_{i\in\omega} \mid (u_i)_{i\in\omega} \in U\}$. This, in turn, yields an F-coalgebra structure $\sigma : U \to FU$ on U:

$$U \xrightarrow{\langle \pi_{i+1}\rangle_{i\in\omega}} U_- \xrightarrow[\simeq]{\varphi} FU$$

The trace map $\mathsf{tr}_\gamma : X \to \mathcal{P}^+Z$ is now given by:

$$\mathsf{tr}_\gamma(x) = \{z \in Z \mid z = !_U(u) \text{ for some } u \in U \text{ with } \pi_0(u) = x\} \tag{3}$$

where $!_U : (U, \sigma) \to (Z, \zeta)$ is the unique F-coalgebra homomorphism arising from the finality of (Z, ζ).

Path-Based Coalgebraic Temporal Logics. The path-based coalgebraic temporal logics defined in [2] are interpreted over coalgebras of type $T \circ F$, with T and F as before, and are parameterised by sets Λ and Λ_F of monotone predicate liftings for the endofunctors T and F, respectively. In [2], a *(unary)*[3]

[3] To simplify the presentation, we assume all predicate liftings to be either unary or nullary; however, our results hold for predicate liftings with arbitrary finite arities.

predicate lifting for an endofunctor $F : \mathsf{C} \to \mathsf{C}$ is taken to be a natural transformation $\lambda : P \Longrightarrow P \circ F$, with $P : \mathsf{C} \to \mathsf{Set}$ a subfunctor of $\hat{\mathcal{P}} \circ U$, $\hat{\mathcal{P}} : \mathsf{Set} \to \mathsf{Set}^{\mathrm{op}}$ the contravariant powerset functor, and $U : \mathsf{C} \to \mathsf{Set}$ mapping a state space to its underlying set. The syntax and semantics of path-based coalgebraic temporal logics are summarised next.

Definition 2 ([2]). *Let Λ and Λ_F denote sets of monotone predicate liftings for T and F, respectively. The language $\mu\mathcal{L} ::= \mu\mathcal{L}_{\Lambda}^{\Lambda_F}(\mathcal{U}, \mathcal{V})$ over a 2-sorted set $(\mathcal{U}, \mathcal{V})$ of propositional variables (with sorts for* paths *and respectively* states*) is defined by the following grammar:*

$$\mu\mathcal{L}_F \ni \varphi ::= \mathsf{tt} \mid \mathsf{ff} \mid q \mid \Phi \mid \varphi \wedge \varphi \mid \varphi \vee \varphi \mid [\lambda_F]\varphi \mid \eta q.\varphi$$
$$\mu\mathcal{L} \ni \Phi ::= \mathsf{tt} \mid \mathsf{ff} \mid p \mid [\lambda]\varphi \mid \Phi \wedge \Phi \mid \Phi \vee \Phi$$

where $q \in \mathcal{U}$, $p \in \mathcal{V}$, $\eta \in \{\mu, \nu\}$, $\lambda_F \in \Lambda_F$ and $\lambda \in \Lambda$.

That is, *path formulas* φ are built from propositional variables q and *state formulas* Φ using positive boolean operators, modal operators $[\lambda_F]$ and fixpoint operators, while state formulas Φ are built from atomic propositions p and modal formulas $[\lambda]\varphi$ using positive boolean operators.

Such languages are interpreted over pairs consisting of a $T \circ F$-coalgebra (X, γ) and a 2-sorted valuation $V : (\mathcal{U}, \mathcal{V}) \to (PZ_X, PX)$ (interpreting path and state variables as sets of maximal executions and respectively of states), by making use of the maximal execution map $\mathsf{exec}_\gamma : X \to TZ_X$.

Definition 3 ([2]). *Given a $T \circ F$-coalgebra (X, γ) and a 2-sorted valuation $V : (\mathcal{U}, \mathcal{V}) \to (PZ_X, PX)$, the semantics $(\!|\varphi|\!)_{\gamma,V} \in PZ_X$ of path formulas $\varphi \in \mu\mathcal{L}_F$ and $[\![\Phi]\!]_{\gamma,V} \in PX$ of state formulas $\Phi \in \mu\mathcal{L}$ is defined by:*

$$\begin{aligned}
(\!|q|\!)_{\gamma,V} &= V(q) \\
(\!|\Phi|\!)_{\gamma,V} &= P(\pi_1 \circ \zeta_X)([\![\Phi]\!]_{\gamma,V}) \\
(\!|[\lambda_F]\varphi|\!)_{\gamma,V} &= (P(\pi_2 \circ \zeta_X) \circ (\lambda_F)_{Z_X})((\!|\varphi|\!)_{\gamma,V}) \\
(\!|\mu q.\varphi|\!)_{\gamma,V} &= \mathsf{lfp}((\varphi)_q^{\gamma,V}) \\
(\!|\nu q.\varphi|\!)_{\gamma,V} &= \mathsf{gfp}((\varphi)_q^{\gamma,V}) \\
[\![p]\!]_{\gamma,V} &= V(p) \\
[\![[\lambda]\varphi]\!]_{\gamma,V} &= (P\mathsf{exec}_\gamma \circ \lambda_{Z_X})((\!|\varphi|\!)_{\gamma,V})
\end{aligned}$$

and the usual clauses for the boolean operators, where, for $q \in \mathcal{U}$, $(\varphi)_q^{\gamma,V} : PX \to PX$ denotes the monotone map defined by $(\varphi)_q^{\gamma,V}(Y) = (\!|\varphi|\!)_{\gamma,V'}$ with $V'(p) = V(p)$ for $p \in \mathcal{V}$, $V'(q) = Y$ and $V'(r) = V(r)$ for $r \in \mathcal{U}$, $r \neq q$, whereas $\mathsf{lfp}(\_)$ *and* $\mathsf{gfp}(\_)$ *construct least and respectively greatest fixpoints.*

Thus, a maximal execution satisfies a state formula Φ (regarded as a path formula) precisely when the first state of that execution (obtained by applying $\pi_1 \circ \zeta_X : Z_X \to X$) satisfies Φ. The F-coalgebra structure $\pi_2 \circ \zeta_X : Z_X \to FZ_X$ on the set of maximal executions is used to define when an execution satisfies the path formula $[\lambda_F]\varphi$. Finally, to obtain the set of states satisfying the state formula $[\lambda]\varphi$, with φ a path formula, one uses the map $PZ_X \xrightarrow{(\lambda)_{Z_X}} PTZ_X \xrightarrow{\mathsf{Pexec}_\gamma} PX$ to go from a set of maximal executions (those satisfying φ) to a set of states.

In what follows, we restrict attention to coalgebras over Set and take $U = \mathsf{Id}$ and $P = \hat{\mathcal{P}}$. Moreover, we only consider the case $T = \mathcal{P}^+$ and take $\Lambda = \{\lambda_\Box\}$ with $\lambda_\Box : \hat{\mathcal{P}} \Longrightarrow \hat{\mathcal{P}} \circ \mathcal{P}^+$ given by $(\lambda_\Box)_X(Y) = \mathcal{P}Y$. In this case, one can consider a fragment of the above language where atomic propositions $p \in \mathcal{V}$ are regarded as path formulas (and hold on a path if they hold in its first state), and where state formulas can only be of the form $\Box\varphi$ with φ a path formula, and can not be viewed as path formulas. This yields what we call *linear* coalgebraic temporal logics for $\mathcal{P}^+ \circ F$-coalgebras, where the linearity pertains to the branching structure arising from the presence of $\mathcal{P}^+$: the state formula $\Box\varphi$ requires *all* maximal executions from a particular state to satisfy the path formula φ. The syntax of such a fragment is thus parameterised by a choice Λ_F of monotone predicate liftings for the functor F, and can be described as follows:

$$\varphi ::= \mathsf{tt} \mid \mathsf{ff} \mid q \mid p \mid \varphi \wedge \varphi \mid \varphi \vee \varphi \mid [\lambda_F]\varphi \mid \eta q.\varphi$$
$$\Phi ::= \Box\varphi$$

The semantics inherited by this fragment provides, for each pair consisting of a $\mathcal{P}^+ \circ F$-coalgebra (X, γ) and a 2-sorted valuation $V : (\mathcal{U}, \mathcal{V}) \to (PZ_X, PX)$, an interpretation of path formulas as subsets of Z_X, and an interpretation of state formulas as subsets of X. In particular, the interpretation of atomic propositions $p \in \mathcal{V}$ is given by $(\!|p|\!)_{\gamma,V} = P(\pi_1 \circ \zeta_X)(V(p))$.

Two observations are now worth making. Firstly, the use of the final $X \times F$-coalgebra (Z_X, ζ_X) in defining the semantics of path formulas is not needed anymore: since state formulas cannot be viewed as path formulas, the final F-coalgebra would suffice. Secondly, the syntax of path formulas is that of the coalgebraic μ-calculus considered in [3], after regarding propositional variables in $\mathcal{V}$ as nullary modal operators. Based on these observations, the above syntactic fragment can be given an equivalent semantics that makes direct use of the semantics of the coalgebraic μ-calculus (as given in [3]). This is achieved by regarding a pair consisting of a $\mathcal{P}^+ \circ F$-coalgebra with carrier X and a valuation of type $\mathcal{V} \to \mathcal{P}X$ as a coalgebra of $\mathcal{P}^+(\mathcal{P}\mathcal{V} \times F)$, and by viewing the elements of $\mathcal{V} \cup \Lambda_F$ as predicate liftings for the functor $\mathcal{P}\mathcal{V} \times F$: an atomic proposition $p \in \mathcal{V}$ yields a nullary predicate lifting $\lambda_p : 1 \Longrightarrow \hat{\mathcal{P}}(\mathcal{P}\mathcal{V} \times F)$ for $\mathcal{P}\mathcal{V} \times F$ given by $(\lambda_p)_X = \{(P, y) \in \mathcal{P}\mathcal{V} \times FX \mid p \in P\}$, whereas a predicate lifting $\lambda_F : \hat{\mathcal{P}} \Longrightarrow \hat{\mathcal{P}}F$ for F yields the predicate lifting $\hat{\mathcal{P}}\pi_2 \circ \lambda_F : \hat{\mathcal{P}} \Longrightarrow \hat{\mathcal{P}}(\mathcal{P}\mathcal{V} \times F)$ for $\mathcal{P}\mathcal{V} \times F$.

The next result gives an equivalent semantics for the linear fragments of path-based coalgebraic temporal logics, solely in terms of maximal trace maps:

Proposition 4. *Let $(Z_{\mathcal{P}\mathcal{V}}, \zeta_{\mathcal{P}\mathcal{V}})$ denote the final $\mathcal{P}\mathcal{V} \times F$-coalgebra, let (X, γ) be an arbitrary $\mathcal{P}\mathcal{V} \times (\mathcal{P}^+ \circ F)$-coalgebra (incorporating a $\mathcal{P}^+ \circ F$-coalgebra and a valuation of type $\mathcal{V} \to \mathcal{P}(X)$), and let $(X, \tilde{\gamma})$ be the $\mathcal{P}^+ \circ (\mathcal{P}\mathcal{V} \times F)$-coalgebra given by $\mathsf{st}_{\mathcal{P}\mathcal{V},F} \circ \gamma$, where $\mathsf{st}_{\mathcal{P}\mathcal{V},F} : \mathcal{P}\mathcal{V} \times (\mathcal{P}^+ \circ F) \Longrightarrow \mathcal{P}^+ \circ (\mathcal{P}\mathcal{V} \times F)$ is the strength of the monad $\mathcal{P}^+$:*

$$X \xrightarrow{\gamma} \mathcal{P}\mathcal{V} \times \mathcal{P}^+(FX) \xrightarrow{\mathsf{st}_{\mathcal{P}\mathcal{V},FX}} \mathcal{P}^+(\mathcal{P}\mathcal{V} \times FX)$$

Then $x \models_\gamma \Box\varphi$ iff $z \models_{\zeta_{\mathcal{P}\mathcal{V}}} \varphi$ for all $z \in \mathsf{tr}_{\tilde{\gamma}}(x)$, where $\models_{\zeta_{\mathcal{P}\mathcal{V}}}$ is the satisfaction relation between states of $(Z_{\mathcal{P}\mathcal{V}}, \zeta_{\mathcal{P}\mathcal{V}})$ and variable-free formulas of the coalgebraic μ-calculus induced by $\mathcal{V} \cup \Lambda_F$, and $\mathsf{tr}_{\tilde{\gamma}} : X \to \mathcal{P}^+ Z_{\mathcal{P}\mathcal{V}}$ is the trace map of $(X, \tilde{\gamma})$.

Example 5. 1. Let $F = A \times \mathsf{Id}$ and $\Lambda_F = \{a \mid a \in A\} \cup \{\bigcirc\}$, with the predicate liftings $\lambda_a : 1 \Longrightarrow \hat{\mathcal{P}} \circ (A \times \mathsf{Id})$ for $a \in A$ and $\lambda_\bigcirc : \hat{\mathcal{P}} \Longrightarrow \hat{\mathcal{P}} \circ (A \times \mathsf{Id})$ being given by $(\lambda_a)_X(*) = \{a\} \times X$ and $(\lambda_\bigcirc)_X(Y) = A \times Y$. This yields a linear temporal logic for $\mathcal{P}^+ \circ (A \times \mathsf{Id})$-coalgebras, with, for example, the formula $\Box\, \mu X.(a \vee \bigcirc X)$ stating that the label a occurs along every trace of a pointed $\mathcal{P}^+ \circ (A \times \mathsf{Id})$-coalgebra.

2. For $F = 1 + A \times \mathsf{Id}$, one can obtain a logic that also talks about termination by extending a variant of the previous set of predicate liftings with the nullary predicate lifting $\lambda_\bot : 1 \Longrightarrow \hat{\mathcal{P}} \circ (1 + A \times \mathsf{Id})$ given by $(\lambda_\bot)_X(*) = \{\iota_1(*)\}$. In the resulting logic, the formula $\Box\, \mu X.(\bot \vee \bigcirc X)$ states that all maximal executions of a pointed $\mathcal{P}^+ \circ (1 + A \times \mathsf{Id})$-coalgebra are finite ones.
3. For $F = A + \mathsf{Id} \times \mathsf{Id}$, a natural choice for Λ_F is the set $\{a \mid a \in A\} \cup \{[\pi_1], [\pi_2]\}$, with the predicate liftings $\lambda_a : 1 \Longrightarrow \hat{\mathcal{P}} \circ (A + \mathsf{Id} \times \mathsf{Id})$ for $a \in A$ and $\lambda_{\pi_1}, \lambda_{\pi_2} : \hat{\mathcal{P}} \Longrightarrow \hat{\mathcal{P}} \circ (A + \mathsf{Id} \times \mathsf{Id})$ being given by $(\lambda_a)_X(*) = \{\iota_1(a)\}$, $(\lambda_{\pi_1})_X(Y) = \iota_2(Y \times X)$ and $(\lambda_{\pi_2})_X(Y) = \iota_2(X \times Y)$. In the resulting temporal logic, the formula $\Box\, \mu X.(a \vee [\pi_1]\, X \vee [\pi_2]\, X)$ requires all maximal traces of a pointed $\mathcal{P}^+ \circ (A + \mathsf{Id} \times \mathsf{Id})$-coalgebra to have a leaf labelled by a.

Linear coalgebraic temporal logics can thus be viewed as coalgebraic generalisations of Linear Temporal Logic [7]. In what follows, we consider $\mathcal{P}^+ \circ F$-coalgebras instead of $\mathcal{P}^+ \circ (\mathcal{P}\mathcal{V} \times F)$-coalgebras, and assume that the interpretation for any atomic propositions has already been incorporated into the functor F.

Coalgebra Automata. We now recall the various coalgebraic notions of automaton that are required for the subsequent development. The definitions that follow assume that the functor $F : \mathsf{Set} \to \mathsf{Set}$ preserves weak pullbacks.

Definition 6 ([5]). *A* (parity) alternating F-automaton *is a tuple (A, a_0, δ, Ω) with A a finite set of* states, *a_0 the* initial state, *$\delta : A \to \mathcal{P}\mathcal{P}FA$ the* transition function *and $\Omega : A \to \omega$ a* parity map. *If for all $a \in A$, all elements of $\delta(a)$ are singletons, the automaton is called* non-deterministic. *Also, if for all $a \in A$, $\delta(a)$ is of the form $\{\{\varphi_a\}\}$ for some $\varphi_a \in FA$, the automaton is called* deterministic.

Definition 7 ([5]). *Given an F-automaton $\mathbb{A} = (A, a_0, \delta, \Omega)$ and an F-coalgebra $\mathbb{S} = (S, \gamma)$, the* acceptance game $\mathcal{G}(\mathbb{A}, \mathbb{S})$ *is the parity game defined by:*

Position: b	*Player: P(b)*	*Admissible moves: E[b]*	*Priority*
$(s, a) \in S \times A$	$\exists$	$\{(s, \Phi) \in S \times \mathcal{P}(FA) \mid \Phi \in \delta(a)\}$	$\Omega(a)$
$(s, \Phi) \in S \times \mathcal{P}(FA)$	$\forall$	$\{(s, \varphi) \in S \times FA \mid \varphi \in \Phi\}$	0
$(s, \varphi) \in S \times FA$	$\exists$	$\{(Z \subseteq S \times A \mid (\gamma(s), \varphi) \in \mathsf{Rel}(F)(Z)\}$	0
$Z \in \mathcal{P}(S \times A)$	$\forall$	Z	0

The automaton $\mathbb{A}$ *is said to* accept a pointed coalgebra $(\mathbb{S}, s_0)$ *if* (s_0, a_0) *is a winning position for* $\exists$ *in* $\mathcal{G}(\mathbb{A}, \mathbb{S})$.

Positions $(s, a) \in S \times A$ of the acceptance game are called *basic* positions. Here, the aim of $\exists$ is to show that the state s of the coalgebra fits the description provided by the state a of the automaton. The game also contains intermediary positions of three other types. Before arriving at a basic position again, $\exists$ must provide a *witnessing relation* $Z \subseteq S \times A$ in a position of type $S \times FA$.

The game $\mathcal{G}(\mathbb{A}, \mathbb{S})$ is closely related to the *bisimulation game* of [1].

Definition 8 ([1]). *Let* $\mathbb{S} = (S, \gamma)$ *and* $\mathbb{S}' = (S', \gamma')$ *be two* F*-coalgebras. The* bisimulation game $\mathcal{B}(\mathbb{S}, \mathbb{S}')$ *is the graph game defined by:*

Position: b	*Player:* $P(b)$	*Admissible moves:* $E[b]$
$(s, s') \in S \times S'$	$\exists$	$\{Z \subseteq S \times S' \mid (\gamma(s), \gamma(s')) \in \mathsf{Rel}(F)(Z)\}$
$Z \subseteq S \times S'$	$\forall$	Z

Finite plays of $\mathcal{B}(\mathbb{S}, \mathbb{S}')$ *are lost by the player who can not move, whereas infinite plays are won by* $\exists$.

[1] shows that the bisimilarity of two states $s \in S$ and $s' \in S'$ is equivalent to the existence of a winning strategy for $\exists$ from position (s, s') in $\mathcal{B}(\mathbb{S}, \mathbb{S}')$.

For a *deterministic* automaton $\mathbb{A}$, the acceptance game $\mathcal{G}(\mathbb{A}, \mathbb{S})$ is essentially the bisimulation game $\mathcal{B}(\mathbb{S}, \mathbb{A}')$, where $\mathbb{A}' = (A, \alpha, a_0)$ is the F-coalgebra given by $\alpha(a) = \varphi_a$ for $a \in A$. However, for arbitrary alternating automata, $\exists$ and $\forall$ must play a small sub-game in each round starting in a basic position, in order to arrive at a position where $\exists$ must provide a witnessing relation Z. The first occurrence of the powerset functor in the codomain of the transition function of an alternating automaton indicates a choice for $\exists$ in this sub-game, whereas the second occurrence of $\mathcal{P}$ indicates a choice for $\forall$. In the case of non-deterministic automata, $\forall$ has no real choice while playing the previously-mentioned sub-game.

Remark 9. Any finite, pointed F-coalgebra $\mathbb{S} = (S, \gamma, s_0)$ can be turned into an F-automaton $\mathbb{A}_{\mathbb{S}, s_0} = (S, s_0, \Delta_\gamma, \Omega_0)$, where $\Delta_\gamma(s) = \{\{\gamma(s)\}\}$ and $\Omega_0(s) = 0$ for $s \in S$. Moreover, this automaton accepts a pointed F-coalgebra (D, δ, d_0) if and only if the states s_0 and d_0 are bisimilar (see e.g. [8] for details).

Automata that account for the satisfaction of coalgebraic μ-calculus formulas by states of F-coalgebras were considered in [3]. For a set Λ of monotone predicate liftings for F, Λ*-automata* differ from F-automata in that their transition functions have type $A \to \mathcal{L}_0\Lambda(A)$, with $\mathcal{L}_0(X)$ the set of lattice terms over X. In the acceptance game for a Λ-automaton, the aim of $\exists$ is to show that a pointed coalgebra satisfies the formula encoded by the automaton. Thus, occurrences of conjunction in the lattice terms used to define the transition function of a Λ-automaton correspond to choices of $\forall$ in the associated acceptance game, whereas occurrences of disjunction correspond to choices of $\exists$. These choices are, however, made implicit by the presentation of the acceptance game given below.

Definition 10 ([3]). *Let Λ be a set of monotone predicate liftings for F. A* Λ-automaton *is a tuple (A, a_0, δ, Ω) with A a finite set of* states, *a_0 the* initial state, *$\delta : A \to \mathcal{L}_0\Lambda(A)$ the* transition function *and $\Omega : A \to \omega$ a parity map.*

The following is a slight reformulation of the acceptance game for a Λ-automaton:

Definition 11 ([3]). *Given a Λ-automaton $\mathbb{A} = (A, a_0, \delta, \Omega)$ and an F-coalgebra $\mathbb{S} = (S, \gamma)$, the* acceptance game *$\mathcal{G}(\mathbb{A}, \mathbb{S})$ is the parity game defined by:*

Position: b	*Player: P(b)*	*Admissible moves: E[b]*	*Priority*
$(s, a) \in S \times A$	$\exists$	$\{(Z \subseteq S \times A \mid \gamma(s) \in [\![\delta(a)]\!]^1_Z\}$	$\Omega(a)$
$Z \in \mathcal{P}(S \times A)$	$\forall$	Z	0

where for $Z \subseteq S \times A$ and $\varphi \in \mathcal{L}_0\Lambda(A)$, $[\![\varphi]\!]^1_Z \in \mathcal{P}(FS)$ is defined inductively by

$$[\![[\lambda](a_1, \ldots, a_n)]\!]^1_Z = \lambda_S([\![a_1]\!]_Z, \ldots, [\![a_n]\!]_Z)$$

and the usual clauses for conjunction and disjunction. Here, $[\![a]\!]_Z \in \mathcal{P}S$ is given by $[\![a]\!]_Z := \{s \in S \mid (s, a) \in Z\}$, for $a \in A$. The automaton $\mathbb{A}$ is said to accept a pointed coalgebra *$(\mathbb{S}, s_0)$* if *(s_0, a_0) is a winning position for $\exists$ in $\mathcal{G}(\mathbb{A}, \mathbb{S})$.*

As shown in [3], any coalgebraic μ-calculus formula φ can be mapped to a Λ-automaton accepting precisely the pointed F-coalgebras that satisfy φ.

3 Automata-Based Characterisation of Maximal Traces

We now provide an automata-theoretic characterisation of the set of maximal traces of a pointed $\mathcal{P}^+ \circ F$-coalgebra (X, γ, x_0) with finite carrier, by defining a non-deterministic F-automaton that accepts exactly those traces (elements of the final F-coalgebra) which belong to $\mathsf{tr}_\gamma(x_0)$. The construction of this automaton is straightforward: we simply regard the coalgebra map γ as the transition function of a *non-deterministic* F-automaton with a trivial parity map. As a result, for each trace of x_0, $\exists$'s moves in basic positions (z, x) of the acceptance game for this F-automaton can be chosen so as to select precisely the values $y \in \gamma(x)$ that generate the given trace.

In what follows, F is assumed to be a polynomial and standard functor (hence pullback and non-empty intersection preserving), and $\lambda : F\mathcal{P}^+ \Longrightarrow \mathcal{P}^+F$ is assumed to be as in (2). Both restrictions are required in order to make use of the alternative definition of the trace map of a $\mathcal{P}^+ \circ F$-coalgebra given in (3).

Definition 12. *The* trace automaton induced by a pointed $\mathcal{P}^+ \circ F$-coalgebra *(X, γ, x_0) is the non-deterministic F-automaton $\mathbb{X}_\gamma = (X, \delta_\gamma, x_0, \Omega_0)$ where:*

- *$\delta_\gamma : X \to \mathcal{P}\mathcal{P}FX$ is given by $\delta_\gamma(x) = \{\{y\} \mid y \in \gamma(x)\}$ for $x \in X$,*
- *$\Omega_0 : X \to \omega$ is given by $\Omega_0(x) = 0$ for $x \in X$.*

This definition should be compared with that of the automaton in Remark 9 – the difference is that the non-determinism provided by the coalgebra γ is now used to provide a *non-deterministic F-automaton*, as opposed to a *deterministic $\mathcal{P}^+ \circ F$-automaton.* Our first main result can now be stated as follows.

Theorem 13. *Let (X, γ, x_0) be a pointed $\mathcal{P}^+ \circ F$-coalgebra, and let (Z, ζ) denote the final F-coalgebra. Then, for $z_0 \in Z$, $z_0 \in \mathsf{tr}_\gamma(x_0)$ iff $\mathbb{X}_\gamma$ accepts the pointed coalgebra (Z, ζ, z_0).*

To prove this result we need the following definition.

Definition 14. *For $y \in FY$, let*

$$\mathsf{Base}(y) := \bigcap \{Y' \subseteq Y \mid y \in FY'\}$$

For $(z, y) \in FZ \times FY$ such that $(z, y) \in \mathsf{Rel}(F)(R)$ for some $R \subseteq Z \times Y$, let

$$\mathsf{Base}(z, y) := \bigcap \{R \subseteq Z \times Y \mid (z, y) \in \mathsf{Rel}(F)(R)\}$$

The preservation of non-empty intersections by F gives $y \in F\mathsf{Base}(y)$. Thus, $\mathsf{Base}(y)$ is the smallest set Y' with the property that $y \in FY'$. Similarly, one can show by induction on the structure of the polynomial functor F that $\mathsf{Rel}(F)$ preserves non-empty intersections. As a result, whenever $\mathsf{Base}(z, y)$ exists, it is the smallest relation R with the property that $(z, y) \in \mathsf{Rel}(F)(R)$. We now return to the proof of Theorem 13 which, due to space limitations, is only outlined here.

Proof (sketch). For the *if* direction, let $\mathcal{S}$ denote a winning strategy for $\exists$ in $\mathcal{G}(\mathbb{X}_\gamma, (Z, \zeta, z_0))$, and assume w.l.o.g. that $\mathcal{S}$ prescribes the smallest possible witnessing relations, namely $\mathsf{Base}(\zeta(z), y)$, in positions $(z, y) \in Z \times FX$. The $\mathcal{S}$-conform $\mathcal{G}(\mathbb{X}_\gamma, (Z, \zeta, z_0))$-plays can now be visualised as branches of the following tree (where positions in which $\forall$ has no choice have been omitted):

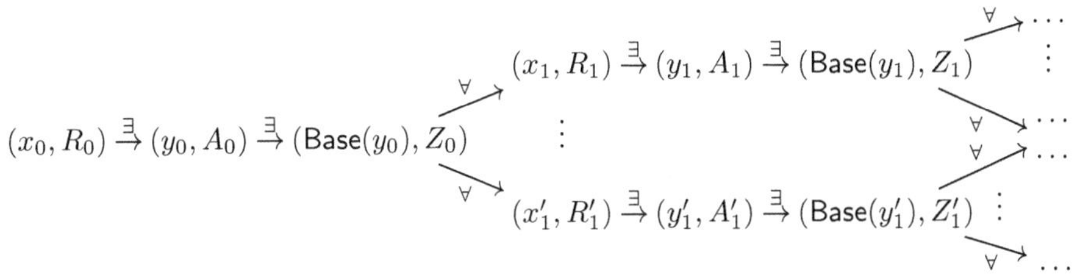

$\exists$'s moves provided by $\mathcal{S}$ can then be used to define an F-coalgebra structure $\xi : D \to FD$ on the set D of basic positions reachable from (z_0, x_0) through an $\mathcal{S}$-conform $\mathcal{G}(\mathbb{X}_\gamma, (Z, \zeta, z_0))$-play. Moreover, the winning strategy $\mathcal{S}$ can be mirrored in the bisimulation game $\mathcal{B}((Z, \zeta, z_0), (D, \xi, (z_0, x_0)))$, by letting $\exists$ move in a position $(z, (z, x))$ to the position $\{(z', (z', x')) \mid (z', x') \in \mathsf{Base}(\zeta(z), y)\}$, where (z, y) is the move prescribed by $\mathcal{S}$ in position (z, x). This yields a winning strategy for $\exists$ in the bisimulation game, and hence $(Z, \zeta, z_0) \sim (D, \xi, (z, x_0))$. Using this and the definition of λ given in (2), once can show that $\pi_i(z_0) \in \gamma_i(x_0)$ for $i \in \omega$, which finally yields $z_0 \in \mathsf{tr}_\gamma(x_0)$ by the definition of tr_γ in (1).

For the *only if* direction one can prove more generally that, for $x \in X$ and $z \in \mathsf{tr}_\gamma(x)$, $\exists$ has a (history-free) winning strategy from position (z, x) in $\mathcal{G}(\mathbb{X}_\gamma, (Z, \zeta, z_0))$. The key idea is to make use of the following property of the trace map, as proved in [4]:

$$\{\zeta(z) \mid z \in \mathsf{tr}_\gamma(x)\} = \bigcup \{\lambda_Z((F\mathsf{tr}_\gamma)(y)) \mid y \in \gamma(x)\}$$

for each $x \in X$. Using the above, $z \in \mathsf{tr}_\gamma(x)$ yields some $y \in \gamma(x)$ such that $\zeta(z) \in \lambda_Z((F\mathsf{tr}_\gamma)(y))$. This, in turn, yields a suitable choice of a trace $z_{x'} \in \mathsf{tr}_\gamma(x')$ for each $x' \in \mathsf{Base}(y)$. Our winning strategy now requires $\exists$ to move in position (z, x) to (z, y) and then immediately to the relation $\{(z_{x'}, x') \mid x' \in \mathsf{Base}(y)\}$. Clearly this guarantees that $\exists$ will never be stuck in a play, and since all infinite $\mathcal{G}(\mathbb{X}_\gamma, (Z, \zeta, z_0))$-plays are won by $\exists$, it follows that the proposed strategy is a winning strategy for $\exists$ in $\mathcal{G}(\mathbb{X}_\gamma, (Z, \zeta, z_0))$. □

Remark 15. Under the assumption that the coalgebra (X, γ) contains no duplicates (see Definition 16), an alternative proof of the previous result can be given by showing that the coalgebra (D, ξ) constructed in the above proof is a sub-coalgebra of the coalgebra (U, σ) defined in Section 2.

4 Model Checking Linear Coalgebraic Temporal Logics

This section defines a regular graph game, called the *non-emptiness game*, for deciding whether a formula $\Box\varphi$ of a linear coalgebraic temporal logic (induced by a set Λ of predicate liftings for F) holds in a finite, pointed $\mathcal{P}^+ \circ F$-coalgebra $\mathbb{X}$. We represent the *negation* of the formula φ as a Λ-automaton[4], and use the non-deterministic F-automaton induced by $\mathbb{X}$ as a description of its traces. The game we define has the property that winning strategies of $\exists$ correspond to traces of $\mathbb{X}$ that are accepted by the Λ-automaton. Thus, the existence of a winning strategy for $\exists$ is equivalent to the formula $\Box\varphi$ not holding in $\mathbb{X}$, while the winning strategy itself provides a counter-example for the statement $\mathbb{X} \models \Box\varphi$.

Although such a game can be defined for any pointed $\mathcal{P}^+ \circ F$-coalgebra, the game has a much simpler presentation for coalgebras *with no duplicates*. We will therefore only define the game for such coalgebras, and show that any $\mathcal{P}^+ \circ F$-coalgebra with finite carrier can be transformed into a $\mathcal{P}^+ \circ F$-coalgebra with no duplicates, and whose carrier is still finite. The next definition formalises the idea of a state x occurring more than once in some $y \in FX$.

Definition 16. *A $\mathcal{P}^+ \circ F$-coalgebra (X, γ)* contains duplicates *if there exist $u \in X$, $y \in \gamma(u)$, $x \in \mathsf{Base}(y)$ and $y' \in F(X + 1)$ such that:*

- $\iota_1(x), \iota_2(*) \in \mathsf{Base}(y')$,
- $y = F[1_X, x](y')$ *(where $x : 1 \to X$ maps $*$ to x).*

We then call $y \in FX$ a duplicate type of *(X, γ), and x* a duplicate state of *y.*

Thus, an F-coalgebra (X, γ) contains duplicates if there exist $x, u \in X$ and $y \in \gamma(u)$ such that x occurs at least twice in y.

Example 17. Let $F = \{0, 1\} + \mathsf{Id} \times \mathsf{Id}$. The elements of the final F-coalgebra are (possibly infinite) binary trees with leaves labelled by either 0 or 1. Now let (X, γ) be the $\mathcal{P}^+ \circ F$-coalgebra given by $X = \{x_0, x\}$, $\gamma(x_0) = \{(x, x)\}$, $\gamma(x) = \{0, 1\}$. In this case, (x, x) is a duplicate type of (X, γ), and x is a duplicate state of (x, x). We also note in passing that in this case, $\mathsf{tr}_\gamma(x) = \{0, 1\}$, whereas $\mathsf{tr}_\gamma(x_0)$ consists of all four binary trees of depth 1 with leaves labelled by either 0 or 1.

[4] We assume that Λ contains enough predicate liftings to encode negations of formulas.

Lemma 18. *Assume that F preserves finite sets*[5]*. There exists an effective procedure translating any $\mathcal{P}^+ \circ F$-coalgebra (X, γ) with finite carrier into a $\mathcal{P}^+ \circ F$-coalgebra (X', γ') with finite carrier and no duplicates, and with a surjective $\mathcal{P}^+ \circ F$-coalgebra homomorphism $\pi : (X', \gamma') \to (X, \gamma)$.*

The next definition will allow the formulation of the winning condition for the non-emptiness game.

Definition 19. *Let $\Omega : A \to \omega$ be a parity map. A* trace through a sequence of relations *$(R_i)_{i\in\omega}$ with $R_i \subseteq A \times A$ for $i \in \omega$ is an infinite sequence $(a_i)_{i\in\omega} \in A^\omega$ such that $(a_i, a_{i+1}) \in R_i$ for each $i \in \omega$. A trace $(a_i)_{i\in\omega}$ through $(R_i)_{i\in\omega}$ is called* bad w.r.t. *Ω if $max\{k \mid k = \Omega(a_i)$ for infinitely many $i \in \omega\}$ is odd.*

Definition 20. *Let $\mathbb{X} = (X, \gamma, x_0)$ be a pointed $\mathcal{P}^+ \circ F$-coalgebra with no duplicates, and let $\mathbb{A} = (A, a_0, \delta, \Omega)$ be a Λ-automaton. The* non-emptiness game *$\mathcal{G}_{\neq\emptyset}(\mathbb{X}, \mathbb{A})$ is the graph game defined by:*

Position	*Player*	*Admissible moves*
$(x, R) \in X \times \mathcal{P}(A \times A)$	$\exists$	$\{(y, \mathsf{ran}(R)) \in FX \times \mathcal{P}A \mid y \in \gamma(x)\}$
$(y, A') \in FX \times \mathcal{P}A$	$\exists$	$\{(\mathsf{Base}(y), Z) \mid Z : A' \to \mathcal{P}(\mathsf{Base}(y) \times A)$ *s.t.* $y \in [\![\delta(a)]\!]_{Z_a}$ *for* $a \in A'\}$
$(B, Z) \in \mathcal{P}(X) \times \mathcal{P}(X \times A)^{A'}$	$\forall$	$\{(x, \bigcup_{a\in A'}\{(a, a') \mid (x, a') \in Z_a\}) \mid x \in B\}$

with initial position $(x_0, \{(a_0, a_0)\})$, where $\mathsf{ran}(R)$ *denotes the range of the relation R. An infinite match with basic positions $(x_0, R_0), (x_1, R_1), \ldots$ is won by $\exists$ if and only if no trace through the sequence of relations $R_0 R_1 \ldots$ is bad w.r.t. Ω.*

The basic positions of $\mathcal{G}_{\neq\emptyset}(\mathbb{X}, \mathbb{A})$ are given by pairs consisting of a state $x \in X$ and a *relation* $R \subseteq A \times A$, whereas the witnessing relations for positions of type (y, A') can be regarded as families of functions of type $Z_a : X \to \mathcal{P}(A)$ (one for each $a \in A'$). To explain the reasons behind this definition and the assumption that $\mathbb{X}$ has no duplicates, let us imagine that the initial position of $\mathcal{G}_{\neq\emptyset}(\mathbb{X}, \mathbb{A})$ was (x_0, a_0) (the obvious choice, given that $\exists$'s goal in this game is to prove that x_0 admits a trace that "satisfies the formula a_0"). In this position, $\exists$ would have to provide a choice of $y_0 \in \gamma(x_0)$ that "makes the formula a_0 true". This move would be followed by $\exists$ providing a witnessing relation $Z \subseteq X \times A$ for (y_0, a_0). However, the play would not be able to continue with $\forall$ choosing an element of this relation, which would then result in a new basic position (x_1, a_1), since in situations where *several* pairs in Z have the *same* first component $x_1 \in X$, the second components of those pairs may or may not need to be satisfied by the same trace of x_1. If x_1 is not a duplicate state of y_0, it is clear that a *single* trace of x_1 should make all formulas a_1 with $(x_1, a_1) \in Z$ true. If, on the other hand, x_1 is a duplicate state of y_0, different occurrences of x_1 could (and might need to) be unfolded in different ways in order to satisfy a_0 in y_0. For instance, given the pointed coalgebra (X, γ, x_0) of Example 17, a Λ-automaton could be

[5] This further restricts polynomial functors by only allowing finite exponents and finite constant functors. However, the restriction to finite constant functors is superfluous.

devised that accepts only traces with both 0 and 1 as leaves, and to obtain a trace of x_0 with this property, different traces of x would need to be considered for the two occurrences of x in $\gamma(x_0)$. In the presence of duplicate states, the witnessing relation Z does not provide sufficient information to decide which formulas should hold for which unfoldings. On the other hand, by assuming that (X, γ) has no duplicates, it becomes much easier to define $\exists$'s possible moves in positions given by witnessing relations Z: *all* pairs in Z with the same first component x_1 must be witnessed by the *same* choice of $y_1 \in \gamma(x_1)$. Now to accommodate this, $\exists$'s moves in basic positions would have to be of type $Z : X \to \mathcal{P}(A)$. However, to define a winning condition for infinite games, we would have to assign priorities to basic positions of type (x, A') with $A' \in \mathcal{P}(A)$, which is not possible in a meaningful way. We therefore take the fairly standard approach of using traces through sequences of relations for defining the winning condition. This leads to basic positions of type $X \times \mathcal{P}(A \times A)$, where in a position (x, R), $\exists$'s goal is to prove true all formulas in $\mathsf{ran}(R)$. Finally, the presence of $\mathsf{Base}(y)$ in the definition of admissible moves for $\exists$ in positions (y, A') is justified by the fact that only pairs (x, a') with $x \in \mathsf{Base}(y)$ are relevant to the satisfaction by $y \in FX$ of the formula represented by a.

Remark 21. A version of the non-emptiness game that also applies to coalgebras with duplicates could be defined by moving from witnessing relations of type $Z_a \in \mathcal{P}(X \times A)$ (or $Z_a : X \to \mathcal{P}(A)$) to witnessing relations of type $Z : X \to \mathcal{P}(\mathcal{P}(A))$, where elements of A that must be simultaneously satisfied on some trace of x are grouped appropriately in $Z_a(x)$. However, the conditions specifying that Z is a witnessing relation would become much more complex.

The non-emptiness game proceeds as follows:

- in a basic position (x, R) (in which $\exists$ must show that some trace of x makes all formulas in $\mathsf{ran}(R)$ true), $\exists$ chooses $y \in \gamma(x)$ and moves to $(y, \mathsf{ran}(R))$;
- in a position (y, A') (in which $\exists$ must show that a suitable choice of trace for each $x \in \mathsf{Base}(y)$ makes all formulas in A' true), $\exists$ provides a suitable witnessing relation $Z_a \subseteq X \times A$ for each formula $a \in A'$;
- in a position (B, Z) with $Z : A' \to \mathcal{P}(B \times A)$, $\forall$ chooses some $x \in B$ and collects all second components of pairs (x, a') in one of the Z_as – these formulas must all be satisfied by the *same* trace of x. The resulting position records the corresponding a for a pair (x, a'), to be used in the formulation of the winning condition. The game is now again in a basic position.

The game $\mathcal{G}_{\neq\emptyset}(\mathbb{X}, \mathbb{A})$ is ω-regular: there exists a parity $\mathcal{P}(A \times A)$-word automaton accepting exactly those sequences of relations which do not contain a bad trace.

The second main result of the paper now states that winning strategies for $\exists$ in $\mathcal{G}_{\neq\emptyset}(\mathbb{X}, \mathbb{A})$ correspond to traces of $\mathbb{X}$ that are accepted by $\mathbb{A}$. This is proved with the help of the following lemma.

Lemma 22. *Let (X, γ) be a $\mathcal{P}^+ \circ F$-coalgebra with no duplicates, and let (U, σ) be the F-coalgebra defined in Section 2. Then, states $u \in U$ are in one-to-one correspondence with infinite trees of the following shape:*

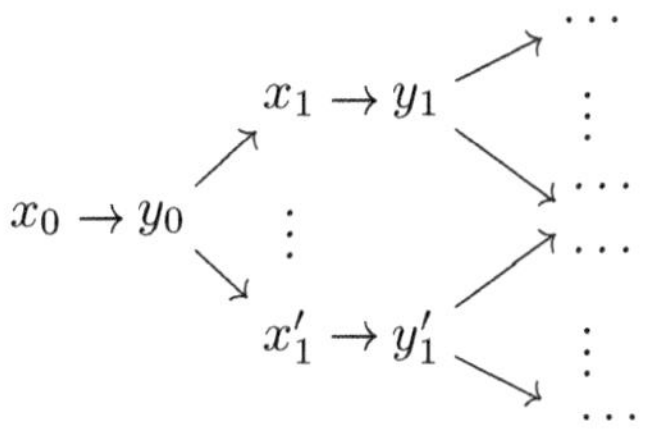

$$(4)$$

with $y_i \in \gamma_i(x_i)$ and with one child of y_i for each $x_{i+1} \in \mathsf{Base}(y_i)$.

Theorem 23. *Let $\mathbb{X} = (X, \gamma, x_0)$ be a pointed $\mathcal{P}^+ \circ F$-coalgebra and let $\mathbb{A} = (A, a_0, \delta, \Omega)$ be a Λ-automaton. Then $\exists$ has a winning strategy in $\mathcal{G}_{\neq\emptyset}(\mathbb{X}, \mathbb{A})$ iff there exists a trace $z \in \mathsf{tr}_\gamma(x_0)$ such that (Z, ζ, z) is accepted by $\mathbb{A}$.*

Proof (sketch). For the *only if* direction, let $\mathcal{S}$ denote a winning strategy for $\exists$ in $\mathcal{G}_{\neq\emptyset}(\mathbb{X}, \mathbb{A})$, and observe that $\mathcal{S}$-conform $\mathcal{G}_{\neq\emptyset}(\mathbb{X}, \mathbb{A})$-plays can be visualised as branches of the following tree:

$$(x_0, R_0) \overset{\exists}{\to} (y_0, A_0) \overset{\exists}{\to} (\mathsf{Base}(y_0), Z_0) \quad \begin{matrix} \overset{\forall}{\to} (x_1, R_1) \overset{\exists}{\to} (y_1, A_1) \overset{\exists}{\to} (\mathsf{Base}(y_1), Z_1) \\ \vdots \\ \overset{\forall}{\to} (x_1', R_1') \overset{\exists}{\to} (y_1', A_1') \overset{\exists}{\to} (\mathsf{Base}(y_1'), Z_1') \end{matrix}$$

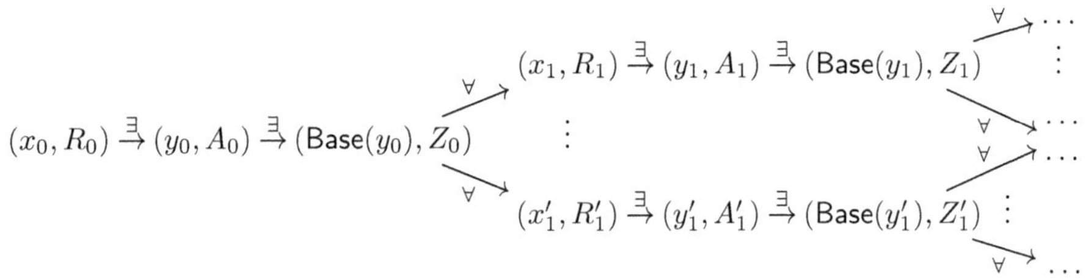

where each node of type $(\mathsf{Base}(y_i), Z_i)$ has exactly one child for each $x_{i+1} \in \mathsf{Base}(y_i)$. A subtree of this tree rooted in some (x_i, R_i) determines an infinite tree of the type required by Lemma 22, and this, in turn, yields a state $u_i \in U$. Moreover, u_0 describes the desired behaviour of a trace of x_0 that is accepted by the automaton $\mathbb{A}$, since the above tree provides winning strategies for $\exists$ in each of the acceptance games $\mathcal{G}(\mathbb{X}, (U, \sigma, u_0))$ and $\mathcal{G}(\mathbb{A}, (U, \sigma, u_0))$. On the one hand, $\mathcal{S}$-conform $\mathcal{G}_{\neq\emptyset}(\mathbb{X}, \mathbb{A})$-plays can be mirrored in the acceptance game $\mathcal{G}(\mathbb{X}, (U, \sigma, u_0))$ by letting $\exists$ move in a position (x_i, u_i) to the witnessing relation $\{(x_{i+1}, u_{i+1}) \mid x_{i+1} \in \mathsf{Base}(y_i)\}$, and this yields a winning strategy for $\exists$ in $\mathcal{G}(\mathbb{X}, (U, \sigma, u_0))$. On the other hand, $\mathcal{S}$-conform $\mathcal{G}_{\neq\emptyset}(\mathbb{X}, \mathbb{A})$-plays can be mirrored in the acceptance game $\mathcal{G}(\mathbb{A}, (U, \sigma, u_0))$: pairs consisting of a path through the above tree and a trace through the sequence of relations determined by that path correspond to $\mathcal{G}(\mathbb{A}, (U, \sigma, u_0))$-plays in which $\exists$ plays essentially the witnessing relations Z_{a_i} (but with x_i substituted by u_i) in positions of type (u_i, a_i) with $a_i \in A_i$. Moreover, the property that any such path is winning for $\exists$ in $\mathcal{G}_{\neq\emptyset}(\mathbb{X}, \mathbb{A})$ translates to any $\mathcal{G}(\mathbb{A}, (U, \sigma, u_0))$-play that is played according to the proposed strategy being winning for $\exists$ in $\mathcal{G}(\mathbb{A}, (U, \sigma, u_0))$. Thus, u_0 defines a trace of $\mathbb{X} = (X, \gamma, x_0)$ that satisfies the property described by the Λ-automaton $\mathbb{A}$.

For the *if* direction, note that a trace $z \in \mathsf{tr}_\gamma(x_0)$ such that (Z, ζ, z) is accepted by $\mathbb{A}$ yields an element $u_0 \in U$ that is accepted by $\mathbb{A}$ (as acceptance by $\mathbb{A}$

is invariant under bisimulation), as well as a winning strategy $\mathcal{S}$ for $\exists$ in the acceptance game $\mathcal{G}(\mathbb{A}, (U, \sigma, u_0))$. Lemma 22 can now be used to obtain an infinite tree similar to that in (4), which, in turn, yields a winning strategy $\mathcal{S}'$ for $\exists$ in $\mathcal{G}_{\neq\emptyset}(\mathbb{X}, \mathbb{A})$: the choices of $y_i \in \gamma(x_i)$ made in the definition of u_0 provide the choices required in basic positions (x_i, R_i) reached from $(x_0, \{(a_0, a_0)\})$ through $\mathcal{S}'$-conform plays, whereas in positions of the form $(y_i, A_i) \in FX \times \mathcal{P}A$ with $A_i = \mathsf{ran}(R_i)$, $\mathcal{S}'$ prescribes that $\exists$ moves to $(\mathsf{Base}(y_i), (Z_a)_{a \in A_i})$ with $Z_a \in \mathcal{P}(\mathsf{Base}(y_i) \times A)$ being obtained from $\exists$'s $\mathcal{S}$-conform move in the position (u_i, a) of $\mathcal{G}(\mathbb{A}, (U, \sigma, u_0))$, by replacing any $u_{i+1} \in \mathsf{Base}(u_i)$ with the corresponding x_{i+1}. The fact that $\mathcal{S}$ is winning for $\exists$ in $\mathcal{G}(\mathbb{A}, (U, \sigma, u_0))$ then results in $\mathcal{S}'$ being winning for $\exists$ in $\mathcal{G}_{\neq\emptyset}(\mathbb{X}, \mathbb{A})$. □

5 Concluding Remarks

We provided an automata-theoretic characterisation of the set of traces of a finite, pointed $\mathcal{P}^+ \circ F$-coalgebra, with F a polynomial endofunctor. Next, we defined a regular graph game that can be used to decide whether a formula of a linear coalgebraic temporal logic (also introduced in this paper) holds in a finite, pointed $\mathcal{P}^+ \circ F$-coalgebra that contains no duplicates.

Future work includes generalising these results to non-polynomial functors F, studying model-checking algorithms based on such regular games, and extending the techniques proposed here to more general coalgebraic types and path-based temporal logics, as considered in [2].

References

1. Baltag, A.: A logic for coalgebraic simulation. In: Proc. CMCS 2000. Electr. Notes Theor. Comput. Sci., vol. 33, pp. 42–60 (2000)
2. Cîrstea, C.: Generic infinite traces and path-based coalgebraic temporal logics. In: Proc. CMCS 2010. Electr. Notes Theor. Comput. Sci., vol. 264, pp. 83–103 (2010)
3. Fontaine, G., Leal, R., Venema, Y.: Automata for coalgebras: an approach using predicate liftings. In: Abramsky, S., Gavoille, C., Kirchner, C., Meyer auf der Heide, F., Spirakis, P.G. (eds.) ICALP 2010. LNCS, vol. 6199, pp. 381–392. Springer, Heidelberg (2010)
4. Jacobs, B.: Trace semantics for coalgebras. In: Proc. CMCS 2004. Electr. Notes Theor. Comput. Sci., vol. 106, pp. 167–184 (2004)
5. Kupke, C., Venema, Y.: Coalgebraic automata theory: basic results. Logical Methods in Computer Science 4(4) (2008)
6. Mazala, R.: 2 infinite games. In: Grädel, E., Thomas, W., Wilke, T. (eds.) Automata, Logics, and Infinite Games. LNCS, vol. 2500, pp. 23–38. Springer, Heidelberg (2002)
7. Vardi, M.: Automata-theoretic techniques fo temporal reasoning. In: Handbook of Modal Logic, pp. 971–990. Addison Wesley, Reading (2005)
8. Venema, Y.: Automata and fixed point logics: a coalgebraic perspective. Information and Computation 204, 637–678 (2006)

Refinement Trees: Calculi, Tools, and Applications

Mihai Codescu and Till Mossakowski

DFKI GmbH Bremen
{Mihai.Codescu,Till.Mossakowski}@dfki.de

Abstract. We recall a language for refinement and branching of formal developments. We introduce a notion of refinement tree and present proof calculi for checking correctness of refinements as well as their consistency. Both calculi have been implemented in the Heterogeneous Tool Set (Hets), and have been integrated with other tools like model finders and conservativity checkers. This technique has already been applied for showing the consistency of a first-order ontology that is too large to be tackled directly by model finders.

1 Introduction

Consider the task of providing an implementation (or finding a model) for a specification SP. The classical theory of refinement [2] provides the means for vertically decomposing this task into a sequence of refinement steps:

$$SP \rightsquigarrow SP_1 \rightsquigarrow \ldots \rightsquigarrow SP_n \rightsquigarrow P$$

Here, $SP_1, \ldots, SP_n$ are intermediate specifications and P is an implementation or a model description. For large specifications, this quickly becomes unmanageable. Horizontal decomposition allows the splitting of an implementation task into manageable subtasks [18] that can be implemented independently (by separate people). This leads to a refinement tree, with leaves being specifications that can be implemented directly.

$$SP \rightsquigarrow \begin{cases} SP_1 \rightsquigarrow P_1 \\ \vdots \\ SP_n \rightsquigarrow \begin{cases} SP_{n1} & \rightsquigarrow \quad \{ SP_{n11} \rightsquigarrow P_{n11} \\ \cdots & \\ SP_{nm} & \underset{\kappa_{nm}}{\rightsquigarrow} P_{nm} \end{cases} \end{cases}$$

This approach is not only applicable for formal software development (where the leaves of the tree are programs), but also for finding models of larger theories like upper ontologies. Note that the latter is a challenge: recently a contest for showing (in)consistency of the ontology SUMO has been set up [16]. We propose using refinement trees as a way for managing this task; the leaves of the trees

A. Corradini, B. Klin, and C. Cîrstea (Eds.): CALCO 2011, LNCS 6859, pp. 145–160, 2011.

then are small theories whose consistency can be proved using an automated model finder.

We recall a language for expressing refinements and branching of formal developments from [13] in Sect. 2 using an example application from [3]. Section 3 formally introduces refinement trees, and section 4 recalls the semantics of refinements and their parts. Section 5 contains our main contribution, a calculus for checking correctness of refinement trees. Moreover, we implement the calculus in a tool that also integrates model finders and other tools. Section 6 provides a calculus for consistency of (possibly architectural) refinements, which already has been successfully applied for showing the consistency of the DOLCE [7] upper ontology. Section 7 concludes the paper.

2 The CASL Refinement Language: Syntax

The Common Algebraic Specification Language CASL [15] has been designed by the "Common Framework Initiative for Algebraic Specification and Development" with the goal to unify the many previous algebraic specification languages and to provide a standard language for the specification and development of modular software systems. CASL has been designed in orthogonal layers:

1. **basic specifications** provide means for writing specifications of the individual software modules. The underlying logic of CASL is multi-sorted first-order logic with equality, partial functions and induction principles for datatypes;
2. **structured specifications** allow to organize large specifications in a modular way;
3. **architectural specifications** [4] describe, in contrast to the previous layer, the structure of the *implementation* of a software system, given by a (possible structured) specification of the requirements.
4. **libraries of specifications** allow storage and retrieval of named specifications.

The orthogonality of layers means that the syntax and semantics of each layer are independent of those of the others. In particular, this allows to replace the layer of CASL basic specifications with a completely different logic without having to modify the other layers. This is achieved in a mathematically sound way by using the formalization of the notion of logical system as institutions [8] and defining the semantics of structured and architectural specifications in an institution-independent way. CASL is supported by the Heterogeneous Tool Set (Hets) [14], that collects parsing, static analysis and proof management tools for *heterogeneous* specifications. Heterogeneity is achieved by formalizing logics as institutions and adding them to the graph of logics that parameterizes Hets. Moreover, Hets interfaces various theorem provers, model finders or consistency checkers.

Example 1. We will illustrate the CASL architectural and refinement language with the help of an industrial case study: specification of a steam boiler control system for controlling the water level in a steam boiler. The problem has been formulated in [1] as a benchmark for specification languages; [3] gives a complete solution using CASL, including architectural design and refinement of components. However, the refinement steps were presented in an informal way and only the refinement language of [13] makes it possible to formally write down the refinement steps using CASL refinement specifications.

The specifications involved can be briefly explained as follows. VALUE specifies in a very abstract way a sort *Value* and some operations and predicates on values. This specification can be regarded as a parameter of the entire design. PRELIMINARY gathers the messages in the system, both sent and received, and also defines a series of constants characterizing the steam boiler. SBCS_STATE introduces observers for the system states, while SBCS_ANALYSIS extends this to an analysis of the messages received, failure detection and computation of messages to be sent. Finally, STEAM_BOILER_CONTROL_SYSTEM specifies the initial state and the reachability relation between states. [1]

The initial design for the architecture of the system is recorded by the architectural specification below:

arch spec ARCH_SBCS =
  units P : VALUE → PRELIMINARY;
  S : PRELIMINARY → SBCS_STATE;
  A : SBCS_STATE → SBCS_ANALYSIS;
  C : SBCS_ANALYSIS → STEAM_BOILER_CONTROL_SYSTEM
  result λ *V* : VALUE • C [A [S [P [V]]]]

Here, the units P, S, A and C are all *generic units*, which denote partial functions taking compatible models of the argument specifications to models of the result specification in a persistent way (that is, the arguments are protected). The compatibility of arguments means that the models can be amalgamated to a model of the union of all the signatures. Moreover, the components are combined in the way prescribed in the result unit of ARCH_SBCS; notice that a model of VALUE is required to able to provide a model of the entire system. □

$ASP ::= S \mid$ **units** $UDD_1 \dots UDD_n$ **result** UE
$UDD ::= UDEFN \mid UDECL$
$UDECL ::= USP$ **given** $UT_1, \dots, UT_n$
$USP ::= SP \mid SP_1 \times \dots \times SP_n \to SP$
$UDEFN ::= A = UE$
$UE ::= UT \mid \lambda\, A_1 : SP_1, \dots,\ A_n : SP_n \bullet UT$
$UT ::= A \mid A\, [FIT_1] \dots [FIT_n] \mid UT$ **and** $UT \mid UT$ **with** $\sigma : \Sigma \to \Sigma' \mid$
  UT **hide** $\sigma : \Sigma \to \Sigma' \mid$ **local** $UDEFN_1 \dots UDEFN_n$ **within** UT
$FIT ::= UT \mid UT$ **fit** $\sigma : \Sigma \to \Sigma'$

Fig. 1. Syntax of the CASL architectural language

[1] The complete specification of the SBCS example can be found at `https://svn-agbkb.informatik.uni-bremen.de/Hets-lib/trunk/UserManual/Sbcs.casl`

Fig. 1 presents the syntax of the architectural language of CASL. Here, S stands for an architectural specification name, A for a component name, Σ and Σ' denote signatures and σ denotes a signature morphism. The architectural language can be shortly explained as follows: an architectural specification *ASP* consists of a list of unit declarations *UDECL* and unit definitions *UDEFN* (where declarations assign unit specifications *USP* to units and definitions assign unit expressions *UE* to units) and a result unit expression formed with the units declared/defined. Unit expressions are used to give definitions for generic units, while unit terms define non-generic units. When the result unit of an architectural specification is generic, the system is 'open', requiring some parameters to provide an implementation. We would like to point out a construction in the architectural language, namely units declared with an optional list of imported unit terms, specified using the **given** clause. This construction has been explained in literature as generic units instantiated once:

<table>
<tr>
<td>units M : SP1;
    N : SP2 given M;
    ...</td>
<td>is equivalent to</td>
<td>units M : SP1;
    N : arch spec {
        units F : SP1 → SP2
        result F[M]};
    ...</td>
</tr>
</table>

This equivalence has been made formal in [6] and thus allows us to treat the first syntactic construction as a "syntactic sugar" for the second, reducing thus the complexity of semantics and verification of architectural specifications. Notice that the name F of the generic unit is chosen arbitrarily and in order to be able to refine the unit N we must slightly adapt the original semantics rules for refinements from [13]. We will address this in more detail in Sections 4 and 5.

In [13], the architectural layer has been complemented with a *refinement* language, which allows to formalize developments as refinement trees. The language provides syntactic constructs (Fig. 2) for expressing refinement between specifications, starting with the simplest form which denotes just model class inclusion between unit specifications: *USP* **refined via** σ **to** *USP'* is a correct refinement when $M|_\sigma \in Mod(USP)$ for any model M of *USP'* (here, $M|_\sigma$ denotes the reduct of M against the signature morphism σ). We denote this $USP \rightsquigarrow_\sigma USP'$ and when σ is identity or clear from the context we omit it. This grows in complexity to compositions of refinements (written as *USP* **refined to** *RSP*, where *RSP* is now an arbitrarily complex refinement or as *RSP* **then** *RSP'*), branching introduced by architectural design, with the specifications of the components being now refinement specifications and finally refinement of components of architectural specifications (written $\{UN_i$ **to** $RSP_i\}_{i \in \mathcal{J}}$, where UN_i stands for a unit name). Notice that refinement specifications subsume architectural specifications and we can therefore speak of a refinement level subsuming the architectural level in the CASL language. In Fig. 2 R stands for the name of a refinement specification.

Example 2. We can then write the initial refinement of the steam boiler system as

RSP := **refinement** R = *RSP* | *USP* | **arch spec** *ASP* | *RSP* **then** *RSP* | *SP* **refined** [**via** σ] **to** *RSP* | $\{A_1$ **to** $RSP_1, \ldots, A_n$ **to** $RSP_n\}$ *UDECL* ::= A : *RSP* | A : *USP* **given** $UT_1, \ldots, UT_n$

Fig. 2. Syntax of the CASL refinement language

refinement REF_SBCS =
STEAM_BOILER_CONTROL_SYSTEM **refined to arch spec** ARCH_SBCS

We further proceed with refining the individual units [2]. The specifications of C and S in ARCH_SBCS above do not require further architectural decomposition. The specification of S, recorded in the unit specification STATE_ABSTR, can be refined by providing an implementation of states as a record of all observable values. This is done in SBCS_STATE_IMPL, assuming an implementation of PRELIMINARY; we record this development in the unit specification UNIT_SBCS_STATE. The refinement of S is then written in STATE_REF.

unit spec STATE_ABSTR = PRELIMINARY → SBCS_STATE
unit spec UNIT_SBCS_STATE =
    PRELIMINARY → SBCS_STATE_IMPL
refinement STATE_REF =
        STATE_ABSTR **refined to** UNIT_SBCS_STATE

For the units P and A, we proceed with designing their architecture. This is recorded in the architectural specifications ARCH_PRELIMINARY and ARCH_ANALYSIS (omitted here). We can now record the component refinement:

refinement REF_SBCS' = REF_SBCS **then**
        {P **to arch spec** ARCH_PRELIMINARY,
        S **to** STATEREF,
        A **to arch spec** ARCH_ANALYSIS}

Moreover, the components of ARCH_ANALYSIS are further refined:

refinement REF_SBCS" =
        REF_SBCS'
        then {A **to**
            {FD **to arch spec** ARCH_FAILURE_DETECTION,
            PR **to arch spec** ARCH_PREDICTION }}

□

3 Refinement Trees

We now give a formal definition of the concept of *refinement trees*. Refinement trees provide visualization means for the structure of the development and access

[2] We tacitly correct some rather minor mistakes in the specifications in [13] that were revealed while testing the implementation of the static analysis rules of refinement in Hets.

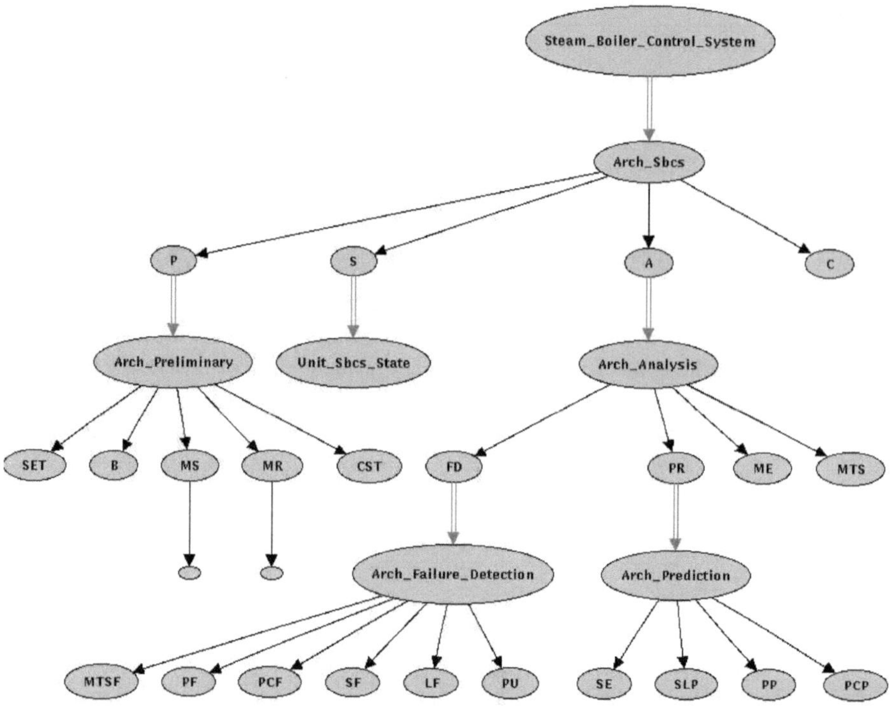

Fig. 3. The refinement tree of the steam boiler control system

points for the logical properties of architectural and refinement specifications. They can be regarded as a counterpart of the development graphs [12] constructed for structured specifications. Notice that the proof calculus for refinements presented in Section 5 produces a development graph for proving the verification conditions, but the components of the system decomposition are quite difficult to observe, since e.g. the nodes introduced in applications of generic units and edges for dependencies between units in architectural specifications are present. The approach that we propose separates the diagram of dependencies from the tree-like representation of the development process.

While intuitively clear, refinement trees are built in a stepwise manner and must be *combined* in the way prescribed by the refinement specifications: composition of refinements gives rise to composition of refinement trees, and we need a mechanism for keeping track of the branches and nodes in the tree corresponding to units and connection points between trees. Moreover, in the case of component refinement, each component produces a (sub)tree.

Example 3. Fig. 3 presents the refinement tree of the specifications in Example 1. Single arrows denote components, while double arrows denote refinements. Notice that in the case of e.g. REF_SBCS'', we need to build the trees of the architectural specifications ARCH_FAILURE_DETECTION and ARCH_PREDICTION, store them as corresponding to the units FD and PR in a component refinement, obtaining thus a set of trees, then connect them to the corresponding components of the unit A, which must be correctly identified.

The example shows that refinement trees should consist of a collection of trees and that they can grow not only at the leaves, but also at the root, thus becoming subtrees. This leads us to the following definition.

Definition 4. *A refinement tree $\mathcal{RT}$ consists of a set of trees with nodes labelled with unit specifications and edges that can be either (i)* refinement links $n_1 \Rightarrow n_2$ *to denote refinement of specifications or (ii)* component links $n_1 \rightarrow n_2$*, where n_1, n_2 are nodes in $\mathcal{RT}$ to denote architectural decomposition.*

We need to define an auxiliary structure to keep track of the roots, leaves and nodes of the branching decompositions; this will make possible to compose refinement trees. We would like to stress that this is only done for book-keeping and it is not visible to the user at all. Let us define *refinement tree pointers* in a refinement tree $\mathcal{RT}$ as either (i) *simple refinement pointers* of form (n_1, n_2) with n_1, n_2 nodes in $\mathcal{RT}$, with the intuition that the first node is the root and the second node is the leaf of a chain[3], (ii) *branching refinement pointers* of form (n, f), where n is a node and f is a map assigning refinement tree pointers to unit names, or (iii) *component refinement pointers* which are maps assigning refinement tree pointers to unit names.

Let us give a series of notations and operations with refinement trees. We denote $\mathcal{RT}_\emptyset$ the empty tree. If $\mathcal{RT}$ is a refinement tree, $\mathcal{RT}[USP]$ is obtained by adding to it a new isolated node (also returned as result) labelled with *USP*. For $\mathcal{RT}_1, \ldots, \mathcal{RT}_k$ refinement trees, $\mathcal{RT}[n_1 \rightarrow \mathcal{RT}_1, \ldots, \mathcal{RT}_k]$ denotes the tree obtained by inserting component links from the node n_1 of $\mathcal{RT}$ to the roots of each of the argument trees. Moreover, for two refinement trees $\mathcal{RT}_1$ with pointer p_1 and $\mathcal{RT}_2$ with pointer p_2, we denote $(p, \mathcal{RT})$ the composition $\mathcal{RT}_1 \circ p_1, p_2 \mathcal{RT}_2$ defined as follows:

- if p_1 is a simple refinement pointer (n_1, n_2) and p_2 is a simple refinement pointer (m_1, m_2), $\mathcal{RT}$ is obtained by adding a refinement link from n_2 to the root of the subtree of m_1 in $\mathcal{RT}_2$. The pointer p is then (n_1, m_2). This corresponds to STATE_REF in Example 1, see branch for unit S in Fig. 3.
- if p_1 is a simple refinement pointer (n_1, n_2) and p_2 is a branching refinement pointer (m_1, f), $\mathcal{RT}$ is obtained by adding a refinement link from n_2 to m_1. The pointer p is (n_1, f). This corresponds to REF_SBCS in Example 1, see the initial refinement link in Fig. 3.
- if p_1 is a branching refinement pointer (n_1, f_1) and p_2 is a component refinement pointer f_2, $\mathcal{RT}$ is obtained by making for each X in $dom(f_2)$ the composition of the subtree pointed by $f_1(X)$ with the tree $f_2(X)$, which also returns a pointer p_X. The pointer p is $(n_1, f_1[f_2])$, where $f_1[f_2]$ updates the value of X in f_1 with the pointer p_X. This corresponds to REF_SBCS' in Example 1 and introduces the second level of refinement links in Fig. 3.

[3] Notice that the two nodes can coincide.

– if p_1 is a component refinement pointer f_1 and p_2 is a component refinement pointer p_2, then the refinement tree is obtained by making for each X in $dom(f_2)$ the composition of the subtree pointed by $f_1(X)$ with the tree $f_2(X)$, which also returns a pointer p_X. The pointer p is $f_1[f_2]$. We can obtain an example for this last case by writing the refinement of the components in a slightly different way:

refinement R = {
P **to arch spec** ARCH_PRELIMINARY, S **to** STATEREF,
A **to arch spec** ARCH_ANALYSIS }
then {
A **to** {FD **to arch spec** ARCH_FAILURE_DETECTION,
PR **to arch spec** ARCH_PREDICTION }}

and the corresponding refinement trees are obtained by removing the first three levels from the tree in Fig. 3.

The composition is undefined otherwise. Notice that the specification of a unit, needed to label the nodes of refinement trees, will be obtained in some cases with the proof calculus for refinement specifications that we present in Section 5 and therefore the refinement trees will be built with the proof calculus rules.

4 The CASL Refinement Language: Semantics

The semantics of refinement specifications is, as usual in CASL, model-theoretic: a specification denotes a signature (static semantics) and a class of models (model semantics). We briefly recall the semantics of architectural specifications (see [15] for details). Unit specifications describe self-contained units (models) or generic units, mapping compatible models of the arguments to models of the result specification, such that the arguments are preserved under reduct. Their static semantics is given by unit signatures, which are themselves either plain signatures Σ or generic unit signatures $\Sigma_1 \times \cdots \times \Sigma_n \to \Sigma$ where Σ extends the union of $\Sigma_1, \ldots, \Sigma_n$. Plain CASL static semantics of architectural specifications consist of a unit signature for the result unit and a mapping giving a unit signature for any unit component, while the model semantics consists of a class of architectural models, which themselves pair a model over the unit signature of the result with a mapping giving a model for each component.

For the static semantics of refinements, [13] introduces *refinement signatures*, $R\Sigma$, which take one of the following forms: (i) *unit refinement signatures* $(U\Sigma, U\Sigma')$ which consist of two unit signatures and correspond to simple refinements (ii) *branching refinement signatures* $(U\Sigma, B\Sigma')$ which consist of a unit signature $U\Sigma$ and a *branching signature* $B\Sigma'$, which is either a unit signature $U\Sigma'$ (in which case the branching refinement signature is a unit refinement signature) or a *branching static context* $BstC'$, which is in turn a (finite) map assigning branching signatures to unit names, and corresponds to architectural decompositions, and finally (iii) *component refinement signatures* which are (finite) maps $\{UN_i \mapsto R\Sigma_i\}_{i\in\mathcal{J}}$ from unit names to refinement signatures and give static semantics for refinements of components.

The rules for static and model semantics of refinements are introduced in [13] as well; due to space limitations, we do not repeat them here. The only modification that we bring handles the case of units with imports, as mentioned in Section 2. The refinement signature of a unit with imports is a branching refinement signature with the branching static context having only one entry; since the architectural specification that equivalently expresses units with imports is generated, the name of the generic unit is not available to the user. The convention that we propose is to adjust the definition of *composition of refinement signatures* given in [13].

Given refinement signatures $R\Sigma_1$ and $R\Sigma_2$, their *composition* $R\Sigma_1\,;\,R\Sigma_2$ is defined inductively on the form of the first argument. We extend the definition by making the composition $R\Sigma_1;\ R\Sigma_2$ also defined when $R\Sigma_1$ is of form $(U\Sigma, \mathit{BstC})$ with only one unit name UN in the domain of BstC and the composition $R\Sigma'$ of $\mathit{BstC}(UN)$ with $R\Sigma_2$ is defined . In this case, $R\Sigma_1;\ R\Sigma 2 = (U\Sigma, \mathit{BstC}[UN/R\Sigma'])$.

We will also make use of the *refinement relations* introduced in the paper cited above, which provide a notion of refinement model. Given a refinement signature $R\Sigma$, *refinement relations*, $\mathcal{R}$, are classes of *assignments*, R, which take the following forms:

- *unit assignments*, for $R\Sigma = (U\Sigma, U\Sigma')$, are pairs (U, U') of units over unit signatures $U\Sigma$ and $U\Sigma'$, respectively;
- *branching assignments*, for $R\Sigma = (U\Sigma, B\Sigma')$, are pairs (U, BM'), where U is a unit over the unit signature $U\Sigma$ and BM' is a *branching model* over the branching signature $B\Sigma'$, which is either a unit over $B\Sigma'$ when $B\Sigma'$ is a unit signature (in which case the branching assignment is a unit assignment), or a *branching environment* BE' that fits $B\Sigma'$ when $B\Sigma'$ is a branching static context. Branching environments are (finite) maps assigning branching models to unit names, with the obvious requirements to ensure compatibility with the branching signatures indicated in the corresponding branching static context.
- *component assignments*, for $R\Sigma = \{UN_i \mapsto R\Sigma_i\}_{i\in\mathcal{J}}$, are (finite) maps $\{UN_i \mapsto R_i\}_{i\in\mathcal{J}}$ from unit names to assignments over the respective refinement signatures. When $R\Sigma$ is a refined-unit static context (and so each R_i, $i \in \mathcal{J}$, is a branching assignment) we refer to $RE = \{UN_i \mapsto (U_i, BM_i)\}_{i\in\mathcal{J}}$ as a *refined-unit environment.* Any such refined-unit environment can be naturally coerced to a unit environment $\pi_1(RE) = \{UN_i \mapsto U_i\}_{i\in\mathcal{J}}$ of the plain CASL semantics, as well as to a branching environment $\pi_2(RE) = \{UN_i \mapsto BM_i\}_{i\in\mathcal{J}}$.

Again, the only change to the rules of model semantics is to adapt in the expected way the *composition* of refinement relations to accommodate refinements of unit with imports, in symmetry with the change of static semantics.

The static semantic rules are of form $\vdash SPR \rhd R\Sigma$ while the model semantic rules are of form $\vdash SPR \Rightarrow \mathcal{R}$; when we only want to state that SPR has a denotation w.r.t. the static or model semantics, the result is replaced with $\square$.

5 Proof Calculus for the CASL Refinement Language

The main motivation for formalizing the development process using CASL architectural specifications and refinements is that one can then *prove* that the process is correct. Verification of correctness for a refinement tree is presented as a *proof calculus*. In [15], Section 4:5, a proof calculus for verification of architectural specifications (in a slightly restricted variant) was introduced as an algorithm for checking whether the resulting units of an architectural specification satisfy a given unit specification. This is denoted $\vdash ASP :: USP$, where ASP is an architectural specification and USP is a unit specification. Since architectural specifications are now a particular case of refinements, we will extend this calculus to support the whole refinement language.

For space limitation reasons, we omit the complete presentation of the proof calculus of architectural specifications mentioned above and just explain the rough intuition. It can be regarded as having two components. The first one is a *constructive* component, building a graph Γ with nodes labeled with non-generic unit signatures and edges labeled with signature morphisms where additionally some of the nodes may be labeled with sets of specifications. Moreover, generic units declared in ASP are stored in a generic context Γ_{gen}. The second component is *deductive* and it uses the diagram built with the constructive component to check whether models of a unit expression satisfy a given unit specification SP. Fig. 4 presents one of the interesting rules of the calculus, namely those for unit expressions (only non-generic). What is particular about the rules for unit expressions is that here is where the two components of the calculus meet: the proof calculus rules for unit terms (of format $\Gamma_{gen}, \Gamma \vdash UT :: \Gamma', A$) modify the context by adding the diagram of the unit term and also return the node of the unit term, and this node is further used for checking in Γ' whether the given specification actually holds or not. On the notations, R is an institution translation (formalised as so-called comorphism) that translates from the institution of interest to another with weak amalgamation property (i.e., given a diagram and a family of models compatible with it, they can be combined to a model of a cocone for the diagram) and the refinement condition verifies that for any

$$\frac{\begin{array}{c}\Gamma_{gen}, \Gamma \vdash UT :: \Gamma', A \\ \text{for all } U \in \mathrm{dom}(\Gamma'), \text{ we have } U :_{\Sigma_U} SP_U \text{ in } \Gamma' \\ \Sigma, \{\eta_U\}_{U \in \mathrm{dom}(\Gamma')} \text{ is a weakly amalgamable cocone over } R(\Gamma') \\ \eta_A(R(SP)) \rightsquigarrow^{J}_{\Sigma} \bigcup_{U \in \mathrm{dom}(\Gamma')} \eta_U(\overline{R}(SP_U))\end{array}}{\Gamma_{gen}, \Gamma \vdash UT \text{ qua } UE :: SP}$$

Fig. 4. Architectural proof calculus rule for non-generic unit expressions

family of models compatible with the diagram the model corresponding to the node A satisfies SP.

As a first step, we modify the proof calculus such that it becomes fully constructive: USP is no longer provided, but rather obtained by defining the specification of each unit term inductively on its structure.

Definition 5. *Let ASP be an architectural specification and UE a unit expression. Then the specification of UE, denoted $\mathcal{S}_{ASP}(UE)$ is defined as follows:*

- *if UE is a unit term UT, $\mathcal{S}_{ASP}(UT)$ is defined inductively:*
  - *if UT is a unit name, then $\mathcal{S}_{ASP}(UT) = SP$ where $UT : SP$ is the declaration of UT in ASP;*
  - *if $UT = F[A_1 \textbf{ fit } \sigma_1] \ldots [A_n \textbf{ fit } \sigma_n]$, where $\mathcal{S}_{ASP}(F) = SP_1 \times \cdots \times SP_n \rightarrow SP$ and for any $i = 1, \ldots, n$, $\mathcal{S}_{ASP}(A_i) \textbf{ hide } \sigma_i \models SP_i$, then $\mathcal{S}_{ASP}(UT) = \{SP \textbf{ with } \sigma\} \textbf{ and } \mathcal{S}_{ASP}(A_1) \textbf{ hide } \sigma_1 \textbf{ and} \ldots \textbf{and } \mathcal{S}_{ASP}(A_n) \textbf{ hide } \sigma_n$, where $\sigma = \cup_{i=1,\ldots,n}\sigma_i$;*
  - *if $\mathcal{S}_{ASP}(A_i) = SP_i$ then $\mathcal{S}_{ASP}(A_1 \textbf{ and } \ldots \textbf{ and } A_n) = SP_1 \textbf{ and } \ldots \textbf{ and } SP_n$;*
  - *if $\mathcal{S}_{ASP}(A) = SP$, then $\mathcal{S}_{ASP}(A \textbf{ with } \sigma) = SP \textbf{ with } \sigma$;*
  - *if $\mathcal{S}_{ASP}(A) = SP$, then $\mathcal{S}_{ASP}(A \textbf{ hide } \sigma) = SP \textbf{ hide } \sigma$;*
  - *$\mathcal{S}_{ASP}(\textbf{local } UDEFN \textbf{ within } UT) = \mathcal{S}_{ASP}(UT)$, where UDEFN is used to obtain the specification of the locally defined units.*
- *if UE is a lambda expression $\lambda A : SP . UT$, then $\mathcal{S}_{ASP}(UE) = SP \rightarrow \mathcal{S}_{ASP}(UT)$.*

The specification of a unit term also gets employed in expressing declarations of units with imports as generic units: a declaration $UN : SP$ **given UT** can be written as $UN :$ **arch spec**$\{$**units** $F : \mathcal{S}_{ASP}(UT) \rightarrow SP$; **result F**[**UT**]$\}$. As noticed in [9], it is not always possible to give a precise axiomatization of the class of models produced by the unit expression. However, we will use $\mathcal{S}_{ASP}(UE)$ as an approximation, since models of the unit expression are also models of this specification.

The proof calculus is then turned into a fully constructive variant by removing verification conditions from the rules for unit expressions and returning the specification of the result unit expression of the architectural specification instead of having one as a input parameter of the calculus. Let us denote the constructive version of the architectural proof calculus (built using $\mathcal{S}_{ASP}(UE)$) by $\vdash ASP ::_c USP$.

When the architectural language is restricted by omitting the unit imports , the constructive and the deductive versions of the proof calculus are related by the following straightforward result.

Proposition 6. *If unit imports are omitted, $\vdash ASP \rhd \Box$ and $\vdash ASP ::_c USP$ then $\vdash ASP :: USP$.*

Moreover, in some cases, the obtained unit specification exactly captures the models of the architectural specification, if the latter are projected with the possible semantics for the result unit term. Let therefore *ProjRes* take any model of an architectural specification to the interpretation of its result unit term in this model.

Theorem 7. *Let ASP be a consistent architectural specification without unit imports and unit definitions, where each parametric unit is applied only once. If* $\vdash ASP \rhd \Box$ *and* $\vdash ASP ::_c USP$ *then* $ProjRes(\mathbf{Mod}(ASP)) = \mathbf{Mod}(USP)$.

While the proof calculus of architectural specifications checks that result units satisfy a given unit specification, we introduce a counterpart for the specification that refinements should satisfy, tailored according to the three kinds of refinements. This allows to express the proof calculus rule for compositions of refinements in a more concise manner.

Definition 8. *Let $R\Sigma$ be a refinement signature. A* verification specification *S over $R\Sigma$ is defined as follows:*

- *if $R\Sigma = (U\Sigma_1, U\Sigma_2)$, $S = (USP_1, USP_2)$ such that $\vdash USP_i \rhd U\Sigma_i$, for i=1,2;*
- *if $R\Sigma = (U\Sigma, B\Sigma)$, $S = (USP, BSP)$, where $\vdash USP \rhd U\Sigma$ and BSP is a branching specification, which is in turn either a unit specification USP' such that $\vdash USP' \rhd U\Sigma'$, when $B\Sigma = U\Sigma'$ or a map SPM such that and $SPM(X)$ is a verification specification over $BstC(X)$, for any $X \in dom(BstC)$, when $B\Sigma = BstC$;*
- *if $R\Sigma = \{UN_i \mapsto R\Sigma_i\}_{i \in \mathcal{J}}$, then $S = \{UN_i \mapsto S_i\}_{i \in \mathcal{J}}$, where S_i is a verification specification over $R\Sigma_i$.*

Again, the proof calculus rules rely on a composition operation on verification specifications. The composition $S_1;\ S_2$ is defined inductively as follows:

- if $S_1 = (USP_1, USP_2)$, then $S_1;\ S_2$ is defined only when $S_2 = (USP_3, SPM)$ and moreover $\vdash USP_2 \rightsquigarrow USP_3$. Then $S_1;\ S_2 = (USP_1, SPM)$.
- if $S_1 = (USP_1, SPM_1)$ then S_2 must be of form SPM_2. We define $S_1;\ S_2 = (USP_1, SPM_1[SPM_2])$, where $SPM_1[SPM_2](A) = SPM_1(A)$, if $A \notin dom(SPM_2)$ and $SPM_1[SPM_2](A) = SPM_1(A);\ SPM_2(A)$ otherwise.
- if $S_1 = SPM_1$, then $S_1;\ S_2$ is defined only if $S_2 = SPM_2$. Then $S_1;\ S_2$ modifies the ill-defined union of SPM_1 and SPM_2 by putting $(S_1;\ S_2)(A) = S_1(A);\ S_2(A)$ for any $A \in dom(S_1) \cap dom(S_2)$.

The proof calculus for architectural specifications is complemented at the level of refinement specifications as in Fig. 5. The judgments of the proof calculus for refinements are then of form $\vdash SPR :: S, \mathcal{RT}, p$, where SPR is a refinement specification, S is a verification specification, $\mathcal{RT}$ is a refinement tree and p is a refinement tree pointer. By a slight abuse of notation, when we are not interested in the refinement tree and the pointer, we omit them as results.

Notice that the proof calculus for architectural specifications of [15] only takes into account the case when the specification of a unit is a unit specification. In the context of refinements, the specification of a unit can be an arbitrary refinement specification, and the proof calculus of architectural specifications

$$\frac{(n, \mathcal{RT}) = \mathcal{RT}_{\emptyset}[USP]}{\vdash USP :: (USP, USP), \mathcal{RT}, (n, n)} \qquad \frac{\begin{array}{c}\vdash USP :: (USP, USP), \mathcal{RT}_1, p_1 \\ \vdash SPR :: (USP', BSP), \mathcal{RT}_2, p_2 \\ (\mathcal{RT}, p) = \mathcal{RT}_1 \circ_{p_1, p_2} \mathcal{RT}_2 \\ \vdash USP \rightsquigarrow USP'\end{array}}{\vdash USP \textbf{ refined to } SPR :: (USP', BSP), \mathcal{RT}, p}$$

$$\frac{\begin{array}{c}\vdash ASP ::_c USP \\ \vdash SPR_i :: (USP_i, BSP_i), \mathcal{RT}_i, p_i \\ \text{for any } UN_i : SPR_i \text{ in } ASP \\ SPM(UN_i) = BSP_i \\ (n, \mathcal{RT}') = \mathcal{RT}_{\emptyset}[USP] \\ \mathcal{RT} = \mathcal{RT}'[n \to \mathcal{RT}_1, \ldots, \mathcal{RT}_k] \\ p = \{UN_i \mapsto p_i\}_{i=1,\ldots,k}\end{array}}{\vdash ASP :: (USP, SPM), \mathcal{RT}, p} \qquad \frac{\begin{array}{c}\vdash SPR_i :: S_i, \mathcal{RT}_i, p_i \\ \mathcal{RT} = \cup \mathcal{RT}_i \\ p = (n, \{UN_i \to p_i\})\end{array}}{\vdash \{UN_i \textbf{ to } SPR_i\}_{i \in \mathcal{J}} :: \{UN_i \to S_i\}_{i \in \mathcal{J}}, \mathcal{RT}, p}$$

$$\frac{\begin{array}{c}\vdash SPR_1 :: S_1, \mathcal{RT}_1, p_1 \\ \vdash SPR_2 :: S_2, \mathcal{RT}_2, p_2 \\ S = S_1;\ S_2 \\ (p, \mathcal{RT}) = \mathcal{RT}_1 \circ_{p_1, p_2} \mathcal{RT}_2\end{array}}{\vdash SPR_1 \textbf{ then } SPR_2 :: S, \mathcal{RT}, p}$$

Fig. 5. Proof calculus for CASL refinements

must be therefore adapted. This means that for a unit declaration $UN_i : SPR_i$, we define $\mathcal{S}_{ASP}(UN_i) = USP_i$ if $\vdash SPR_i :: (USP_i, SPM_i)$. Moreover, we can set $SPM(UN_i) = SPM_i$ in the verification specification of ASP. Finally, in the case of named refinement specifications, we need to store their verification specifications at the library level and retrieve them by name, in the usual way.

Definition 9. *Let $R\Sigma$ be a refinement signature, S a verification specification of $R\Sigma$ and $\mathcal{R}$ a refinement relation over $R\Sigma$. We define the satisfaction of a verification specification by a refinement relation, denoted $\mathcal{R} \models S$, inductively as follows:*

- *if $R\Sigma = (U\Sigma, U\Sigma')$, then $S = (USP, USP')$ and $\mathcal{R} = \{(u, u') | u \in Unit(U\Sigma), u' \in Unit(U\Sigma')\}$. Then $\mathcal{R} \models S$ iff $u \in Unit(USP)$ and $u' \in Unit(USP')$ for any $(u, u') \in \mathcal{R}$;*
- *if $R\Sigma = (U\Sigma, B\Sigma)$, then $S = (USP, SPM)$ and $\mathcal{R} = \{(u, bm) | u \in Unit(U\Sigma)$, bm is a branching model over $B\Sigma\}$. Then $\mathcal{R} \models S$ iff for any $(u, bm) \in R$, $u \in Unit(USP)$ and for any $A \in dom(SPM)$ we have that $bm(A) \models S(A)$. (Notice that SPM and bm have the same domain);*
- *if $R\Sigma = SPM$, then $S = \{UN_i \to S_i\}_{i \in \mathcal{J}}$ and $\mathcal{R} = \{UN_i \to \mathcal{R}_i\}_{i \in \mathcal{J}}$. Then $\mathcal{R} \models S$ iff $\mathcal{R}_i \models S_i$ for any i.*

The following lemma can be proven by induction and making a case distinction on refinement signatures.

Lemma 10. *Let $R\Sigma_1, R\Sigma_2$ be two refinement signatures such that their composition is defined. Let S_i be a verification specification over $R\Sigma_i$ and $\mathcal{R}_i$ be a refinement relation over $R\Sigma_i$ such that $\mathcal{R}_i \models S_i$, for $i = 1, 2$ such that $\mathcal{R}_1;\ \mathcal{R}_2$ and $S_1;\ S_2$ are defined. Then $\mathcal{R}_1;\ \mathcal{R}_2 \models S_1;\ S_2$.*

The following result states that if a statically well-formed refinement specification *SPR* can be proven correct w.r.t. a verification specification S using the proof calculus for refinements, then *SPR* has a denotation according to the model semantics and moreover the refinement relation thus obtained satisfies S.

Theorem 11 (Soundness). *Let SPR be a refinement specification such that $\vdash SPR \rhd \Box$. If $\vdash SPR :: S$, then there is $\mathcal{R}$ such that $\vdash SPR \Rightarrow \mathcal{R}$ and $\mathcal{R} \models S$.*

Because we approximate the specification of the result unit of architectural specifications, completeness is much more difficult to obtain and is therefore postponed to future work.

6 Checking Consistency of Refinement Specifications

We introduce a calculus for checking whether a refinement specification is consistent, i.e. it has a refinement model. In [10], we have successfully applied this calculus to verify the consistency of the upper ontology Dolce. Indeed, Dolce is too large for contemporary model finders. Instead of hand-crafting a large and specific model, we have shown the consistency of Dolce using an architectural refinement. This has the advantage of giving a modular model for Dolce, i.e. one that can be changed at various local places (= leaves of the refinement tree) without affecting the possibility to assemble (via the semantics of architectural specifications) a global model of Dolce.

Intuitively, a refinement is consistent if its target is, and an architectural specification is consistent if all its unit specifications are. This makes it clear that our calculus eventually (for checking consistency of the leaves of the refinement tree) needs to be based on a calculus for the consistency of unit specifications, which we denote $\vdash cons(USP)$ and is given by the rules in Fig. 6. Checking consistency of non-parametric unit specification amounts to checking consistency of structured specifications; a calculus for this has been introduced in [17] (this is in turn based on some institution-specific calculus for consistency of basic specifications). Checking consistency of parametric unit specification amounts to checking conservativity of extensions of structured specifications; for the case of first-order logic and CASL basic specifications, a calculus has been developed in [11].

Note that in the case of compositions, if SPR_1 contains a branching, it does not suffice for SPR_2 (which must be a component refinement) to be consistent, because some component of SPR_1 outside the domain of SPR_2 might be inconsistent.

Proposition 12 (Soundness). *If $\vdash SPR \rhd \Box$ and $\vdash cons(SPR)$, there is a refinement relation $\mathcal{R}$ such that $\vdash SPR \Rightarrow \mathcal{R}$.*

$$\frac{\vdash cons(USP)}{\vdash cons(USP\ qua\ \texttt{SPEC-REF})} \qquad \frac{\vdash cons(SPR)}{\vdash cons(USP\ \textbf{refined to}\ SPR)}$$

$$\frac{\vdash cons(SPR)\ for\ any\ UN : SPR\ in\ ASP}{\vdash cons(ASP)} \qquad \frac{\vdash cons(SPR_i)}{\vdash cons(\{U_i\ \textbf{to}\ SPR_i\}_{i \in \mathcal{J}})}$$

$$\frac{\begin{array}{c}\vdash cons(SPR_1)\\ \vdash cons(SPR_2)\end{array}}{\vdash cons(SPR_1\ \textbf{then}\ SPR_2)}$$

Fig. 6. Consistency calculus for refinement specifications

Completeness holds only by restricting the language again to a variant without imports.

Proposition 13 (Completeness). *If unit imports are omitted,* $\vdash SPR \rhd \Box$ *and* $\vdash SPR \Rightarrow \mathcal{R}$*, then* $\vdash cons(SPR)$.

7 Conclusions

We have recalled the language for refinements in CASL and we provided a sound proof calculus for it. Thus we can formalize development process for software systems and prove their correctness. Moreover, we have introduced refinements trees in theory, and also practically implemented them in the Heterogeneous Tool Set Hets, such that browsing through and inspection of complex formal developments becomes possible. An implementation of the proof calculus is currently in progress; the refinement part is already implemented. Note that the proof calculus for architectural specifications of [15] was given for a restricted version of the language; it can be extended to the whole language in a way substantially simplified by the transformation of units with imports into generic units. We also have introduced and implemented a sound and complete calculus for consistency of refinements and architectural specification, which already has been applied for proving the consistency of the upper ontology Dolce in a modular way.

Future work includes extending the language to support *behavioral refinement.* Often, a specification does not satisfy the requirements literally, but only up to some observational equivalence. The standard example is the implementation of stacks as arrays with pointer that may differ (inessentially w.r.t. the behavior) on the entries beyond the pointer position. This has been discussed in the case of CASL in [5]. Another useful addition would be amalgamability checks for other logics in the Hets' logic graph, making thus possible to have architectural specifications in that logic.

Acknowledgement. This work has been supported by the German Research Council (DFG) under grant Mo-971/2 "Logic Atlas and Integration (LATIN)".

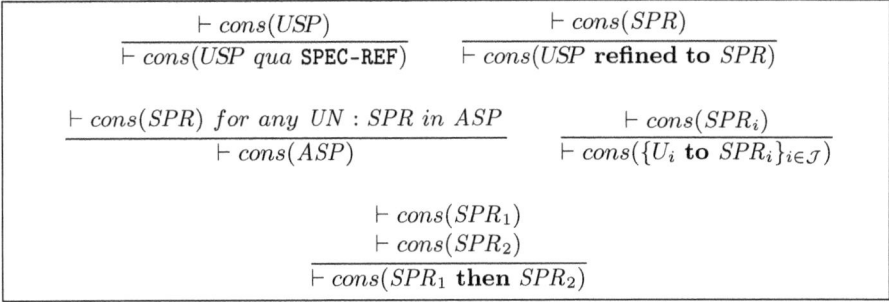

References

1. Abrial, J.-R., Börger, E., Langmaack, H.: Formal Methods for Industrial Applications, Specifying and Programming the Steam Boiler Control. LNCS, vol. 1165. Springer, Heidelberg (1996)
2. Astesiano, E., Kreowski, H.-J., Krieg-Brückner, B.: Algebraic Foundations of Systems Specification. Springer, Heidelberg (1999)
3. Bidoit, M., Mosses, P.D. (eds.): CASL User Manual. LNCS, vol. 2900. Springer, Heidelberg (2004)
4. Bidoit, M., Sannella, D., Tarlecki, A.: Architectural specifications in CASL. Formal Aspects of Computing 13, 252–273 (2002)
5. Bidoit, M., Sannella, D., Tarlecki, A.: Observational interpretation of CASL specifications. Mathematical Structures in Computer Science 18(2), 325–371 (2008)
6. Codescu, M.: Lambda Expressions in CASL Architectural Specifications. In: Mossakowski, T., Kreowski, H.-J. (eds.) 20th International Workshop on Recent Trends in Algebraic Development Techniques, WADT 2010. LNCS. Springer, Heidelberg (2011)
7. Gangemi, A., Guarino, N., Masolo, C., Oltramari, A., Schneider, L.: Sweetening ontologies with DOLCE. In: Gómez-Pérez, A., Benjamins, V.R. (eds.) EKAW 2002. LNCS (LNAI), vol. 2473, pp. 166–181. Springer, Heidelberg (2002)
8. Goguen, J.A., Burstall, R.M.: Institutions: Abstract model theory for specification and programming. Logic of Programs 1983 39, 95–146 (1992); Predecessor in: LNCS. vol. 164, pp. 221–256 (1984)
9. Hoffman, P.: Architectural Specifications and Their Verification. PhD thesis, Warsaw University (2005)
10. Kutz, O., Mossakowski, T.: A modular consistency proof for Dolce. In: 25th Conference on Artificial Intelligence, AAAI 2011 (to appear, 2011)
11. Liu, M.: Konsistenz-Check von CASL-Spezifikationen. Master's thesis, University of Bremen (2008)
12. Mossakowski, T., Autexier, S., Hutter, D.: Development graphs – proof management for structured specifications. Journal of Logic and Algebraic Programming 67(1-2), 114–145 (2006)
13. Mossakowski, T., Sannella, D., Tarlecki, A.: A simple refinement language for CASL. In: Fiadeiro, J.L., Mosses, P.D., Yu, Y. (eds.) WADT 2004. LNCS, vol. 3423, pp. 162–185. Springer, Heidelberg (2005)
14. Mossakowski, T., Maeder, C., Lüttich, K.: The Heterogeneous Tool Set, HETS. In: Grumberg, O., Huth, M. (eds.) TACAS 2007. LNCS, vol. 4424, pp. 519–522. Springer, Heidelberg (2007)
15. Mosses, P.D. (ed.): CASL Reference Manual. LNCS, vol. 2960. Springer, Heidelberg (2004)
16. Pease, A.: The SUMO challenges, `http://www.cs.miami.edu/~tptp/SUMOChallenge/`
17. Roggenbach, M., Schröder, L.: Towards Trustworthy Specifications I: Consistency Checks. In: Cerioli, M., Reggio, G. (eds.) WADT 2001 and CoFI WG Meeting 2001. LNCS, vol. 2267, p. 305. Springer, Heidelberg (2002), `http://www.springer.de`
18. Sannella, D., Tarlecki, A.: Toward formal development of programs from algebraic specifications: implementations revisited. Acta Informatica 25, 233–281 (1988)

On the Fusion of Coalgebraic Logics

Fredrik Dahlqvist and Dirk Pattinson

Dept. of Computing, Imperial College London

Abstract. Fusion is arguably the simplest way to combine modal logics. For normal modal logics with Kripke semantics, many properties such as completeness and decidability are known to transfer from the component logics to their fusion. In this paper we investigate to what extent these results can be generalised to the case of arbitrary coalgebraic logics. Our main result generalises a construction of Kracht and Wolter and confirms that completeness transfers to fusion for a large class of logics over coalgebraic semantics. This result is independent of the rank of the logics and relies on generalising the notions of distance and box operator to coalgebraic models.

Keywords: modal logic, coalgebra, fusion, completeness.

1 Introduction

The most common and simplest way to combine two modal logics L (for 'left') and R (for 'right') that we take as having disjoint sets of modal operators is their *fusion* $L \otimes R$, i.e. the smallest modal logic containing both L and R. In particular $L \otimes R$ does not contain any axioms combining operators from L with operators from R.

For modal logics with relational semantics, a large number of properties such as decidability and completeness are known to transfer from the component logics to their fusion [11,17,8,13]. The situation for (non-normal) logics outside the realm of relational semantics is far less satisfactory, despite the fact that there is an ever-growing class of logics that fall into this category such as probabilistic modal logic [6] or the logic of (monotone) neighbourhood frames [3]. While decidability can be established, also in the non-normal case, by purely algebraic methods [1], transfer of completeness remains largely open.

In fact, the only result we are aware of is negative: it is shown in [7] that the construction known as 'modalising', also discussed in [13] cannot be used to transfer completeness in the case of non-normal modal logics.

The main result of this paper generalises a model-building construction of Kracht and Wolter [11] and confirms that completeness transfers to fusion for a large class of logics over coalgebraic semantics. This technique, also known as *iterated dovetailing* [13,9], uses many familiar concepts such as successor states, distances and necessitation which are readily available in the context of Kripke semantics but need to be suitably adapted to be put to work in the coalgebraic

A. Corradini, B. Klin, and C. Cîrstea (Eds.): CALCO 2011, LNCS 6859, pp. 161–175, 2011.

setting. While all concepts above can be expressed in the coalgebraic framework, we have to assume that the component logics allow to express a form of necessitation.

In contrast to existing work that addresses the combination of modal logics in the framework of coalgebraic semantics [4,15] our result only assumes completeness of the component logics with respect to a subclass of (coalgebraic) frames, whereas *op.cit* assumes completeness with respect to the class of *all* coalgebras. In particular, we do not restrict to logics whose axiomatisation only uses so-called rank-1 axioms and our result can be applied to logics that incorporate arbitrary frame conditions, as long as they come equipped with a complete semantics and possess a necessitation operator.

2 Preliminaries

We fix a countable set V of atomic propositions throughout. A *modal signature* Λ is a set of (modal) operators with associated arities. A modal signature Λ' *extends* a modal signature Λ if $\Lambda' \supseteq \Lambda$. Given two modal signatures Λ_L and Λ_R, their disjoint union is denoted by $\Lambda_L + \Lambda_R$. The set $\mathcal{L}(\Lambda, V_0)$ of Λ-formulae over the set $V_0 \subseteq V$ of propositional variables is given by the grammar

$$\phi ::= p \mid \bot \mid \neg\phi \mid \phi \wedge \psi \mid \heartsuit(\phi_1, \ldots, \phi_n)$$

where $p \in V_0$ ranges over atomic propositions and $\heartsuit \in \Lambda$ is n-ary. If $V_0 = V$ is the set of all propositional variables, we write $\mathcal{L}(\Lambda) = \mathcal{L}(\Lambda, V)$. The set of propositional variables that occur in a formula $\phi \in \mathcal{L}(\Lambda, V_0)$ is denoted by $\mathrm{var}(\phi)$, $\mathrm{sf}(\phi)$ is the set of subformulae of ϕ and $\mathrm{md}(\phi)$ denotes the modal depth, i.e. the maximal nesting depth of modal operators in ϕ.

A Λ-logic L is a set of $\mathcal{L}(\Lambda)$-formulae containing all propositional tautologies, and closed under modus ponens, uniform substitution and the congruence rules

$$\frac{p_1 \leftrightarrow q_1 \wedge \ldots \wedge p_n \leftrightarrow q_n}{\heartsuit(p_1, \ldots, p_n) \leftrightarrow \heartsuit(q_1, \ldots, q_n)}$$

for each n-ary operator $\heartsuit \in \Lambda$. An *L-theorem* is a formula $\phi \in L$ and we write $\vdash_L \phi$ in this case, and a formula $\phi \in \mathcal{L}(\Lambda)$ is *L-consistent* if $\neg\phi \notin L$. The smallest congruential Λ-logic is denoted by $\mathbf{E}_\Lambda$.

Given two modal signatures Λ and Λ' such that Λ' extends Λ, we will say that a Λ'-logic M is an *extension* of a Λ-logic L if $L \subseteq M$. If, additionally, for every Λ-formula ϕ, $\phi \in L$ iff $\phi \in M$, then M is called a *conservative extension* of L. The *lattice of extensions* $\mathscr{E}(L)$ of a coalgebraic logic L is the set of all extensions $L \subseteq M$ of L with the meet and join operations given by the set intersection $\cap$ and union $\cup$ operations respectively.

On the semantical side, formulae are interpreted over T-coalgebras, where $T : \mathbf{Set} \to \mathbf{Set}$ is an endofunctor. A *T-coalgebra* is a pair $F = (W, \gamma)$ where W is a set (of worlds) and $\gamma : W \to TW$ is a (transition) function. Here, T-coalgebras play the role of frames, and we frequently refer to T-coalgebras as

T-frames. A *T-model* is a triple $M = (W, \gamma, \sigma)$ where (W, γ) is a T-frame and $\sigma : V \to \mathcal{P}(W)$ is a valuation (and $\mathcal{P}(W)$ is the powerset of W). A T-model (W, γ, σ) is *based* on the T-frame (W, γ).

Assumption 1. Modulo a modification of its action on the empty set and empty mappings, any **Set**-endofunctor T can be assumed to preserve finite intersections, all monos as well as the inverse image of injective maps [10]. Since modifying T on the empty set only gives an isomorphic category of coalgebras we will assume throughout that all functors do possess these preservation properties.

The interpretation of Λ-formulae over T-models requires that T extends to a *Λ-structure*, i.e. T comes equipped with an interpretation of the operators in Λ. Concretely, a Λ-structure consists of an endofunctor $T : \mathbf{Set} \to \mathbf{Set}$ together with an assignment of an n-ary *predicate lifting*, i.e. a set-indexed family of maps

$$([\![\heartsuit]\!]_X : \mathcal{P}(X)^n \to \mathcal{P}(TX)_{X \in \mathbf{Set}})$$

to every n-ary operator $\heartsuit \in \Lambda$ that satisfies the *naturality condition*

$$[\![\heartsuit]\!]_X \circ (f^{-1})^n = (Tf)^{-1} \circ \heartsuit_Y$$

for all maps $f : X \to Y$. Categorically speaking, $[\![\heartsuit]\!]$ is a natural transformation $2^- \to 2^{T-}$ where $2^- : \mathbf{Set} \to \mathbf{Set}^{\mathrm{op}}$ is the contravariant powerset functor. We will usually keep the assignment of predicate liftings implicit and just refer to Λ-structures by the underlying endofunctor.

Given a modal signature Λ and a Λ-structure T, the satisfaction relation between worlds of T-models and Λ-formulae is given inductively by

$$\begin{array}{ll} \mathcal{M}, w \models p \text{ iff } w \in \sigma(p) & \mathcal{M}, w \models \neg\phi \text{ iff not } \mathcal{M}, w \models \phi \\ \mathcal{M}, w \models \bot \text{ never} & \mathcal{M}, w \models \phi \wedge \psi \text{ iff } \mathcal{M}, w \models \phi \text{ and } \mathcal{M}, w \models \psi \end{array}$$

$$\mathcal{M}, w \models \heartsuit(\phi_1, \ldots, \phi_n) \text{ iff } \gamma(w) \in [\![\heartsuit]\!]_W([\![\phi_1]\!], \ldots, [\![\phi_n]\!])$$

where $\mathcal{M} = (W, \gamma, \sigma)$ is a T-model. We write $[\![\phi]\!]_{\mathcal{M}} = \{w \in W \mid \mathcal{M}, w \models \phi\}$ for the *truth-set* of ϕ relative to $\mathcal{M}$.

A formula $\phi \in \mathcal{L}(\Lambda)$ is *valid* in a T-frame $F = (W, \gamma)$ if $\mathcal{M}, w \models \phi$ for all T-models $\mathcal{M}$ based on F, this is denoted by $F \models \phi$. If $\mathscr{C}$ is a class of T-frames, we say that ϕ is *valid* on $\mathscr{C}$ if $F \models \phi$ for all $F \in \mathscr{C}$, this is denoted $\mathscr{C} \models \phi$. The *logic of* $\mathscr{C}$ is the set of all formulae valid on $\mathscr{C}$, i.e.

$$\mathrm{Log}(\mathscr{C}) = \{\phi \in \mathcal{L}(\Lambda) \mid \mathscr{C} \models \phi\}.$$

It is easy to check that $\mathrm{Log}(\mathscr{C})$ is a Λ-logic.

3 Fusion and Transfer of Soundness and Consistency

Given two modal signatures Λ_L and Λ_R (where the subscripts stand for 'left' and 'right', respectively), the *fusion* of a Λ_L-logic with a Λ_R-logic is the smallest $\Lambda_L + \Lambda_R$-logic that extends both. In other words, we may mix Λ_L and Λ_R operators freely in the fusion, but not stipulate any interaction between them.

Definition 2 (Fusion of Logics). Given a Λ_L-logic L and a Λ_R-logic R where Λ_L and Λ_R are two arbitrary modal signatures, the *fusion* $F = L \otimes R$ is the smallest $\Lambda_L + \Lambda_R$-logic containing both L and R. The fusion is therefore a binary operation $- \otimes - : \mathscr{E}(\mathbf{E}_{\Lambda_L}) \times \mathscr{E}(\mathbf{E}_{\Lambda_R}) \to \mathscr{E}(\mathbf{E}_{\Lambda_L+\Lambda_R})$.

While we consider the modal operators of L and R to be disjoint by assumption, propositional connectives are shared. Given two structures for Λ_L and Λ_R, we can interpret the fusion over a $\Lambda_L + \Lambda_R$-structure as follows:

Definition 3 (Fusion of Structures). Given a Λ_L-structure S and a Λ_R-structure T over modal signatures Λ_L and Λ_R, the *fusion* of S and T is the $\Lambda_L + \Lambda_R$-structure over the endofunctor $S \times T$ where the assignment of predicate liftings is given by

$$\llbracket \heartsuit \rrbracket^1_X = \pi_1^{-1} \circ \llbracket \heartsuit \rrbracket_X \qquad \text{for } \heartsuit \in \Lambda_L$$
$$\llbracket \spadesuit \rrbracket^2_X = \pi_2^{-1} \circ \llbracket \spadesuit \rrbracket_X \qquad \text{for } \spadesuit \in \Lambda_R$$

where π_1 and π_2 are the projections $SX \xleftarrow{\pi_1} SX \times TX \xrightarrow{\pi_2} TX$. Note that both $\llbracket \heartsuit \rrbracket^1$ and $\llbracket \spadesuit \rrbracket^2$ are predicate liftings of type $2^{(-)^n} \to 2^{S-\times T-}$ for n-ary operators $\heartsuit$ and $\spadesuit$.

In other words, the fusion of two structures S and T produces a structure for the disjoint union of operators. In particular $S\times T$-frames carry both the structure of an S-frame and a T-frame, and the interpretation of modalities over the product of S and T is obtained by their interpretation over S and T by just projecting to the respective component. This induces a fusion operation on frame classes.

Definition 4 (Fusion of Frame Classes). Let S and T be two **Set**-endofunctors. If $\mathscr{C}_L$ and $\mathscr{C}_R$ are classes of S and T-frames, respectively, the *fusion* $\mathscr{C}_L \otimes \mathscr{C}_R$ of $\mathscr{C}_L$ and $\mathscr{C}_R$ is the class of $S \times T$-frames given by

$$\mathscr{C}_L \otimes \mathscr{C}_R = \{(X \xrightarrow{\langle \gamma, \delta \rangle} SX \times TX) \mid (X, \gamma) \in \mathscr{C}_L \text{ and } (X, \delta) \in \mathscr{C}_R\}$$

where $\langle \gamma, \delta \rangle(x) = (\gamma(x), \delta(x))$.

The purpose of this paper is essentially to show that for a large class of functors Definitions 2 and 4 are counterparts of one another, i.e. $L \otimes R = \mathrm{Log}(\mathscr{C}_L \otimes \mathscr{C}_R)$ iff $L = \mathrm{Log}(\mathscr{C}_L)$ and $R = \mathrm{Log}(\mathscr{C}_R)$. For now, let us introduce some examples of Λ-logics, their coalgebraic semantics as well as some examples of fusion.

Example 5. 1. For $\Lambda = \{\Box\}$, the logic $\mathbf{E}_{\{\Box\}}$ is the smallest congruential modal logic (classical modal logic in the terminology of [3]). It is sound and complete with respect to neighbourhood frames, i.e. $\mathcal{N}$-coalgebras, where $\mathcal{N} = 2^{2^-}$ is the neighbourhood functor (see also [14]).

2. The smallest extension of $\mathbf{E}$ containing $\Box(p \to q) \to \Box p \to \Box q$ that is closed under the necessitation rule $(p/\Box p)$ is the logic $\mathbf{K}$ which is sound and complete with respect to the class of all $\mathcal{P}$-coalgebras, where $\mathcal{P}$ is the covariant powerset functor and $\llbracket \Box \rrbracket_X(A) = \{B \in \mathcal{P}(X) \mid B \subseteq A\}$. A $\mathcal{P}$-coalgebra γ :

$W \to \mathcal{P}W$ is equivalent to a relation $\mathcal{R}$ on W via $(x, y) \in \mathcal{R}$ iff $y \in \gamma(x)$. It follows that a $\mathcal{P} \times \mathcal{P}$-coalgebra $\langle \gamma, \delta \rangle$ as defined by Def. 4 is equivalent to a pair of relations $\mathcal{R}_L$ and $\mathcal{R}_R$ via $(x, y) \in \mathcal{R}_L$ iff $y \in \pi_1 \circ \langle \gamma, \delta \rangle(x)$ and $(x, y) \in \mathcal{R}_R$ iff $y \in \pi_2 \circ \langle \gamma, \delta \rangle(x)$. We thus recover the standard definition of fusion of Kripke frames [13,11].

3. If we restrict the previous example to the class $\mathscr{C}_{\mathbf{S5}}$ of $\mathcal{P}$-coalgebras corresponding to Kripke frames where the relation is an equivalence, then $\mathrm{Log}(\mathscr{C}_{\mathbf{S5}})$ is the epistemic logic **S5** [2]. Syntactically **S5**, is the smallest extension of **K** containing the axioms for transitivity (4), reflexivity (T) and symmetry (B). It is customary write the operator of **S5** as K and to consider the n-fold fusion $\mathbf{S5}_n = \bigotimes_{i=1}^n \mathbf{S5}$. The modal signature of this logic is $\coprod_{i=1}^n \{K\} \simeq \{K_i\}_{1 \leq i \leq n}$. By Definition 2, $\mathbf{S5}_n$ is thus the smallest $\{K_i\}_{1 \leq i \leq n}$-logic closed under a necessitation rule and a copy of $\Box(p \to q) \to \Box p \to \Box q$, T, B and 4 per operator.

4. The finitely supported probability distribution functor $\mathcal{D}(X) = \{\mu : X \to_f [0,1] \mid \sum_{x \in X} \mu(x) = 1\}$ (where $\to_f$ denotes finite support) extends to a Λ-structure for $\Lambda = \{L_u \mid u \in [0,1] \cap \mathbb{Q}\}$ consisting of unary operators L_u, read as 'with probability at least u'. The functor $\mathcal{D}$ extends to a Λ-structure by stipulating $[\![L_u]\!]_X(A) = \{\mu \in \mathcal{D}(X) \mid \mu(A) \geq p\}$. We therefore obtain *probabilistic epistemic logic* as the fusion of $\mathbf{S5_n} \otimes \mathrm{Log}(\mathscr{D})$ where $\mathscr{D}$ is the class of all $\mathcal{D}$-frames.

Our first goal is to show that consistency transfers under fusion, i.e. $\bot \notin L \otimes R$ whenever $\bot \notin L, R$. We will show this algebraically following [9] using terminology adapted from [5]. The transfer of consistency can be proved using purely coalgebraic arguments, but it requires the assumption that both consitituents of the fusion be sound and complete w.r.t. some classes of coalgebras. Using algebraic arguments we can prove the transfer of consistency under fusion without any extra hypothesis. The proof is also shorter and more elegant.

Definition 6 (Algebraic Semantics). Given a modal signature Λ, a *Λ-modal algebra* $\mathfrak{A}$ is a boolean algebra $(A, 0, 1, \neg^{\mathfrak{A}}, \wedge^{\mathfrak{A}})$ together with an n-ary function $f_\heartsuit : A^n \to A$, for each $\heartsuit \in \Lambda$. The boolean algebra $(A, 0, 1, \neg^{\mathfrak{A}}, \wedge^{\mathfrak{A}})$ is called the *boolean reduct* of $\mathfrak{A}$. A *variable assignment* is a map $\sigma : V \to A$ into the carrier set of $\mathfrak{A}$. Every variable assignment σ extends to an *interpretation* (denoted by the same symbol) $\sigma : \mathcal{L}(\Lambda) \to \mathfrak{A}$ given by

$$\sigma(\phi \wedge \psi) = \sigma(\phi) \wedge^{\mathfrak{A}} \sigma(\psi) \qquad \sigma(\neg\phi) = \neg^{\mathfrak{A}} \sigma(\phi)$$
$$\sigma(\heartsuit(\phi_0, \ldots, \phi_n)) = f_\heartsuit(\sigma(\phi_1), \ldots, \sigma(\phi_n))$$

for $\heartsuit \in \Lambda$ n-ary. An *algebraic model* is a pair $\mathfrak{M} = (\mathfrak{A}, \sigma)$ where $\mathfrak{A}$ is a Λ-modal algebra and σ is an interpretation of $\mathcal{L}(\Lambda)$, $\mathfrak{M}$ is said to be *based on* $\mathfrak{A}$. We say that a formula ϕ is *satisfied* in $\mathfrak{M}$ if $\sigma(\phi) \neq 0$ and we say that ϕ is *true* in $\mathfrak{M}$ (notation $(\mathfrak{A}, \sigma) \Vdash \phi$) when $\sigma(\phi) = 1$. Finally, we say that ϕ is *valid* in $\mathfrak{A}$ if ϕ is true for any model $(\mathfrak{A}, \sigma)$ based on $\mathfrak{A}$.

We will sketch the proof that if L and R are consistent (i.e. $\bot \notin L, R$) then $L \otimes R$ is a conservative extension of both L and R. For more details we refer the reader to [9] where this theorem is proved for normal unary multi-modal logics. However, since the proof does not involve normality or the arity of the operators it extends trivially to our setup. To show that $L \otimes R$ is a conservative extension of both L and R the first step is to build the Lindenbaum-Tarski algebra of L which provides us with a Λ_L-modal algebra $\mathfrak{A}_L$ which validates all formulae of L. Secondly, by contraposition consider a Λ_L-formula ϕ such that $\nvdash_L \neg\phi$. Then $\sigma_{\mathfrak{A}_L}(\neg\phi) \neq 1$, where $\sigma_{\mathfrak{A}_L}$ is the interpretation associated with the Lindenbaum-Tarski algebra of L. Thirdly, we build the Lindenbaum-Tarski algebra $\mathfrak{A}_R$ of R which validates all formulae in R. It is easy to show that both $\mathfrak{A}_L$ and $\mathfrak{A}_R$ are countably infinite atomless provided both similarity types are at most countable [9]. As any countably infinite boolean algebras are isomorphic, their boolean reducts (and thus carrier sets) can therefore be assumed to be equal. The $\Lambda_L + \Lambda_R$-modal algebra consisting of this common boolean algebra plus all operators from $\mathfrak{A}_L$ and $\mathfrak{A}_R$ provides us with a $\Lambda_L + \Lambda_R$-modal algebra $\mathfrak{A}_{L\otimes R}$ which validates L and R and in which $\sigma(\neg\phi) \neq 1$. By the completeness of algebraic semantics we can thus conclude that $\nvdash_{L\otimes R} \neg\phi$.

Theorem 7. *If L and R are consistent logics over at most countable similarity types Λ_L and Λ_R (i.e. $\bot \notin L, R$) then $L \otimes R$ is a conservative extension of both L and R.*

A direct consequence of this proposition is that fusion preserves consistency.

Theorem 8 (Consistency Transfers). *The fusion of two consistent Λ-logics is consistent, i.e. $\bot \notin L \otimes R$ whenever $\bot \notin L, R$.*

While the transfer of consistency can be based on a purely syntactic argument, the transfer of soundness involves the (coalgebraic) semantics, and in particular the fact that a $S\times T$-model can be seen as an S (or T) model by simply forgetting the T (or S) structure.

Theorem 9 (Soundness Transfers). *Let L and R be Λ_L and Λ_R-logics, respectively. If $\mathscr{C}_L$ and $\mathscr{C}_R$ are classes of S-frames and T-frames, respectively, then $L \otimes R \subseteq \mathrm{Log}(\mathscr{C}_L \otimes \mathscr{C}_R)$.*

4 Transfer of Completeness

The remainder of the paper is devoted to establishing the converse of Theorem 9: we establish that completeness transfers to the fusion of two logics over coalgebraic semantics. While an algebraic approach yields transfer of consistency without any further assumptions (in particular without assuming a complete semantics), the situation is different when it comes to transfer of completeness. A naive approach via categorical duality only yields completeness with respect to (the coalgebraic analogue of) descriptive general frames, or coalgebras over Stone spaces [12]. In particular, this form of algebraic completeness does not appear to yield completeness with respect to the fusion of frame classes, mainly because constructions like canonical extensions do not have a coalgebraic counterpart. We therefore adapt a classical construction that witnesses satisfiability of (fusion-) consistent formulas [11] to the coalgebraic setting, and directly build (set-theoretic) models.

For the whole section, we fix two modal signatures Λ_L and Λ_R, two Λ_L- and Λ_R-structures over functors S and T respectively and two logics L and R that are sound and complete with respect to classes $\mathscr{C}_L$ and $\mathscr{C}_R$ of S and T-frames, respectively, that is, $L = \mathrm{Log}(\mathscr{C}_L)$ and $R = \mathrm{Log}(\mathscr{C}_R)$. Our goal is to show that the fusion $L \otimes R$ is complete with respect to the fusion of the corresponding frame classes, i.e. $L \otimes R = \mathrm{Log}(\mathscr{C}_L \otimes \mathscr{C}_R)$. As usual, we consider the contrapositive and show that every $L \otimes R$-consistent formula can be satisfied in a model that is based on a $\mathscr{C}_L \otimes \mathscr{C}_R$-frame. We use the fact that we know how to build coalgebraic models for L- and R-consistent formulae. So the trick is to turn $L \otimes R$-consistent formulae into L- and R-consistent ones and glue the satisfying models together in a suitable way. The passage from $\mathcal{L}(\Lambda_L + \Lambda_R)$ formulae to formulae of the component languages $\mathcal{L}(\Lambda_L)$ and $\mathcal{L}(\Lambda_R)$ is achieved by the following constructions introduced in [11].

4.1 Ersatz and Reconstruction

Definition 10 (Ersatz). Let Λ_L and Λ_R be two similarity types. If ϕ is a formula in the language of the fusion $\mathcal{L}(\Lambda_L + \Lambda_R)$, then its *L-ersatz* (or left-ersatz) ϕ^L is defined by putting $p^L = p$, $(\phi \wedge \psi)^L = \phi^L \wedge \psi^L$, $(\neg\phi)^L = \neg\phi^L$ and

$$(\heartsuit(\phi_1, \dots, \phi_n))^L = \heartsuit(\phi_1^L, \dots, \phi_n^L) \qquad \text{for all } \heartsuit \in \Lambda_L$$
$$(\spadesuit(\phi_1, \dots, \phi_m))^L = q_{\spadesuit(\phi_1, \dots, \phi_m)} \qquad \text{for all } \spadesuit \in \Lambda_R$$

where $p \in V$ and $q_{\spadesuit(\phi_1, \dots, \phi_m)} \in V_R$ is a fresh variable from a set V_R of variables disjoint from V, called an *R-surrogate* (or right-surrogate). The *R-ersatz* (or right-ersatz) ϕ^R of ϕ is defined dually by switching the role of $\heartsuit$ and $\spadesuit$.

The ersatz operations $(.)^L$ and $(.)^R$ are thus an operations of type

$$(.)^L : \mathcal{L}(\Lambda_L + \Lambda_R, V) \to \mathcal{L}(\Lambda_L, V \cup V_R) \qquad (.)^R : \mathcal{L}(\Lambda_L + \Lambda_R, V) \to \mathcal{L}(\Lambda_R, V \cup V_L).$$

Althought not mentioned in [11], the construction relies on the fact that the ersatz operations preserve consistency.

Lemma 11 (Ersatz Perserves Consistency). *If ϕ is an $L \otimes R$-consistent formula then ϕ^L (respectively ϕ^R) is L-consistent (respectively R-consistent).*

Example 12. Consider the formula $\phi = L_u K_i L_v p \wedge L_u q$ in the language of $\mathbf{S}_5^n \otimes \mathbf{Prob}$ introduced in Example 5. We obtain

$$\phi^L = q_{L_u K_i L_v p} \wedge q_{L_u q} \text{ and } \phi^R = L_u q_{K_i L_v p} \wedge L_u q.$$

As can be seen from this example and from the definition, the left-ersatz construction transforms a subformula into a surrogate variable as soon as it sees an R-operator. The nesting of operators following the outermost R-operator is therefore lost. One of the key ideas of the completeness transfer theorem is to alternate left and right ersatz constructions, building a model at each stage until we have drilled down to the level of propositional variables. In order to do this we need to be able to build ersatz formulae deeper inside the nesting of operators.

Definition 13 (Reconstruction). Let ϕ be free of left (respectively right) surrogate variables. We define the reconstruction operators $\uparrow_L$ and $\uparrow_R$ by

$$\uparrow_R\phi = \phi\left(\spadesuit((\psi_1^i)^R, ..., (\psi_m^i)^R)/q_{\spadesuit(\psi_1^i,...,\psi_m^i)}, p_j\right) \text{ for all } \spadesuit \in \Lambda_R$$

$$\uparrow_L\phi \ = \phi\left(\heartsuit((\psi_1^i)^L, ..., (\psi_n^i)^L)/q_{\heartsuit(\psi_1^i,...,\psi_n^i)}, p_j\right) \text{ for all } \heartsuit \in \Lambda_L$$

where i ranges over the set of right (resp. left) surrogate variables in ϕ and j ranges over the set of non-surrogate variables in ϕ. These operations have type

$$\uparrow_L : \mathcal{L}(\Lambda_L + \Lambda_R, V \cup V_L) \to \mathcal{L}(\Lambda_L + \Lambda_R, V \cup V_R)$$
$$\uparrow_R : \mathcal{L}(\Lambda_L + \Lambda_R, V \cup V_R) \to \mathcal{L}(\Lambda_L + \Lambda_R, V \cup V_L),$$

that is $\uparrow_L$ maps formulae free of R-surrogates to formulae free of L-surrogates, dually for $\uparrow_R$. As a consequence, left and right reconstructions can be alternated. To simplify notation we write $\uparrow$ for both and $\uparrow^n$ for the n-fold iteration of $\uparrow$.

Example 14. Let ϕ be as in Example 12. Then we have the following:

$$\begin{aligned}
\uparrow\phi^L &= \uparrow(q_{L_uK_iL_vp} \wedge q_{L_uq}) = L_u(K_iL_vp)^R \wedge (L_uq)^R &&= L_uq_{K_iL_vp} \wedge L_uq \\
\uparrow^2\phi^L &= \uparrow(L_uq_{K_iL_vp} \wedge L_uq) = \quad L_uK_i(L_vp)^L \wedge L_uq &&= L_uK_iq_{L_vp} \wedge L_uq \\
\uparrow^3\phi^L &= \uparrow(L_uK_iq_{L_vp} \wedge L_uq) = \quad L_uK_iL_v(p)^R \wedge L_uq &&= L_uK_iL_vp \wedge L_uq = \phi
\end{aligned}$$

It is intuitively clear that repeated reconstruction reconstructs the original formula step-by-step. We note this as:

Lemma and Definition 15. *Let ϕ be a formula on which the reconstruction operator $\uparrow$ is defined. Then there exists $n \in \omega$ such that $\uparrow^n\phi = \uparrow^{n+1}\phi$ This is the case exactly when the $\uparrow^n\phi$ has no surrogate variables (i.e. has the same variables as ϕ). We call $\uparrow^n\phi$ the* total reconstruction *of ϕ, denoted by $\phi^\uparrow$.*

4.2 Consistency Sets and Consistency Formulae

We are now equiped with two constructions that allow us to 'project' an $L \otimes R$-consistent formula ϕ onto L- and R-consistent formulae (by Lemma 11) for which we know how to build coalgebraic models. Let us for instance start by building ϕ^L. Since it is L-consistent and L is complete with respect to a class $\mathscr{C}_L$ of S-coalgebras, we can build a coalgebraic model for ϕ^L. However, we quickly run into trouble: any model for ϕ^L cannot take into account the actual meaning of its surrogate variables, i.e. of their indices. In particular, it can assign to two surrogate variables q_ϕ, q_ψ a truth value of *true* at the same point even if $\phi \wedge \psi$ is $L \otimes R$-inconsistent. To avoid this problem we need some way of enforcing that a model of ϕ^L gives $L \otimes R$-consistent valuations to surrogate variables. This is achieved by using the following constructions. By convention we will always consider reconstructions based on the left-ersatz (this can be done without loss of generality, since $L \otimes R \simeq R \otimes L$). To simplify the notation we will write the ith reconstruction of ϕ^L as $\phi_i = \uparrow^i(\phi^L)$, and we define for each i the *index set* $\mathbb{S}_i$ of indices of surrogate variables of the ith reconstruction and their subformulae. Formally:

$$\mathbb{S}_i = \mathrm{sf}\{\psi \mid q_\psi \in \mathrm{var}(\phi_i)\} \cup \mathrm{var}(\phi)\}$$

$\mathbb{S}_0$ for example, regroups the indices of all surrogate variables of ϕ^L as well as their subformulae. To enforce $L \otimes R$-consistent valuations of the surrogate variables of ϕ^L, the idea is to first list all the possible $L \otimes R$-consistent combinations of formulae in $\mathbb{S}_0$ into a 'consistency set'.

Definition 16 (Consistency Sets). Let L be a logic and let Δ be a finite set of formulae in the language of L. The *L-consistency set* $\Sigma(\Delta)$ is defined by

$$\Sigma(\Delta) = \{\bigwedge M \mid M \subseteq \Delta \cup \neg\Delta \mid M \text{ maximally consistent}\}$$

where $\neg\Delta = \{\neg\phi \mid \phi \in \Delta\}$ and $s^L_\Delta = \bigvee \Sigma(\Delta)$ is the *L-consistency formula* of Δ.

We read 'maximally consistent' above as maximal among the subsets of $\Delta \cup \neg\Delta$, and $\Sigma(\Delta)$ contains all different possible realisations of combinations of formulae in Δ and the consistency formula s_Δ amounts to requiring that one such combination can be satisfied in a model. Returning to our problem and setting $\Delta = \mathbb{S}_0$, it is easy to see that if the consistency formula $s^L_{\mathbb{S}_0}$ is true at a certain point of an L-model, then we are guaranteed that the combination of surrogate variables which are valuated as 'true' at that point stand for an $L \otimes R$-consistent combination of formulae of $\mathbb{S}_0$. We crucially have that:

Lemma 17. *L-consistency formulae are L-theorems.*

Since $s_{\mathbb{S}_0}$ is an $L \otimes R$-theorem, $\phi \wedge s_{\mathbb{S}_0}$ is $L \otimes R$-consistent, thus by Lemma 11, $\phi^L \wedge s^L_{\mathbb{S}_0}$ is L-consistent, and by completeness of L a model can thus be build for it. Such a model of ϕ^L will necessarily have an $L \otimes R$-consistent valuation for surrogate variables.

4.3 Necessity Operators and Distances

We have just solved a problem in the construction of our model, but we are almost immediately confronted by another one. Indeed, we may have avoided $L \otimes R$-inconsistencies at *one* point in the L-model of ϕ^L (namely the point w making ϕ^L true), but $L \otimes R$-inconsistent valuations of surrogate variables could still happen elsewhere in the model. We therefore need to 'propagate' consistency. In Kripke frames we can simply use the necessitation rule and the box operator $\Box$ to propagate $s^L_{\mathbb{S}_0}$, but in a coalgebraic interpretation this is in general not possible. This problem is the biggest hurdle, but also the most interesting, in generalising Kracht and Wolter's construction [11] to coalgebraic semantics.

Definition 18 (One-step Successors). Given a T-coalgebra $W \xrightarrow{\gamma} TW$, and an element $w \in W$, we define the *1-step successors* $S_1(w)$ of w to be the set

$$S_1(w) = \bigcap\{U \subseteq W \mid \gamma(w) \in Ti[TU]\}$$

where $i : U \hookrightarrow W$ is inclusion and $Ti[TU]$ is the direct image of TU under Ti.

Intuitively, $S_1(w)$ is the smallest subset of W providing a 'support' for $\gamma(w)$. Note however, that Definition 18 is in general not well defined, as arbitrary intersections need not be preserved by **Set**-endofunctors (unlike finite intersections, see Assumption 1). The filter functor provides an easy example of such a behaviour. In fact, [10, Corollary 4.8] provides an elegant criterion for the the existence of the set of 1-step successors of a point in a coalgebra.

Proposition 19. *T preserves infinite intersections iff for any $u \in TW$ there is a smallest $U \subseteq W$ with $u \in Ti[TU]$.*

Assumption 20. From now on we will therefore assume that we are dealing with **Set**-endofunctors that preserve arbitrary intersections. Note that for logics with the finite model property, since we can always assume that the carrier set of any coalgebraic model is finite and since all **Set**-endofunctor preserve finite intersections, we drop this assumption. In particular, all complete rank-1 coalgebraic logic have the finite model property.

The notion of 1-step successor allows us to define a notion of distance on the points of a coalgebraic model $W \xrightarrow{\gamma} TW$.

Definition 21 (Distance). We say that there is *a path of length n* between $x, y \in W$ if there is a sequence of elements $(x_i)_{1\leq i\leq n}$ such that $x = x_1$, $x_n = y$ and $x_{i+1} \in S_1(x_i)$ for all $1 \leq i < n$. The *distance* $dist(x, y)$ between $x, y \in W$ is the length of the shortest path between x and y, or ∞ if no such path exists. Any T-coalgebra (W, γ) induces a distance function $dist : W \times W \to \mathbb{N} \cup \{\infty\}$. Based on the notion of distance we can generalise the set of one-step successors to the following sets of *n-step successors* of w (ball and sphere of radius n around w):

$$S_n(w) = \{x \in W \mid dist(w, x) = n\} \quad B_n(w) = \{x \in W \mid dist(w, x) \leq n\}$$

Finally, we will say that there is a path between two points w and x and write $w \rightsquigarrow x$ if there is a path of finite length between w and x.

Remark 22. 1. The notion of one-step successor is not symmetric and the notion of distance defined above is therefore not a true metric. However, it is an (extended) quasimetric, i.e. it is non-negative, satisfies the triangle inequality and is zero iff the two arguments are equal.

2. For an $S \times T$-coalgebra $W \xrightarrow{\langle\gamma,\delta\rangle} SW \times TW$, there are three notions of distance: the S-distance based on the notion of S-successors $B_1^S(w) := \bigcap\{U \subseteq W \mid \gamma(w) \in Si[SU]\}$, the T-distance based on the notion of T-successors $B_1^T(w) := \bigcap\{U \subseteq W \mid \delta(w) \in Ti[TU]\}$ and the combined $S \times T$-distance based on the notion of $S \times T$-successors $B_1^{S\times T}(w) := \bigcap\{U \subseteq W \mid (\gamma(w), \delta(w)) \in Si[SU] \times Ti[TU]\}$.

A key feature of Kripke semantics is the local aspect of truth, i.e. that the truth of a formula of modal depth n at a world w depends only on points at most n-steps away from w. Similarly, in the coalgebraic semantics it is intuitively clear that if $\phi \in L$ is of modal depth $\mathrm{md}(\phi) = n$ and $\gamma : W \to TW$ is a coalgebraic frame, then the truth value of ϕ at $w \in W$ will also only depends of the valuations of propositional variables at points $x \in B_n(w)$.

Theorem 23 (Coalgebraic Semantics is Local). *Let ϕ be a Λ-formula of modal depth $\mathrm{md}(\phi) = n$. If $M = (W, \gamma, \sigma)$ and $M' = (W, \gamma, \sigma')$ are T-models based on the same frame and T preserves arbitrary intersections, then we have for all $w \in W$ that*

$$M, w \models \phi \iff M', w \models \phi$$

whenever $\sigma(p) \cap B_n(w) = \sigma'(p) \cap B_n(w)$ for all $p \in \mathrm{var}(\phi)$.

Returning to our problem, this result tells us how far we need to propagate the truth of our consistency formula $s^L_{\mathbb{S}_0}$: if the modal depth of ϕ^L is n, we only need to concentrate our efforts on enforcing $s^L_{\mathbb{S}_0}$ in a ball of radius n around the point w where ϕ^L will be satsified. But how can this be done in an coalgebraic model? Over relational semantics we can use the box operator $\Box$ to enforce the truth of a formula ψ at all 1-step away successors but this ability of Kripke semantics to enforce truth on successor states is in general not available in the coalgebraic framework. We will therefore need a coalgebraic generalisation of the necessitation operator.

Definition 24 (Necessity and Necessity Operators). Let L be a Λ-logic. Then L has *weak necessity* over a Λ-structure T if, for every L-consistent formula ϕ there exists an L-consistent formula $\mathrm{nec}(\phi)$ such that

$$S_1(w) \subseteq [\![\phi]\!]_{\mathcal{M}} \text{ whenever } \mathcal{M}, w \models \mathrm{nec}(\phi)$$

for all T-models $\mathcal{M} = (W, \gamma, \sigma)$ and all $w \in W$. We will say that L has *strong necessity* over T if

$$S_1(w) \subseteq [\![\phi]\!]_{\mathcal{M}} \text{ iff } \mathcal{M}, w \models \mathrm{nec}(\phi)$$

A unary operator $\heartsuit \in \Lambda$ is a *necessity operator* over T if

$$S_1(w) \subseteq [\![\phi]\!]_{\mathcal{M}} \text{ whenever } \mathcal{M}, w \models \heartsuit\phi$$

for every T-model $\mathcal{M} = (W, \gamma, \sigma)$ and all $w \in W$. We usually use the symbol $\Box$ for necessity operators.

We will solely focus on notions of necessity arising from necessity operators. In most practical cases such an operator can either be found directly in the logic itself or can be *simulated* within the logic as a boolean combination of existing operators. The following result shows how frequent necessity operators are.

Proposition 25. *Suppose that T : **Set** $\to$ **Set** preserves weak pullbacks and Λ is a modal signature containing $\Box$. Then there exists a predicate lifting $\lambda : 2^- \to 2^{T-}$ making $\Box$ a necessity operator over T.*

In other words, a necessity operator $\Box$ exists for any endofunctor $T : \mathbf{Set} \to \mathbf{Set}$ that preserves weak-pullbacks. It is the predicate lifting associated with the subset $T1$ of $T2$ as described in [16].

Example 26. Many logics have a necessitation operator. This is evident for extensions of (multi-modal) K. It is easy to see that probabilistic modal logic also has a necessitation operator. In the terminology of Example 5, it can be seen easily that $\Box = L_0$ is a necessitation operator over all extensions of probabilistic modal logic, as long as the latter is interpreted over the structure presented in *loc.cit.* An easy calculation shows that this operator indeed arises from the set $\mathcal{D}(1)$ via the construction described in [16].

In the construction of a satisfiable model for $L \otimes R$-consistent formulae, we will use necessitation operators to propagate consistency formulae.

Lemma 27 (Necessity Operators Satisfy Necessitation). *Suppose that L is a Λ-logic and $\Box \in \Lambda$ is a necessity operator over a Λ-structure T, $\phi \in \mathcal{L}(\Lambda)$ and $\mathcal{M}$ is a T-model. Then $\mathcal{M} \models \Box\phi$ whenever $\mathcal{M} \models \phi$.*

We now return to the construction of satisfying $S \times T$-models. We impose $L \otimes R$-consistency at all relevant points in an L-model of ϕ^L as follows: since $s_{\mathbb{S}_0}$ is a $L \otimes R$-theorem, so is $\Box^{\leq md(\phi^L)} s_{\mathbb{S}_0}$ (by Lemma 27) and by using the same reasoning as earlier we can thus build an L-model

$$(W_0, \gamma_0, \sigma_0), w \models \phi^L \wedge \Box^{\leq md^L(\phi)} s^L_{\mathbb{S}_0} \tag{1}$$

that satisfies the consistency formula at all points that influence the interpretation of ϕ (see Theorem 23).

4.4 Generated Submodels

What do we do with the points of the model in Equation 1 that cannot affect the truth of ϕ at w? They may not affect ϕ at w but we still need to build an $L \otimes R$-consistent model. We deal with these points in two ways: first we ensure that our model has no truly excessive points by using generated submodels, secondly we will use non-standard valuations during the construction of the model for points that cannot influence ϕ at w. Only at the last step of the construction will we return to standard (boolean) valuations. Let us first deal with the first point.

Definition 28 (Generated Submodels). Given a coalgebra $W \xrightarrow{\gamma} TW$, we define the *subcoalgebra generated* by $w \in W$ as the set of points reachable via a $\rightsquigarrow$-trace from w, i.e.

$$\mathrm{Tr}(w) = \{x \in W \mid w \rightsquigarrow x\}$$

together with the map $\delta : \mathrm{Tr}(w) \to T(\mathrm{Tr}(w)), x \mapsto Ti^{-1}(\gamma(x))$ where i is the injection of $\mathrm{Tr}(w)$ in W.

It follows from Proposition 19 that the above is well-defined. In fact, we can show slightly more:

Proposition 29. *Given a coalgebra $W \xrightarrow{\gamma} TW$ and $w \in W$,* $\mathrm{Tr}(w)$ *is the smallest subcoalgebra containing w, i.e.*

$$\mathrm{Tr}(w) = \bigcap\{S \subseteq W \mid (S,\delta) \subseteq (W,\gamma) \text{ is a subcoalgebra for some } \delta : S \to TS\}.$$

Clearly the passage from points in satisfying models to generated submodels does not change the validity of formulae at that point.

Proposition 30. *Let ϕ be a Λ-formula for some modal signature Λ for which we assume a Λ-structure has been fixed and let $\mathcal{M} = (W,\gamma,\sigma)$ be a coalgebraic model. Then for any $w \in W$*

$$\mathcal{M}, w \models \phi \Leftrightarrow \mathrm{Tr}(w), w \models \phi.$$

Once we have pruned our model of irrelevant points by considering only generated submodels we deal with the remaining points that cannot influence ϕ^L by weakening the valuations at these points with a non-standard valuation [11].

4.5 Characteristic Sets and Formulae

Finally, how do we give the surrogate variables of ϕ^L an interpretation that reflects their index? The idea is to build at each $x \in W_0$ an R-model which will provide an R-interpretation of the index of all surrogate variables true at x. We must therefore 'sum-up' all that is true at a certain point.

Definition 31 (Characteristic Sets). Given an L-model $\mathcal{M} = (W,\sigma,\gamma)$ and a set of $L \otimes R$-formulae Δ, the *L-characteristic set* $X_L^{V,\Delta}(t)$ at $t \in W$ is

$$X_L^{\sigma,\Delta}(t) = \{\psi \mid \psi \in \Delta \text{ and } \mathcal{M}, t \models \psi^L\} \cup \{\neg\psi \mid \psi \in \Delta \text{ and } \mathcal{M}, t \not\models \psi^L\}$$

The *L-characteristic formula* $\chi_L^{\sigma,\Delta}(t)$ is defined as:

$$\chi_L^{\sigma,\Delta}(t) = \bigwedge X_L^{\sigma,\Delta}(t)$$

The R-characteristic sets and formulae are defined dually.

It is easy to see that by construction characteristic formulae are consistent and sum-up all the formulae of Δ that are true at a point. By taking Δ to be the index set $\mathbb{S}_0$ we can now give a meaning to the surrogate variables in ϕ^L, for if $(\chi^{\sigma_0,\mathbb{S}_0}(t))^R$ is satisfied at a point x_1^t in an R-model $(W_1^t, \delta_1^t, \sigma_1^t)$, then a surrogate variable q_ψ is true at $t \in W_0$ iff ψ^R is true at $x_1^t \in W_1^t$.

4.6 Transfer of Completeness

We have just seen how we can unravel the R-meaning of surrogate variables of ϕ^L in our original model $(W_0, \gamma_0, \sigma_0)$. The next step in the construction of the model is to perform this operation at every point of W_0, i.e. for each $t \in W_0$ we build an R-model (a 'fibre') of $(\chi^{\sigma_0,\mathbb{S}_0}(t))^R$. We then glue all our models together by identifying t and x_1^t (which as mentioned above is legitimate from the point of view of truth values). This gives us a model for the first reconstruction of ϕ which contains a host of R-surrogate variables which much be given a proper L-interpretation and the process we have described is thus iterated. By alternating and gluing these $L \otimes R$-consistent L and R-models in the way we described we eventually reach a model of the final reconstruction of ϕ, i.e. ϕ itself.

Theorem 32 (Completeness Transfer). *Let Λ_L and Λ_R be two modal signatures and let L and R be consistent Λ_L and Λ_R-logic respectively. If both L and R have a necessitation operator over their respective structures, then*

$$L = \mathrm{Log}(\mathscr{C}_L) \text{ and } R = \mathrm{Log}(\mathscr{C}_R) \text{ iff } L \otimes R = \mathrm{Log}(\mathscr{C}_L \otimes \mathscr{C}_R)$$

that is, completeness transfers to the fusion of coalgebraic logics.

Example 33. Given that both **S5** and **Prob** have necessitation over their respective structures (Example 26), we may apply Theorem 32 to show that $\mathbf{S5}_n \otimes \mathbf{Prob}$ is sound and complete with respect to the class $\bigotimes_{i=1}^{n} (\mathscr{C}_{\mathbf{S5}})_i \otimes \mathscr{D}$.

Note that if L and R have the finite model property, then at any stage of our construction, the L- and R-models can be chosen to be finite, and since the total number of steps in the construction is finite (bounded by the modal depth of the formula), the final model is also finite.

Theorem 34. *Under the assumptions of Theorem 32, the finite model property transfers.*

5 Discussion and Future Work

Several questions emerge from our generalisation of Kracht and Wolter's construction. Firstly, can we drop the assumption that functors need to preserve all intersections? As we mentioned above, the case of logics with the finite model property offers a partial solution to this problem which includes any complete rank-1 logic, and in particular the classical logic **E** for which the completenss transfer problem is still open.

Secondly, in order to deal with logics whose semantics is given by non weak-pullback preserving functors we cannot use notions of necessity that are given by simply applying a unary necessity operator to formulae. Instead we may need more complex formulae, i.e. our general notion of necessity (Definition 24). But can the formula nec(ϕ) be constructed or be proven to exist in general, and if not what are the restrictions that prevent it from happening? Less ambitiously, but perhaps more realistically, is there a generalization of our unary necessity operators to the n-ary case for weak-pullback preserving functors?

Finally, is the Kracht and Wolter construction necessary at all? Could some duality argument interpret the syntactic fusion as a binary operation on algebras/theories whose dual would be the semantic fusion operation on (classes of) coalgebras/models. Syntactically, there is a strong connection between the fusion and co-product constructions. The signature of a fusion can be seen as the co-product $\Lambda_L + \Lambda_R$ of its constituent signatures. Moreover, if we view the algebraic models of a Λ_L-logic L as models of a Lawvere theory T_L (the theory of Λ_L modal algebras) and the algebraic models of a Λ_R-logic R as models of a Lawvere theory T_R, then the models of $L \otimes R$ are models of the pushout $T_L \stackrel{i_1}{\leftarrow} T_B \stackrel{i_2}{\rightarrow} T_R$ where T_B is the Lawvere theory of boolean algebras and i_1, i_2 are just the inclusion as boolean reducts. Dually, the fusion of models is based on products, but the correct categorical framework in which to view the operation of fusion on classes of coalgebras as a kind of product or pullback is not clear.

References

1. Baader, F., Ghilardi, S., Tinelli, C.: A new combination procedure for the word problem that generalizes fusion decidability results in modal logics. Information and Computation 204(10), 1413–1452 (2006)
2. Blackburn, P., de Rijke, M., Venema, Y.: Modal Logic. Cambridge Tracts in Theoretical Computer Science, vol. 53. Cambridge University Press, Cambridge (2001)
3. Chellas, B.: Modal Logic, Cambridge (1980)
4. Cirstea, C., Pattinson, D.: Modular proof systems for coalgebraic logics. Theoret. Comput. Sci. 388, 83–108 (2007)
5. Došen, K.: Duality between modal algebras and neighbourhood frames. Studia Logica 48(2), 219–234 (1989)
6. Fagin, R., Halpern, J.: Reasoning about knowledge and probability. J. ACM 41, 340–367 (1994)
7. Fajardo, R., Finger, M.: How not to combine modal logics. In: Prasad, B. (ed.) IICAI, pp. 1629–1647 (2005)
8. Fine, K., Schurz, G.: Transfer theorems for stratified modal logics. In: Copeland, J. (ed.) Logic and Reality, Essats in Pure and Applied Logic. In Memory of Arthur Prior, pp. 169–213. Oxford Univeristy Press, Oxford (1996)
9. Gabbay, D.M., Kurucz, A., Wolter, F., Zakharyaschev, M.: Many-Dimensional Modal Logics: Theory and Applications. Elsevier, Amsterdam (2003)
10. Gumm, H.P.: From T-coalgebras to filter structures and transition systems. In: Fiadeiro, J.L., Harman, N.A., Roggenbach, M., Rutten, J. (eds.) CALCO 2005. LNCS, vol. 3629, pp. 194–212. Springer, Heidelberg (2005)
11. Kracht, M., Wolter, F.: Properties of independently axiomatizable bimodal logics. J. Symb. Logic 56(4), 1469–1485 (1991)
12. Kupke, C., Kurz, A., Venema, Y.: Stone coalgebras. Theor. Comput. Sci. 327(1-2), 109–134 (2004)
13. Kurucz, A.: Combining modal logics. In: Patrick Blackburn, J.V.B., Wolter, F. (eds.) Handbook of Modal Logic. Studies in Logic and Practical Reasoning, vol. 3, pp. 869–924. Elsevier, Amsterdam (2007)
14. Schröder, L., Pattinson, D.: Rank-1 logics are coalgebraic. Journal of Logic and Computation 20(5), 1113–1147 (2010)
15. Schröder, L., Pattinson, D.: Modular algorithms for heterogeneous modal logics via multi-sorted coalgebra. Math. Struct. Comput. Sci. 21 (2011)
16. Schröder, L.: Expressivity of coalgebraic modal logic: The limits and beyond. Theor. Comput. Sci. 390(2-3), 230–247 (2008)
17. Wolter, F.: Fusions of modal logics revisited. In: Kracht, M., de Rijke, M., Wansing, H., Zakharyaschev, M. (eds.) Advances in Modal Logic 1096. CSLI Lecture Notes, pp. 361–379. CSLI Publications, Stanford (1998)

Indexed Induction and Coinduction, Fibrationally

Clément Fumex, Neil Ghani, and Patricia Johann

University of Strathclyde, Scotland

Abstract. This paper extends the fibrational approach to induction and coinduction pioneered by Hermida and Jacobs, and developed by the current authors, in two key directions. First, we present a sound coinduction rule for any data type arising as the final coalgebra of a functor, thus relaxing Hermida and Jacobs' restriction to polynomial data types. For this we introduce the notion of a *quotient category with equality* (QCE), which both abstracts the standard notion of a fibration of relations constructed from a given fibration, and plays a role in the theory of coinduction dual to that of a comprehension category with unit (CCU) in the theory of induction. Second, we show that indexed inductive and coinductive types also admit sound induction and coinduction rules. Indexed data types often arise as initial algebras and final coalgebras of functors on slice categories, so our key technical results give sufficent conditions under which we can construct, from a CCU (QCE) $U : \mathcal{E} \to \mathcal{B}$, a fibration with base $\mathcal{B}/I$ that models indexing by I and is also a CCU (QCE).

1 Introduction

Iteration operators provide a uniform way to express common and naturally occurring patterns of recursion over inductive data types. Categorically, iteration operators arise from initial algebra semantics: the constructors of an inductive data type are modelled as a functor F, the data type itself is modelled as the carrier μF of the initial F-algebra $in : F(\mu F) \to \mu F$, and the iteration operator $fold : (FA \to A) \to \mu F \to A$ for μF maps an F-algebra $h : FA \to A$ to the unique F-algebra morphism from in to h. Initial algebra semantics provides a comprehensive theory of iteration which is i) *principled*, in that it ensures that programs have rigorous mathematical foundations that can be used to give them meaning and prove their soundness; ii) *expressive*, in that it is applicable to *all* inductive types — i.e., all types that are carriers of initial algebras — rather than just to syntactically defined classes of data types such as polynomial ones; and iii) *sound*, in that it is valid in any model — set-theoretic, domain-theoretic, realizability, etc. — interpreting data types as carriers of initial algebras.

Final coalgebra semantics gives an equally comprehensive understanding of coinductive types: the destructors of a coinductive data type are modelled as a functor F, the data type itself is modelled as the carrier νF of the final F-coalgebra $out : \nu F \to F(\nu F)$, and the coiteration operator $unfold : (A \to FA) \to A \to \nu F$ for νF maps an F-coalgebra $k : A \to FA$ to the unique F-coalgebra morphism from k to out. Final coalgebra semantics thus provides a theory of coiteration which is as principled, expressive, and sound as that for induction.

A. Corradini, B. Klin, and C. Cîrstea (Eds.): CALCO 2011, LNCS 6859, pp. 176–191, 2011.

Since induction and iteration are closely linked, we might expect initial algebra semantics to give a principled, expressive, and sound theory of induction as well. However, most theories of induction for a data type μF, where $F : \mathcal{B} \to \mathcal{B}$, are sound only under significant restrictions on the category $\mathcal{B}$, the functor F, or the property to be established. Recently a conceptual breakthrough in the theory of induction was made by Hermida and Jacobs [7]. They show how to lift an arbitrary functor F on a base category $\mathcal{B}$ of types to a functor $\hat{F}$ on a category of properties over those types. They take the premises of an induction rule for μF to be an $\hat{F}$-algebra, and their main theorem shows that such a rule is sound if the lifting $\hat{F}$ preserves truth predicates. Hermida and Jacobs work in a fibrational, and hence axiomatic, setting and treat *any* notion of property that can be suitably fibred over $\mathcal{B}$. Moreover, they place no stringent requirements on $\mathcal{B}$. Thus, they overcome two of the aforementioned limitations. But since they give sound induction rules only for polynomial data types, the limitation on the functors treated remains in their work. The current authors [3] subsequently removed this final restriction to give sound induction rules for all inductive types on the underlying fibration under conditions commensurate with those in [7].

In this paper, we extend the existing body of work in three key directions. First, Hermida and Jacobs developed a fibrational theory of coinduction to complement their theory of induction. But this theory, too, is sound only for polynomial data types, and so does not apply to final coalgebras of some key functors, such as the finite powerset functor. In this paper, we derive a sound fibrational coinduction rule for *every* coinductive data type, i.e., for every type that is the carrier of a final coalgebra. Second, data types arising as initial algebras of functors are fairly simple. More sophisticated data types — e.g., untyped lambda terms and red-black trees — are often modelled as inductive indexed types arising as initial algebras of functors on slice categories, presheaf categories, and similar structures. In this paper, we derive sound induction rules for such inductive indexed types. Finally, since we can derive sound induction rules for inductive types and inductive indexed types, and sound coinduction rules for coinductive types, we might expect to be able to derive sound coinduction rules for coinductive indexed types, too. In this paper, we confirm that this is the case.

This rest of this paper is structured as follows. In Section 2 we recall the fibrational approach to induction pioneered in [7] and extended in [3]. In Section 3 we extend the results of [7] to derive sound coinduction rules for *all* functors with final coalgebras. We give sound induction (coinduction) rules for inductive (resp., coinductive) indexed types in Section 4 (resp., Section 5). Section 6 summarises our conclusions, and discusses related work and possibilities for future work.

2 Induction in a Fibrational Setting

Fibrations support a uniform, axiomatic approach to induction and coinduction that is widely applicable and abstracts over the specific choices of category, functor, and predicate. This is advantageous because i) the semantics of data types in languages involving recursion and other effects usually involves categories other than Set; ii) in such circumstances, the standard set-based interpretations of predicates are no longer germane; iii) in any setting, there can be more than one reasonable notion of predicate; and iv) fibrations allow induction and coinduction rules for many classes of data types to be obtained by instantiation of a single, generic theory rather than developed an ad hoc, case-by-case basis.

2.1 Fibrations in a Nutshell

We begin with fibrations. More details can be found in, e.g., [9,14].

Definition 2.1. *Let $U : \mathcal{E} \to \mathcal{B}$ be a functor. A morphism $g : Q \to P$ in $\mathcal{E}$ is* cartesian *over a morphism $f : X \to Y$ in $\mathcal{B}$ if $Ug = f$ and, for every $g' : Q' \to P$ in $\mathcal{E}$ with $Ug' = fv$ for some $v : UQ' \to X$, there exists a unique $h : Q' \to Q$ in $\mathcal{E}$ such that $Uh = v$ and $gh = g'$.*

The cartesian morphism $f^{\S}_P$ over a morphism f with codomain UP is unique up to isomorphism. We write f^*P for the domain of $f^{\S}_P$, and omit the subscript P when it can be inferred from context.

Definition 2.2. *Let $U : \mathcal{E} \to \mathcal{B}$ be a functor. Then U is a* fibration *if for every object P of $\mathcal{E}$ and every morphism $f : X \to UP$ in $\mathcal{B}$ there is a cartesian morphism $f^{\S}_P : f^*P \to P$ in $\mathcal{E}$ over f.*

If $U : \mathcal{E} \to \mathcal{B}$ is a fibration, we call $\mathcal{B}$ the *base category* of U and $\mathcal{E}$ its *total category*. Objects of $\mathcal{E}$ are thought of as properties, objects of $\mathcal{B}$ are thought of as types, and U is thought to map each property P in $\mathcal{E}$ to the type UP about which it is a property. An object P in $\mathcal{E}$ is said to be *above* its image UP under U, and similarly for morphisms. For any object X of $\mathcal{B}$, we write $\mathcal{E}_X$ for the *fibre above* X, i.e., the subcategory of $\mathcal{E}$ comprising objects above X and morphisms above id_X. Morphisms within a fibre are said to be *vertical.* If $f : X \to Y$ is a morphism in $\mathcal{B}$, then the function mapping each object P of $\mathcal{E}$ to f^*P extends to a functor $f^* : \mathcal{E}_Y \to \mathcal{E}_X$ called the *reindexing functor induced by f*.

Example 2.3. The category Fam(Set) has as objects pairs (X, P) with X a set and $P : X \to \text{Set}$. We call X the *domain* of (X, P), and write P for (X, P) when convenient. A morphism from $P : X \to \text{Set}$ to $P' : X' \to \text{Set}$ is a pair $(f, f^{\sim})$ of functions $f : X \to X'$ and $f^{\sim} : \forall x : X.\, P\,x \to P'(f\,x)$. The functor $U : \text{Fam(Set)} \to \text{Set}$ mapping (X, P) to X is called the *families fibration.*

Example 2.4. The *arrow category* of $\mathcal{B}$, denoted $\mathcal{B}^{\to}$, has morphisms of $\mathcal{B}$ as its objects. A morphism from $f : X \to Y$ to $f' : X' \to Y'$ in $\mathcal{B}^{\to}$ is a pair (α_1, α_2) of morphisms in $\mathcal{B}$ such that $f'\alpha_1 = \alpha_2 f$. The codomain functor $cod : \mathcal{B}^{\to} \to \mathcal{B}$ maps an object $f : X \to Y$ of $\mathcal{B}^{\to}$ to the object Y of $\mathcal{B}$. If $\mathcal{B}$ has pullbacks, then *cod* is a fibration, called the *codomain fibration over* $\mathcal{B}$. Indeed, given an object $f : X \to Y$ in the fibre above Y and a morphism $f' : X' \to Y$ in $\mathcal{B}$, the pullback of f along f' gives a cartesian morphism over f'.

We say $U : \mathcal{E} \to \mathcal{B}$ is an opfibration, if $U^{op} : \mathcal{E}^{op} \to \mathcal{B}^{op}$ is a fibration. Concretely:

Definition 2.5. *Let $U : \mathcal{E} \to \mathcal{B}$ be a functor. A morphism $g : P \to Q$ in $\mathcal{E}$ is* opcartesian *over a morphism $f : X \to Y$ in $\mathcal{B}$ if $Ug = f$ and, for every $g' : P \to Q'$ in $\mathcal{E}$ with $Ug' = vf$ for some $v : Y \to UQ'$, there exists a unique $h : Q \to Q'$ in $\mathcal{E}$ such that $Uh = v$ and $hg = g'$.*

As for cartesian morphisms, the opcartesian morphism $f_{\S}^P$ over a morphism f with codomain UP is unique up to isomorphism. We write $\Sigma_f P$ for the domain of $f_{\S}^P$, and omit the superscript P when it can be inferred from context.

Definition 2.6. *If $U : \mathcal{E} \to \mathcal{B}$ is a functor, then U is an* opfibration *if for every object P of $\mathcal{E}$ and every morphism $f : UP \to Y$ in $\mathcal{B}$ there is an opcartesian morphism $f_{\S}^P : P \to \Sigma_f P$ in $\mathcal{E}$ over f. A functor U is a* bifibration *if it is simultaneously a fibration and an opfibration.*

If $f : X \to Y$ is a morphism in the base of an opfibration, then the function mapping each object P of $\mathcal{E}_X$ to $\Sigma_f P$ extends to a functor $\Sigma_f : \mathcal{E}_X \to \mathcal{E}_Y$ called the *opreindexing functor induced by* f. The following useful result is from [10].

Lemma 2.7. *Let $U : \mathcal{E} \to \mathcal{B}$ be a fibration. Then U is a bifibration iff, for every morphism $f : X \to Y$ in $\mathcal{B}$, f^* is right adjoint to Σ_f.*

2.2 Fibrational Induction in Another Nutshell

At the heart of Hermida and Jacobs' approach to induction is the observation that if $U : \mathcal{E} \to \mathcal{B}$ is a fibration and $F : \mathcal{B} \to \mathcal{B}$ is a functor, then F can be lifted to a functor $\hat{F} : \mathcal{E} \to \mathcal{E}$ and the premises of the induction rule for μF can be taken to be an $\hat{F}$-algebra. Crucially, for this induction rule to be sound, the lifting must be truth-preserving. These terms are defined as follows.

Definition 2.8. *Let $U : \mathcal{E} \to \mathcal{B}$ be a fibration and $F : \mathcal{B} \to \mathcal{B}$ be a functor. A* lifting *of F with respect to U is a functor $\hat{F} : \mathcal{E} \to \mathcal{E}$ such that $U\hat{F} = FU$. If each fibre $\mathcal{E}_X$ has a terminal object, and if reindexing preserves terminal objects, then we say that U* has fibred terminal objects. *In this case, the map assigning to every X in $\mathcal{B}$ the terminal object in $\mathcal{E}_X$ defines a functor K_U which is called the* truth functor for U *and is right adjoint to U. We omit the subscript on K_U when this can be inferred. A lifting $\hat{F}$ of F is called* truth-preserving *if $KF \cong \hat{F}K$.*

The codomain fibration *cod* from Example 2.4, for instance, has fibred terminal objects: the terminal object in $\mathcal{E}_X$ is id_X. A truth-preserving lifting $F^{\to}$ of F with respect to *cod* is given by the action of F on morphisms.

Definition 2.9. *A* comprehension category with unit *(CCU) is a fibration $U : \mathcal{E} \to \mathcal{B}$ with a truth functor K_U which itself has a right adjoint $\{-\}_U$. In this case, $\{-\}_U$ is called the* comprehension functor *for U.*

We omit the subscript on $\{-\}_U$ when this can be inferred. The fibration *cod* is the canonical CCU: the comprehension functor is the domain functor $dom : \mathcal{B}^{\to} \to \mathcal{B}$ mapping $f : X \to Y$ in $\mathcal{B}^{\to}$ to X. Truth-preserving liftings with respect to CCUs are used in [7] to state and prove soundness of induction rules.

Theorem 2.10. *Let $U : \mathcal{E} \to \mathcal{B}$ be a CCU, let $F : \mathcal{B} \to \mathcal{B}$ be a functor with initial algebra μF, and let $\hat{F}$ be a truth-preserving lifting of F. Then the following induction rule for F is sound:*

$$ind_F : \forall(P : \mathcal{E}).\ (\hat{F}P \to P) \to \mu F \to \{P\}$$

Proof. Because $\hat{F}$ is truth-preserving, the initial $\hat{F}$-algebra exists and has carrier $K(\mu F)$. Thus, for any $\hat{F}$-algebra $h : \hat{F}P \to P$, we have $fold\, h : K(\mu F) \to P$. Since $K \dashv \{-\}$, this map in turn gives the desired map from μF to $\{P\}$.

This very elegant theorem shows that fibrations provide just the right structure to derive sound induction rules for inductive data types whose underlying functors have truth-preserving liftings. Although Hermida and Jacobs gave such liftings only for polynomial functors, [3] showed that every functor has a truth-preserving lifting with respect to certain bifibrations, called *Lawvere categories.*

Definition 2.11. *A fibration $U : \mathcal{E} \to \mathcal{B}$ is a* Lawvere category *if it is a CCU which is also a bifibration.*

If ϵ is the counit of the adjunction $K \dashv \{-\}$ for a CCU U, then $\pi_P = U\epsilon_P$ defines a natural transformation $\pi : \{P\} \to UP$. (The domain of π_P really is $\{P\}$ since $UK = Id$.) Moreover, π extends to a functor $\pi : \mathcal{E} \to \mathcal{B}^{\to}$ in the obvious way.

Lemma 2.12. *Let $U : \mathcal{E} \to \mathcal{B}$ be a Lawvere category. Then π has a left adjoint $I : \mathcal{B}^{\to} \to \mathcal{E}$ defined by $I\,(f : X \to Y) = \Sigma_f\,(KX)$.*

For any functor F, the composition $\hat{F} = IF^{\to}\pi : \mathcal{E} \to \mathcal{E}$ defines a truth-preserving lifting with respect to the Lawvere category U [4]. Here, $F^{\to}$ is the lifting given after Definition 2.8. If F also has an initial algebra, then Theorem 2.10 guarantees that it has a sound induction rule as well.

If $\mathcal{B}$ has pullbacks, the following diagram shows that we have actually given a modular construction of a lifting with respect to a Lawvere category by factorisation through the lifting for the codomain fibration:

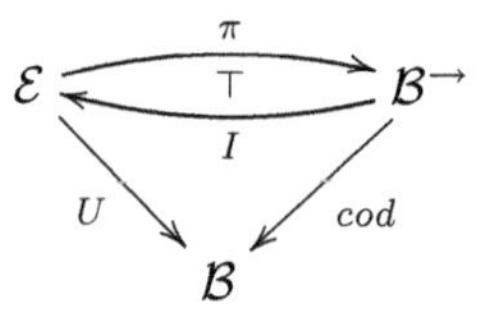

3 Coinduction

In [7], a sound fibrational coinduction rule is given for final coalgebras of polynomial functors. The development is based on a fibration U, but since coinduction is concerned with relations, a new fibration $Rel(U)$ of relations is first constructed.

Definition 3.1. *Let $U : \mathcal{E} \to \mathcal{B}$ be a fibration, assume $\mathcal{B}$ has products, and let $\Delta : \mathcal{B} \to \mathcal{B}$ be the diagonal functor sending an object X to $X \times X$. Then the fibration $Rel(U) : Rel(\mathcal{E}) \to \mathcal{B}$ is obtained by the pullback of U along Δ.*

That the pullback of a fibration along any functor is a fibration is well-known [11]. The process of pulling back a fibration along a functor F, called *change of base along F*, is also well-known to preserve fibred terminal objects [6]. The fibration $Rel(U)$ therefore has a truth functor. Below we denote the pullback of *any* functor $F : \mathcal{A} \to \mathcal{B}$ along a functor $G : \mathcal{B}' \to \mathcal{B}$ by $G^*F : G^*\mathcal{A} \to \mathcal{B}'$. The objects of $G^*\mathcal{A}$ are pairs (X, Y) such that $GX = FY$, and G^*F maps the pair (X, Y) to the object X. We write Y for (X, Y) in G^*F when convenient. Definition 3.1 entails that the fibre of $Rel(\mathcal{E})$ above X is the fibre $\mathcal{E}_{X \times X}$. A morphism from (X, Y) to (X', Y') in $Rel(\mathcal{E})$ consists of a pair of morphisms $\alpha : X \to X'$ and $\beta : Y \to Y'$ such that $U\beta = \alpha \times \alpha$. Finally, if U has truth functor K, then the truth functor for $Rel(U)$ is given by $K_{Rel(U)}X = K(X \times X)$.

In the inductive setting, truth-preserving liftings were needed. In the coinductive setting, we need equality-preserving liftings, where equality is given by:

Definition 3.2. *Let $U : \mathcal{E} \to \mathcal{B}$ be a bifibration with a truth functor and assume $\mathcal{B}$ has products. The* equality functor *for U is the functor $Eq_U : \mathcal{B} \to Rel(\mathcal{E})$ mapping an object X to $\Sigma_\delta K_{Rel(U)} X$ and a morphism $f : X \to I$ to the unique morphism above $f \times f$ induced by the naturality of δ at f and the opcartesian map $\delta_\S^{KX}$. Here, $\delta : Id_\mathcal{B} \to \Delta$ is the diagonal natural transformation with components $\delta_X : X \to X \times X$, and $\Sigma_\delta : \mathcal{E} \to Rel(\mathcal{E})$ maps an object P in $\mathcal{E}_X$ to an object $\Sigma_{\delta_X} P$ in $\mathcal{E}_{X \times X}$. If Eq_U has a left adjoint Q_U, then Q_U is called the* quotient functor *for U. We suppress the subscripts on Eq_U and Q_U when convenient.*

Definition 3.3. *Let $U : \mathcal{E} \to \mathcal{B}$ be a bifibration which has a truth functor, assume $\mathcal{B}$ has products, and let $F : \mathcal{B} \to \mathcal{B}$ be a functor. A lifting $\check{F}$ of F with respect to $Rel(U)$ is called* equality-preserving *if $Eq\, F \cong \check{F}\, Eq$.*

That every polynomial functor has an equality-preserving lifting is shown in [7]. Theorem 3.4 is Hermida and Jacobs' main theorem about coinduction. Note the duality: in the inductive, setting the truth functor K must have a right adjoint, while in the coinductive one, the equality functor Eq must have a left adjoint.

Theorem 3.4. *Let $U : \mathcal{E} \to \mathcal{B}$ be a fibration which has a truth functor, assume $\mathcal{B}$ has products, let $F : \mathcal{B} \to \mathcal{B}$ be a functor with final coalgebra νF, and let $\check{F}$ be an equality-preserving lifting of F. If Eq has a left adjoint Q, then the following coinduction rule for F is sound:*

$$coind_F : \forall (R : Rel(\mathcal{E})).\ (R \to \check{F} R) \to QR \to \nu F$$

Proof. Because $\check{F}$ is equality-preserving, the final $\check{F}$-coalgebra exists and has carrier $Eq(\nu F)$. Thus, for any $\check{F}$-coalgebra $k : R \to \check{F}R$, we have *unfold* $k : R \to Eq(\nu F)$. Since $Q \dashv Eq$, this map in turn gives the desired map from QR to νF.

3.1 Generic Coinduction for All Final Coalgebras

The first contribution of this paper is to give a sound coinduction rule for any functor with a final coalgebra. To do so, we show how to generate liftings that can be instantiated to give both the truth-preserving liftings required for induction and, by duality, the equality-preserving liftings required for coinduction.

Lemma 3.5. *Consider a quotient category with equality (QCE) over $\mathcal{B}$, i.e., a fibration $U : \mathcal{E} \to \mathcal{B}$ with a full and faithful functor $E : \mathcal{B} \to \mathcal{E}$ such that $UE = Id_{\mathcal{B}}$ and E has left adjoint Q with unit η. Define functors ρ, J, and $\check{F}$ by*

$$\begin{array}{lll} \rho : \mathcal{E} \to \mathcal{B}^{\to} & J : \mathcal{B}^{\to} \to \mathcal{E} & \check{F} : \mathcal{E} \to \mathcal{E} \\ \rho P = U\eta_P & J\,(f : X \to Y) = f^* EY & \check{F} = J\,F^{\to}\,\rho \end{array}$$

Then $U\check{F} = FU$ and $\check{F}E \cong EF$.

Proof. To prove $U\check{F} = FU$, note that the morphisms ρP each have domain UP, that $dom\, F^{\to}\rho = FU$, and that $UJ = dom$. Together these give $U\check{F} = UJF^{\to}\rho = FU$. To prove $\check{F}E \cong EF$, we first assume that i) for every X in $\mathcal{B}$, ρEX is an isomorphism in $\mathcal{B}$, and ii) for every isomorphism f in $\mathcal{B}$, $Jf \cong E(dom\, f)$. Then since $UE = Id_{\mathcal{B}}$, i) and ii) imply that $\check{F}E = JF^{\to}\rho E \cong E\, dom\, F^{\to}\rho E = EFUE = EF$. To discharge i), note that, since E is full and faithful, $\eta E : E \to EQE$ is $E\kappa$ for a natural transformation $\kappa : Id_{\mathcal{E}} \to QE$, where each κ_X is an isomorphism with inverse ϵ_X and ϵ is the counit of $Q \dashv E$. Then $\rho EX = U\eta_{EX} = UE\kappa_X = \kappa_X$, so that ρEX is indeed an isomorphism. To discharge ii), let f be an isomorphism in $\mathcal{B}$. Since cartesian morphisms over isomorphisms are isomorphisms, we have $Jf = f^*(E\,(cod f)) \cong E\,(cod\, f) \cong E\,(dom\, f)$. Here, the first isomorphism is witnessed by $f^{\S}$ and the second by Ef^{-1}.

The lifting $\check{F}$ has as its dual the lifting $\hat{F}$ given in the following lemma.

Lemma 3.6. *Let $U : \mathcal{E} \to \mathcal{B}$ be an opfibration, let $K : \mathcal{B} \to \mathcal{E}$ a full and faithful functor such that $UK = Id_{\mathcal{B}}$, and let $C : \mathcal{E} \to \mathcal{B}$ be a right adjoint to K with counit ϵ. Define functors π, I, and $\hat{F}$ by*

$$\begin{array}{lll} \pi : \mathcal{E} \to \mathcal{B}^{\to} & I : \mathcal{B}^{\to} \to \mathcal{E} & \hat{F} : \mathcal{E} \to \mathcal{E} \\ \pi P = U\epsilon_P & I\,(f : X \to Y) = \Sigma_f KY & \hat{F} = I\,F^{\to}\,\pi \end{array}$$

Then $U\hat{F} = FU$ and $\hat{F}K \cong KF$.

Proof. By dualisation of Lemma 3.5. The setting on the left below with U an opfibration is equivalent to the setting on the right with U a fibration.

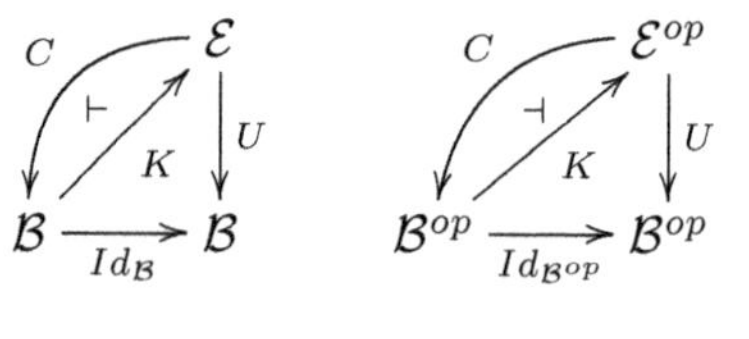

We can instantiate Lemmas 3.5 and 3.6 to derive both the truth-preserving lifting for all functors from [3] and an equality-preserving lifting for all functors. The former gives the sound induction rule for all inductive types presented in [3], and the latter gives a sound coinduction rule for all coinductive types. To obtain the lifting for induction, let $U : \mathcal{E} \to \mathcal{B}$ be a Lawvere category, K be the truth functor for U, and C be the comprehension functor for U. Since a Lawvere category is an opfibration, Lemma 3.6 ensures that any functor $F : \mathcal{B} \to \mathcal{B}$ lifts to a truth-preserving lifting $\hat{F} : \mathcal{E} \to \mathcal{E}$. This is exactly the lifting of [3]. To obtain the lifting for coinduction, let $U : \mathcal{E} \to \mathcal{B}$ be a bifibration with a truth functor and products in $\mathcal{B}$, and let E be the equality functor $Eq = \Sigma_\delta K_{Rel(U)}$ for U. Since both $K_{Rel(U)}$ and Σ_δ are full and faithful, so is Eq. Moreover, since EqX is in the fibre of $Rel(U)$ above X we have $Rel(U)\, Eq = Id_{\mathcal{B}}$. We can therefore take E to be Eq in Lemma 3.5 provided Eq has a left adjoint Q. In this case, every functor $F : \mathcal{B} \to \mathcal{B}$ has an equality-preserving lifting $\check{F} : Rel(\mathcal{E}) \to Rel(\mathcal{E})$. Thus, if F has a final coalgebra, then νF has a sound coinduction rule.

The domain functor $dom : \mathcal{B}^{\to} \to \mathcal{B}$ is actually a fibration called the *domain fibration over* $\mathcal{B}$. No conditions on $\mathcal{B}$ are required. Just as *cod* is the canonical CCU, *dom* is the canonical QCE. A QCE $Rel(U)$ over $\mathcal{B}$ which is obtained by change of base along Δ, and for which the functor E is the equality functor for U, is called a *relational QCE.*

Example 3.7. We can take U to be $dom : \mathcal{B}^{\to} \to \mathcal{B}$, E to map each X in $\mathcal{B}$ to id_X, and Q to be *cod* in Lemma 3.5. Then $\check{F}$ is exactly $F^{\to}$, so that $F^{\to}$ and $\check{F}$ are interdefinable. Thus, just as the lifting $\hat{F}$ with respect to an arbitrary fibration U satisfying the hypotheses of Lemma 3.6 can be modularly constructed from the specific lifting $F^{\to}$ with respect to *cod* [3], so the lifting $\check{F}$ with respect to an arbitrary fibration U satisfying the hypotheses of Lemma 3.5 can be modularly constructed from the specific lifting $F^{\to}$ with respect to *dom*.

So *dom* plays a role role in the coinductive setting similar to that played by *cod* in the inductive one. We think of a morphism $f : X \to Y$ in the total category of *cod* as a predicate on Y whose proofs constitute X. Intuitively, f maps each p in X to the element y in Y about which it is a proof. Similarly, we think of a morphism $f : X \to Y$ in the total category of *dom* as a relation on X, the quotient of X by which has equivalence classes comprising Y. Intuitively, f maps each x in X to its equivalence class in that quotient.

Example 3.8. If U is the families fibration, then the fibre above X in $Rel(U)$ consists of functions $R : X \times X \to Set$. We think of these as constructive relations, where $R(x, x')$ gives the set of proofs that x is related to x'. In Lemma 3.5 we can take U to be the families fibration, E to map each set X to the relation eq_X defined by $eq_X(x, x') = 1$ if $x = x'$ and $eq_X(x, x') = 0$ otherwise, and Q to map each relation $R : X \times X \to Set$ to the quotient X/R of X by the least equivalence relation containing R. We get this instantiation of the definition of $\check{F}$, for $F : Set \to Set$, from Lemma 3.5: $\rho : Rel(U) \to Set^{\to}$ maps a relation $R : X \times X \to$ Set to the quotient map $\rho_R : X \to X/R$, $F^{\to}$ maps f to Ff, and

$J : Set^{\rightarrow} \rightarrow Rel(U)$ maps $f : X \rightarrow Y$ to the relation $\bar{f}$ mapping (x, x') to 1 if $fx = fx'$ and to 0 otherwise. Thus $\check{F} : FA \times FA \rightarrow Set$ is given by $\check{F}R = \overline{F\rho_R}$.

We now derive the coinduction rule, for the functor $\mathscr{P}_{\mathit{fin}}$, which maps a set to its finite powerset, with respect to the fibration of relations constructed from the families fibration in Example 3.8. Since $\mathscr{P}_{\mathit{fin}}$ is not polynomial, it lies outside the scope of [7], but it is important, since a number of canonical coalgebras are built from it. For example, a finitely branching labelled transition system with state space S and labels from an alphabet A is a coalgebra $S \rightarrow \mathscr{P}_{\mathit{fin}}(A \times S)$.

Example 3.9. By Example 3.8, the lifting $\check{\mathscr{P}}_{\mathit{fin}}$ maps a relation $R : A \times A \rightarrow Set$ to the relation $\check{\mathscr{P}}_{\mathit{fin}}R : \mathscr{P}_{\mathit{fin}}A \times \mathscr{P}_{\mathit{fin}}A \rightarrow Set$ defined by $\check{\mathscr{P}}_{\mathit{fin}}R = \overline{\mathscr{P}_{\mathit{fin}}\rho_R}$. Thus, if X and Y are finite subsets of A, then $(X, Y) \in \check{\mathscr{P}}_{\mathit{fin}}R$ iff $\mathscr{P}_{\mathit{fin}}\rho_R X = \mathscr{P}_{\mathit{fin}}\rho_R Y$. Since the action of $\mathscr{P}_{\mathit{fin}}$ on a morphism f maps any subset of the domain of f to its image under f, $\mathscr{P}_{\mathit{fin}}\rho_R X = \mathscr{P}_{\mathit{fin}}\rho_R Y$ iff $(\forall x : X).(\exists y : Y).\, xRy \wedge (\forall y : Y).(\exists x : X).\, xRy$. From $\check{\mathscr{P}}_{\mathit{fin}}$ we have that the resulting coinduction rule has as its premises a $\check{\mathscr{P}}_{\mathit{fin}}$-coalgebra, i.e., a relation $R : A \times A \rightarrow Set$ and a map from R to $\check{\mathscr{P}}_{\mathit{fin}}R$ in $Rel(U)$. An object of $Rel(U)$ is a pair $(X, (Y, P))$ where X is a set, (Y, P) is an object of $Fam(Set)$, and $Y = X \times X$. A morphism in $Rel(U)$ from $(X, (Y, P))$ to $(X', (Y', P'))$ consists of a morphism $\phi : X \rightarrow X'$ in Set and a morphism $(\psi, \psi^{\sim}) : (Y, P) \rightarrow (Y', P')$ in $Fam(Set)$ such that $\psi = \phi \times \phi$. Thus, a $\check{\mathscr{P}}_{\mathit{fin}}$-coalgebra consists of a function $\alpha : A \rightarrow \mathscr{P}_{\mathit{fin}}A$ together with a function $\alpha^{\sim} : (\forall a, a' : A).\, aRa' \rightarrow (\alpha a)\ \check{\mathscr{P}}_{\mathit{fin}}R\ (\alpha a')$. If we regard $\alpha : A \rightarrow \mathscr{P}_{\mathit{fin}}A$ as a transition function, i.e., if we define $a \rightarrow b$ iff $b \in \alpha a$, then $\alpha^{\sim}$ captures the condition that R is a bisimulation over α. The coinduction rule thus asserts that any two bisimilar states have the same interpretation in the final coalgebra.

4 Indexed Induction

Data types arising as initial algebras and final coalgebras on traditional semantic categories such as Set and $\omega cpo_{\perp}$ are of limited expressivity. More sophisticated data types arise as initial algebras of functors on their indexed versions. To build intuition about the resulting *inductive indexed types*, first consider the inductive type $\mathsf{List}\,X$ of lists of X. It is clear that the definition of $\mathsf{List}\,X$ does not require an understanding of $\mathsf{List}\,Y$ for for any $Y \neq X$. Since, each type $\mathsf{List}\,X$ is, in isolation, inductive, List can be considered a *family of inductive types*. By contrast, for each n in Nat, let $\mathsf{Fin}\,n$ be the data type of n-element sets, and consider the inductive definition of the Nat-indexed type $\mathsf{Lam} : \mathsf{Nat} \rightarrow Set$ of untyped λ-terms up to α-equivalence with free variables in $\mathsf{Fin}\,n$ given by

$$\frac{i : \mathsf{Fin}\,n}{\mathit{Var}\,i : \mathsf{Lam}\,n} \qquad \frac{f : \mathsf{Lam}\,n \quad a : \mathsf{Lam}\,n}{\mathit{App}\,f\,a : \mathsf{Lam}\,n} \qquad \frac{b : \mathsf{Lam}\,(n+1)}{\mathit{Abs}\,b : \mathsf{Lam}\,n}$$

Unlike $\mathsf{List}\,X$, the type $\mathsf{Lam}\,n$ cannot be defined in isolation using only the elements of $\mathsf{Lam}\,n$ that have already been constructed. Indeed, elements of $\mathsf{Lam}\,(n+1)$ are needed to construct elements of $\mathsf{Lam}\,n$ so that, in effect, all of the types

$\mathsf{Lam}\, n$ must be inductively constructed simultaneously. The indexed type Lam is thus an *inductive family of types*, rather than a family of inductive types.

There is considerable interest in inductive and coinductive indexed types. If types are interpreted in a category $\mathcal{B}$, and if I is a set of indices considered as a discrete category, then an inductive I-indexed type can be modelled by the initial algebra of a functor on the functor category $I \to \mathcal{B}$. Alternatively, indices can be modelled by objects of $\mathcal{B}$, and inductive I-indexed types can be modelled by initial algebras of functors on slice categories $\mathcal{B}/I$. Coinductive indexed types can similarly be modelled by final colagebras of functors on slice categories.

Initial algebra semantics for inductive indexed types has been developed extensively [2,12]. Pleasingly, no fundamentally new insights were required: the standard initial algebra semantics needed only to be instantiated to categories such as $\mathcal{B}/I$. By contrast, the theory of induction for inductive indexed types has received comparatively little attention. The second contribution of this paper is to use our fibrational framework to derive sound induction rules for such types by similarly instantiating initial algebra semantics to appropriate categories. The key technical question to be solved turns out to be: given a Lawvere category of properties fibred over types, can we construct a new Lawvere category fibred over indexed types from which induction rules for the indexed types can be derived? To answer it, we make the simplifying assumption that the inductive indexed types of interest arise as initial algebras of functors over slice categories, i.e., of functors $F : \mathcal{B}/I \to \mathcal{B}/I$, where I is an object of $\mathcal{B}$. Let U/I denote the Lawvere category to be constructed. We conjecture that the total category of U/I should be a slice category of $\mathcal{E}$, and so make the canonical choice to slice over KI, where K is the truth functor for U. We then define $U/I : \mathcal{E}/KI \to \mathcal{B}/I$ by $U/I\,(f : P \to KI) = Uf : UP \to I$. Here, *cod* (Uf) really is I because $UK = Id$.

We first show that U/I is indeed a bifibration. We give a concrete proof before indicating how the same result can be derived from a more abstract treatment.

Lemma 4.1. *If $U : \mathcal{E} \to \mathcal{B}$ is a fibration (bifibration) and I is an object of $\mathcal{B}$, then U/I is a fibration (resp., bifibration).*

Proof. Let $\alpha : Y \to I$ and $\beta : X \to I$ be objects of $\mathcal{B}/I$, and let $\phi : Y \to X$ be a morphism in $\mathcal{B}/I$ from α to β, i.e., be such that $\alpha = \beta\phi$. First, let $f : P \to KI$ be an object of $\mathcal{E}/KI$ such that $(U/I)f = Uf = \beta$, and let $\phi_P^\S : \phi^* P \to P$ be the cartesian morphism in $\mathcal{E}$ over ϕ with respect to U. Then $\phi_P^\S$ is a morphism in $\mathcal{E}/KI$ with domain $f\phi_P^\S$ and codomain f, and it is cartesian over ϕ with respect to U/I. Thus, U/I is a fibration if U is. Now, let $g : Q \to KI$ be an object of $\mathcal{E}/KI$ such that $(U/I)g = Ug = \alpha$, and let $\phi_\S^Q : Q \to \Sigma_\phi Q$ be the opcartesian morphism in $\mathcal{E}$ over ϕ with respect to U. Since $\alpha = \beta\phi$, the opcartesianness of $\phi_\S^Q$ ensures that there is a unique map $k : \Sigma_\phi\, Q \to KI$ in $\mathcal{E}$ above β such that $g = k\phi_\S^Q$. Then $\phi_\S^Q$ is a morphism in $\mathcal{E}/KI$ with domain g and codomain k, and it is opcartesian over ϕ with respect to U/I. Thus, U/I is an opfibration if U is. Combining these results gives that if U is a bifibration then so is U/I.

There is an alternative characterisation of $\mathcal{E}/KI$ which both clarifies the conceptual basis of our treatment of indexed induction and simplifies our calculations. The next lemma is the key observation underlying this characterisation.

Lemma 4.2. *Let $U : \mathcal{E} \to \mathcal{B}$ be a fibration with truth functor K, let I be an object of $\mathcal{B}$, and let $\alpha : X \to I$. Then $(\mathcal{E}/KI)_\alpha \cong \mathcal{E}_X$.*

Proof. One half of the isomorphism maps the object $f : P \to KI$ of $(\mathcal{E}/KI)_\alpha$ to P. For the other half, note that since truth functors map objects to terminal objects, and since reindexing preserves terminal objects, we have $KX \cong \alpha^* KI$. Thus, for any object Q above X, we get a morphism from Q to KI by composing $\alpha^\S_{KI}$ and the unique morphism ! from Q to KX. Since ! is vertical and $\alpha^\S_{KI}$ is above α, this composition is above α. Thus each object Q in $\mathcal{E}_X$ maps to an object of $(\mathcal{E}/KI)_\alpha$. It is routine to verify that these maps constitute an isomorphism.

By Lemma 4.2 we may identify objects (morphisms) of $(\mathcal{E}/KI)_\alpha$ and objects (resp., morphisms) of $\mathcal{E}_X$. This gives our abstract characterisation of U/I:

Lemma 4.3. *Let $U : \mathcal{E} \to \mathcal{B}$ be a fibration and I be an object of $\mathcal{B}$. Then U/I can be obtained by change of base by pulling U back along $dom : \mathcal{B}/I \to \mathcal{B}$.*

Proof. As noted in Section 3, the pullback of a fibration along a functor is a fibration. The objects (morphisms) of the fibre above $\alpha : X \to I$ of the pullback of U along *dom* are the objects (resp., morphisms) of $\mathcal{E}_X$. By Lemma 4.2, the pullback of U along *dom* is therefore U/I.

As observed in Section 3, pulling back a fibration along a functor preserves fibred terminal objects, so U/I has fibred terminal objects if U does by Lemma 4.3. Concretely, the truth functor $K_{U/I} : B/I \to \mathcal{E}/KI$ maps an object $f : X \to I$ to $Kf : KX \to KI$. To see that U/I is a Lawvere category if U is, we thus need to show that $K_{U/I}$ has a right adjoint. For this, we use an abstract theorem due to Hermida [5] to transport adjunctions to pullbacks along fibrations.

Lemma 4.4. *Let $F \dashv G : \mathcal{A} \to \mathcal{B}$ be an adjunction with counit ϵ, and let $U : \mathcal{E} \to \mathcal{B}$ be a fibration. Then the functor $U^*F : U^*\mathcal{A} \to \mathcal{E}$ has a right adjoint $G_U : \mathcal{E} \to U^*\mathcal{A}$ whose action maps each object E to the object $(\epsilon^*_{UE}E, GUE)$.*

Lemma 4.5. *Change of base along a fibration preserves CCUs, i.e., if $U : \mathcal{E} \to \mathcal{B}$ is a CCU and $U' : \mathcal{E}' \to \mathcal{B}$ is a fibration, then the pullback U'^*U is a CCU.*

Proof. We already have that U'^*U is a fibration with fibred terminal objects. To see that $K_{U'^*U}$ has a right adjoint, consider the pullback of U^*U' and K_U. This pullback is given by $\mathcal{E}'$, $K_{U'^*U} : \mathcal{E}' \to U'^*\mathcal{E}$, and $U' : \mathcal{E}' \to \mathcal{B}$. Note that U^*U' is a fibration since it is obtained by pulling U' back along U. Lemma 4.4 then ensures that, since K_U has a right adjoint, so does $K_{U'^*U}$. Thus U'^*U is a CCU.

When $U : \mathcal{E} \to \mathcal{B}$, I is an object of $\mathcal{B}$, and U' is $dom : \mathcal{B}/I \to \mathcal{B}$, the comprehension functor for U'^*U — i.e., for U/I — maps an object P in $\mathcal{E}_X$ above $\alpha : X \to I$ to $\alpha\pi_P : \{P\} \to I$. Combining Lemmas 4.1 and 4.5 and the fact that U^{op} is a fibration if U is an opfibration, we have

Lemma 4.6. *Let $U : \mathcal{E} \to \mathcal{B}$ be a Lawvere category and $U' : \mathcal{E}' \to \mathcal{B}$ be a fibration. Then U'^*U is a Lawvere category.*

Thus, if F is a functor on $\mathcal{E}'$ with initial algebra μF, then Theorem 2.10 guarantees the existence of a sound induction rule for μF. We use this observation to derive an induction rule for the indexed containers of Morris and Altenkirch [12].

Example 4.7. If I is a set, then the *category of I-indexed sets* is the fibre $\mathrm{Fam}(\mathrm{Set})_I$. An *$I$-indexed set* is thus a function $X : I \to \mathrm{Set}$, and a morphism h from X to X', written $h : X \to_I X'$, is a function of type $(\Pi i : I).\, Xi \to X'i$. Morris and Altenkirch denote this category $I \to \mathrm{Set}$ and define an *I-indexed container* to be a pair (S, P) with $S : I \to \mathrm{Set}$ and $P : (\Pi i : I).\, Si \to I \to \mathrm{Set}$. An I-indexed container defines a functor $[S, P] : (I \to \mathrm{Set}) \to I \to \mathrm{Set}$ by $[S,P]Xi = (\Sigma s : Si).\, P\, i\, s \to_I X$. Thus, if $t : [S,P]\, X\, i$, then t is of the form (s, f), with projections ρ_0 and ρ_1 defined by $\rho_0\, t = s$ and $\rho_1\, t = f$. The action of $[S,P]$ on a morphism $g : X \to_I Y$ maps a pair (s,f) to (s, gf). The initial algebra of $[S,P]$ is denoted $in : [S,P]W_{S,P} \to_I W_{S,P}$. Since $I \to \mathrm{Set}$ is equivalent to Set/I, we can use the results of this section to extend those of [12] by deriving an induction rule for $W_{S,P}$. A predicate over an I-indexed set X is a function $Q : (\Pi i : I).\, Xi \to \mathrm{Set}$. To simplify notation, this is written $Q : X \to_I \mathrm{Set}$. The lifting $\widehat{[S,P]}$ of $[S,P]$ maps each $Q : X \to_I \mathrm{Set}$ to the predicate $\widehat{[S,P]}Q : [S,P]X \to_I \mathrm{Set}$ defined by $\widehat{[S,P]}\, Q\, i\, (s,f) = (\Pi j : I).\, (\Pi p : P\, i\, s\, j).\, Q\, j\, (f\, j\, p)$. Altogether, this gives the following induction rule for establishing a predicate $Q : W_{S,P} \to_I \mathrm{Set}$:

$$
\begin{aligned}
&(\Pi i{:}I).\,(\Pi(s,f){:}[S,P]\,W_{S,P}\,i).\,((\Pi j{:}I).\,(\Pi p{:}P\,i\,s\,j).\,Q\,j(f\,j\,p) \to Q\,i(in\,i\,(s,f))))\\
&\quad \to (\Pi i{:}I).\,(\Pi t{:}W_{S,P}\,i).\,Q\,i\,t
\end{aligned}
$$

5 Indexed Coinduction

We now present our third contribution: we derive coinduction rules for coinductive indexed types. Examples of such types are infinitary versions of inductive indexed types, such as infinitary untyped lambda terms and the interaction structures of Hancock and Hyvernat [8]. If $U : \mathcal{E} \to \mathcal{B}$ supports coinduction for the final coalgebra of any functor on $\mathcal{B}$ having one, and if $U' : \mathcal{E}' \to \mathcal{B}$ gives a change of base to an indexed notion of data described by $\mathcal{E}'$, then is there a fibration over $\mathcal{E}'$ supporting indexed coinduction for the final coalgebra of any functor on $\mathcal{E}'$ having one. However, the details in the coinductive setting are much more involved than in the inductive one, here we present only the following simpler result, showing that for any relational QCE over a base category $\mathcal{B}$ and object I of $\mathcal{B}$, change of base along $dom : \mathcal{B}/I \to \mathcal{B}$ yields a relational QCE over $\mathcal{B}/I$.

If $\mathcal{B}$ has products and $U : \mathcal{E} \to \mathcal{B}$ is a bifibration with truth functor K, then the equality functor Eq for U is given by $Eq = \Sigma_\delta K$. Let $Rel(U) : Rel(E) \to \mathcal{B}$ be a QCE, i.e., let Eq have a left adjoint Q. To define a relational QCE over $\mathcal{B}/I$ we must first see that $\mathcal{B}/I$ has products. But the product of f and g in $\mathcal{B}/I$ is determined by their pullback: if W, $j : W \to Z$, and $i : W \to X$ give the pullback of f and g, then their product in $\mathcal{B}/I$ is the morphism fi or, equivalently, gj. Below, we write f^2 for the product of f with itself in $\mathcal{B}/I$ and X_fX for the domain of f^2. Then, if $\mathcal{B}$ has pullbacks, we can construct the relation fibration $Rel(U/I) : Rel(\mathcal{E}/KI) \to \mathcal{B}/I$ from the pullback of U/I along the product functor $\Delta/I : \mathcal{B}/I \to \mathcal{B}/I$ mapping f to f^2. Concretely, an object of $Rel(\mathcal{E}/KI)$ above $f : X \to I$ is an object of $\mathcal{E}/KI$ above f^2 with respect to U/I. This is, in turn, equivalent to an object of $\mathcal{E}$ above X_fX with respect to U.

5.1 The Equality Functor for U/I

If U is a bifibration with a truth functor then, for any object I of $\mathcal{B}$, U/I is as well, and so U/I has an equality functor $Eq_{U/I}$. To define this functor concretely, note that the component of the diagonal natural transformation $\delta/I : Id \to \Delta/I$ at $f : X \to I$ is given by the diagram on the left. Thus, $Eq_{U/I}$ maps an object $f : X \to I$ of $\mathcal{B}/I$ to the unique morphism above f^2 in the diagram on the right induced by the opcartesian map m above $(\delta/I)_f$:

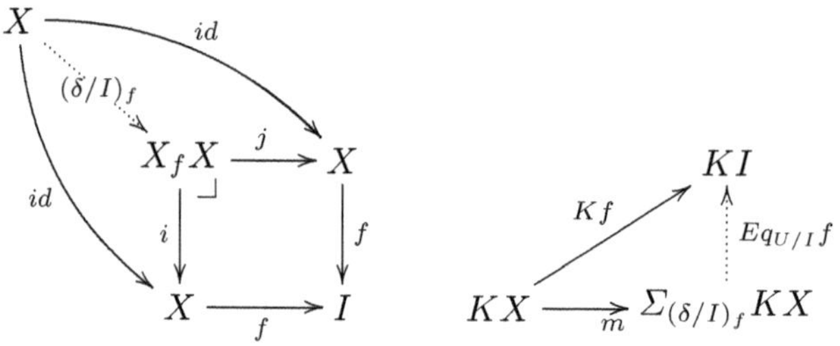

5.2 The Quotient Functor for U/I

Whereas defining the equality functor for U/I was straightforward, defining its quotient functor is actually tricky. We have not (yet!) found any abstract fibrational results to deliver it, so we give a concrete construction. For each object I of $\mathcal{B}$, we define another fibration, denoted $Rel(U)/I : Rel(\mathcal{E})/Eq\,I \to \mathcal{B}/I$, where $Eq : \mathcal{B} \to Rel(\mathcal{E})$ is the equality functor for U. The objects of $Rel(\mathcal{E})/Eq\,I$ above $f : X \to I$ are morphisms $\alpha : P \to Eq\,I$ for some object P of $Rel(\mathcal{E})$ such that $U\alpha = \Delta f$. Our first result identifies conditions under which $Rel(U)/I$ is a QCE.

Lemma 5.1. *Let $\mathcal{B}$ have pullbacks, let I be an object of $\mathcal{B}$, and let $Rel(U)$: $Rel(\mathcal{E}) \to \mathcal{B}$ be a relational QCE. Then $Rel(U)/I$ is a QCE.*

Proof. Let $Eq : \mathcal{B} \to Rel(\mathcal{E})$ and $Q : Rel(\mathcal{E}) \to \mathcal{B}$ be the equality and quotient functors for U, respectively. We construct a full and faithful functor $E' : \mathcal{B}/I \to Rel(\mathcal{E})/Eq\,I$ such that $(Rel(U)/I)\,E' = Id_{\mathcal{B}/I}$ and a left adjoint Q' for E' as follows. Take E' to be Eq. Then E' is full and faithful since Eq is. Moreover, for any $f : X \to I$, Definition 3.2 ensures that $Eq\,f$ is above $f \times f$ with respect to U, so $(Rel(U)/I)\, E'f = f$, and thus $(Rel(U)/I)\, E' = Id_{\mathcal{B}/I}$. Finally, we define Q' to map each object $\alpha : P \to Eq\, I$ of $Rel(\mathcal{E})/Eq\, I$ to its transpose $\alpha' : QP \to I$ under the adjunction $Q \dashv Eq$. That $Q' \dashv E'$ follows directly from $Q \dashv Eq$.

We can now define the quotient functor for $Rel(U/I)$ using the functor Q' from the proof of Lemma 5.1. The key step is to define an adjunction $\tau \dashv \sigma$ so that the diagram below commutes. Then if E' and Q' are the functors witnessing the fact that $Rel(U)/I$ is a QCE, the compositions $\sigma E'$ and $Q'\tau$ give equality and quotient functors for $Rel(U/I)$.

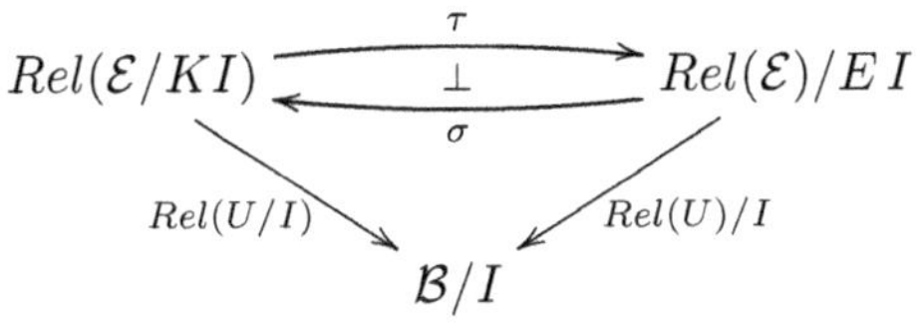

To define τ and σ, let $f : X \to I$, let i and j be the projections for the pullback square defining $X_f X$. The universal property of the product $X \times X$ ensures the existence of a morphism $v : X_f X \to X \times X$ such that $\pi_1 v = i$ and $\pi_2 v = j$. By the universal property of the pullback of f along itself, v is a monomorphism; we will use this in the proof of Lemma 5.3. From the diagram on the left below, we have that $\delta_X = v\,(\delta/I)_f$, which also gives the diagram on the right:

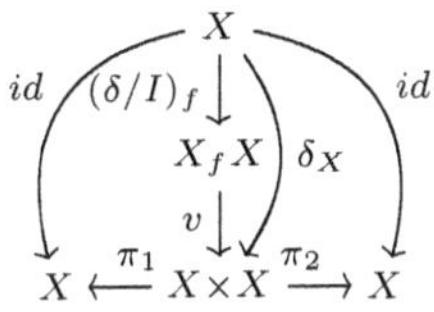

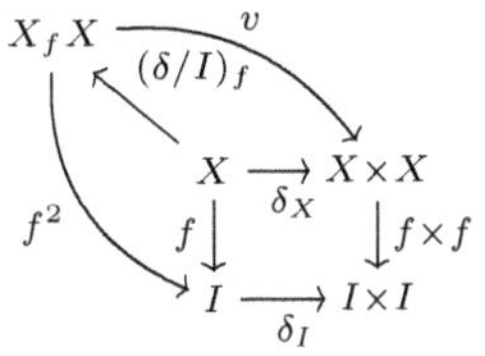

We use the right diagram and the opcartesianness of $v_\S^A$ to define the functor τ. First recall that, if $R : A \to KI$ is an object of $Rel(\mathcal{E}/KI)$ above $f : X \to I$ with respect to $Rel(U/I)$, then $Rel(U/I)R = (U/I)R = UR = f^2$. The following diagram then defines the morphism h above $f \times f$ in $\mathcal{E}$ to which τ maps R:

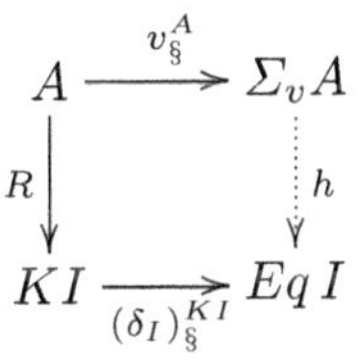

So, assuming U satisfies the Beck-Chevalley condition [9], this all gives:

Definition 5.2. *The functors τ and σ are given by:*

$$\tau : Rel(\mathcal{E}/KI) \to Rel(\mathcal{E})/Eq\, I \qquad \sigma : Rel(\mathcal{E})/Eq\, I \to Rel(\mathcal{E}/KI)$$
$$\tau(R : A \to KI) = h \qquad \sigma(S : B \to Eq\, I) = v^* B$$

Lemma 5.3. *τ is a full and faithful left adjoint to σ.*

Proof. We exhibit the unit η of the adjunction $\tau \dashv \sigma$ and show that it is an isomorphism. Because v is a monomorphism, the unit η' of the adjunction $\Sigma_v \dashv v^*$ is an isomorphism. Moreover, since objects of $Rel(\mathcal{E}/KI)$ above $f : X \to I$ in the fibration $Rel(U/I)$ can be seen as objects of $\mathcal{E}$ above $X_f X$ in U, η must assign to every object R above $X_f X$ in U a morphism from R to $v^* \Sigma_v A$. We define $\eta = \eta'$. The universality of η follows from the fact that η' is an isomorphism.

Recall that our candidate for the quotient functor for $Rel(U/I)$ is $Q'\tau$. To see that $Q'\tau \dashv Eq_{U/I}$, note that $Q'\tau \dashv \sigma E'$, so we need only verify that $Eq_{U/I}$ is $\sigma E'$. It is routine to check that $\tau Eq_{U/I} = E'$, from which $Eq_{U/I} = \sigma E'$ follows.

We now use the results of this section to give a coinduction rule for final coalgebras of indexed containers that is dual to the induction rule of Example 4.7.

Example 5.4. Let (S, P) be an I-indexed container with final coalgebra $out : M_{S,P} \to_I [S, P]M_{S,P}$. A relation over an I-indexed set $X : I \to \mathrm{Set}$ is an I-indexed family of relations Ri on Xi. The relational lifting of $[S, P]$ maps a relation R over an I-indexed set X to the relation R' over the I-indexed set $[S, P]X$ that relates $(s, f) \in [S, P]Xi$ and $(s', f') \in [S, P]Xi$ iff $s = s'$ and, for all $j : I$ and $p : P\, i\, s\, j$, $f\, j\, p$ is related in Rj to $f'\, j\, p$. This gives the following notion of bisimulation for $[S, P]$ coalgebras $k : X \to_I [S, P]X$: if $x, x' \in Xi$, then $x \sim_i x'$ iff $\rho_0(kx) = \rho_0(kx')$ and, for all $j : I$ and $p : P\, i\, (\pi_0(kx))\, j$, $\rho_1(fx)p \sim_j \rho_1(fx')p$.

6 Conclusions, Related Work, and Future Work

In this paper, we have extended the fibrational approach to induction and coinduction pioneered by Hermida and Jacobs, and further developed by the current authors, in three key directions: We gave sound coinduction rules for all functors having final coalgebras provided the fibration interpreting them is a QCE, and we gave similarly sound generic induction and coinduction rules for all functors over slice categories having initial algebras and final coalgebras.

The work of Hermida and Jacobs is most closely related to ours, but there is, of course, a large body of work on induction and coinduction. in a broader setting. In dependent type theory, for example, data types are usually presented with elimination rules that are exactly induction rules. Along these lines, [13] has heavily influenced the development of induction in Coq. Another important strand of related work concerns inductive families and their induction rules [2]. On the coinductive side, papers such as [1,15,16] have had immense impact in bringing bisimulation into the mainstream of theoretical computer science.

There are several evident directions for future work. The most immediate is showing that change of base along a fibration preserves QCEs, just as it does CCUs; this would yield a compact derivation of the results in Section 5 analogous to that in Section 4. We also expect to exploit the predictive power of our theory to provide induction and coinduction rules for advanced data types — such as inductive recursive types — whose rules are not discernible by sheer intuition. In such circumstances our generic fibrational approach should provide rules whose use is justified by their soundness proofs. Finally, we would like to see our induction and coinduction rules for advanced data types incorporated into implementations such as Agda and Coq.

References

1. Aczel, P., Mendler, P.: A Final Coalgebra Theorem. In: Proceedings, Category Theory and Computer Science, pp. 357–365 (1989)
2. Dybjer, P.: Inductive Families. Formal Aspects of Computing 6(4), 440–465 (1994)
3. Ghani, N., Johann, P., Fumex, C.: Fibrational Induction Rules for Initial Algebras. In: Proceedings, Computer Science Logic, pp. 336–350 (2010)
4. Ghani, N., Johann, P., Fumex, C.: Generic Fibrational Induction (2011) (submitted)
5. Hermida, C.: Some properties of Fib as a Fibred 2-Category. Journal of Pure and Applied Algebra 134(1), 83–109 (1993)
6. Hermida, C.: Fibrations, Logical Predicates and Related Topics. Dissertation, University of Edinburgh (1993)
7. Hermida, C., Jacobs, B.: Structural Induction and Coinduction in a Fibrational Setting. Information and Computation 145, 107–152 (1998)
8. Hancock, P.G., Hyvernat, P.: Programming Interfaces and Basic Topology. Annals of Pure and Applied Logic 137(1-3), 189–239 (2006)
9. Jacobs, B.: Categorical Logic and Type Theory. Studies in Logic and the Foundations of Mathematics, vol. 141 (1999)
10. Jacobs, B.: Comprehension Categories and the Semantics of Type Dependency. Theoretical Computer Science 107, 169–207 (1993)
11. Jacobs, B.: Quotients in Simple Type Theory. Mathematics Institute (1994)
12. Morris, P., Altenkirch, T.: Indexed Containers. In: Proceedings, Logic in Computer Science, pp. 277–285 (2009)
13. Pfenning, F., Paulin-Mohring, C.: Inductively Defined Types in the Calculus of Constructions. In: Proceedings, Mathematical Foundations of Programming Semantics, pp. 209–228 (1989)
14. Pavlovič, D.: Predicates and Fibrations. Dissertation, University of Utrecht (1990)
15. Rutten, J.: Universal Coalgebra: A Theory of Systems. Theoretial Computer Science 249(1), 3–80 (2000)
16. Turi, D., Rutten, J.: On the Foundations of Final Coalgebra Semantics. Mathematical Structures in Computer Science 8(5), 481–540 (1998)

Stone Duality for Nominal Boolean Algebras with И

Murdoch J. Gabbay, Tadeusz Litak, and Daniela Petrişan

Abstract. We define Boolean algebras over nominal sets with a function-symbol и mirroring the И 'fresh name' quantifier (Banonas), and dual notions of nominal topology and Stone space. We prove a representation theorem over fields of nominal sets, and extend this to a Stone duality.

1 Introduction

We study Boolean logic with И over nominal sets. $\text{И}a.\phi(a)$ means 'for *fresh* a, $\phi(a)$ holds'. Here a is intended to be an atomic resource: examples are a variable symbol, memory location, channel name, or (in security) nonce.

We introduce Banonas (Boolean Algebra, NOminal with Иa) and nominal Stone spaces, and prove representation and duality theorems. Stone's representation does not generalise easily to nominal sets: the ultrafilter lemma breaks down (since a union of finitely-supported sets need not be finitely-supported). To 'fix' this, something corresponding to И seems to be required, so the mathematics leads us to Banonas even if we started off just with Boolean algebras in nominal sets (without И).

И is interesting from both a mathematical and computational point of view. It is *self-dual* ($\neg\text{И}a.\phi(a) \Leftrightarrow \text{И}a.\neg\phi(a)$) and this gives a computationally useful *some/any property*: to prove И we check $\phi(a)$ for *some* fresh name, but we can then use *any* fresh b. (Definitions below or in [20,16].)

И also appears to be useful. It has been applied to syntax-with-binding [20], resource generation in functional programming [32], local names in incomplete trees [6], π-calculus name-restriction [3], game semantics [1,36], and nominal domains [31,35]. A categorical analysis of self-duality has been suggested [27], and several proof-theories for И [13,17,8,11].

Equational axiomatisation, a representation theorem, and topology with И have not been studied. As we shall see, algebra and topology with И are non-trivial and interesting. For more discussion, see Example 3.6.

2 Background on Nominal Sets and the И-quantifier

A nominal set is a 'set with names'. The notion of a name being 'in' an element is given by support $supp(x)$ (Definition 2.7). For more details of nominal sets, see [20,16]. The reader can think of $supp(x)$ as 'the free names of x', but without committing to x being syntax—on the contrary, x could have any structure provided it admits a permutation action.

A. Corradini, B. Klin, and C. Cîrstea (Eds.): CALCO 2011, LNCS 6859, pp. 192–207, 2011.

Definition 2.1. Fix a countably infinite set of **atoms** $\mathbb{A}$. We use a **permutative convention** that $a, b, c, \ldots$ range over *distinct* atoms.

Definition 2.2. A **(finite) permutation** π is a bijection on atoms such that $nontriv(\pi) = \{a \mid \pi(a) \neq a\}$ is finite.

Write id for the **identity** permutation such that $\mathrm{id}(a) = a$ for all a. Write $\pi' \circ \pi$ for composition, so that $(\pi' \circ \pi)(a) = \pi'(\pi(a))$. Write π^{-1} for inverse, so that $\pi^{-1} \circ \pi = \mathrm{id} = \pi \circ \pi^{-1}$. Write $(a\ b)$ for the **swapping** (terminology from [20]) mapping a to b, b to a, and all other c to themselves, and take $(a\ a) = \mathrm{id}$.

Definition 2.3. If $A \subseteq \mathbb{A}$ define $fix(A) = \{\pi \mid \forall a \in A. \pi(a) = a\}$.

Definition 2.4. A **set with a permutation action** $\mathcal{X}$ is a pair $(|\mathcal{X}|, \cdot)$ of an **underlying set** $|\mathcal{X}|$ and a **permutation action** written $\pi \cdot x$ which is a group action on $|\mathcal{X}|$, so that $\mathrm{id} \cdot x = x$ and $\pi \cdot (\pi' \cdot x) = (\pi \circ \pi') \cdot x$ for all $x \in |\mathcal{X}|$ and permutations π and π'.

Say that $A \subseteq \mathbb{A}$ **supports** $x \in |\mathcal{X}|$ when $\forall \pi. \pi \in fix(A) \Rightarrow \pi \cdot x = x$. If a finite A supporting x exists, call x **finitely-supported**.

Definition 2.5. Call a set with a permutation action $\mathcal{X}$ a **nominal set** when every $x \in |\mathcal{X}|$ has finite support. $\mathcal{X}$, $\mathcal{Y}$, $\mathcal{Z}$ will range over nominal sets.

$\mathbb{A}$ forms a nominal set where $\pi \cdot a = \pi(a)$, and $\mathcal{X} \times \mathcal{Y}$ is a nominal set with underlying set $\{(x, y) \mid x \in |\mathcal{X}|, y \in |\mathcal{Y}|\}$ and action $\pi \cdot (x, y) = (\pi \cdot x, \pi \cdot y)$.

Definition 2.6. Call a function $f \in |\mathcal{X}| \to |\mathcal{Y}|$ **equivariant** when $\pi \cdot f(x) = f(\pi \cdot x)$ for all permutations π and $x \in |\mathcal{X}|$. In this case write $f : \mathcal{X} \to \mathcal{Y}$.

Definition 2.7. Suppose $\mathcal{X}$ is a nominal set and $x \in |\mathcal{X}|$. Define the **support** of x by $supp(x) = \bigcap\{A \mid A \text{ supports } x\}$. Write $a\#x$ as shorthand for $a \notin supp(x)$ and read this as a is **fresh for** x.

Theorem 2.8. *Suppose $\mathcal{X}$ is a nominal set and $x \in |\mathcal{X}|$. Then $supp(x)$ is the unique least finite set of atoms that supports x. (Proofs in [20,16].)*

Definition 2.9. Write $\pi|_A$ for the partial function which is π restricted to A.

Corollary 2.10. *1. If $\pi(a) = a$ for all $a \in supp(x)$ then $\pi \cdot x = x$.*
2. If $\pi|_{supp(x)} = \pi'|_{supp(x)}$ then $\pi \cdot x = \pi' \cdot x$.
3. $a\#x$ if and only if $\exists b. b\#x \wedge (b\ a) \cdot x = x$.

Proposition 2.11. *$supp(\pi \cdot x) = \{\pi(a) \mid a \in supp(x)\}$ (cf. Definition 4.1).*

Our reasoning can be formalised in first-order logic with axioms of set theory with atoms (the name for this is Zermelo-Fraenkel with atoms, or **ZFA**). Because one atom will do as well as any other, we obtain Theorem 2.12, from which concisely follow results about equivariance and support.[1] See e.g. proofs of Lemmas 4.6, 5.8, 5.13, 5.14, 5.19, and Proposition 5.23.

[1] Nominal sets can be implemented in ZFA sets such that nominal sets map to equivariant elements (elements with empty support) and the permutation action maps to 'real' permutation of atoms in the model. See [16, Subsection 9.3] and [16, Section 4].

Theorem 2.12. *If $\overline{x}$ is a list $x_1, \dots, x_n$, write $\pi\cdot\overline{x}$ for $\pi\cdot x_1, \dots, \pi\cdot x_n$. Suppose $\phi(\overline{x})$ is a first-order logic predicate with free variables $\overline{x}$. Suppose $\chi(\overline{x})$ is a function specified using a first-order predicate with free variables $\overline{x}$. Then we have the following principles:*

1. **Equivariance of predicates.** $\phi(\overline{x}) \Leftrightarrow \phi(\pi\cdot\overline{x})$.[2]
2. **Equivariance of functions.** $\pi\cdot\chi(\overline{x}) = \chi(\pi\cdot\overline{x})$.
3. **Conservation of support.** *If $\overline{x}$ denotes elements with finite support then $supp(\chi(\overline{x})) \subseteq supp(x_1) \cup \dots \cup supp(x_n)$.*

Definition 2.13. Write $\text{И}a.\phi(a)$ for '$\{a \mid \neg\phi(a)\}$ is finite'.

Remark 2.14. We can read И as 'for all but finitely many a', 'for fresh a', or 'for new a'. It captures a *generative* aspect of names, that for any x we can find plenty of atoms a such that $a \notin supp(x)$. И was designed in [20] to model the quantifier being used when we informally write "rename x in $\lambda x.t$ to be fresh", or "emit a fresh channel name" or "generate a fresh memory cell".

И is a 'for most' quantifier [38], and is a *generalised quantifier* [21, Section 1.2.1]. But importantly, И over nominal sets satisfies the *some/any property* that to prove a И-quantified property we test it for *one* fresh atom; we may then use it for *any* fresh atom. This is Theorem 2.15, which we use implicitly when later we choose a 'fresh atom' without proving that it does not matter which one we choose. See e.g. proofs of Proposition 4.11 and Lemmas 4.7, 5.10, and 6.6.

Theorem 2.15. *Suppose $\phi(\overline{z}, a)$ is a predicate with free variables $\overline{z}, a$.*[3] *Suppose $\overline{z}$ denotes elements with finite support. Then the following are equivalent:*

$$\forall a.(a \in \mathbb{A} \wedge a\#\overline{z}) \Rightarrow \phi(\overline{z}, a) \qquad \text{И}a.\phi(\overline{z}, a) \qquad \exists a.a \in \mathbb{A} \wedge a\#\overline{z} \wedge \phi(\overline{z}, a)$$

3 Nominal Boolean Algebra with и

Definition 3.1. A **nominal Boolean algebra with** и (with a new-binder) or **Banona** is a tuple $(\mathcal{B}, \wedge, \neg, \text{и})$ of a nonempty nominal set $\mathcal{B}$, and equivariant functions (Definition 2.6)

- *conjunction* $\wedge : \mathcal{B} \times \mathcal{B} \to \mathcal{B}$ written $x \wedge y$ (for $\wedge(x, y)$),
- *negation* $\neg : \mathcal{B} \to \mathcal{B}$ written $\neg x$, and
- *new* $\text{и} : \mathbb{A} \times \mathcal{B} \to \mathcal{B}$ written $\text{и}a.x$ (for $\text{и}(a, x)$),

such that the equalities in Figure 1 hold.

Banonas arise naturally as nominal powersets (Section 4), just as Boolean algebras arise naturally as 'ordinary' powersets. See also Example 3.6.

2 $\overline{x}$ must contain *all* the variables mentioned in the predicate. It is not the case that $a = a$ if and only if $a = b$—but it is the case that $a = b$ if and only if $b = a$.

3 ϕ should be provable without the axiom of choice. Every ϕ used in this paper will satisfy this property.

(**Commute**)		$x \wedge y = y \wedge x$
(**Assoc**)		$(x \wedge y) \wedge z = x \wedge (y \wedge z)$
(**Huntington**)		$x = \neg(\neg x \wedge \neg y) \wedge \neg(\neg x \wedge y)$
(**Swap**)		$\text{И}a.\text{И}b.x = \text{И}b.\text{И}a.x$
(**Garbage**)	$a\#x \Rightarrow$	$\text{И}a.x = x$
(**Distrib**)		$\text{И}a.(x \wedge y) = (\text{И}a.x) \wedge (\text{И}a.y)$
(**SelfDual**)		$\neg\text{И}a.x = \text{И}a.\neg x$
(**Alpha**)	$b\#x \Rightarrow$	$\text{И}a.x = \text{И}b.(b\ a)\cdot x$

Fig. 1. Axioms of Banonas

Remark 3.2. (**Commute**), (**Assoc**), and (**Huntington**) axiomatise Boolean algebra (see, e.g., [26]). We write $\bot$ for $x \wedge \neg x$, $x \vee y$ for $\neg(\neg x \wedge \neg y)$, and $x \leq y$ for $x \wedge y = x$. With these definitions, the axioms ensure standard properties of Boolean algebra including: absorption, distributivity, and poset properties.

The other axioms describe properties of the И-quantifier [20] (Definition 2.13, in this paper). See Theorem 4.10 for a proof.[4]

Lemma 3.3. *If $x \leq y$ then $\text{И}b.x \leq \text{И}b.y$. In words: И is monotone.*

Proof. By $x \leq y$ we mean $x \wedge y = x$. So $\text{И}b.(x \wedge y) = \text{И}b.x$. We use (**Distrib**).

Lemma 3.4. *$a\#\text{И}a.x$. As a corollary, $\text{И}a.\text{И}a.x = \text{И}a.x$.*

Proof. Choose a fresh b (so $b\#x$). By (**Alpha**) $\text{И}a.x = \text{И}b.(b\ a)\cdot x$. By equivariance of $\text{И} : \mathbb{A} \times \mathcal{B} \to \mathcal{B}$, $\text{И}b.(b\ a)\cdot x = (b\ a)\cdot\text{И}a.x$. By Proposition 2.11 $a\#(b\ a)\cdot\text{И}a.x$. The result follows. The corollary follows using (**Garbage**).

Definition 3.5. Call a function $f \in |\mathcal{B}'| \to |\mathcal{B}|$ a (Banona) **morphism** when:

$$f(x \wedge y) = f(x) \wedge f(y) \quad f(\neg x) = \neg f(x) \quad f(\text{И}a.x) = \text{И}a.f(x) \quad f(\pi\cdot x) = \pi\cdot f(x)$$

Write BAnona for the category of Banonas and Banona morphisms.

Example 3.6.
- Call a set $X \subseteq \mathbb{A}$ **cofinite** when $\mathbb{A} \setminus X$ is finite. The set of finite or cofinite sets of atoms is a Banona where conjunction and negation are interpreted as intersection and complement respectively, and $\text{И}a.X = X \setminus \{a\}$ if X is finite, and $\text{И}a.X = X \cup \{a\}$ if X is cofinite (cf. Example 4.5). Section 4 exhibits nominal powersets as Banonas. Section 5 proves them complete for all possible examples, in a certain formal sense.
- The discrete *nominal restriction set* $\mathbb{B} = \{\mathit{True}, \mathit{False}\}$ from [29, Section 3.2] is also a Banona.

[4] The axioms are *nominal* algebraic because of their freshness side-conditions. In a sense which has been made formal, nominal algebra is an equational logic [19,14,22], and is sound and complete for nominal sets.

- The term model for nominal logic constructed by Cheney and Urban in [9] is a Banona where we interpret conjunction as conjunction, negation as negation, and $\text{и}a.\phi$ as 'for fresh a, ϕ' as specified in Figure 9 of [9].
- Predicates of Cardelli and Gordon's *ambient calculus* up to logical equivalence form a (very richly-structured) Banona where и is interpreted as И as defined in Definition 4.3 of [7] (see [7, Corollary 4.5]).
- Formulas of hybrid logic with downarrow $\downarrow x.\phi$ (see, e.g, [2]) up to logical equivalence, where atoms serve as *world variables*, permutation is syntactic permutation and freshness is 'not free in', are a Banona where и maps to $\downarrow$. A broader class of Banonas (with additional operators) is obtained by adapting an abstract equational class of algebras including sets denotations for hybrid logic defined in [23] to a nominal setting.

4 The Nominal Powerset as an Algebra

The main (and by Theorem 5.25 canonical) example of a Banona, is a nominal powerset:

Definition 4.1. Define the **nominal powerset** $pow(\mathcal{X})$ by:

- $U \subseteq |\mathcal{X}|$ has the **pointwise** action $\pi{\cdot}U = \{\pi{\cdot}x \mid x \in U\}$.
- $|pow(\mathcal{X})|$ is the set of finitely-supported $U \subseteq |\mathcal{X}|$.

Call $U \in |pow(\mathcal{X})|$ **equivariant** when $supp(U) = \varnothing$.

Lemma 4.2. *U is equivariant if and only if $\pi{\cdot}x \in U$ if and only if $x \in U$.*

Definition 4.3. If $X \in |pow(\mathcal{X})|$ then define $\mathsf{n}a.X = \{x \mid \text{И}b.(b\ a){\cdot}x \in X\}$.

Remark 4.4. Definition 4.3 is the sets-based interpretation of И which we will prove sound and complete for the axioms in Definition 3.1. Visibly, n is defined using И. Conversely, $\text{И}a.\phi(a)$ if and only if $a \in \mathsf{n}a.\{x \in \mathbb{A} \mid \phi(x)\}$.

Intuitively, $\mathsf{n}a.X$ adds/removes just those elements of X which are 'responsible' for a being in the support of X, and $\mathsf{n}a.X$ is a 'nearest set' to X for which a is fresh. We see this in Example 4.5, and in a sense part 1 of Lemma 4.7 and Proposition 4.11 are corollaries of this intuition.

$\mathsf{n}a.$ is equal to the operation $X{-}a$ described in [15, Subsection 4.2], used there for different purposes. The elements added/removed by n are called *crucial elements* in [15, Subsection 4.2] and correspond to adding/removing elements from a-orbits [15, Subsection 4.1].

Example 4.5. $|pow(\mathbb{A})|$ is the set of finite and cofinite subsets of atoms and $\mathsf{n}a.X$ is characterised on $pow(\mathbb{A})$ by $\mathsf{n}a.X = X \setminus \{a\}$ if X is finite, and $\mathsf{n}a.X = X \cup \{a\}$ if X is cofinite.

Lemma 4.6. *If $X \in |pow(\mathcal{X})|$ then $\mathsf{n}a.X \in |pow(\mathcal{X})|$.*

Proof. By Theorem 2.12 $supp(\mathsf{n}a.X) \subseteq supp(X) \cup \{a\}$.

Lemma 4.7. *1. If $a\#X$ then* $\mathsf{n}a.X = X$.
2. $\mathsf{n}a.(X \cap Y) = (\mathsf{n}a.X) \cap (\mathsf{n}a.Y)$.
3. $\mathsf{n}a.(|\mathfrak{X}| \setminus X) = |\mathfrak{X}| \setminus \mathsf{n}a.X$.
As a corollary, if $X \subseteq Y$ then $\mathsf{n}a.X \subseteq \mathsf{n}a.Y$.

Proof (Sketch). Suppose $a\#X$. Choose $x{\in}|\mathfrak{X}|$ and fresh b (so $b\#x, X$). By the pointwise action and Corollary 2.10 $(b\ a){\cdot}x{\in}X$ if and only if $x{\in}X$.

Choose $x \in |\mathfrak{X}|$ and fresh b (so $b\#x, X, Y$). Then $(b\ a){\cdot}x \in X \cap Y$ if and only if $(b\ a){\cdot}x \in X$ and $(b\ a){\cdot}x \in Y$.

Choose $x \in |\mathfrak{X}|$ and fresh b (so $b\#x, X, |\mathfrak{X}|\backslash X$). Then $(b\ a){\cdot}x \in |\mathfrak{X}| \setminus X$ if and only if $(b\ a){\cdot}x \notin X$.

Lemma 4.8. *$b\#X$ implies* $\mathsf{n}a.X = \mathsf{n}b.(b\ a){\cdot}X$. *As a corollary,* $a\#\mathsf{n}a.X$.

Proof (Sketch). Choose fresh b (so $b\#X$) and $x{\in}|\mathfrak{X}|$. By the pointwise action $(b\ c){\cdot}x{\in}(a\ b){\cdot}X$ if and only if $x{\in}(b\ c){\cdot}(a\ b){\cdot}X$. By Corollary 2.10 $(b\ c){\cdot}(a\ b){\cdot}X = (a\ c){\cdot}X$. The result follows.

Lemma 4.9. $\mathsf{n}a.\mathsf{n}b.X = \mathsf{n}b.\mathsf{n}a.X$.

Proof. By routine calculations using the fact that $(a'\ a){\cdot}(b'\ b){\cdot}x = (b'\ b){\cdot}(a'\ a){\cdot}x$.

Theorem 4.10. $(pow(\mathfrak{X}), \cap, |\mathfrak{X}|\backslash\text{-}, \mathsf{n})$ *is an object of* BAnona.

Proof. Validity of (**Commute**), (**Assoc**), and (**Huntington**) is by routine sets calculations. Validity of (**Alpha**) and (**Swap**) is by Lemmas 4.8 and 4.9. Validity of (**Garbage**), (**Distrib**), and (**SelfDual**) is by Lemma 4.7.

Proposition 4.11. *If $a\#x$ then $x \in X$ if and only if* $x \in \mathsf{n}a.X$.

Proof. Choose b fresh (so $b\#x, X$). By definition $x \in \mathsf{n}a.X$ when $(b\ a){\cdot}x \in X$. By Corollary 2.10 $(b\ a){\cdot}x = x$. The result follows.

5 A Representation Theorem

We introduce n-*filters* of a Banona $\mathcal{B}$ (Definition 5.1). We define the *canonical extension* $\mathcal{B}^\bullet$ as the nominal powerset of its maximal n-filters (Definition 5.20). Finally we prove $\mathcal{B}$ isomorphic to a subalgebra of $\mathcal{B}^\bullet$ (Theorem 5.25).

5.1 n-Filters

Definition 5.1. An **n-filter** is a finitely-supported subset $p \subseteq |\mathcal{B}|$ such that:

1. $\bot \notin p$	2. $\forall x, y.(x \in p \wedge y \in p) \Leftrightarrow (x \wedge y \in p)$	3. $\text{И}a.\forall x.(x{\in}p \Rightarrow \text{и}a.x{\in}p)$

Remark 5.2. We are proving a representation theorem, so the game we play is to convert algebra—like $\wedge$, $\neg$, and и—into operations on (nominal) sets. Conditions 1 and 2 of Definition 5.1 correspond to properties of the empty set and sets intersection. The third condition corresponds to a property of *nominal* sets: Proposition 4.11; as made formal in the proof of Lemma 5.13.

Definition 5.3. Given $C \subseteq \mathbb{A}$ and $x \in |\mathcal{B}|$ define $иC.x$ by:

$$иC.x = иc_1 \ldots иc_n.x \quad \text{where} \quad C \cap supp(x) = \{c_1, \ldots, c_n\}$$

By assumption x has finite support so $C \cap supp(x)$ is finite even if C is not. Also, by (**Swap**) the order of the c_i does not matter.

Lemma 5.4. $supp(\bot) = \varnothing$.

Proof. We defined $\bot = x \wedge \neg x$. It follows from the axioms of Boolean algebra in Figure 1 that $x \wedge \neg x = y \wedge \neg y$ for any other y, and in particular by equivariance $x \wedge \neg x = \pi \cdot (x \wedge \neg x)$. The result follows.

Lemma 5.5. *Suppose* $C \subseteq \mathbb{A}$ *and* $a \in C$. *Then:*

1. $иC.\bot = \bot$ 2. $иC.(x \wedge y) = (иC.x) \wedge иC.y$ 3. $иC.x = иC.иa.x$

Proof. 1. Follows by (**Garbage**) and Lemma 5.4. 2. Follows by (**Distrib**),(**Swap**) and (**Garbage**). 3. If $a\#x$ then $x = иa.x$ by Lemma 3.4. If $a \in supp(x)$, then $a \in C \cap supp(x)$ and the result follows by part 2 of Lemma 3.4 and (**Swap**).

Definition 5.6. If $z \in |\mathcal{B}|$ write $C = \mathbb{A} \backslash supp(z)$ and define $z{\uparrow} = \{x \mid z \leq иC.x\}$.

Remark 5.7. The standard definition of $z{\uparrow}$ is $\{x \mid z \leq x\}$. This definition seems to *not work*, and proofs based on it break. We can view Definition 5.6 as elaborating the standard definition to respect property 3 of Definition 5.1.

Lemma 5.8. *If* $z \in |\mathcal{B}|$ *and* $z \neq \bot$ *then* $supp(z{\uparrow}) \subseteq supp(z)$ *and* $z{\uparrow}$ *is an* n-*filter.*

Proof. Write $C = \mathbb{A} \backslash supp(z)$. The first part is by Theorem 2.12 and the fact that $supp(C) = supp(z)$.[5] That $z{\uparrow}$ is an n-filter follows from Lemma 5.5.

Definition 5.9. Call an n-filter $p \subseteq |\mathcal{B}|$ **maximal** when for all n-filters $p' \subseteq |\mathcal{B}|$ if $p \subseteq p'$ then $p' = p$.

We now show that every n-filter is contained in a maximal n-filter (Theorem 5.17). We use Zorn's lemma, but this requires a bound on support. Thus we consider n-filters maximal amongst n-filters with no greater support. Surprisingly, these n-filters are the maximal n-filters (Lemma 5.14).

Lemma 5.10. *Suppose* q *is an* n-*filter and suppose* $x \notin q$. *Write* q' *for the set* $q' = \{z \mid z \vee x \in q\}$. *Then* q' *is an* n-*filter.*

Proof. That $\bot \notin q'$ and $\forall z_1, z_2.(z_1 \in q' \wedge z_2 \in q') \Leftrightarrow (z_1 \wedge z_2 \in q')$ is routine. $Иa.\forall z.(z \in q' \Rightarrow иa.z \in q')$ follows from the fact that if a is fresh (so $a\#q, x$) then by (**Garbage**), (**SelfDual**), and (**Distrib**), $иa.(z \vee x) = (иa.z) \vee иa.x = (иa.z) \vee x$.

Proposition 5.11. q *is maximal if and only if* $\forall x. \neg x \in q \Leftrightarrow x \notin q$.

---

[5] In fact it can be proved by a further calculation that $supp(z{\uparrow}) = supp(z)$.

Proof. For q maximal, suppose $\neg x \notin q$ and $x \notin q$. By Lemma 5.10, $q' = \{z \mid z\vee x\in q\}$ is an n-filter. Also, $\neg x\in q'$ whereas $\neg x\notin q$, so $q \subsetneq q'$, contradicting maximality of q. The rest follows from conditions 1 and 2 of Definition 5.1.

Lemma 5.12. *If q is an* n*-filter and $C=\mathbb{A}\setminus supp(q)$ then so is $q' = \{z \mid \text{и}C.z\in q\}$.*

Proof. Follows from Lemma 5.5.

Lemma 5.13. *Suppose q is an* n*-filter such that for all* n*-filters q', $q \subseteq q'$ and $supp(q') \subseteq supp(q)$ imply $q = q'$. Then $\text{И}a.(\forall x.x \in q \Leftrightarrow \text{и}a.x \in q)$.*

Proof. The left-to-right implication is condition 3 of Definition 5.1. Write $C = \mathbb{A} \setminus supp(q)$. The set $q' = \{z \mid \text{и}C.z \in q\}$ (Definition 5.3) is an n-filter by Lemma 5.12. By condition 3 of Definition 5.1 $q \subseteq q'$, by Theorem 2.12 $supp(q') \subseteq supp(q)$, hence $q' = q$ (note that $supp(C) = supp(q)$). The right-to-left implication follows.

Lemma 5.14. *q is a maximal* n*-filter if and only if it is maximal amongst* n*-filters with no greater support (that is, if and only if for all* n*-filters q', $q \subseteq q'$ and $supp(q') \subseteq supp(q)$ imply $q = q'$).*

Proof. The left-to-right implication is trivial. Now suppose q is maximal amongst n-filters with no greater support. By Proposition 5.11 it suffices to show that $\neg x \in q$ if and only if $x \notin q$.

$\neg x \in q$ and $x \in q$ is impossible by conditions 1 and 2 of Definition 5.1.

Now suppose $\neg x \notin q$ and also $x \notin q$. Write $C = \mathbb{A} \setminus supp(q)$ and $x' = \text{и}C.x$. By definition, (**SelfDual**), and Lemma 5.13 $x' \notin q$ and $\neg x' \notin q$.

By Lemma 5.10, $q' = \{z \mid z \vee x' \in q\}$ is an n-filter. Also, $\neg x' \in q'$ whereas $\neg x' \notin q$, so $q \subsetneq q'$. By Theorem 2.12 $supp(q') \subseteq supp(q)$ and so $q = q'$ by assumption, a contradiction.

Remark 5.15. Given a nominal set $\mathcal{X}$, call $Y \subseteq |\mathcal{X}|$ **bounded-supported** when $\bigcup\{supp(x) \mid x \in Y\}$ is finite. By [16, Theorem 2.29] if Y is bounded-supported then Y is finitely-supported and $supp(Y) = \bigcup\{supp(x) \mid x \in Y\}$. See also [35, Definition 3.4.2.3] and subsequent discussion.

Lemma 5.16. *Consider a nominal set $\mathcal{X}$ and $\leq$ a partial order on $|\mathcal{X}|$. If every chain $C \in pow(\mathcal{X})$ has an upper bound $b(C)$ with $supp(b(C)) \subseteq supp(C)$ then for every $x \in |\mathcal{X}|$ the set $x^\circ = \{y \in |\mathcal{X}| \mid y \geq x,\ supp(y) \subseteq supp(x)\}$ has a maximal element.*

Proof. By Remark 5.15 every chain C in x° is finitely-supported and $supp(C) \subseteq supp(x)$. Then C has an upper bound $b(C)$ such that $supp(b(C)) \subseteq supp(x)$. Thus $b(C) \in x^\circ$. The result follows using Zorn's lemma for x°.

Theorem 5.17. *For every* n*-filter p, there is a maximal* n*-filter q with $p \subseteq q$.*

Proof. If C is a finitely-supported chain in the nominal set of n-filters on $\mathcal{B}$ ordered by subset inclusion, then $\bigcup C = \{x \mid \exists p' \in C.x \in p'\}$ is an upper bound for C. By Theorem 2.12 $supp(\bigcup C) \subseteq supp(C)$. By Lemma 5.16 the set p° of n-filters p' such that $p \subseteq p'$ and $supp(p') \subseteq supp(p)$ has a maximal element q with respect to inclusion. By Lemma 5.14 q is a maximal n-filter (Definition 5.9).

5.2   The Canonical Extension -•

Definition 5.18. Define $points(\mathcal{B}) = \{p \subseteq |\mathcal{B}| \mid p \text{ is a maximal n-filter}\}$.

Lemma 5.19. $points(\mathcal{B})$ *with the pointwise action (Defn. 4.1) is a nominal set.*

Proof. By Theorem 2.12 the predicate 'p is a maximal n-filter' holds if and only if the predicate '$\pi{\cdot}p$ is a maximal n-filter' holds.

Definition 5.20. Define   the   **canonical   extension** $\mathcal{B}^{\bullet}=(pow(points(\mathcal{B})), \wedge, \neg, \text{и})$:

$$A \wedge B = A \cap B \qquad \neg A = |\mathcal{B}^{\bullet}| \setminus A \qquad \text{и}a.A = \text{na}.A$$

A and B will range over elements of $|\mathcal{B}^{\bullet}|$. ($\text{na}.A$ defined in Definition 4.3.)

Proposition 5.21. $\mathcal{B}^{\bullet}$ *is a nominal Boolean algebra with* и.

Proof. From Theorem 4.10.

Definition 5.22. Define a map $-^{\bullet} \in |\mathcal{B}| \to |\mathcal{B}^{\bullet}|$ by:

$$x^{\bullet} = \{p \in |points(\mathcal{B})| \mid x \in p\}$$

We need to check that $-^{\bullet}$ does map to $|\mathcal{B}^{\bullet}|$. It suffices to show that $supp(x^{\bullet})$ is finite. But it follows from Theorem 2.12 that $supp(x^{\bullet}) \subseteq supp(x)$.

Proposition 5.23. $-^{\bullet}$ *is an arrow in* BAnona *(Definition 3.5).*

Proof. Equivariance is by Theorem 2.12.

1. $(x \wedge y)^{\bullet} = \{p \in |points(\mathcal{B})| \mid x \wedge y \in p\}$. By assumption in Definition 5.1 $x \wedge y \in p$ if and only if $x \in p$ and $y \in p$ and it follows that $(x \wedge y)^{\bullet} = x^{\bullet} \wedge y^{\bullet}$.
2. $(\neg x)^{\bullet} = \{p \in |points(\mathcal{B})| \mid \neg x \in p\}$. We use Proposition 5.11.
3. $(\text{и}a.x)^{\bullet} = \{p \in |points(\mathcal{B})| \mid \text{и}a.x \in p\}$. Suppose $p \in (\text{и}a.x)^{\bullet}$. Choose fresh a' (so $a'\#x, x^{\bullet}, p$). By (**Alpha**) $\text{и}a.x = \text{и}a'.(a'\ a){\cdot}x$. By definition $\text{и}a'.(a'\ a){\cdot}x \in p$. By Lemma 5.13 $(a'\ a){\cdot}x \in p$. By definition $p \in (a'\ a){\cdot}x^{\bullet}$. By Proposition 4.11 $p \in \text{na}'.(a'\ a){\cdot}x^{\bullet}$. By Lemma 4.8 $\text{na}'.(a'\ a){\cdot}x^{\bullet} = \text{na}.x^{\bullet}$.

Remark 5.24. Note the 'internal' and 'external' names in part 3 of the proof of Proposition 5.23. We begin with a 'internally restricted' with и. We use (**Alpha**) to apply an 'external' renaming of a to 'externally' fresh a'. This is picked up by the definition of n-filter and Proposition 4.11 moves to the 'external' И.

In part 3 $a\#x$ implies $a\#x^{\bullet}$—but this does not matter; we choose a *fresh*.

Theorem 5.25. $-^{\bullet}$ *is injective, thus* $\mathcal{B}$ *is isomorphic to a subalgebra of* $\mathcal{B}^{\bullet}$.

Proof. Suppose $x \in |\mathcal{B}|$ and $y \in |\mathcal{B}|$ are distinct. Suppose without loss of generality that $x \not\leq y$, so that $x \wedge \neg y \neq \bot$. By Lemma 5.8 $(x \wedge \neg y){\uparrow}$ is an n-filter. By Theorem 5.17 there exists a point q containing $(x \wedge \neg y){\uparrow}$. Then $x \wedge \neg y \in q$, hence $q \in x^{\bullet}$ and $q \notin y^{\bullet}$. The result follows by Proposition 5.23.

6 Nominal Stone Duality

Nominal topological spaces (Definition 6.1) are just topology over nominal sets (finite support can consierably restrict the open sets; see Example 4.5). We cannot take arbitrary unions of open sets, but only finitely-supported unions. Banonas correspond to nominal topological spaces with additional properties which we call nominal Stone spaces with и (Definition 6.7). To the new-binder и corresponds on the topological side the semantic n (Definition 4.3); elements correspond to clopen sets (sets that are both open and closed), so clopens must be closed under n. The notion of compactness must also be subtly tweaked to take into account the role of n. We conclude with a duality theorem.

Definition 6.1. A nominal topological space $\mathcal{T}$ is a pair $(|\mathcal{T}|, \mathcal{O}_\mathcal{T})$ of a **carrier** nominal set $|\mathcal{T}|$ and equivariant set of **open sets** $\mathcal{O}_\mathcal{T} \subseteq pow(|\mathcal{T}|)$ such that:

- $\emptyset \in \mathcal{O}_\mathsf{T}$ and $|\mathcal{T}| \in \mathcal{O}_\mathsf{T}$
- $U \in \mathcal{O}_\mathcal{T} \wedge V \in \mathcal{O}_\mathcal{T}$ implies $U \cap V \in \mathcal{O}_\mathcal{T}$.
- $\mathcal{U} \in pow(\mathcal{O}_\mathcal{T})$ implies $\bigcup \mathcal{U} \in \mathcal{O}_\mathcal{T}$; we call this a **finitely-supported union**.

Call equivariant $f \in |\mathsf{T}_1| \to |\mathsf{T}_2|$ **continuous** when $V \in \mathcal{O}_{\mathsf{T}_2}$ implies $f^{-1}(V) \in \mathcal{O}_{\mathsf{T}_1}$. Write nTop for the category of nominal topological spaces and continuous maps.

Definition 6.2. Given $\mathcal{B}$ in BAnona define $F(\mathcal{B})$ in nTop by:

- $|F(\mathcal{B})| = points(\mathcal{B})$ (Definition 5.18).
- $\mathcal{O}_{F(\mathcal{B})}$ is the closure of $\{x^\bullet \mid x \in |\mathcal{B}|\}$ (Definition 5.22) under finitely-supported unions. So $U \in \mathcal{O}_{F(\mathcal{B})}$ when $\exists M \in pow(|\mathcal{B}|).U = \bigcup\{x^\bullet \mid x \in M\}$.

Given $f : \mathcal{B} \to \mathcal{B}'$ in BAnona define $F(f) : F(\mathcal{B}') \to F(\mathcal{B})$ by $F(f)(p) = f^{-1}(p)$.

$F(\mathcal{B})$ is indeed a nominal topological space, for if $\mathcal{U} \in pow(\mathcal{O}_{F(\mathcal{B})})$ then $\bigcup \mathcal{U} = \bigcup\{x^\bullet \mid \exists U \in \mathcal{U}. x^\bullet \subseteq U\}$ is open. Lemma 6.3 proves $F(f)$ well-defined. We then pin down the properties that completely characterise nominal topological spaces arising from banonas.

Lemma 6.3. *For* $f : \mathcal{B} \to \mathcal{B}'$ *in* BAnona *and* $p \in points(\mathcal{B}')$, $f^{-1}(p) \in points(\mathcal{B})$.

Proof. It is not hard to use the homomorphism properties of f and Theorem 2.12 to show that $f^{-1}(p)$ is an n-filter. Using Proposition 5.11 $\neg x \in f^{-1}(p)$ if and only if $x \notin f^{-1}(p)$, and it follows that $f^{-1}(p)$ is also maximal.

Definition 6.4. Call $\mathcal{U} \in pow(\mathcal{O}_\mathcal{T})$ **n-closed** when $\text{И}a.\forall U.(U \in \mathcal{U} \Rightarrow \mathsf{n}a.U \in \mathcal{U})$. Call $\mathcal{U} \in pow(\mathcal{O}_\mathcal{T})$ a **cover** when $\bigcup \mathcal{U} = |\mathcal{T}|$. If $\mathcal{U}$ is a cover and is n-closed then call $\mathcal{U}$ an **n-cover**. Call $\mathcal{T}$ **n-compact** when every n-cover has a finite subcover.

Lemma 6.5. *If* $\mathcal{U} \in pow(\mathcal{O}_\mathcal{T})$ *is finite then* $\text{И}a.\forall U.U \in \mathcal{U} \Rightarrow a\#U$. *As a corollary,* $\mathcal{T}$ *is* n*-compact when every* n*-cover has an* n*-closed finite subcover.*

Proof. For finite $\mathcal{U}$, $supp(\mathcal{U}) = \bigcup\{supp(U) \mid U \in \mathcal{U}\}$ [16, Theorem 2.29]. We use part 1 of Lemma 4.7.

Lemma 6.6. *If $\mathcal{U}\in pow(\mathcal{O}_{F(\mathcal{B})})$ is* n*-closed then so is* $\mathcal{V} = \{x^\bullet \mid \exists U{\in}\mathcal{U}.x^\bullet{\subseteq}U\}$.

Proof. Choose b fresh (so $b\#\mathcal{U},\mathcal{V}$). If $x^\bullet \in \mathcal{V}$ then $x^\bullet \subseteq U \in \mathcal{U}$. By Lemma 4.7 $\mathsf{n}b.(x^\bullet) \subseteq \mathsf{n}b.U$ and by assumption $\mathsf{n}b.U \in \mathcal{U}$.

Definition 6.7. Call U **closed** when $|\mathcal{T}|\setminus U \in \mathcal{O}_\mathsf{T}$, and **clopen** when U is open and closed. Call $\mathcal{T}$ **totally separated** when for every $x, y \in |\mathcal{T}|$ there is a clopen U with $x \in U$ and $y \notin U$.

Say $\mathcal{T}$ is a **nominal topological space with** И when $\mathsf{n}a.U \in \mathcal{O}_\mathsf{T}$ for every clopen U.[6] Write $\mathsf{nTop}_{И}$ for the full subcategory of nTop on $\mathcal{T}$ with И.

A **nominal Stone space with** И is a totally separated n-compact nominal topological space with И. Write $\mathsf{nStone}_{И}$ for the full subcategory of $\mathsf{nTop}_{И}$ on nominal Stone spaces with И.

Proposition 6.8. *$F(\mathcal{B})$ is totally separated and* n*-compact.*

Proof. Consider distinct $p, q \in |F(\mathcal{B})|$. Without loss of generality take $x \in p \setminus q$. By Definition 5.20, $p \in x^\bullet$ and $q \notin x^\bullet$. So $x^\bullet$ is an open set separating p and q. By Proposition 5.11 $points(\mathcal{B}) \setminus x^\bullet = (\neg x)^\bullet$, so $x^\bullet$ is also closed.

Consider an n-cover $\mathcal{U} \in pow(\mathcal{O}_{F(\mathcal{B})})$. It suffices to find a finite subcover of $\mathcal{V} = \{x^\bullet \mid \exists U{\in}\mathcal{U}.x^\bullet \subseteq U\}$, since for every $x^\bullet \in \mathcal{V}$ there exists $x^\bullet \subseteq U \in \mathcal{U}$.

Write $X = \bigcap_{fin}\{x' \mid \exists x.\neg x{\leq}x'{\wedge}x^\bullet{\in}\mathcal{V}\}$ where $\bigcap_{fin}$ denotes closure under finite intersections. So $\mathcal{V}$ has a finite subcover if and only if $\bot \in X$. By Proposition 5.23 and Lemma 6.6 X satisfies conditions 2 and 3 of Definition 5.1. If X satisfied also $\bot \notin X$ then X would be an n-filter; by Theorem 5.17 $X \subseteq p$ for some point p; it would follow that $p \notin \bigcup\mathcal{V}$, a contradiction. Therefore $\bot \in X$.

Lemma 6.9. *If $U \in F(\mathcal{B})$ is clopen then $U = x^\bullet$ for some $x \in |\mathcal{B}|$.*

Proof. By assumption $U = \bigcup\{x^\bullet \mid x^\bullet \subseteq U\}$ and $|F(\mathcal{B})| \setminus U = \bigcup\{x^\bullet \mid x^\bullet \subseteq |F(\mathcal{B})| \setminus U\}$. It follows that $\{x^\bullet \mid x^\bullet \subseteq U \vee x^\bullet \cap U = \varnothing\}$ covers $F(\mathcal{B})$. This cover is also n-closed, by part 1 and the corollary in Lemma 4.7. So it has a finite subcover by Proposition 6.8. The result follows by Proposition 5.23.

Proposition 6.10. *F is a functor from* BAnona *to* $\mathsf{nStone}_{И}^{op}$.

Proof. If $U{\in}\mathcal{O}_{\mathsf{F}(\mathcal{B})}$ is clopen then $U{=}x^\bullet$ for some $x{\in}|\mathcal{B}|$ by Lemma 6.9. By part 3 of Proposition 5.23 $\mathsf{n}a.(x^\bullet){=}(И a.x)^\bullet{\in}\mathcal{O}_{\mathsf{F}(\mathcal{B})}$. By Proposition 6.8 $F(\mathcal{B})$ is a nominal Stone space with И.

Consider $f : \mathcal{B} \to \mathcal{B}'$ in BAnona. By Lemma 6.3 $F(f)$ maps $|F(\mathcal{B}')|$ to $|F(\mathcal{B})|$. Continuity of $F(f)$ follows using the fact that $F(f)^{-1}(x^\bullet) = (f(x))^\bullet$.

Definition 6.11. Map $\mathcal{T} \in \mathsf{nStone}_{И}$ to a $G(\mathcal{T}) \in \mathsf{BAnona}$ defined by:

- $|G(\mathcal{T})| = \{U \in \mathcal{O}_\mathcal{T} \mid U \text{ is clopen}\}$.
- $\wedge$, $\neg$, and И are interpreted as intersection, complement, and n.

Given $f : \mathcal{T} \to \mathcal{T}'$ in $\mathsf{nStone}_{И}$ define $G(f) : G(\mathcal{T}') \to G(\mathcal{T})$ by $G(f)(U) = f^{-1}(U)$.

6 We do not require $\mathsf{n}a.U$ to be open for all U. More on this in the Conclusions.

Lemma 6.12. *G is a functor from* $\mathsf{nStone}_{\text{И}}^{op}$ *to* BAnona.

Proof. If U is clopen then $\mathsf{n}a.U$ is open by Definition 6.7. $\mathsf{n}a.U$ is closed from the fact that $\mathsf{n}a.(|\mathcal{T}| \setminus U) = |\mathcal{T}| \setminus \mathsf{n}a.U$, immediate from part 3 of Lemma 4.7. It is routine to check that $G(\mathcal{T})$ is an object of BAnona.

Consider continuous $f : \mathcal{T} \to \mathcal{T}'$. If U is clopen then so is $f^{-1}(U)$. $G(f)$ is equivariant because f is, and preserves intersections and complements. Given $U \in \mathcal{O}_{\mathcal{T}'}$ a clopen, $G(f)(\mathsf{n}a.U) = \mathsf{n}a.G(f)(U)$ follows from Definitions 4.3 and 6.11 and the equivariance of f. Thus $G(f)$ is a morphism in BAnona.

Lemma 6.13. *Suppose $\mathcal{T} \in \mathsf{nTop}_{\text{И}}$ is* n*-compact and $\mathcal{U}$ is a finitely-supported set of closed sets with the* **finite intersection property***: the intersection of finitely many sets in $\mathcal{U}$ is nonempty. Then* И$b.\forall U.U \in \mathcal{U} \Rightarrow \mathsf{n}b.U \in \mathcal{U}$ *implies* $\bigcap \mathcal{U} \neq \emptyset$.

Proof. If $\bigcap \mathcal{U} = \emptyset$ then $\{\mathcal{T} \setminus U \mid U \in \mathcal{U}\}$ is an n-cover of $\mathcal{T}$. Since $\mathcal{T}$ is n-compact it has a finite subcover. This contradicts the finite intersection property.

Theorem 6.14. *G defines an equivalence between* BAnona *and* $\mathsf{nStone}_{\text{И}}^{op}$.

Proof. We use [24, Theorem 1, Chapter IV, Section 4].
G is essentially surjective on objects. Given $\mathcal{B}$ in BAnona and $x \in |\mathcal{B}|$, $x^\bullet \in \mathcal{O}_{F(\mathcal{B})}$ is clopen. By Lemma 6.9 if $U \in \mathcal{O}_{F(\mathcal{B})}$ is clopen then $U = x^\bullet$ for some $x \in |\mathcal{B}|$. By Theorem 5.25, the map $\text{-}^\bullet$ defines an isomorphism between $GF(\mathcal{B})$ and $\mathcal{B}$.
G is faithful. From the fact that nominal Stone spaces are totally separated.
G is full. Given $\mathcal{T}, \mathcal{T}'$ in $\mathsf{nStone}_{\text{И}}$ and $u : G(\mathcal{T}') \to G(\mathcal{T})$ in BAnona we construct a morphism $v : \mathcal{T} \to \mathcal{T}'$ in $\mathsf{nStone}_{\text{И}}$, such that $G(v) = u$.

Define $\alpha_{\mathcal{T}} : \mathcal{T} \to FG(\mathcal{T})$ by $t \mapsto \{U \in G(\mathcal{T}) \mid t \in U\}$.

$\alpha_{\mathcal{T}}$ is well defined: $\alpha_{\mathcal{T}}(t)$ is supported by $supp(t)$, and is indeed a maximal n-filter. We must also show that $a\#t$ and $U \in \alpha(t)$ imply $\mathsf{n}a.U \in \alpha(t)$; this follows from Proposition 4.11.

That $\alpha_{\mathcal{T}}$ is injective follows, as in the classical Stone duality, from total separation of the spaces. For surjectivity, consider a maximal n-filter $\mathcal{U} \in FG(\mathcal{T})$. This is a finitely-supported set of closed sets of $\mathcal{T}$ with the finite intersection property such that $U \in \mathcal{U}$ and $b\#\mathcal{U}$ imply $\mathsf{n}b.U \in \mathcal{U}$. By Lemma 6.13 there exists some $t \in \bigcap \mathcal{U} \subseteq |\mathcal{T}|$. That $\mathcal{U} = \alpha(t)$ follows from maximality of $\mathcal{U}$. $\alpha_{\mathcal{T}}$ and $\alpha_{\mathcal{T}}^{-1}$ are continuous. The proof is analoguous to the classical case, see [5].

We set $v = \alpha_{\mathcal{T}'}^{-1} \circ F(u) \circ \alpha_{\mathcal{T}}$. This is continuous and $G(v) = u$.

7 Conclusions

Boolean algebras in nominal sets naturally support an operation corresponding to the И-quantifier from [20]. Nominal sets are different enough that this paper is not a pure 'replay' of the standard proofs—the fine detail can be subtle.

As an empirical observation, our proofs 'want' и. They break for nominal Boolean algebras *without* и (remove и and its axioms from Definition 3.1, presumably to obtain a *Bano*). Without и we would need to use finitely-supported filters, rather than n-filters. But then the proofs of Lemma 5.14 and Theorem 5.17 would break. A representation theorem for Banos is future work.

We should also try to adapt the results to subreducts like Heyting algebras or distributive lattices (in this paper, we use negation in Proposition 5.11) and expansions with operations/operators. A concrete example of a *Hanona* (Heyting algebras with ᴎ; weaken the initial three axioms appropriately and add $\text{ᴎ}a.(x{\Rightarrow}y) = (\text{ᴎ}a.x){\Rightarrow}(\text{ᴎ}a.y)$ in Figure 1) is a intuitionistic hybrid logic with $\downarrow$ (cf. Example 3.6 and [30]). We believe that, modulo an easy syntactic translation, the logic *LG* by Tiu [34] is another.

On $pow(\mathbb{A})$, n is *unique*; the only function satisfying the axioms of Definition 3.1. We do not know whether n is determined by the axioms for arbitrary nominal powersets. (The map $X \mapsto$ *if* $a\#X$ *then* X *else* $\varnothing$ fails (**SelfDual**).)

At the start of this paper we wanted to represent ᴎ as an operation on nominal sets, just as $\neg$ and $\wedge$ are represented as complement and intersection. We found our representation in n, but this has some curious consequences. Notably, there is a mismatch between the internal notion of topological space in nominal sets (Definition 6.1) and the natural notion of nominal Stone space with И: namely, we have to insist that if U is clopen then so is $\mathsf{n}.U$ (because clopens come from the algebra), but we do not insist this of all open sets (because open sets come from the internal notion of topology in nominal sets). If add to Definition 6.1 that "U open implies $\mathsf{n}a.U$ open" then Definition 6.2 would break: we would have to insist on closure under n, to obtain a topology with И. It remains to check in future work whether the proofs using Definition 6.2 can be sensibly modified.

ᴎ is a nominal *name-restriction* (or *-generation*) operator. The premier such is И itself, which acts on predicates. Fernández and the first author added И to nominal rewriting [12], and Pitts added it to system T [28, Definition 1]. Name-restriction on nominal sets $X{-}a$ was considered in [15, Subsection 4.2] and is equal to $\mathsf{n}a$, and another operation νa on nominal sets (distinct from the $\mathsf{n}a$ of this paper) is studied in [18]. This paper locates the original И-quantifier in this landscape: it can always be represented as $\mathsf{n}a$ on nominal sets, which is equal to the $X{-}a$ of [15].

Sections 5.3 and 5.4 of [37] outline the Jónsson-Tarski representation of Boolean algebras with *operators*, i.e., *normal modalities*. Does it suffice to consider ᴎ and permutations as families of modalities? Yes, but our paper gives a strictly stronger result. Nominal sets have an 'external' theory of names and freshness given by permutations and support (extending the 'external' set intersection and complement corresponding to 'internal' conjunction and negation of Boolean algebra). An 'internal' notion of freshness is $x = \text{ᴎ}a.x$; a similar idea is used in cylindric/polyadic algebras and Lambda Abstraction Algebras [25]. Our challenge has been to make 'internal' and 'external' theories coincide for Boolean connectives, *and* to represent ᴎ as n (Definition 4.3), permutation as permutation, and to satisfy e.g. $a\#x$ implies $x = \text{ᴎ}a.x$.

Nominal powersets (Definition 4.1) have rich structure and can be characterised as nominal complete atomic Boolean algebras. There are *two* natural name-restrictions on nominal sets: *freshness* $X_{\#a} = \{x \in X \mid a\#x\}$ and *restriction* $\nu a.X = \{\pi{\cdot}x \mid x \in X, \pi \in \mathit{fix}(\mathit{supp}(X) \setminus \{a\})\}$ [18]. We do not believe that these can be represented using n. Note that n does not equal ν. For instance $X \subseteq \nu a.X$ is true, but $X \subseteq \mathsf{n}a.X$ is false in general.

Another approach to nominal Stone dualities is to consider the connection between nominal sets and functor categories. Lifting Stone type dualities pointwise to functor categories has been used in [4] to give sound and complete logics for π-calculus. The indexing category $\mathbb{I}$ of finite sets and injective maps between them is closely related to nominal sets: Nom is equivalent to a full reflective subcategory of $\mathsf{Set}^{\mathbb{I}}$. However the category of many-sorted algebras used for their logical interpretation of π-calculus, $\mathbf{DDL}^{\mathbb{I}^{op}}$, is not nominal in spirit.

The duality of this paper is related to a different indexing category $\mathsf{p}\mathbb{I}$—the category of finite sets of names and partial injective maps between them—which is self-dual. According to [29], Staton observed that the category of nominal restriction sets *Res* is equivalent to $\mathsf{Set}^{\mathsf{p}\mathbb{I}}$. Since Banonas are Boolean algebra objects in *Res*, BAnona is equivalent to $\mathsf{BA}^{\mathsf{p}\mathbb{I}}$. Lifting Stone duality pointwise yields a duality between $\mathsf{BA}^{\mathsf{p}\mathbb{I}}$ and $\mathsf{Stone}^{\mathsf{p}\mathbb{I}^{op}} \simeq \mathsf{Stone}^{\mathsf{p}\mathbb{I}}$. Nominal Stone spaces with И are equivalent to $\mathsf{Stone}^{\mathsf{p}\mathbb{I}}$ but they have an underlying nominal set, rather than a nominal restriction set. This gives an adjunction between Banonas and nominal sets which may be useful for a duality-based approach to *nominal coalgebraic logic* with built-in name generation. Coalgebras on nominal sets do provide a natural semantics for name-passing calculi such as π-calculus [10],[33].

Acknowledgements. We would like to thank Alexander Kurz and the anonymous referees for very useful comments.

References

1. Abramsky, S., Ghica, D.R., Murawski, A.S., Luke Ong, C.-H., Stark, I.D.B.: Nominal games and full abstraction for the nu-calculus. In: Proceedings of the 19th IEEE Symposium on Logic in Computer Science, LICS 2004, pp. 150–159. IEEE Computer Society Press, Los Alamitos (2004)
2. Areces, C., ten Cate, B.: Hybrid logics. In: Blackburn, P., van Benthem, J., Wolter, F. (eds.) Handbook of Modal Logic. Elsevier, Amsterdam (2007)
3. Bengtson, J., Parrow, J.: Formalising the π-Calculus Using Nominal Logic. In: Seidl, H. (ed.) FOSSACS 2007. LNCS, vol. 4423, pp. 63–77. Springer, Heidelberg (2007)
4. Bonsangue, M., Kurz, A.: Pi-calculus in logical form. In: Proceedings of the 22nd Annual IEEE Symposium on Logic in Computer Science, LICS 2007, pp. 303–312. IEEE Computer Society Press, Los Alamitos (2007)
5. Burris, S., Sankappanavar, H.: A Course in Universal Algebra. Graduate Texts in Mathematics. Springer, Heidelberg (1981)
6. Caires, L., Cardelli, L.: A spatial logic for concurrency (part I). Information and Computation 186(2), 194–235 (2003)
7. Cardelli, L., Gordon, A.: Logical Properties of Name Restriction. In: Abramsky, S. (ed.) TLCA 2001. LNCS, vol. 2044, pp. 46–60. Springer, Heidelberg (2001)
8. Cheney, J.: A simpler proof theory for nominal logic. In: Sassone, V. (ed.) FOSSACS 2005. LNCS, vol. 3441, pp. 379–394. Springer, Heidelberg (2005)

9. Cheney, J., Urban, C.: Nominal logic programming. ACM Transactions on Programming Languages and Systems (TOPLAS) 30(5), 1–47 (2008)
10. Cîrstea, C., Kurz, A., Pattinson, D., Schröder, L., Venema, Y.: Modal logics are coalgebraic. The Computer Journal (2009)
11. Dowek, G., Gabbay, M.J.: Permissive Nominal Logic. In: Proceedings of the 12th International ACM SIGPLAN Symposium on Principles and Practice of Declarative Programming, PPDP 2010, pp. 165–176 (2010)
12. Fernández, M., Gabbay, M.J.: Nominal rewriting with name generation: abstraction vs. locality. In: Proceedings of the 7th ACM SIGPLAN International Symposium on Principles and Practice of Declarative Programming, PPDP 2005, pp. 47–58. ACM Press, New York (2005)
13. Gabbay, M.J.: Fresh Logic. Journal of Applied Logic 5(2), 356–387 (2007)
14. Gabbay, M.J.: Nominal Algebra and the HSP Theorem. Journal of Logic and Computation 19(2), 341–367 (2009)
15. Gabbay, M.J.: A study of substitution, using nominal techniques and Fraenkel-Mostowski sets. Theoretical Computer Science 410(12-13), 1159–1189 (2009)
16. Gabbay, M.J.: Foundations of nominal techniques: logic and semantics of variables in abstract syntax. Bulletin of Symbolic Logic (2011) (in press)
17. Gabbay, M.J., Cheney, J.: A Sequent Calculus for Nominal Logic. In: Proceedings of the 19th IEEE Symposium on Logic in Computer Science, LICS 2004, pp. 139–148. IEEE Computer Society, Los Alamitos (2004)
18. Gabbay, M.J., Ciancia, V.: Freshness and Name-Restriction in Sets of Traces with Names. In: Hofmann, M. (ed.) FOSSACS 2011. LNCS, vol. 6604, pp. 365–380. Springer, Heidelberg (2011)
19. Gabbay, M.J., Mathijssen, A.: Nominal universal algebra: equational logic with names and binding. Journal of Logic and Computation 19(6), 1455–1508 (2009)
20. Gabbay, M.J., Pitts, A.M.: A New Approach to Abstract Syntax with Variable Binding. Formal Aspects of Computing 13(3-5), 341–363 (2001)
21. Keenan, E., Westerståhl, D.: Generalized quantifiers in linguistics and logic. In: Van Benthem, J., Ter Meulen, A. (eds.) Handbook of Logic and Language, pp. 837–894. Elsevier, Amsterdam (1996)
22. Kurz, A., Petrişan, D.: On universal algebra over nominal sets. Mathematical Structures in Computer Science 20, 285–318 (2010)
23. Litak, T.: Algebraization of Hybrid Logic with Binders. In: Schmidt, R. (ed.) RelMiCS/AKA 2006. LNCS, vol. 4136, pp. 281–295. Springer, Heidelberg (2006)
24. Lane, S.M.: Categories for the Working Mathematician. Graduate Texts in Mathematics, vol. 5. Springer, Heidelberg (1971)
25. Manzonetto, G., Salibra, A.: Applying universal algebra to lambda calculus. Journal of Logic and Computation 20(4), 877–915 (2010)
26. McCune, W.: Solution of the Robbins problem. Journal of Automated Reasoning 19, 263–276 (1997)
27. Menni, M.: About И-quantifiers. Applied Categorical Structures 11(5), 421–445 (2003)
28. Pitts, A.M.: Nominal system T. In: Proceedings of the 37th ACM SIGACT-SIGPLAN Symposium on Principles of Programming Languages, POPL 2010, pp. 159–170. ACM Press, New York (2010)
29. Pitts, A.M.: Structural recursion with locally scoped names (September 2010) (submitted for publication)
30. Reed, J.: Hybridizing a logical framework. Electronic Notes in Theoretical Computer Science 174(6), 135–148 (2006); Proceedings of the International Workshop on Hybrid Logic (HyLo 2006)

31. Shinwell, M.R., Pitts, A.M.: On a monadic semantics for freshness. Theoretical Computer Science 342(1), 28–55 (2005)
32. Shinwell, M.R., Pitts, A.M., Gabbay, M.J.: FreshML: Programming with Binders Made Simple. In: Proceedings of the 8th ACM SIGPLAN International Conference on Functional Programming, ICFP 2003, vol. 38, pp. 263–274. ACM Press, New York (2003)
33. Staton, S.: Name-passing process calculi: operational models and structural operational semantics. Technical Report UCAM-CL-TR-688, University of Cambridge, Computer Laboratory (June 2007)
34. Tiu, A.: A logic for reasoning about generic judgments. Electronic Notes in Theoretical Computer Science 174(5), 3–18 (2007)
35. Turner, D.C.: Nominal Domain Theory for Concurrency. PhD thesis, University of Cambridge (2009)
36. Tzevelekos, N.: Full abstraction for nominal general references. In: Proceedings of the 22nd IEEE Symposium on Logic in Computer Science, LICS 2007, pp. 399–410. IEEE Computer Society Press, Los Alamitos (2007)
37. Venema, Y.: Algebras and coalgebras. In: Blackburn, P., Van Benthem, J., Wolter, P. (eds.) Handbook of Modal Logic. Studies in logic and practical reasoning, ch. 6, vol. 3. Elsevier, Amsterdam (2007)
38. Westerståhl, D.: Quantifiers in formal and natural languages. In: Handbook of Philosophical Logic. Synthèse, ch. 2, vol. 4, pp. 1–131. Reidel, Dordrechtz (1989)

A Counterexample to Tensorability of Effects

Sergey Goncharov[1] and Lutz Schröder[1,2,⋆]

[1] Safe and Secure Cognitive Systems, DFKI Bremen
[2] Department of Computer Science, Universität Bremen

Abstract. Monads are widely used in programming semantics and in functional programming to encapsulate notions of side-effect, such as state, exceptions, input/output, or continuations. One of their advantages is that they allow for a modular treatment of effects, using composition operators such as sum and tensor. Here, the sum represents the non-interacting combination of effects, while the tensor imposes a high degree of interaction in the shape of a commutation law. Although many important effects are ranked, i.e. presented by algebraic operations of bounded arity, there is also a range of relevant unranked effects, with prominent examples including continuations and unbounded non-determinism. While the sum and tensor of ranked effects always exist, this is not so clear already when one of the components is unranked, in which case size problems come into play. In contrast to the case of sums where a counterexample can be constructed rather trivially, the general existence of tensors has, so far, been an open issue — as the tensor identifies more terms than the sum, it does exist in many cases where the sum fails to exist. As a possible counterexample, tensors of continuations with unranked effects have been discussed; however, we have disproved that possibility in recent work. In the present work, we nevertheless settle the question in the negative by presenting a well-order monad whose tensor with a simple ranked monad fails to exist; as a consequence, we show also that there is an unranked monad whose tensor with the finite list monad fails to exist.

1 Introduction

In theory and practice of programming, monads are commonly recognized as a standard formal abstraction for computational effects. An important feature of the monad-based approach to effects is its native support for modularity, i.e. the assembly of more complex effects from more primitive ones. One possible way to implement such modularity is by means of *monad transformers* [4]. Essentially, a monad transformer is a function that sends monads to monads (additional properties, such as functoriality, are not in general imposed). In practice, monad transformers usually enrich the given monad with certain additional features, thus enabling an *incremental approach* in denotational semantics and side-effecting programs [16]. Monad transformers have been very successful in practice; in particular, they form the core of the treatment of side-effects in the functional programming language Haskell [17].

Nevertheless, monad transformers have been criticized for their asymmetry [10] which treats one set of effects as the monad transformer and the other set of effects as an

⋆ Research supported by the DFG project *A construction kit for program logics* (LU 707/9-1)

A. Corradini, B. Klin, and C. Cîrstea (Eds.): CALCO 2011, LNCS 6859, pp. 208–221, 2011.

argument of the latter. E.g. the approach via monad transformers hides the symmetry in the combination of exceptions and I/O: this combination can be obtained either by applying the I/O monad transformer to the exception monad or by applying the exception monad transformer to the I/O monad, but this equivalence is not apparent from the corresponding monad transformer expressions. Moreover, monad transformers are ad hoc in character, and in particular have been known only for a limited number of effects, a prominent negative example being nondeterminism. An elegant solution to these problems was the introduction of binary composition operators over monads, such as *sum* and *tensor* [10, 9]. Whereas the sum of two monads is the simplest monad supporting both given effects without any interaction between them, the tensor (whose definition goes back to [6]) moreover requires distribution of these effects over each other, e.g. in case of tensoring statefulness with finite nondeterminism one has

$$(\mathtt{x} := \mathtt{a}; (\mathtt{b} + \mathtt{c})) \;=\; ((\mathtt{x} := \mathtt{a}; \mathtt{b}) + (\mathtt{x} := \mathtt{a}; \mathtt{c}))$$

and the like. We refer to the general form of this condition as the *tensor law*.

An important example of the tensor product is tensoring with global state, in which case the result is equivalent to application of the state monad transformer [19], e.g. $\mathcal{P} \otimes T_S = S \to \mathcal{P}(S \times \_)$ where $\mathcal{P}$ is the powerset monad and $T_S = S \to (S \times \_)$ is the (global) state monad. One can look at this case from the opposite perspective and consider it as an application of a *nondeterminism monad transformer* $T \mapsto \mathcal{P} \otimes T$. This transformer yields the universal *completely additive monad* over T [8], which therefore allows for a generalized Fischer-Ladner decomposition of control operators, i.e. roughly the translation

$$\mathtt{if}\; b\; \mathtt{then}\; p\; \mathtt{else}\; q := b?; p + (\neg b)?; q$$
$$\mathtt{while}\; b\; \mathtt{do}\; p := (b?; p)^\star; (\neg b)?$$

The catch in all this is that unless one requires both component monads to be *ranked*, i.e. generated by algebraic operations of bounded arity, there is no guarantee that sum and tensor exist [10]. Unranked monads arise as soon as the number of values that can participate in a computation is unbounded. Unbounded non-determinism and continuations are prominent examples of unranked monads; in particular, it has long been unclear that the above-mentioned non-determinism monad transformer actually exists. It is comparatively easy to see that the sum of simple ranked monads with most unranked monads will typically fail to exist (see, e.g., [11]). The case of tensoring is more subtle. For the specific example of the (unranked) continuation monad, it has been shown in [9] that the tensor does exist if the partner monad is ranked. It has been conjectured in *op. cit.* (p. 30) that the tensor of unranked monads does not exist in general, and it has been implicitly indicated (*op. cit.*, p. 33) that the tensor of continuations with a suitable unranked monad might serve as a counterexample, which seemed reasonable insofar as continuations generally constitute a good source of counterexamples (see, e.g., [20]).

However, we have recently proved that the tensor of two monads always exists if one of them is *uniform*, a natural criterion that ensures sufficiently pervasive applicability of the tensor law [7]. The class of uniform monads is surprisingly broad and includes not only several variants of non-determinism (which implies that the above-mentioned non-determinism monad transformer does after all exist), but also continuations, thus discharging the latter as a suspect for a potential counterexample to existence of tensors. In summary, prior to the current work, the question of universal existence of monad tensors was open, and no good candidates for possible counterexamples were known.

Having said this, we do settle the question in the negative in the present paper. Specifically, we define a quite natural *well-order monad* (on the category of sets) whose tensor with a simple free algebra monad (with two binary operations) fails to exist. By an ensuing non-constructive argument, our result implies that already for the non-empty finite list monad, not all tensors exist (so that, w.r.t. tensorability, unbounded unordered non-determinism behaves better than finite ordered non-determinism).

The paper is organised as follows. We give an introduction to monads in Section 2, and recall their equivalence with large Lawvere theories; we then actually base much of the technical development on Lawvere theories. In Section 3, we discuss tensors and *tensor algebras*, a notion that goes back to [14]. To set the mood, we summarize known existence results for tensors in Section 4 and state a few (easy) new ones, and then proceed to prove our negative result in Section 5, where tensor algebras appear as a key technical tool in that we show non-existence of the tensor by exhibiting a family of reachable tensor algebras of unbounded cardinality.

2 Monads vs. Large Lawvere theories

In a nutshell, the principle of monadic encapsulation of side-effects originally due to Moggi [15] and subsequently introduced into the functional programming language Haskell as the principal means of dealing with impure features [21] consists in moving the side effect from the function arrow into the result type of a function: a side-effecting function $X \to Y$ becomes a pure function $X \to TY$, where TY is a type of side-effecting computations over Y; the base example is $TY = S \to (S \times Y)$ for a fixed set S of states, so that functions $X \to TY$ are functions that may read and update a global state. Formally, a *monad* on the category of sets, presented as a *Kleisli triple* $\mathbb{T} = (T, \eta, \_^\star)$, consists of a function T mapping sets X (of values) to sets TX (of computations), a family of functions $\eta_X : X \to TX$, and a map assigning to every function $f : X \to TY$ a function $f^\star : TX \to TY$ that lifts f from X to computations over X. These data are subject to the equations

$$\eta^\star = \mathrm{id} \qquad f^\star\eta = f \qquad (f^\star g)^\star = f^\star g^\star$$

(for $g : Z \to TX$) which ensure that the *Kleisli category* of $\mathbb{T}$, which has sets as objects and maps $X \to TY$ as morphisms, is actually a category, with identities $\eta : X \to TY$ and composition $(f, g) \mapsto f^\star g$. We are mainly interested in monads on **Set**, as these are exactly equivalent to large Lawvere theories. On **Set**, all monads are *strong*, i.e. come with a natural transformation $X \times TY \to T(X \times Y)$ satisfying a number of coherence conditions [15].

Monads were originally intended as abstract presentations of algebraic theories, with TX abstracting the free algebra over X, i.e. terms over X, thought of as generated from basic operations, modulo provable equality. The notion of *rank* explained in more detail below refers to the arity of the involved algebraic operations. It has been shown that the algebraic view of monads gives rise to computationally natural operations for effects;

e.g. the state monad (with state set $S = V^L$ for sets V of values and L of locations) can be algebraically presented in terms of operations *lookup* and *update* [18]. Categorically, this shift of viewpoint amounts to generating monads from Lawvere theories. To cover unranked theories, we use the notion of large Lawvere theory [12], introduced into the theory of generic effects in [9]. Generally, we denote hom-sets of a category $\mathbf{C}$ in the form $\mathbf{C}(A, B)$.

Definition 1 (Large Lawvere theory). A *large Lawvere theory* is given by a locally small category L with small (i.e. set-indexed) products, together with a strict product preserving identity-on-objects functor $I : \mathbf{Set}^{op} \to L$. We call I the *indexing functor*, and denote If by $[f]$ for a map f. A morphism of large Lawvere theories $L_1 \to L_2$ is a functor $L_1 \to L_2$ that commutes with the indexing functors (and hence preserves small products). A *model* of a large Lawvere theory L in a category $\mathbf{C}$ with small products is a small product preserving functor $L \to \mathbf{C}$. A *homomorphism* of such models is just a natural transformation (which is determined uniquely by its component at 1).

The algebraic intuition behind these definitions is that the objects of a large Lawvere theory are sets (typically denoted $n, m, k, \dots$) of variables, and morphisms $n \to m$ are m-tuples of terms over n, or substitutions from m into terms over n. The indexing functor prescribes the effect of rearranging variables in terms. The notion of model recalled above implies that Lawvere theories provide a representation of effects that is independent of the base category $\mathbf{C}$, and given enough structure on $\mathbf{C}$ a Lawvere theory will induce a monad on $\mathbf{C}$. E.g., in categories of domains, the theory of finite non-blocking nondeterminism (Example 3.2 below) induces precisely the Plotkin powerdomain monad (while the Hoare and Smyth powerdomains require enriched Lawvere theories) [2].

Notation 2. Let L be a large Lawvere theory. For an object n of L and $i \in n$, we let κ_i denote the map $1 \to n$ that picks i. Thus, the κ_i induce product projections $[\kappa_i] : n \to 1$ in L. Given two sets n and m, their $\mathbf{Set}$-product $n \times m$ can be viewed as the sum of m copies of n in $\mathbf{Set}$, and hence as the m-th power of n in L. This induces for every $f : n \to k$ in L the morphisms $f \otimes m : n \times m \to k \times m$ and $m \otimes f : m \times n \to m \times k$.

It is well-known that large Lawvere theories and monads on $\mathbf{Set}$ form equivalent (overlarge) categories [12, 9]. The equivalence maps a large Lawvere theory L to the monad $T_L X = L(X, 1)$ with unit η^L given by $\eta^L_X(x) = [\kappa_x]$ using notation just introduced. Noting that $f : X \to L(Y, 1)$ induces $\tilde{f} \in L(Y, X)$ via the product structure of L, Kleisli star $f^\star$ is defined by $f^\star(g) = g\tilde{f} \in L(Y, 1)$ for $g \in L(X, 1)$. Conversely, a monad T on $\mathbf{Set}$ is mapped to the dual of its Kleisli category. We therefore largely drop the distinction between monads on $\mathbf{Set}$ and large Lawvere theories, and freely transfer concepts and examples from one setting to the other; occasionally we leave the choice open by just using the term *effect*.

We say that a large Lawvere theory L is *ranked* if it can be presented by operations (and equations) of arity less than κ for some cardinal κ; otherwise, L is *unranked*. Categorically, L having rank κ amounts to preservation of κ-directed colimits by the induced monad.

We give some standard examples of monadic effects, mostly from [15]:

Example 3. 1. *Global state:* as stated initially, $TX = S \to (S \times X)$ is a monad (for this and other standard examples, we omit the description of the remaining data), the well-known *state monad.*

2. *Nondeterminism:* the (unranked) large Lawvere theory $L_{\mathcal{P}}$ for nondeterminism corresponds to the powerset monad $\mathcal{P}$. It has m-tuples of subsets of n as morphisms $n \to m$. Variants arise on the one hand by restricting to nonempty subsets, thus ruling out non-termination, and on the other hand by bounding the cardinality of subsets. We denote nonemptyness by a superscript $\star$, and cardinality bounds by subscripts. E.g., the large Lawvere theory $L_{\mathcal{P}^{\star}_{\omega_1}}$ describes countable non-blocking nondeterminism; its morphisms $n \to m$ are m-tuples of nonempty countable subsets of n. Yet another variant arises by replacing sets with multisets, i.e. maps $X \to (\mathbb{N} \cup \{\infty\})$, thus modelling weighted nondeterminism [5] as a large Lawvere theory L_{mult}.
3. *Continuations:* The continuation monad maps a set X to the set $(X \to R) \to R$, for a fixed set R of results. The corresponding *unranked* large Lawvere theory L^{R}_{cont} has maps $m \to ((n \to R) \to R)$ as morphisms $n \to m$.
4. *Input/Output:* For a given set I of input symbols, the Lawvere theory L_I for input is generated by a single I-ary operation; it is an *absolutely free* theory, i.e. has no equations. Similarly, given a set O of output symbols, the Lawvere theory L_O for output is generated by unary operations f_o for $o \in O$.

Further effects that fit the algebraic framework are exceptions ($TX = X + E$), resumptions ($RX = \mu Y.\, T(X + Y)$ for a given base effect T) and many more.

A convenient way of denoting generic computations is the so-called *computational metalanguage* [15], which has found its way into functional programming in the shape of Haskell's do-notation. We briefly outline the version of the metalanguage we use below; this version is deliberately simplistic, as it serves only to elucidate the definition of tensors.

The metalanguage denotes morphisms in the underlying category of a given monad, using the monadic structure; since large Lawvere theories correspond to monads on **Set**, the metalanguage just denotes maps in our setting. We let a signature Σ consist of a set $\mathcal{B}$ of *base types*, to be interpreted as sets, and a collection of typed *function symbols* $f : A_1 \to A_2$ to be interpreted as functions, where A_1, A_2 are types. Here, we assume that the set $\mathcal{T}$ of *types* is generated from the base types by the grammar

$$\mathcal{T} \ni A_1, A_2 ::= 1 \mid B \mid A_1 \times A_2 \mid TA_1 \qquad (B \in \mathcal{B})$$

where $\times$ is interpreted as set theoretic product, 1 is a singleton set, and T is application of the given monad. We then have standard formation rules for terms-in-context $\Gamma \rhd t : A$, read 'term t has type A in context Γ', where a *context* is a list $\Gamma = (x_1 : A_1, \dots, x_n : A_n)$ of typed variables (later, contexts will mostly be omitted):

$$\frac{x : A \in \Gamma}{\Gamma \rhd x : A} \qquad \frac{f : A \to B \in \Sigma \quad \Gamma \rhd t : A}{\Gamma \rhd f(t) : B} \qquad \frac{}{\Gamma \rhd \star : 1}$$

$$\frac{\Gamma \rhd s : A \quad \Gamma \rhd t : B}{\Gamma \rhd \langle s, t \rangle : A \times B} \qquad \frac{\Gamma \rhd t : A \times B}{\Gamma \rhd \mathsf{fst}\, t : A} \qquad \frac{\Gamma \rhd t : A \times B}{\Gamma \rhd \mathsf{snd}\, t : B}$$

$$\frac{\Gamma \rhd t : A}{\Gamma \rhd \mathsf{ret}\, t : TA} \qquad \frac{\Gamma \rhd p : TA \quad \Gamma, x : A \rhd q : TB}{\Gamma \rhd \mathsf{do}\ x \leftarrow p; q : TB}$$

Only the operations in the last line are specific to monads; they are called *return* and *binding*, respectively. Return is interpreted by the unit η of the monad, and can be thought of as returning a value. A binding $\mathsf{do}\ x \leftarrow p; q$ executes p, binds its result to x, and then executes q, which may use x (if not, mention of x may be omitted). It is interpreted using Kleisli composition and strength, where the latter serves to propagate the context Γ [15]. In consequence, one has the *monad laws*

$$\mathsf{do}\ x \leftarrow p; \mathsf{ret}\, x = p \qquad \mathsf{do}\ x \leftarrow \mathsf{ret}\, a; p = p[a/x]$$

$$\mathsf{do}\ x \leftarrow (\mathsf{do}\ y \leftarrow p; q); r = \mathsf{do}\ x \leftarrow p; y \leftarrow q; r.$$

3 Tensors and Tensor Algebras

We proceed to recall the notion of tensor of effects. Roughly speaking, the tensor starts from the disjoint union of two algebraic theories, i.e. their *sum* in the language of Lawvere theories [10], but then imposes a strong degree of interaction between the component effects, namely that every operation in one of the theories is homomorphic w.r.t. the operations of the other theory. In a setting where we cannot take the existence of tensors for granted, it seems best to start the development coming from the intended models of the tensor, which just formalize the above:

Definition 4. Given two large Lawvere theories L_1, L_2, the category of $L_1 \otimes L_2$-*algebras* and their homomorphisms is the category

$$\mathsf{Mod}(L_1, \mathsf{Mod}(L_2, \mathbf{Set}))$$

of L_1-models in $\mathsf{Mod}(L_2, \mathbf{Set})$.

Rephrasing the above definition, an $L_1 \otimes L_2$-algebra is a set carrying both an L_1-model and an L_2-model in such a way that the operations of L_1 become homomorphic w.r.t. the operations of L_2 as indicated above, and a homomorphism of $L_1 \otimes L_2$-algebras is just a map between the underlying sets that is homomorphic w.r.t. the operations of both L_1 and L_2. Explicitly, if $f : n \to 1$ is an operation of L_1 and $g : m \to 1$ is an operation of L_2, then an $L_1 \otimes L_2$-algebra must satisfy the equation

$$f(g(x_{ij})_{j \in m})_{i \in n} = g(f(x_{ij})_{i \in n})_{j \in m}, \qquad (*)$$

called the *tensor law*. By the symmetry of the tensor law, it is clear that $L_1 \otimes L_2$-algebras are essentially the same as $L_2 \otimes L_1$-algebras.

Now the tensor of Lawvere theories is intended to capture precisely the category of $L_1 \otimes L_2$-algebras:

Definition 5. Let L_1, L_2 be large Lawvere theories. A Lawvere theory L is the *tensor product* of L_1 with L_2, $L = L_1 \otimes L_2$, if the category of L-models is isomorphic to the category of $L_1 \otimes L_2$-algebras:

$$\mathsf{Mod}(L_1 \otimes L_2, \mathbf{Set}) \cong \mathsf{Mod}(L_1, \mathsf{Mod}(L_2, \mathbf{Set})).$$

As Lawvere theories are determined by their categories of algebras, the above defines the tensor product of large Lawvere theories uniquely up to isomorphism whenever it exists. We have the following characterization of tensors:

Proposition 6. [9] *For large Lawvere theories L, L_1, L_2, the following are equivalent:*

1. $L = L_1 \otimes L_2$.
2. *For every category* $\mathbf{C}$ *with small products,*

$$\mathsf{Mod}(L, \mathbf{C}) \cong \mathsf{Mod}(L_1, \mathsf{Mod}(L_2, \mathbf{C})).$$

3. *L is universal w.r.t. having* commuting *morphisms $L_1 \to L \leftarrow L_2$ (elided in the notation). Here, commutation is satisfaction of the* tensor law, *i.e. given $f_1 : n_1 \to m_1$ in L_1 and $f_2 : n_2 \to m_2$ in L_2 we demand commutativity of the diagram*

$$\begin{array}{ccc} n_1 \times n_2 & \xrightarrow{n_1 \otimes f_2} & n_1 \times m_2 \\ {\scriptstyle f_1 \otimes n_2}\downarrow & & \downarrow{\scriptstyle f_1 \otimes m_2} \\ m_1 \times n_2 & \xrightarrow{m_1 \otimes f_2} & m_1 \times m_2. \end{array}$$

For existence of tensors, we have the following criteria:

Proposition 7 (Existence of tensors). [9] *Let L_1, L_2 be large Lawvere theories. The following are equivalent:*

1. *The tensor $L_1 \otimes L_2$ exists.*
2. *The forgetful functor* $\mathsf{Mod}(L_1, \mathsf{Mod}(L_2, \mathbf{Set})) \to \mathbf{Set}$ *is monadic.*
3. *The forgetful functor* $\mathsf{Mod}(L_1, \mathsf{Mod}(L_2, \mathbf{Set})) \to \mathbf{Set}$ *has a left adjoint.*

Remark 8. The existence of a tensor $L_1 \otimes L_2$ is essentially a size issue: one has to prove that the collection of terms generated from a given set of variables by operations from L_1 and L_2 is a set when taken modulo the equations of L_1 and L_2 and the tensor law.

For the sake of completeness, we recall explicit reformulations of the above concepts in terms of monads [14].

Definition 9. Given two monads T and S over $\mathbf{Set}$, *$T \otimes S$-algebras* are triples of the form (X, α, β) where (X, α) is a T-algebra, (X, β) is an S-algebra, and moreover for all sets Y, Z and all $p \in SY$, $q \in TZ$, $f : Y \times Z \to X$,

$$\beta(T(\lambda z.\, \alpha(S f_{\_,z}\, p))q) = \alpha(S(\lambda y.\, \beta(T f_{y,\_}\, q))p)$$

where $f_{\_,z}(y) = f_{y,\_}(z) = f(y, z)$ for $(y, z) \in Y \times Z$. Morphisms of $T \otimes S$-algebras are maps between the respective carriers which are homomorphic for both T and S.

(We avoid the term *bialgebra* as used in [14], as this has since come to signify simultaneous algebra and coalgebra structure.) Then the tensor $T \otimes S$ of monads T, S on $\mathbf{Set}$

is defined, uniquely up to isomorphism, as the monad whose Eilenberg-Moore algebras are precisely the $T \otimes S$-algebras, if such a monad exists. The computational meaning of the tensor law becomes clearer in the computational metalanguage: if we extended the metalanguage with subtypes $T_i A$ of TA interpreted using the component monads T_1, T_2 of the tensor $T = T_1 \otimes T_2$, it amounts to the equality

$$\mathsf{do}\ x_1 \leftarrow p_1; x_2 \leftarrow p_2; \mathsf{ret}\langle x_1, x_2\rangle = \mathsf{do}\ x_2 \leftarrow p_2; x_1 \leftarrow p_1; \mathsf{ret}\langle x_1, x_2\rangle$$

in context Γ_1, Γ_2, where $\Gamma_i \rhd p_i : T_i A_i$ for $i = 1, 2$; i.e. programs having effects only from T_1 do not interfere with programs having effects only from T_2.

Finally, we also have an obvious translation of the existence criterion for tensors (Proposition 7) into the language of monads, which we record explicitly:

Proposition 10. *The tensor $T \otimes S$ of monads T, S on* **Set** *exists iff the forgetful functor from $T \otimes S$-algebras to sets is monadic, equivalently has a left adjoint.*

Our counterexample will be based on the following simple consequence of this:

Corollary 11. *If the tensor $T \otimes S$ of monads T, S on* **Set** *exists, then there exists an initial $T \otimes S$-algebra.*

4 Existence of Tensor Products

We now recall some positive results on existence of tensor products, and establish some basic additional ones. On the one hand, this helps appreciate that although it has always been believed that tensors of monads on **Set** do not exist in general, large classes of monads are in fact ruled out as counterexamples. On the other hand, some of the results presented here enable us to derive additional counterexamples from the one we present in the next section.

It is well-known that the tensor product of two large Lawvere theories with rank does exist [9], so that any counterexample needs to involve at least one unranked monad. The following result on a class of monads where the tensor law is applicable in sufficiently many cases rules out many of the obvious candidates.

Definition 12 (Uniformity). [7] Let L be a large Lawvere theory. The *constants* of L are the elements of $c_L := L(0, 1)$. For every set n we denote by $\mathsf{c}_L^n : n \to n + c_L$ the morphism $[\mathrm{id}] \times \prod_{f \in c_L} f$. We say that L is *uniform* if for every L-morphism $f : n \to m$ there exists a *generic morphism*, i.e. a morphism $\hat{f} : k \to 1$ for some set k such that there exists a set-function $\mathsf{u} : k \times m \to n + c_L$ with $f = (\hat{f} \otimes m) \circ [\mathsf{u}] \circ \mathsf{c}_L^n$.

Definition 13 (Tensorability). A large Lawvere theory L_1 is *tensorable* if the tensor $L_1 \otimes L_2$ exists for all large Lawvere theories L_2.

Theorem 14. [7] *Uniform large Lawvere theories are tensorable.*

As indicated in the introduction, continuations were previously named as a candidate for a counterexample to tensorability [9]. However:

Example 15. Powerset, non-empty powerset, countable powerset, countable non-empty powerset, multisets, and continuations (i.e. the theories $L_{\mathcal{P}}$, $L_{\mathcal{P}^{\star}}$, $L_{\mathcal{P}_{\omega_1}}$, $L_{\mathcal{P}^{\star}_{\omega_1}}$, L_{mult}, and L^R_{cont} of Example 3) are uniform and, consequently, tensorable.

Remark 16. While tensoring with unbounded non-determinism typically forms a well-behaved completion of a given Lawvere theory [7], the general effect of tensoring with continuations is less easy to grasp. However, there are known cases where tensoring with continuations is a reasonable operation, e.g. in case the partner effect is global state [9] — in this case, one obtains the underlying monad of Scheme [1].

Sometimes existence of tensors can be inherited along the structure of theories:

Definition 17. A morphism $L_1 \to L_2$ of large Lawvere theories is *full* if the underlying functor $L_1 \to L_2$ is full.

Intuitively, if $L_1 \to L_2$ is a full morphism, then L_2 has the same operations as L_1 but more equations.

Lemma 18. *Let L_1, L_2, and L be large Lawvere theories and let $F : L_1 \to L_2$ be a full morphism of large Lawvere theories. Then existence of $L_1 \otimes L$ implies existence of $L_2 \otimes L$.*

Proof. Precomposition of models with F yields a full embedding

$$\alpha : \mathsf{Mod}(L_2, \mathsf{Mod}(L, \mathbf{Set})) \hookrightarrow \mathsf{Mod}(L_1, \mathsf{Mod}(L, \mathbf{Set})).$$

This embedding is, in fact, reflective, i.e. has a left adjoint. To see this, first note that $\mathbf{Set}$ has a factorization structure (surjective, jointly injective) for sources [3], which lifts to a factorization structure (surjective, jointly injective) for sources on $\mathsf{Mod}(L_1, \mathsf{Mod}(L, \mathbf{Set}))$. Therefore, it suffices to show that $\mathsf{Mod}(L_2, \mathsf{Mod}(L, \mathbf{Set}))$ is closed under jointly injective sources in $\mathsf{Mod}(L_1, \mathsf{Mod}(L, \mathbf{Set}))$ (see *op. cit.*). Thus, put $\mathbf{C} = \mathsf{Mod}(L, \mathbf{Set})$, let $M : L_1 \to \mathbf{C}$ be a model of L_1 in $\mathbf{C}$, let $M_i : L_2 \to \mathbf{C}$ be a model of L_2 in $\mathbf{C}$ for all i in some index set I, and let $(f_i : M \to M_i F)_{i \in I}$ be a jointly injective source of homomorphisms in $\mathsf{Mod}(L_1, \mathbf{C})$. We have to show that M factors through an L_2-model $\bar{M}$, i.e. $M = \bar{M}F$. We hence define $\bar{M}$ by $\bar{M}Fg = Mg$ for morphisms g in L_1 (which defines $\bar{M}$ on all of L_2 because F is full). This is well-defined by joint injectivity of the f_i; it is then straightforward to show that $\bar{M}$ is a model of L_2.

Now observe that the forgetful functor from $\mathsf{Mod}(L_2, \mathsf{Mod}(L, \mathbf{Set}))$ to $\mathbf{Set}$ is precisely the composition of α with the forgetful functor from $\mathsf{Mod}(L_1, \mathsf{Mod}(L, \mathbf{Set}))$ to $\mathbf{Set}$. If $L_1 \otimes L$ exists then the former functor has a left adjoint, hence the composition will also have a left adjoint, which by Proposition 7 means that $L_2 \otimes L$ also exists. □

Given a set E, we denote by L_E the large Lawvere theory generated by taking constants from E as the only operations, and no equations (thus, $L_E(n, m) = (n + E)^m$ for all n). The sum $L + L_E$ exists for every L and E, and in terms of monads yields exactly the well-known exception monad transformer which maps a monad T to $T(\_ + E)$ [13].

Lemma 19. *Given large Lawvere theories L_1, L_2, if $L_1 \otimes L_2$ exists then $(L_1 + L_E) \otimes L_2$ exists for every E.*

Proof. We have a full embedding

$$\alpha : \mathsf{Mod}(L_1 + L_E, \mathsf{Mod}(L_2, \mathbf{Set})) \hookrightarrow \mathsf{Mod}((L_1 \otimes L_2) + L_E, \mathbf{Set}),$$

which essentially just forgets that the operations of L_E are homomorphic w.r.t. the operations of L_2 in $\mathsf{Mod}(L_1 + L_E, \mathsf{Mod}(L_2, \mathbf{Set}))$. As in the proof of Lemma 18, it suffices to show that α has a left adjoint. This is shown similarly as for Lemma 18 by exploiting the (surjective, jointly injective) factorization structure for sources on $\mathsf{Mod}((L_1 \otimes L_2) + L_E, \mathbf{Set})$ and showing that $\mathsf{Mod}(L_1 + L_E, \mathsf{Mod}(L_2, \mathbf{Set}))$ is closed under jointly injective sources in $\mathsf{Mod}((L_1 \otimes L_2) + L_E, \mathbf{Set})$, i.e. that the tensor law for L_E vs. L_2 is inherited along jointly injective sources. □

5 Nonexistence of Tensor Products

We now present our main (negative) result, an example of two monads whose tensor fails to exist. Necessarily, one of these must be unranked; we describe this monad first.

We define a *$\mathcal{W}$-algebra* to be a set X equipped with an ordinal-indexed family of operations $\iota_\kappa : X^\kappa \to X$ satisfying the following conditions.

1. *Strictness:* $\iota_\kappa(w) = \iota_0$ whenever $w(\alpha) = \iota_0$ for some $\alpha < \kappa$.
2. *Non-repetitiveness:* $\iota_\kappa(w) = \iota_0$ whenever $w(\alpha_1) = w(\alpha_2)$ for some $\alpha_1 < \alpha_2 < \kappa$.
3. *Associativity:* For every ordinal-indexed family $(\kappa_\mu)_{\mu<\nu}$ of ordinals $\kappa_\mu > 0$,

$$\iota_\kappa(w) = \iota_\nu(\lambda\mu < \nu.\, \iota_{\kappa_\mu}(w_\mu))$$

where $\kappa = \sum_{\mu<\nu} \kappa_\mu$ is regarded as having elements $\langle\mu, \alpha\rangle$ with $\mu < \nu$ and $\alpha < \kappa_\mu$, and for every such $\langle\mu, \alpha\rangle$, $w(\mu, \alpha) = w_\mu(\alpha)$.

Since $\iota_0 : 1 \to X$ is essentially a constant, we omit its argument as above. Moreover, we regard an ordinal κ as the set of all ordinals $\alpha < \kappa$ unless we explicitly specify otherwise, as in the associativity law above where we use a more convenient isomorphic representation of ordinal sums. Notice that although the above formulations of strictness and non-repetitiveness employ the word 'whenever', these conditions are effectively equational axioms. By non-repetitiveness, for every κ whose cardinality exceeds $|X|$, ι_κ is identically ι_0, which means that the class of operations of a given $\mathcal{W}$-algebra is effectively a set. As usual, a homomorphism of two $\mathcal{W}$-algebras (X, ι_κ) and (Y, ι_κ) is a map $f : X \to Y$ that commutes with the operations:

$$f(\iota_\kappa(w)) = \iota_\kappa(f \circ w) \quad \text{for } w \in X^\kappa.$$

Lemma and Definition 20. *The forgetful functor from $\mathcal{W}$-algebras to* **Set** *is monadic. The associated monad $\mathcal{W}$, the* well-order monad, *maps a set X to the set*

$$\mathcal{W}X = \{(Y, \rho) \mid Y \subseteq X, \rho \text{ a well-order on } Y\}.$$

One may alternatively think of the well-order monad as a monad of infinite non-repetitive non-empty lists, with the empty well-order $(\varnothing, \varnothing)$ playing the role of an error element that is thrown in case of repetitions, and is propagated through composition by the strictness law. The monad structure of $\mathcal{W}$ will become apparent in the proof below.

Proof. The left adjoint maps a set X to the obvious $\mathcal{W}$-algebra on $\mathcal{W}X$, also denoted $\mathcal{W}X$; explicitly, for $w \in (\mathcal{W}X)^{\kappa}$, $\iota_{\kappa}(w)$ is the concatenation of the well-orderings $w(\alpha)$, $\alpha < \kappa$, if the domains of the $w(\alpha)$ are non-empty and disjoint, and $\iota_{\kappa}(w) = (\varnothing, \varnothing)$ otherwise. The unit $\eta_X : X \to \mathcal{W}X$ maps x to the unique well-ordering on $\{x\}$; the universal property is straightforward.

Monadicity now follows by Beck's monadicity theorem, as creation of split coequalizers (and in fact, more generally, coequalizers of congruences) by the forgetful functor is straightforward thanks to the equational character of $\mathcal{W}$-algebras. □

The second monad for our example is very simple, and has finite rank: Let $\Sigma^{\star}_{2,2}$ be the free monad generated by the signature functor $\Sigma_{2,2} = \lambda X.\, 2 \times X \times X$, i.e. by two binary operations and no equations.

Lemma 21. *For every infinite cardinal κ, there exists a $(\mathcal{W}(\_ + 2) \otimes \Sigma^{\star}_{2,2})$-algebra W_{κ} such that W_{κ} is* reachable, *i.e.* generated from the empty set of generators, *and $|W_{\kappa}| > \kappa$.*

Proof. The domain of W_{κ} is the union $\{\bot, 0, 1\} \cup U^0_{\kappa} \cup U^1_{\kappa} \cup L_{\kappa}$ where the U^i_{κ} and L_{κ} are sets of terms defined by infinitary mutual recursion according to the the rules

$$\frac{t \in W_{\kappa} - \{0\}}{\langle i, 0, t\rangle \in U^i_{\kappa}}$$

where $i \in \{0, 1\}$, and

$$\frac{t : \nu \hookrightarrow U^0_{\kappa} \cup U^1_{\kappa} \qquad \forall \mu.\, \mu + 1 < \nu \implies (t(\mu) \in U^0_{\kappa} \iff t(\mu+1) \in U^1_{\kappa})}{t \in L_{\kappa}}$$

where ν is an ordinal such that $1 < |\nu| \leqslant \kappa$ and $\hookrightarrow$ is read as t being injective (not a subset inclusion). Notice that $U^0_{\kappa} \cap U^1_{\kappa} = \varnothing$, so the second premise says that $t(\mu)$ alternates between U^0_{κ} and U^1_{κ}. Let us define a length map $\#$ from W_{κ} to ordinals as follows: we put $\#t = 1$ for $t \in \{\bot, 0, 1\} \cup U^0_{\kappa} \cup U^1_{\kappa}$, and $\#t = \nu$ whenever $t : \nu \hookrightarrow U^0_{\kappa} \cup U^1_{\kappa} \in L_{\kappa}$. Note that this implies $\#t > 1$ iff $t \in L_{\kappa}$.

To give a $\Sigma^{\star}_{2,2}$-algebra structure over W_{κ} is the same as to define two binary maps $u_0, u_1 : W_{\kappa} \times W_{\kappa} \to W_{\kappa}$. For $i = 0, 1$ we put by definition

- $u_i(t, t) = t$ if $t \in \{0, 1\}$;
- $u_i(0, t) = \langle i, 0, t\rangle \in U^i_{\kappa}$ whenever $t \in \{1\} \cup L_{\kappa}$;
- $u_i(s, t) = \bot$ in the remaining cases.

Defining a $\mathcal{W}(\_ + 2)$-algebra structure on W_{κ} amounts to giving two constants $c_0, c_1 \in W_{\kappa}$ and ordinal-indexed operations $\iota_{\nu} : (W_{\kappa})^{\nu} \to W_{\kappa}$ satisfying the conditions (1)–(3) above. Let $c_0 = 0$, $c_1 = 1$, $\iota_0 = \bot$, and $\iota_1 = \mathrm{id}$. For $\nu > 1$ and $t \in (W_{\kappa})^{\nu}$ we define $\iota_{\nu}(t)$ by the clauses

- $\iota_{\nu}(t) = s$, provided the map $s : \zeta \to W_{\kappa}$ on $\varsigma = \sum_{\mu<\nu} \#t(\mu)$ defined as follows is in L_{κ}: We regard ς as consisting of pairs $\langle \mu, \kappa\rangle$ where $\mu < \nu$ and $\kappa < \#t(\mu)$. For every such $\langle \mu, \kappa\rangle$, put $s\langle \mu, \kappa\rangle = t(\mu)(\kappa)$ if $t(\mu) \in L_{\kappa}$, and $s\langle \mu, \kappa\rangle = t(\mu)$ otherwise (in which case necessarily $\kappa = 0$).
- $\iota_{\nu}(t) = \bot$ otherwise.

It is then clear by construction that W_κ is reachable, as the rules defining L_κ and the U^i_κ just amount to closure under the u_i and ι_ν as defined above. Next, we have to check that W_κ is really a $(\mathcal{W}(\_+2) \otimes \Sigma^\star_{2,2})$-algebra. By definition, for every t, $\iota_\nu(t) \in L_\kappa \cup \{\bot\}$, hence the conditions (1) and (2) are ensured automatically. Condition (3) is less trivial, but still routine. Finally we need to verify the tensor law (∗). In the case at hand it amounts to proving the equation

$$u_i(\iota_\nu(t), \iota_\nu(s)) = \iota_\nu(\lambda\mu < \nu.\, u_i(t(\mu), s(\mu)))$$

for every $s, t \in W_\kappa$, $i = 0, 1$. It is immediate by definition that both sides of this equation equal $\bot$ unless $\nu = 1$. In the latter case the equation also follows since, by definition, $\iota_1 = \mathrm{id}$.

Finally, we show that $|W_\kappa| > \kappa$. In order to derive a contradiction, assume that $|W_\kappa| \leqslant \kappa$ and let ς be an ordinal number such that $|W_\kappa| = |\varsigma|$. Let ρ be a bijection $\varsigma \to W_\kappa - \{0\}$. Since $|\varsigma| \leqslant \kappa$ and hence $|\varsigma \cdot 2| \leqslant \kappa$ (since κ is infinite), we can form an element $t_\rho : \varsigma \cdot 2 \hookrightarrow U^0_\kappa \cup U^1_\kappa$ of W_κ by putting $t_\rho(\varsigma', i) = \langle i, 0, \rho(\varsigma')\rangle$ for $\varsigma' < \varsigma$, $i = 0, 1$. By varying ρ, we can produce as many such elements as there are isomorphisms from ς to $W_\kappa - \{0\}$, i.e. strictly more than $|\varsigma| = |W_\kappa|$, contradiction. □

Theorem 22. *The tensor of the well-order monad $\mathcal{W}$ and $\Sigma^\star_{2,2}$ does not exist.*

Proof. By Lemma 19, it suffices to prove that the tensor of $\mathcal{W}(\_+2)$ and $\Sigma^\star_{2,2}$ does not exist. By Lemma 21, there is no initial $(\mathcal{W}(\_+2) \otimes \Sigma^\star_{2,2})$-algebra I, as the unique homomorphism $I \to W_\kappa$ into the algebras constructed in the lemma would have to be surjective for every cardinal κ. Now Corollary 11 implies that the tensor $(\mathcal{W}(\_+2) \otimes \Sigma^\star_{2,2})$ does not exist. □

The above theorem also yields a (nonconstructive) argument showing that not all tensors with the non-empty list monad exist.

Theorem 23. *The non-empty finite list monad fails to be tensorable.*

Proof. Let L be the large Lawvere theory for non-empty finite lists. It is generated by one binary operation u and the associativity axiom:

$$u(u(a, b), c) = u(a, u(b, c)).$$

The tensor product $L \otimes L$ is generated by u and a duplicate u' of u, associativity for both u and u', and the tensor law

$$u'(u(a_1, b_1), u(a_2, b_2)) = u(u'(a_1, a_2), u'(b_1, b_2)).$$

In other words, $L \otimes L$-algebras are precisely the $\Sigma^\star_{2,2}$-algebras satisfying the mentioned equations. According to this, it is easy to see that the algebras W_κ constructed in Lemma 21 are in fact $(L \otimes L) \otimes \mathcal{W}(\_+2)$-algebras. By the same argument as for Theorem 22, the tensor $(L \otimes L) \otimes \mathcal{W}$ does not exist. If $L \otimes \mathcal{W}$ does not exist, then W witnesses non-tensorability of L. Otherwise, $L \otimes \mathcal{W}$ witnesses non-tensorability of L: given that $L \otimes \mathcal{W}$ exists, it is clear from Definition 5 that existence of $L \otimes (L \otimes \mathcal{W})$ would imply existence of $(L \otimes L) \otimes \mathcal{W}$. □

The way we have proved the last counterexample, namely by reducing it to the more complex one from Theorem 22, seems roundabout, but we do not currently see an alternative. A direct approach to proving that the tensor of, say, $\mathcal{W}$ with the non-empty finite list monad need not exist does not appear to be feasible, as it leads to tensor algebras whose cardinality is extremely difficult to estimate due to broad propagation of the tensor law.

6 Conclusion

We have shown that the tensor of the well-order monad and a simple finitary monad, generated by two binary operations and no equations, fails to exist, and we have concluded that the more familiar non-empty finite list monad L fails to be *tensorable*, i.e. there exists a monad whose tensor with L fails to exist. We have thus settled in the negative the open question of universal existence of tensors of set-based monads, which dates back at least to [14] where it appears in the context of early developments of the categorical foundations of universal algebra, and which has recently reemerged in the perspective of work on algebraic effects [10, 9]. The negative answer as such is in accordance with expectations, but the actual counterexample is rather different from what was previously suspected. Nevertheless, it has been possible to keep the counterexample not only natural but also reasonably simple.

The nature of our counterexample seems to indicate that the positive results that complement it, specifically tensorability of uniform monads [7], leave little room for improvement. One specific open question that does remain, however, is whether *finite* non-determinism is tensorable, i.e. whether every effect can be extended with binary choice (and deadlock). We do not venture a guess here — on the one hand, both unbounded non-determinism and countable non-determinism are uniform and hence tensorable by the results of [7], and on the other hand we have shown here that the closely related non-empty finite list monad fails to be tensorable.

Our main motivation for the study of tensors as such is to develop a monadic framework for non-interference of effects, noting that the tensor law precisely amounts to orthogonality of the component effects; these ideas will be further developed in future research.

Acknowledgements. We wish to thank various contributors to the categories mailing list, in particular Paul Levy and Peter Johnstone, for useful insights communicated via the list, and the anonymous referees for valuable pointers to the literature. Erwin R. Catesbeiana has commented on inconsistent monads.

References

[1] Abelson, H., Dybvig, K., Haynes, C., Rozas, G., Adams, N., Friedman, D., Kohlbecker, E., Steele, G., Bartley, D., Halstead, R., Oxley, D., Sussman, G., Brooks, G., Hanson, C., Pitman, K., Wand, M.: Revised report on the algorithmic language Scheme. Higher-Order Symb. Comput. 11, 7–105 (1998)
[2] Abramsky, S., Jung, A.: Domain theory. In: Handbook of Logic in Computer Science, vol. 3, pp. 1–168. Oxford University Press, Oxford (1994)

[3] Adámek, J., Herrlich, H., Strecker, G.: Abstract and concrete categories. John Wiley & Sons Inc., New York (1990)
[4] Cenciarelli, P., Moggi, E.: A syntactic approach to modularity in denotational semantics. In: Category Theory and Computer Science, CTCS 1993 (1993)
[5] Droste, M., Kuich, W., Vogler, H. (eds.): Handbook of Weighted Automata. Springer, Heidelberg (2009)
[6] Freyd, P.: Algebra valued functors in general and tensor products in particular. Colloq. Math. 14, 89–106 (1966)
[7] Goncharov, S., Schröder, L.: Powermonads and tensors of unranked effects. In: Logic in Computer Science, LICS 2011. IEEE Computer Society Press, Los Alamitos , pp. 227–236 (2011)
[8] Goncharov, S., Schröder, L., Mossakowski, T.: Kleene monads: handling iteration in a framework of generic effects. In: Kurz, A., Lenisa, M., Tarlecki, A. (eds.) CALCO 2009. LNCS, vol. 5728, pp. 18–33. Springer, Heidelberg (2009)
[9] Hyland, M., Levy, P., Plotkin, G., Power, J.: Combining algebraic effects with continuations. Theoret. Comput. Sci. 375, 20–40 (2007)
[10] Hyland, M., Plotkin, G., Power, J.: Combining effects: Sum and tensor. Theoret. Comput. Sci. 357, 70–99 (2006)
[11] Hyland, M., Power, J.: The category theoretic understanding of universal algebra: Lawvere theories and monads. ENTCS 172, 437–458 (2007)
[12] Linton, F.: Some aspects of equational categories. In: Proc. Conf. Categor. Algebra, La Jolla, pp. 84–94 (1966)
[13] Lüth, C., Ghani, N.: Composing monads using coproducts. In: International Conference on Functional Programming, ICFP 2002, pp. 133–144. ACM Press, New York (2002)
[14] Manes, E.: A triple theoretic construction of compact algebras. In: Seminar on Triples and Categorical Homology Theory. Lect. Notes Math., vol. 80, pp. 91–118. Springer, Heidelberg (1969)
[15] Moggi, E.: Notions of computation and monads. Inf. Comput. 93, 55–92 (1991)
[16] Moggi, E.: A semantics for evaluation logic. Fund. Inform. 22, 117–152 (1995)
[17] Peyton-Jones, S. (ed.): Haskell 98 Language and Libraries — The Revised Report. Cambridge University Press, Cambridge (2003); Also: J. Funct. Prog. 13 (2003)
[18] Plotkin, G., Power, J.: Notions of computation determine monads. In: Nielsen, M., Engberg, U. (eds.) FOSSACS 2002. LNCS, vol. 2303, pp. 342–356. Springer, Heidelberg (2002)
[19] Power, J., Shkaravska, O.: From comodels to coalgebras: State and arrays. In: Coalgebraic Methods in Computer Science, CMCS 2004. ENTCS, vol. 106, pp. 297–314. Elsevier, Amsterdam (2004)
[20] Schröder, L., Mossakowski, T.: Generic exception handling and the java monad. In: Rattray, C., Maharaj, S., Shankland, C. (eds.) AMAST 2004. LNCS, vol. 3116, pp. 443–459. Springer, Heidelberg (2004)
[21] Wadler, P.: How to declare an imperative. ACM Comput. Surveys 29, 240–263 (1997)

The Microcosm Principle and Compositionality of GSOS-Based Component Calculi

Ichiro Hasuo*

University of Tokyo, Japan

Abstract. In the previous work by Jacobs, Sokolova and the author, synchronous parallel composition of coalgebras—yielding a coalgebra—and parallel composition of behaviors—yielding a behavior, where behaviors are identified with states of the final coalgebra—were observed to form an instance of the *microcosm principle*. The microcosm principle, a term by Baez and Dolan, refers to the general phenomenon of nested algebraic structures such as a monoid in a monoidal category. Suitable organization of these two levels of parallel composition led to a general *compositionality* theorem: the behavior of the composed system relies only on the behaviors of its constituent parts. In the current paper this framework is extended so that it accommodates any process operator—not restricted to parallel composition—whose meaning is specified by means of GSOS rules. This generalizes Turi and Plotkin's bialgebraic modeling of GSOS, by allowing a process operator to act as a connector between components as coalgebras.

1 Introduction

1.1 Structural Operational Semantics and Its Bialgebraic Modeling

Structural operational semantics (SOS) [20] is a well-developed mathematical tool for defining operational semantics of a programming language. It is based on *SOS rules* from which transitions between program terms are derived. SOS has been notably successful for *process calculi*: simple programming languages for concurrent processes. Various *syntactic formats*—syntactic restrictions on SOS rules—have been proposed to ensure good properties of SOS (see [1] for a survey); the *GSOS format* [6] is one of the most common among them.

In SOS for process calculi, dynamic behaviors of processes—such as $(a; b) \parallel \overline{b} \xrightarrow{a} (0; b) \parallel \overline{b}$—are derived by structural induction on the construction of process terms. In categorical terms, dynamics of processes (the former) are modeled by a *coalgebra* while the set of process terms forms an initial *algebra* that supports structural induction (the latter); see e.g. [15]. It is Turi and Plotkin's seminal work [22] that combines these two on a categorical level, resulting in a *bialgebraic modeling* of SOS.

* Thanks are due to Marcelo Fiore, Masahito Hasegawa, Bart Jacobs, Paul-André Melliès, Bartek Klin, John Power, Ana Sokolova and Sam Staton for helpful discussions. Helpful comments from the reviewers for the earlier versions of this paper are gratefully acknowledged. Supported by PRESTO Promotion Program, Japan Science and Technology Agency.

A. Corradini, B. Klin, and C. Cîrstea (Eds.): CALCO 2011, LNCS 6859, pp. 222–236, 2011.

Basic Bialgebraic Modeling. In the simplest setting in [22], a bialgebra is a carrier X equipped with both algebra and coalgebra structures: $\Sigma X \to X \to FX$. Typically, the functor Σ for the algebra part is $\Sigma = \coprod_{\sigma \in \Sigma}(\_)^{\mathsf{arity}(\sigma)} : \mathbf{Sets} \to \mathbf{Sets}$, a functor that represents a process algebra signature Σ; and the other functor is $F = (\mathcal{P}_{\omega}\_)^{A} : \mathbf{Sets} \to \mathbf{Sets}$, a functor for coalgebraic modeling of labeled transition systems (LTSs). Here $\mathcal{P}_{\omega}$ denotes the finite powerset functor, and A is the set of labels. The algebra and coalgebra structures are further subject to a certain compatibility condition via a natural transformation $\Sigma F \Rightarrow F\Sigma$; this natural transformation is what represents SOS rules.

The following two are almost only examples of such bialgebras in the literature.

- The one induced by the initial algebra $\Sigma I \to I$ (or, slightly more generally, a free algebra). Here I is the set of Σ-terms. The induced coalgebra $I \to FI$ is the transition structure between process terms, which is what is derived in the conventional SOS framework [20].
- The one induced by the final coalgebra $Z \to FZ$ (or, slightly more generally, a cofree coalgebra). Here Z is the set of "behaviors"—specifically bisimilarity classes of states of LTSs (see e.g. [11, §1.3]).[1] Existence of the induced algebra structure $\Sigma Z \to Z$ implies that the process operators are well-defined modulo bisimilarity, that is, *bisimilarity is a congruence*. As laid out shortly in §1.2, this induced algebraic structure is what is generalized in the current paper.

Bialgebraic Modeling of GSOS Rules. It turns out, however, that only a very limited class of SOS rules can be represented by a natural transformation of the form $\Sigma F \Rightarrow F\Sigma$. Therefore in [22] a couple of extensions of the above basic scheme are proposed; the most notable among which is for the GSOS format. It is such that: a natural transformation representing GSOS rules is of the form $\Sigma F_{\bullet} \Rightarrow F\Sigma^{*}$. Here Σ^{*} is the *free monad* over Σ, with $\Sigma^{*}X$ being the set of Σ-terms that can contain elements of X as variables. The functor $F_{\bullet}$ is the *cofree copointed functor* over F, concretely: $F_{\bullet}X = X \times FX$. In [22] it is shown that any *GSOS specification*—a set of SOS rules compliant with the GSOS format—can be represented by a natural transformation of the form $\Sigma F_{\bullet} \Rightarrow F\Sigma^{*}$ and vice versa. See [16] for an introduction to the development.

The idea of bialgebraic modeling of SOS has been further pursued by many authors. See [16] and the references therein.

1.2 Parallel Composition of Coalgebras and the Microcosm Principle

In bialgebraic modeling, it is the elements of the carrier of a bialgebra that are combined using process operators. As described above, typical examples of such are: process terms (combined syntactically); and "behaviors," i.e. bisimilarity classes of LTS states (combined thanks to 'bisimilarity is a congruence').

However, the rise of *component calculi* as a foundation of component-based system design (see e.g. [4, 7]) poses a new challenge. In component-based design it is existing

[1] "Bisimilarity" here is more precisely *behavioral equivalence*; they coincide for functors F that weakly preserve pullbacks (see e.g. [15]). This is the case with $F = (\mathcal{P}_{\omega}\_)^{A}$.

systems that are to be composed; and this lies out of the realm of bialgebraic modeling. Specifically, 'existing systems' do not always mean 'process terms': one may be given two LTSs $\mathcal{S}_1$ and $\mathcal{S}_2$ that are generated from two process terms written in two different process calculi. 'Existing systems' do not necessarily mean their 'behaviors' either: given two LTSs $\mathcal{S}_1$ and $\mathcal{S}_2$, calculating their full behaviors is usually expensive. Therefore it is nice to be able to combine LTSs *as they are*—much like a product of two automata (e.g. in [4]), where one takes the product of two state spaces.

This idea of combining LTSs as they are, using a process operator whose meaning is specified by SOS rules, has been in the literature implicitly or explicitly (e.g. in [8, 5]). However this idea is often regarded as a "cosmetic" extension to SOS and its mathematical/categorical foundation has not been systematically pursued.

In our previous work [14] we formalized combination of LTSs as: a functor $\|$: $\mathbf{Coalg}(F) \times \mathbf{Coalg}(F) \to \mathbf{Coalg}(F)$ arising from a natural transformation $\mathsf{sync}_{X,Y} : FX \times FY \to F(X \times Y)$. If $F = (\mathcal{P}_{\omega}\_)^A$ this allows us to model synchronous parallel composition of LTSs (but hardly any other operator). There the carrier of the LTS $(X \stackrel{c}{\to} FX) \parallel (Y \stackrel{d}{\to} FY)$ is $X \times Y$. The natural transformation sync specifies a "synchronization mechanism," which represents a very limited class of SOS rules.

The operation $\|$ for composing LTSs yields a canonical operation $\|$ for composing "behaviors" like the one in the bialgebraic modeling (§1.1). Namely, behaviors are identified with elements of the final coalgebra $\zeta : Z \stackrel{\cong}{\to} FZ$ and $\|$ is induced by the coinduction diagram on the right.

$$\begin{array}{ccc} F(Z \times Z) & \xrightarrow{F\|} & FZ \\ \uparrow{\scriptstyle \zeta\|\zeta} & & \uparrow{\scriptstyle \zeta} \\ Z \times Z & \xrightarrow[\|]{} & Z \end{array}$$

Furthermore in [14], the two composition operators

$$\| : \mathbf{Coalg}(F) \times \mathbf{Coalg}(F) \longrightarrow \mathbf{Coalg}(F) \quad \text{and} \quad \| : Z \times Z \longrightarrow Z$$

with the latter being a coalgebra morphism $\|: \zeta \parallel \zeta \to \zeta$, are identified as an instance of so-called the *microcosm principle*. It is a term coined in [2], referring to the phenomenon that: a category $\mathbb{C}$ and its object $X \in \mathbb{C}$ have the "same" algebraic structures, with X's *inner* algebraic structure depending on $\mathbb{C}$'s *outer* one. A prototypical example is a monoid object X in a monoidal category $\mathbb{C}$: they both have a multiplication operator ($\mathsf{m} : X \otimes X \to X$ and $\otimes : \mathbb{C} \times \mathbb{C} \to \mathbb{C}$); and the definition of m uses $\otimes$ in it. In [14] we formalized what the microcosm principle means—especially what is meant by the "same" algebraic structures on different levels $\mathbb{C}$ and $X \in \mathbb{C}$—using Lawvere theories.

As an application we proved a *compositionality* result: for any coalgebras $X \stackrel{c}{\to} FX$ and $Y \stackrel{d}{\to} FY$ we have the top diagram on the right commute, where maps like $\mathsf{beh}(c)$ are by coinduction (the bottom diagram). This reads: the behavior of the composed system $c \parallel d$ can be computed from the behaviors of c and d, using $\|$. In particular—denoting bisimilarity by $\sim$—we have that $c \sim c'$ and $d \sim d'$ implies $c \parallel d \sim c' \parallel d'$, with regards to an appropriate choice of initial states.

$$\begin{array}{ccc} X \times Y & \xrightarrow{\mathsf{beh}(c\|d)} & Z \\ \downarrow{\scriptstyle \mathsf{beh}(c) \times \mathsf{beh}(d)} & \nearrow{\scriptstyle \|} & \\ Z \times Z & & \end{array}$$

$$\begin{array}{ccc} FX & \xrightarrow{F\,\mathsf{beh}(c)} & FZ \\ \uparrow{\scriptstyle c} & & \uparrow{\scriptstyle \zeta} \\ X & \xrightarrow[\mathsf{beh}(c)]{} & Z \end{array}$$

1.3 Microcosm Interpretation of Full GSOS Rules

The state of art in [14] roughly corresponds to the basic bialgebraic modeling in §1.1 where an SOS specification is represented by a natural transformation $\Sigma F \Rightarrow F\Sigma$ (see the table). The current work extends [14] by accommodating any process operator whose meaning is specified by GSOS rules. Some GSOS rules are shown below; we assume that the set of labels is $N \cup \overline{N} \cup \{\tau\}$, consisting of names, conames and the internal action.

	composing behaviors	composing both LTSs & behaviors
sync. $\parallel$	[22], $\Sigma F \Rightarrow F\Sigma$	[14]
full GSOS	[22], $\Sigma F_\bullet \Rightarrow F\Sigma^*$	current work

$$\frac{x \xrightarrow{a} x'}{x \parallel y \xrightarrow{a} x' \parallel y}\,(\|\mathrm{L}) \quad \frac{y \xrightarrow{a} y'}{x \parallel y \xrightarrow{a} x \parallel y'}\,(\|\mathrm{R}) \quad \frac{x \xrightarrow{a} x' \quad y \xrightarrow{\overline{a}} y'}{x \parallel y \xrightarrow{\tau} x' \parallel y'}\,(\|\mathrm{SYNC}) \quad \frac{}{a \xrightarrow{a} 0}\,(\mathrm{AT})$$

$$\frac{x \xrightarrow{a} x'}{x ; y \xrightarrow{a} x' ; y}\,(;\mathrm{L}) \quad \frac{x \not\xrightarrow{a} \ (\forall a \in A) \quad y \xrightarrow{b} y'}{x ; y \xrightarrow{b} y'}\,(;\mathrm{R}) \quad \frac{x \xrightarrow{a} x'}{!x \xrightarrow{a} x' \parallel !x}\,(!) \quad \frac{x \xrightarrow{a} x'}{x^* \xrightarrow{a} x' ; x^*}\,((\_)^*)$$

We use these rules for constructing a new LTS from given LTSs. This means:

$$(\|\mathrm{SYNC}) \text{ is read as } \quad \frac{x \xrightarrow{a} x' \ \boxed{\text{in } \mathcal{S}} \qquad y \xrightarrow{\overline{a}} y' \ \boxed{\text{in } \mathcal{T}}}{x \parallel y \xrightarrow{\tau} x' \parallel y' \ \boxed{\text{in } \mathcal{S} \parallel \mathcal{T}}}\,(\|\mathrm{SYNC}) \tag{1}$$

with an additional class of variables ($\mathcal{S}$ and $\mathcal{T}$) that tells in which LTS each transition takes place. The variables x, y here designate states of LTSs—unlike in the usual reading where they designate process terms. Different variables can designate states of distinct LTSs: in (1), x and x' are states of $\mathcal{S}$, while y and y' are of $\mathcal{T}$.

1.4 A Technical Challenge: State Spaces

Think of the new reading of the rules (;R) and (!). The challenge is: what is the state space of the sequential composition $\mathcal{S};\mathcal{T}$? We can start with $X \times Y$—where X and Y are the state spaces of $\mathcal{S}$ and $\mathcal{T}$, respectively—denoting its element by $x; y$. According to the rule (;R), however, a state $x; y$ can "evolve" into y', which is no longer in $X \times Y$. So the answer seems to be $X \times Y + Y$. But how about the replication $!\mathcal{S}$? One would think of $X^+ = \coprod_{1 \leq n} X^n$ or X^ω; both seem plausible.

In this paper we introduce a uniform, syntactic and modular way of constructing such a state space, so that it is compatible with the given set $\mathcal{R}$ of GSOS rules. We shall call it an *$\mathcal{R}$-state space*. The construction is syntactic in the sense that an $\mathcal{R}$-state space consists of rather simple sets like $X_1 \times \cdots \times X_m$, summed up over all the relevant algebraic terms (Def. 2.10). The construction is modular because the $\mathcal{R}$-state space for a composed term t can be calculated using the state spaces for t's subterms as building blocks ((4) later). Such modularity is an essential property of an "algebra."

As a technical tool in this construction we introduce the notion of *term lineage graph (TLG)*. A TLG is roughly a graph between two terms (thought of as parse trees) that keeps track of evolution of terms (like $x; y \mapsto y'$ in (; R) above).

Organization of the Paper. In §2, after fixing notations for GSOS rules we describe the construction of $\mathcal{R}$-state spaces. In §3 we present a GSOS specification as a natural transformation: this generalizes $\Sigma F_{\bullet} \Rightarrow F\Sigma^*$ in [22] (for composing different LTSs) as well as $\mathsf{sync}_{X,Y} : FX \times FY \to F(X \times Y)$ in [14] (for the full GSOS format). This results in the two interpretations of process operators in §4, acting on LTSs and on behaviors. We prove a compositionality result, and show that our framework indeed generalizes the GSOS fragment of [22]. In §5 we conclude and discuss related work.

2 State Space Compatible with GSOS Rules

2.1 GSOS Specification

We first fix a signature Σ of a process calculus. For each $n \in \mathbb{N}$ it determines the set Σ_n of *n-ary operators.*

Definition 2.1 (GSOS). A *GSOS rule* R (as in [22]) over a signature Σ is a syntactic expression of the following form.

$$\frac{\{x_i \xrightarrow{a} y_i^{a,j}\}_{i\in[1,m]}^{a\in A, j\in[1,N_i^a]} \quad \{x_i \stackrel{b}{\not\to}\}_{i\in[1,m]}^{b\in B_i}}{\sigma(x_1,\dots,x_m) \xrightarrow{e} t} \tag{2}$$

Here A is a fixed set of *labels*; $x_i, y_i^{a,j}$ are distinct variables; N_i^a is a natural number that is 0 for almost every i and a; B_i is a (possibly infinite) subset of A; σ is an m-ary operator in Σ; and t is a Σ-term where only x_i and $y_i^{a,j}$ occur as variables. For a GSOS rule R like (2), the operator σ on the left in the conclusion shall be denoted by σ_R; the term t on the right is denoted by t_R.

A *GSOS specification* is a pair $(\Sigma, \mathcal{R})$ of a signature Σ and an *image-finite* set $\mathcal{R}$ of GSOS rules over Σ. Here image-finiteness means that there are only finitely many rules in $\mathcal{R}$, once $\sigma \in \Sigma$ and $c \in A$ are fixed.

Our another reading of (2)—in the sense of (1)—is as follows, deriving a transition in a new LTS $\sigma(\mathcal{S}_1,\dots,\mathcal{S}_m)$.

$$\frac{\{x_i \xrightarrow{a} y_i^{a,j} \text{ in } \mathcal{S}_i\}_{i\in[1,m]}^{a\in A, j\in[1,N_i^a]} \quad \{x_i \stackrel{b}{\not\to} \text{ in } \mathcal{S}_i\}_{i\in[1,m]}^{b\in B_i}}{\sigma(x_1,\dots,x_m) \xrightarrow{e} t \quad \text{in } \sigma(\mathcal{S}_1,\dots,\mathcal{S}_m)}$$

Here $\mathcal{S}_1,\dots,\mathcal{S}_m$ are a new class of variables that designate LTSs; variables x_i and $y_i^{a,j}$ with a subscript i therefore designate states of $\mathcal{S}_i$.

We will need a careful inspection of the structure of Σ-terms. We fix some notations.

Notation 2.2. We assume that a Σ-term t always comes with an explicit *context* of variables: $x_1,\dots,x_m \vdash t$. Any occurrence of a variable in t must be that of $x_1, x_2, \dots$ or x_m; each variable x_i can have multiple or no occurrences in t. For example, we distinguish two terms $x_1 \vdash x_1 \parallel x_1$ and $x_1, x_2 \vdash x_1 \parallel x_1$ because of different contexts. In the sequel, however, we suppress the context of a Σ-term when it is obvious.

Definition 2.3 (i_t, $|t|$ **and** t^ℓ). Given a Σ-term $x_1, \ldots, x_m \vdash t$, the number of *occurrences* of variables in t is denoted by $|t|$. Note that this is not necessarily the same as m. Then the term t induces an "indexing" function $\mathsf{i}_t : [1, |t|] \to [1, m]$ such that: the i-th occurrence of variables (counted from left to right) in t is that of $x_{\mathsf{i}_t(i)}$.

The term obtained from t, by replacing all the occurrences of variables by those of distinct variables $x_1, \ldots, x_{|t|}$ from left to right, is denoted by t^ℓ. This is the *linear term* induced by t. An easy consequence is: $t = t^\ell[x_{\mathsf{i}_t(1)}/x_1, \ldots, x_{\mathsf{i}_t(|t|)}/x_{|t|}]$.

For example, let t be the term $x_1, x_2 \vdash x_2 \parallel (x_1 \parallel !x_1)$. Then $|t| = 3$, $\mathsf{i}_t : 1 \mapsto 2, 2 \mapsto 1, 3 \mapsto 1$. The term t^ℓ is $x_1, x_2, x_3 \vdash x_1 \parallel (x_2 \parallel !x_3)$.

2.2 Requirements on State Spaces

Shortly we will introduce the notion of *term lineage graph (TLG)*. Since it is technically rather involved, we shall first lay out what we aim to achieve using TLGs.

We construct state spaces for LTSs like $\mathcal{S}_1 \parallel \mathcal{S}_2$, $\mathcal{S}_1; \mathcal{S}_2$, $!\mathcal{S}$, etc., deriving from given GSOS rules. Let $x_1, \ldots, x_m \vdash t$ be a Σ-term and $X_1, \ldots, X_m$ be sets, with the idea that X_i is the state space of the LTS $\mathcal{S}_i$. We shall define an $\mathcal{R}$-*state space* $(\!|t|\!)(X_1, \ldots, X_m)$. The following requirements are essential to to base the framework in [14] on it; these will be exploited in the technical development later in §3.

Requirements 2.4. 1. The set $(\!|t|\!)(\vec{X})$ should accommodate *initial* states that are there prior to any evolution—such as $x_1 \parallel x_2$ in $\mathcal{S}_1 \parallel \mathcal{S}_2$ or $!x$ in $!\mathcal{S}$ (see §1.4). The set of such initial states are given by

$$|t|(X_1, \ldots, X_m) \;:=\; X_{\mathsf{i}_t(1)} \times \cdots \times X_{\mathsf{i}_t(|t|)} \;, \tag{3}$$

where i_t and $|t|$ are from Def. 2.3. Note that $|t|(\vec{X})$ need not be the same as $X_1 \times \cdots \times X_m$. Our definition (3) implies, for example, that we allow $x_1 \parallel x_1'$ (with $x_1, x_1' \in X_1$ and $x_1 \neq x_1'$) as an initial state in an LTS $\mathcal{S}_1 \parallel \mathcal{S}_1$.
2. The set $(\!|t|\!)(\vec{X})$ should also accommodate those states which would arise through "evolution" specified by GSOS rules (see §1.4).
3. *Functoriality*: the operation $(\!|t|\!)$ extends to a functor $\mathbf{Sets}^m \to \mathbf{Sets}$.
4. *Modularity*: the operation $(\!|\_|\!)$—applied to a term t and yielding a functor $(\!|t|\!) : \mathbf{Sets}^m \to \mathbf{Sets}$—is compatible with substitution of terms. That is,

$$(\!|\, t[t_i/x_i] \,|\!)\,(X_1, \ldots, X_m) \;\cong\; (\!|t|\!)\,\big(\, (\!|t_1|\!)(\vec{X}), \ldots, (\!|t_n|\!)(\vec{X}) \,\big) \;. \tag{4}$$

2.3 Term Lineage Graph (TLG)

First we note that a Σ-term t can be identified with its *parse tree*. Its leaves are variables and 0-ary operators; and its internal nodes are operators with positive arities with a suitable branching degree. For example the term $x_1 \parallel !x_1$ is understood as on the right.

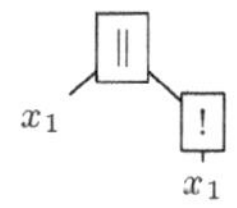

Definition 2.5 (Term Lineage Graph). Let s, t be Σ-terms. We require them to be in the same variable context (Notation 2.2): $x_1, \ldots, x_m \vdash s, t$. A *term lineage graph (TLG)* ρ from s to t, denoted by $\rho : s \Rightarrow t$, is an unlabeled directed graph (like in (5)) whose nodes are nodes of s and t (seen as parse trees), such that:

- any edge is from a node in the *domain term* s to a node in the *codomain term* t;
- each node in s has exactly one outgoing edge;
- the edges are *monotone*: assume that the origin of one edge is a descendant (in the parse tree s) of the origin of another edge. Then the target of the former is also a descendant of (or the same as) that of the latter;
- an edge from an operator symbol σ goes into a (not necessarily the same) operator symbol;
- an edge from a variable x_i in s goes into the same variable x_i in t.

A TLG is sometimes denoted by $x_1, \ldots, x_m \vdash \rho : s \Rightarrow t$, making its context explicit.

Definition 2.6 (The TLG ρ_R). Let R be a GSOS rule of the form (2). It induces a *TLG* ρ_R defined as follows. Its type is $\rho_R : t_R[x_i/y_i^{a,j}] \Rightarrow \sigma_R(x_1, \ldots, x_m)$. Concretely

- each node in the domain term $t_R[x_i/y_i^{a,j}]$ that is labeled with an operator symbol is tied to the root node (labeled with σ_R) of the codomain term $\sigma_R(x_1, \ldots, x_m)$;
- each node in the domain term that is labeled with a variable x_i is tied to the unique occurrence of the same variable x_i in the codomain $\sigma_R(x_1, \ldots, x_m)$.

Intuitively, the substitution $y_i^{a,j} \mapsto x_i$ in the domain term $t_R[x_i/y_i^{a,j}]$ forces every occurrence of a state of $\mathcal{S}_i$ to be denoted by x_i.

Here are some examples of TLGs induced by GSOS rules. The solid lines represent the order in each parse tree; the dotted lines are the edges of the TLGs.

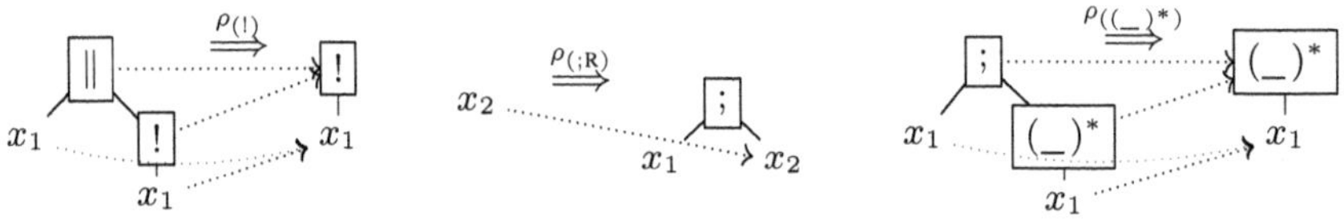

(5)

Definition 2.7 (Operations on TLGs).

- (Identity) For each Σ-term s, the *identity TLG* $\mathrm{id}_s : s \Rightarrow s$ is the one in which each node in the domain term is tied to the same node of the codomain term.
- (Composition) Given two successive TLGs $\rho : t \Rightarrow s$ and $\rho' : s \Rightarrow u$, the *composition* $\rho' \bullet \rho : t \Rightarrow u$ is obtained by composing two successive edges of ρ and ρ', and then forgetting about the mediating term s.
- (Substitution) Let $x_1, \ldots, x_n \vdash \rho : t \Rightarrow t'$ be a TLG. Assume further that we have a TLG $x_1, \ldots, x_m \vdash \rho_i : t_i \Rightarrow t'_i$, for each $i \in [1, n]$.[2] We define the *substitution* $\rho[\rho_i/x_i]$, which is a TLG between substituted terms from $t[t_i/x_i]$ to $t'[t'_i/x_i]$, as follows. Pictorially:

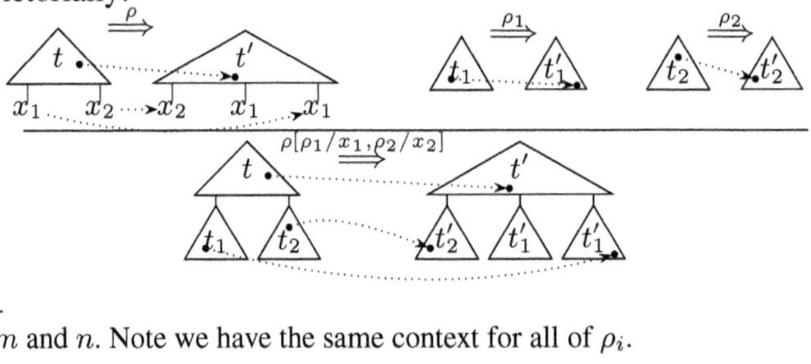

That is,

[2] Distinguish m and n. Note we have the same context for all of ρ_i.

- Each node in the (upper) t-part of the domain term $t[t_i/x_i]$ has the same outgoing edge as in ρ, into the t'-part of the codomain term.
- As for a node in the (lower) t_i-part of the domain term: basically the outgoing edge is the same as the corresponding one in $\rho_i : t_i \Rightarrow t'_i$. However, like t'_1 in the above example, there can be multiple occurrences of t'_i in the codomain term; we must decide which occurrence the edge points to. There we use the information in $\rho : t \Rightarrow t'$: we follow the edge in ρ that starts from the corresponding occurrence of a variable.

Let us now illustrate the operations on TLGs. Using three TLGs $\rho_{(;\mathrm{R})} : x_2 \Rightarrow x_1; x_2$, $\mathrm{id}_{x_1} : x_1 \Rightarrow x_1$ and $\mathrm{id}_{x_1^*} : x_1^* \Rightarrow x_1^*$, we can form the substitution $\rho_{(;\mathrm{R})}[\mathrm{id}_{x_1}/x_1, \mathrm{id}_{x_1^*}/x_2]$ that is shown below on the left. The equality below asserts that, if we pre-compose $\rho_{((\_)^*)}$ to that substitution, then it is the same as the identity TLG.

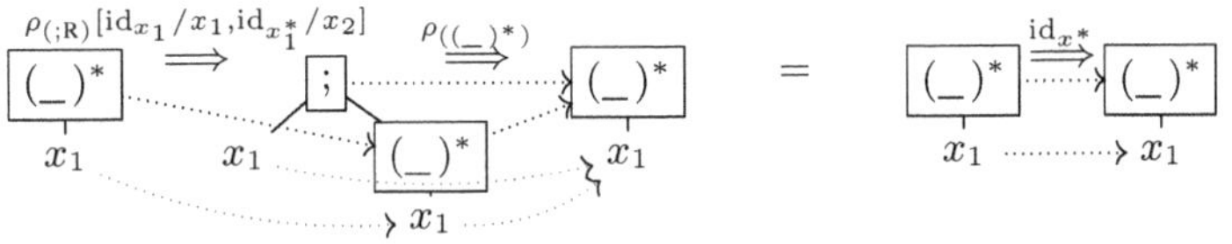

(6)

Finally, a given set $\mathcal{R}$ of GSOS rules determines a class of TLGs that are relevant to it.

Definition 2.8 ($\mathcal{R}$-TLG). Let $(\Sigma, \mathcal{R})$ be a GSOS specification (Def. 2.1). The class of *$\mathcal{R}$-TLGs* is a subclass of TLGs between Σ-terms, defined inductively as follows: 1) for each GSOS rule $R \in \mathcal{R}$, the induced TLG ρ_R (Def. 2.6) is an $\mathcal{R}$-TLG; 2) the class of $\mathcal{R}$-TLGs is closed under identities, composition, and substitution (Def. 2.7).

2.4 $\mathcal{R}$-State Space

We define an *$\mathcal{R}$-state space* operator $(\!|t|\!)$. Let us fix a GSOS specification $(\Sigma, \mathcal{R})$.

Definition 2.9 (Initial State Space). Given a Σ-term $x_1, \dots, x_m \vdash t$, we define the *initial state space functor* $|t| : \mathbf{Sets}^m \to \mathbf{Sets}$ by (3) in §2.2. Its functoriality is obvious. We are overriding the notation $|t|$ (cf. Def. 2.3); this will not cause any confusion.

Definition 2.10 ($\mathcal{R}$-state Space). Let t be a Σ-term. We define the *$\mathcal{R}$-state space functor* $(\!|t|\!) : \mathbf{Sets}^m \to \mathbf{Sets}$ by:

$$(\!|t|\!)(X_1, \dots, X_m) := \coprod \{ \, |s|(X_1, \dots, X_m) \mid \rho : s \Rightarrow t, \ \mathcal{R}\text{-TLG} \, \} \ . \qquad (7)$$

Here $|s|$ is the initial state space functor (Def. 2.9); the sum $\coprod$ is taken over all the $\mathcal{R}$-TLGs ρ with a codomain term t. We have one summand for each ρ, not for each s.

Let us now calculate some $\mathcal{R}$-state spaces. We take as $\mathcal{R}$ the set of rules presented in §1.3; we take the corresponding signature $\Sigma = \{\|, !, ;, (\_)^*\} \cup \{a \mid a \in A\}$.

For a rule $R \in \mathcal{R}$ whose *principal operator* is $\|$—namely ($\|$L), ($\|$R) and ($\|$SYNC)—the induced TLG ρ_R coincides with the identity (Def. 2.7). It follows that any $\mathcal{R}$-TLG of the form $\cdot \Rightarrow x_1 \parallel x_2$ is the identity TLG. Hence by (7) we have $(\!|x_1 \parallel x_2|\!)(X_1, X_2) = |x_1 \parallel x_2|(X_1, X_2) = X_1 \times X_2$.

For the operator ; , an $\mathcal{R}$-TLG whose codomain term is $x_1, x_2 \vdash x_1; x_2$ is either the identity or $\rho_{(;\mathrm{R})}$ (see (5)). Hence $(\!|x_1; x_2|\!)(X_1, X_2) = |x_1; x_2|(X_1, X_2) + |x_2|(X_1, X_2) = X_1 \times X_2 + X_2$. Here the term x_2 inside the expression $|x_2|$ is $x_1, x_2 \vdash x_2$, to be precise about its context.

Regarding !, we have countably many $\mathcal{R}$-TLGs whose codomain is $!x_1$:

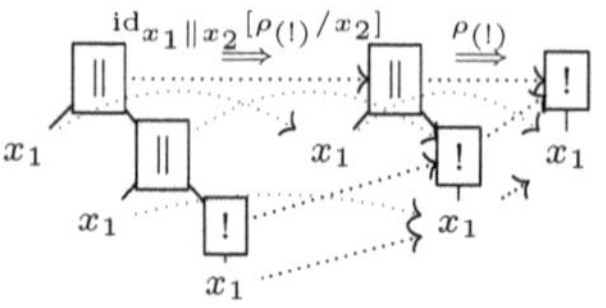

$$\begin{aligned} \mathrm{id} :&\quad !x_1 \Rightarrow !x_1 \ , \\ \rho_{(!)} :&\quad x_1 \parallel !x_1 \Rightarrow !x_1 \ , \\ \rho_{(!)} \bullet (\mathrm{id}_{x_1\|x_2}[\rho_{(!)}/x_2]) :&\quad x_1 \parallel (x_1 \parallel !x_1) \Rightarrow !x_1 \ , \text{ etc.} \end{aligned}$$

The third one is depicted above. Therefore we have:

$$(\!|\,!x_1\,|\!)(X_1) = \coprod_{i \geq 0} \Big| \underbrace{x_1 \parallel (\cdots \parallel (x_1}_{i} \parallel !x_1) \cdots) \Big| (X_1) = \coprod_{i \geq 0} X_1^{i+1} = X_1^+ \ .$$

Regarding $(\_)^*$, from the equality (6) it follows that the situation is similar to ! above. Namely, we have one $\mathcal{R}$-TLG for each of the following types: $x_1^* \Rightarrow x_1^*$, $x_1; x_1^* \Rightarrow x_1^*$, $x_1; (x_1; x_1^*) \Rightarrow x_1^*$, etc. Hence we have $(\!|x_1^*|\!)(X_1) = X_1^+$.

We now verify that Requirements 2.4 are indeed fulfilled. The item 3 is obvious. The proof is in [12, Appendix A].

Proposition 2.11. *1. There is a canonical embedding $(\nu_t)_{\vec{X}} : |t|(\vec{X}) \to (\!|t|\!)(\vec{X})$. This extends to a natural transformation $\nu_t : |t| \Rightarrow (\!|t|\!) : \mathbf{Sets}^m \to \mathbf{Sets}$.*

2. Given an $\mathcal{R}$-TLG $\rho : s \Rightarrow t$, it induces a canonical map $(\!|\rho|\!)_{\vec{X}} : (\!|s|\!)(\vec{X}) \to (\!|t|\!)(\vec{X})$. It also extends to a natural transformation $(\!|\rho|\!) : (\!|s|\!) \Rightarrow (\!|t|\!)$.

4. The operation $(\!|\_|\!)$ is compatible with substitution: for Σ-terms $x_1, \ldots, x_n \vdash t$ and $x_1, \ldots, x_m \vdash t_i$ for each $i \in [1, n]$, we have a canonical isomorphism (4). □

Using the embedding $(\nu_t)_{\vec{X}}$ in Prop. 2.11.1, we can also embed the set $X_1 \times \cdots \times X_m$ in the $\mathcal{R}$-state space $(\!|t|\!)(X_1, \ldots, X_m)$. This is by the following $\theta_{X_1,\ldots,X_m}$:

$$\theta_{\vec{X}} := \Big(X_1 \times \cdots \times X_m \overset{(\psi_t)_{X_1,\ldots,X_m}}{\longrightarrow} X_{\mathrm{i}_t(1)} \times \cdots \times X_{\mathrm{i}_t(|t|)} = |t|(\vec{X}) \overset{(\nu_t)_{\vec{X}}}{\longrightarrow} (\!|t|\!)(\vec{X}) \Big) \ , \quad (8)$$

where $(\psi_t)_{\vec{X}}$ is a function such that $(x_1, \ldots, x_m) \mapsto (x_{\mathrm{i}_t(1)}, \ldots, x_{\mathrm{i}_t(|t|)})$ (cf. Def. 2.3). It rearranges arguments $x_1, \ldots, x_m$ according to the occurrences of variables in t.

Remark 2.12. Our choice of state spaces (Def. 2.10) is in fact one out of a spectrum. The smallest extreme in the spectrum is obtained by: singling out the "reachable" part of our $\mathcal{R}$-state space and then quotienting it out by the bisimilarity. Although this yields a smaller state space, calculating such is expensive. Moreover it does not satisfy Requirements 2.4.4, breaking modularity/algebraicity of the whole framework.

The biggest extreme is to take, as the state space $(\!|t|\!)(X_1, \ldots, X_m)$, the whole set of Σ-terms with all states $x_i \in X_i$ as variables. This satisfies Requirements 2.4 so we could develop the theory on top of it. However, we claim it to be our finding that we can trim down this big state space into the one in Def. 2.10. In particular, our refined definition yields $(\!|\,\|\,|\!)(X_1, X_2) = X_1 \times X_2$ which matches with intuition as well as with the usual definitions of product of automata in e.g. [4].

In-between is the "Kleene-style" composition of automata. The classic result on regular language and DFA/NFA has recently seen its vast generalization in coalgebraic terms; see e.g. [21]. In this approach the Kleene star $(\_)^*$ preserves finiteness of state spaces, whereas our approach yields the state space X^* that is inevitably infinite. However the definition of Kleene-style composition calls for different ingenuity for each operator; it is not clear if that can be done uniformly for any operation defined in GSOS.

We further note that our choice goes well with *first-order representation* of LTSs. (Variants of) the latter are now widely used (e.g. CARML in [3]) because it enables BDD-based representation and symbolic model checking [9]. In first-order representation, a state is represented by an assignment of values to variables such as $[\mathtt{agent1} \mapsto \mathtt{critical}, \mathtt{agent2} \mapsto \mathtt{noncritical}]$; hence a state space is the *product* of the value domains for the variables. Combining such state spaces by means of products and coproducts (7) could be done in a programming language with advanced type constructors.

3 Categorical GSOS Specification

We introduced appropriate state spaces for $\mathcal{S}_1 \parallel \mathcal{S}_2, \mathcal{S}_1; \mathcal{S}_2$, etc. in §2; we wish to derive transition structures on those state spaces, making them LTSs. In concrete terms it is via derivations using GSOS rules, with their reading like (1). In categorical terms, GSOS rules are first translated into a natural transformation ξ_t (for each term t, this section); which gives rise to transition structures (§4).

3.1 Copointed Functors and Copointed Coalgebras

We will need the following notion of copointedness—a technical but standard one in the field (see e.g. [22, 18]). For the functor $F = (\mathcal{P}_\omega\_)^A : \mathbf{Sets} \to \mathbf{Sets}$, we set a functor $F_\bullet : \mathbf{Sets} \to \mathbf{Sets}$ by $F_\bullet X := X \times FX = X \times (\mathcal{P}_\omega X)^A$. We denote the first projection $F_\bullet X (= X \times FX) \to X$ by ε_X.

An $F_\bullet$-coalgebra $c : X \to F_\bullet X$ is said to be *copointed* if the diagram on the right commutes. Intuitively, the additional component X in $F_\bullet X = X \times FX$ records the "original" state before a transition; indeed for a copointed $F_\bullet$-coalgebra $c : X \to X \times (\mathcal{P}_\omega X)^A$ we have $c(x) = (\, x,\, \lambda a.\, \{x' \mid x \xrightarrow{a} x'\}\,)$. The next result is standard and easy; see [18].

$$\begin{array}{ccc} X \times FX & \xrightarrow{\varepsilon_X} & X \\ {\scriptstyle c}\uparrow & \nearrow_{=} & \\ X & & \end{array}$$

Lemma 3.1. *Let us denote the category of F-coalgebras by* $\mathbf{Coalg}(F)$*; and the category of copointed $F_\bullet$-coalgebras by* $\mathbf{Coalg}_\bullet(F_\bullet)$*. The two categories are isomorphic:* $\mathbf{Coalg}(F) \cong \mathbf{Coalg}_\bullet(F_\bullet)$*. In particular,* $\mathbf{Coalg}_\bullet(F_\bullet)$ *has a final object (a* final copointed $F_\bullet$-coalgebra*) that corresponds to a final object in* $\mathbf{Coalg}(F)$. □

Due to this result, in the sequel we identify an LTS with a copointed coalgebra for the functor $F_\bullet X = X \times (\mathcal{P}_\omega X)^A$. Bisimilarity in LTSs can then be captured by the final copointed $F_\bullet$-coalgebra—which we denote by $\zeta_\bullet : Z \to F_\bullet Z$. Note that the coalgebra $\zeta_\bullet$ has the same carrier set Z as the final F-coalgebra; this is also a standard fact.

3.2 GSOS Rules, Categorically

We shall translate a set $\mathcal{R}$ of GSOS rules into functions of the following type:

$$(\xi_t)_{\vec{X}} \;:\; (\!|t|\!)(\,F_\bullet X_1,\,\ldots,\,F_\bullet X_m\,) \longrightarrow F_\bullet(\,(\!|t|\!)(X_1,\ldots,X_m)\,)\;, \tag{9}$$

defined for each Σ-term $x_1,\ldots,x_m \vdash t$ and sets $X_1,\ldots,X_m$. Here $F = (\mathcal{P}_{\omega}\_)^A$, and m is the length of t's context $x_1,\ldots,x_m \vdash t$. These functions $(\xi_t)_{\vec{X}}$ form a natural transformation ξ_t, defined for each term t. The difference from the natural transformation $\Sigma F_\bullet X \Rightarrow F\Sigma^* X$ in [22] is that ours $(\xi_t)_{\vec{X}}$ is required to be natural in multiple sets $X_1,\ldots,X_m$. This is because we deal with state spaces of multiple LTSs.

Towards the natural transformation ξ_t in (9) we proceed step by step: we first derive from a GSOS rule R a natural transformation $\xi^{(1)}_R$, from which we derive $\xi^{(2)}_R, \xi^{(3)}_\sigma, \ldots,$ finally reaching ξ_t of the desired form.

Step 1. Let $R \in \mathcal{R}$ be a GSOS rule in (2). It induces the following function $(\xi^{(1)}_R)_{X_1,\ldots,X_m}$:

$$\begin{aligned}
&|\sigma_R|\,(F_\bullet X_1,\cdots,F_\bullet X_m) \overset{(\xi^{(1)}_R)_{\vec{X}}}{\longrightarrow} F(\,|t_R[x_i/y_i^{a,j}]|\,(X_1,\ldots,X_m)\,)\;,\\
\text{i.e. }\; &\textstyle\prod_{i\in[1,m]}(\,X_i\times(\mathcal{P}_\omega X_i)^A\,) \overset{(\xi^{(1)}_R)_{\vec{X}}}{\longrightarrow} \Big(\mathcal{P}_\omega\Big(\,|\,t_R[x_i/y_i^{a,j}]\,|(\vec{X})\Big)\Big)^A\\
\text{by }\; &(\,(\xi^{(1)}_R)_{\vec{X}}(\mathtt{x}_1,\varphi_1,\ldots,\mathtt{x}_m,\varphi_m)\,)(e)\\
&= \left\{\, t_R[\mathtt{x}_i/x_i, \mathtt{y}_i^{a,j}/y_i^{a,j}] \;\middle|\; \begin{array}{l}\forall i.\,\forall b\in B_i.\;\; \varphi_i(b)=\emptyset, \text{ and}\\ \forall i.\,\forall a.\,\forall j\in[1,N_i^a].\;\; \mathtt{y}_i^{a,j}\in\varphi_i(a).\end{array}\right\}\;.
\end{aligned}$$

Here $\mathtt{x}_i$ and $\mathtt{y}_i^{a,j}$ denote elements of X_i; φ_i belongs to $(\mathcal{P}_\omega X_i)^A$; and $e \in A$ is a label. Note that $|\sigma_R|(F_\bullet X_1,\cdots,F_\bullet X_m) = F_\bullet X_1\times\cdots\times F_\bullet X_m$ (Def. 2.9). The expression $t_R[\mathtt{x}_i/x_i, \mathtt{y}_i^{a,j}/y_i^{a,j}]$ denotes the obvious element of the set $|\,t_R[x_i/y_i^{a,j}]\,|(X_1,\ldots,X_m)$. To be precise, the latter set is of the form $X_{k_1}\times\cdots\times X_{k_{|t_R|}}$, with each component X_{k_l} coming from an occurrence of either x_i or $y_i^{a,j}$ in t_R. If X_{k_l} is from x_i then the l-th component of $t_R[\mathtt{x}_i/x_i, \mathtt{y}_i^{a,j}/y_i^{a,j}]$ is $\mathtt{x}_i$; if X_{k_l} is from $y_i^{a,j}$ then it is $\mathtt{y}_i^{a,j}$.

Step 2. We exploit properties of $(\!|\_|\!)$ (Prop. 2.11.1–2) to obtain the following two successive functions. Here ρ_R is the $\mathcal{R}$-TLG induced by the GSOS rule R (Def. 2.6).

$$\begin{aligned}
(\nu_{t_R[x_i/y_i^{a,j}]})_{\vec{X}} &: |\,t_R[x_i/y_i^{a,j}]\,|(\vec{X}) \longrightarrow (\!|\,t_R[x_i/y_i^{a,j}]\,|\!)(\vec{X})\;,\\
(\!|\rho_R|\!)_{\vec{X}} &: (\!|\,t_R[x_i/y_i^{a,j}]\,|\!)(\vec{X}) \longrightarrow (\!|\sigma_R|\!)(\vec{X})\;.
\end{aligned}$$

The functions are natural in $X_1,\ldots,X_m$. We compose these arrows, apply the functor F, and then post-compose it to $(\xi^{(1)}_R)_{\vec{X}}$ in Step 1. The outcome is a natural transformation $(\xi^{(2)}_R)_{X_1,\ldots,X_m} \;:\; |\sigma_R|\,(F_\bullet X_1,\cdots,F_\bullet X_m) \longrightarrow F(\,(\!|\sigma_R|\!)\,(X_1,\ldots,X_m)\,)$.

Step 3. The natural transformation $\xi^{(2)}_R$ has been defined for each rule $R \in \mathcal{R}$; we shall take their "union" to define $\xi^{(3)}_\sigma$, now for each operator $\sigma \in \Sigma$. Specifically, for each operator $\sigma \in \Sigma$ and $e \in A$, let $\mathcal{R}_{\sigma,e}$ denote the collection of those rules $R \in \mathcal{R}$ whose conclusion is of the form $\sigma(x_1,\ldots,x_m) \overset{e}{\to} \cdot$. By the image finiteness assumption (Def. 2.1) the set $\mathcal{R}_{\sigma,e}$ is finite. We define a function

$$\begin{aligned}
&(\xi^{(3)}_\sigma)_{\vec{X}} \;:\; |\sigma|\,(F_\bullet X_1,\cdots,F_\bullet X_m) \longrightarrow F(\,(\!|\sigma|\!)\,(X_1,\ldots,X_m)\,)\\
&\text{by }\,(\,\xi^{(3)}_\sigma(\mathtt{x}_1,\varphi_1,\ldots,\mathtt{x}_m,\varphi_m)\,)(e)\\
&\quad := \textstyle\bigcup_{R\in\mathcal{R}_{\sigma,e}}(\,\xi^{(2)}_R(\mathtt{x}_1,\varphi_1,\ldots,\mathtt{x}_m,\varphi_m)\,)(e)\;.
\end{aligned}$$

Recall that $FX = (\mathcal{P}_\omega X)^A$ and $F_\bullet X = X \times (\mathcal{P}_\omega X)^A$; the union is in $\mathcal{P}_\omega((\!|\sigma|\!)(\vec{X}))$.

Step 4. We shall extend F in the codomain of $\xi_\sigma^{(3)}$ to $F_\bullet$, so that we obtain $(\xi_\sigma^{(4)})_{\vec{X}} : |\sigma|(\overrightarrow{F_\bullet X}) \to F_\bullet((\!|\sigma|\!)(\vec{X}))$. What we need is a function of the type $|\sigma|(\overrightarrow{F_\bullet X}) \to (\!|\sigma|\!)(\vec{X})$; then we can tuple it with $\xi_\sigma^{(3)}$ and obtain $\xi_\sigma^{(4)}$. Such a function is given by

$$|\sigma|(\overrightarrow{F_\bullet X}) \stackrel{|\sigma|(\varepsilon_{\vec{X}})}{\longrightarrow} |\sigma|(\vec{X}) \stackrel{(\nu_\sigma)_{\vec{X}}}{\longrightarrow} (\!|\sigma|\!)(\vec{X}) \ .$$

Here $\varepsilon_{X_i} : F_\bullet X_i \to X_i$ is the first projection (§3.1); ν_σ is from Prop. 2.11.1. We used the functoriality of the operation $|\sigma|$ (Def. 2.9).

We shall further extend this definition of $\xi_\sigma^{(4)}$ to $(\xi_t^{(4)})_{\vec{X}} : |t|(\overrightarrow{F_\bullet X}) \to F_\bullet((\!|t|\!)(\vec{X}))$, now defined for each Σ-term t. This is by induction on the formation of Σ-terms.

If t is a variable x_i (i.e. $x_1, \ldots, x_m \vdash x_i$), we have $|x_i|(\overrightarrow{F_\bullet X}) = F_\bullet X_i$ (Def. 2.9) and $(\!|x_i|\!)(\vec{X}) = X_i$ (Def. 2.10, the only $\mathcal{R}$-TLG into x_i is the identity). Hence the function $(\xi_{x_i}^{(4)})_{\vec{X}}$ we are defining is of the type $F_\bullet X_i \to F_\bullet X_i$; it is defined to be the identity. If t is a composed term $\sigma(s_1, \ldots, s_n)$, we define $(\xi_t^{(4)})_{\vec{X}}$ by the composite:

$$\begin{array}{l}
|\sigma(s_1,\ldots,s_n)|(F_\bullet X_1,\ldots,F_\bullet X_m) \\
\quad \xrightarrow{\qquad\qquad\cong\qquad\qquad} |\sigma|(\,|s_1|(\overrightarrow{F_\bullet X}),\ \ldots,\ |s_n|(\overrightarrow{F_\bullet X})\,) \\
\quad \xrightarrow{|\sigma|(\,(\xi_{s_1}^{(4)})_{\vec{X}},\ \ldots,\ (\xi_{s_n}^{(4)})_{\vec{X}}\,)} |\sigma|(\,F_\bullet((\!|s_1|\!)(\vec{X})),\ \ldots,\ F_\bullet((\!|s_n|\!)(\vec{X}))\,) \\
\quad \xrightarrow{(\xi_\sigma^{(4)})_{(\!|s_1|\!)(\vec{X}),\ldots,(\!|s_n|\!)(\vec{X})}} F_\bullet(\,(\!|\sigma|\!)(\,(\!|s_1|\!)(\vec{X}),\ \ldots,\ (\!|s_n|\!)(\vec{X})\,)\,) \\
\quad \xrightarrow{\qquad\qquad\cong\qquad\qquad} F_\bullet(\,(\!|\sigma(s_1,\ldots,s_n)|\!)(X_1,\ldots,X_m)\,)\ .
\end{array}$$

Here the first isomorphism is a variant of (4), for $|\_|$ instead of $(\!|\_|\!)$, which we readily obtain. The second one uses functoriality of $|\sigma|$ applied to functions $(\xi_{s_i}^{(4)})_{\vec{X}}$; the latter are available by the induction hypothesis. The third function is what we defined in the first half of the current Step 4; the last isomorphism is by Prop. 2.11.4.

Step 5. Finally we shall obtain the goal (9) of the current §3.2, by extending $|t|$ into $(\!|t|\!)$ in the domain of the previous $(\xi_t^{(4)})_{\vec{X}} : |t|(\overrightarrow{F_\bullet X}) \to F_\bullet((\!|t|\!)(\vec{X}))$. By Def. 2.10 of $(\!|t|\!)$, such an extension can be done through finding, for each $\mathcal{R}$-TLG $\rho : s \Rightarrow t$, a function of the type $|s|(\overrightarrow{F_\bullet X}) \to F_\bullet((\!|t|\!)(\vec{X}))$. The following composite does this job.

$$|s|(F_\bullet X_1, \ldots, F_\bullet X_m) \stackrel{(\xi_s^{(4)})_{\vec{X}}}{\longrightarrow} F_\bullet((\!|s|\!)(\vec{X})) \stackrel{F_\bullet((\!|\rho|\!)_{\vec{X}})}{\longrightarrow} F_\bullet((\!|t|\!)(\vec{X}))$$

Here $(\!|\rho|\!)$ is from Prop. 2.11.2. We take the cotuple of these functions (i.e. definition by cases) to obtain $(\xi_t)_{\vec{X}}$.

4 The Microcosm Interpretation of GSOS Rules

Let $(\Sigma, \mathcal{R})$ be a GSOS specification. We shall define its *microcosm interpretation*. It consists of, for each Σ-term $x_1, \ldots, x_m \vdash t$,

the *outer* interpretation $[\![t]\!] : \big(\mathbf{Coalg}_\bullet(F_\bullet)\big)^m \to \mathbf{Coalg}_\bullet(F_\bullet)$; and
the *inner* interpretation $[t] : Z^m \to Z$.

Recall (§3.1) that we identify an LTS with $c \in \mathbf{Coalg}_\bullet(F_\bullet)$. Hence the outer interpretation is a "process operator" t acting on LTSs. Therefore the two interpretations combined constitute an instance of the microcosm principle; see §1.2.

Definition 4.1. Let $(\Sigma, \mathcal{R})$ be a GSOS specification, and t be a Σ-term.

1. We define t's *outer interpretation* $[\![t]\!] : (\mathbf{Coalg}_\bullet(F_\bullet))^m \to \mathbf{Coalg}_\bullet(F_\bullet)$ as follows, using functoriality of $(\!|t|\!)$. The function $(\xi_t)_{\vec{X}}$ was introduced in §3.2 as "categorical GSOS rules." Functoriality of thus defined $[\![t]\!]$ is easy.

$$
\left(\begin{array}{c} F_\bullet X_1 \\ c_1\uparrow \\ X_1 \end{array}, \ldots, \begin{array}{c} F_\bullet X_m \\ c_m\uparrow \\ X_m \end{array}\right) \stackrel{[\![t]\!]}{\longmapsto} \begin{array}{c} F_\bullet\big(\,(\!|t|\!)(X_1,\ldots,X_m)\big) \\ \uparrow(\xi_t)_{X_1,\ldots,X_m} \\ (\!|t|\!)(F_\bullet X_1,\ldots,F_\bullet X_m) \\ \uparrow(\!|t|\!)(c_1,\ldots,c_m) \\ (\!|t|\!)(X_1,\ldots,X_m) \end{array}
$$

The copointedness (§3.1) of the resulting $F_\bullet$-coalgebra is easy, too. It has $(\!|t|\!)(\vec{X})$—the $\mathcal{R}$-state space from §2—as a carrier.

2. We define t's *inner interpretation* $[t] : Z^m \to Z$ as follows. Once we have the outer interpretation $[\![t]\!]$, we can apply it to the m-tuple $(\zeta_\bullet, \ldots, \zeta_\bullet)$ of the final copointed coalgebra $\zeta_\bullet : Z \to F_\bullet Z$. Then the resulting coalgebra $[\![t]\!](\vec{\zeta_\bullet})$ induces a unique "behavior" map into the final one, as below. This behavior map is almost what we want; we pre-compose $\theta_{Z,\ldots,Z} : Z^m \to (\!|t|\!)(\vec{Z})$ (from (8)) to it and obtain $[t]$. That is, $[t] := \mathsf{beh}([\![t]\!])(\vec{\zeta_\bullet}) \circ \theta_{\vec{Z}} : Z^m \to Z$.

$$
\begin{array}{ccc}
F_\bullet((\!|t|\!)(\vec{Z})) & \dashrightarrow & F_\bullet Z \\
{\scriptstyle [\![t]\!](\vec{\zeta_\bullet})}\uparrow & & \zeta_\bullet\uparrow{\scriptstyle \text{final}} \\
(\!|t|\!)(\vec{Z}) & \underset{\mathsf{beh}([\![t]\!](\vec{\zeta_\bullet}))}{\dashrightarrow} & Z
\end{array}
$$

Proposition 4.2 (Modularity). *Assume Σ-terms t and t_i are as in Prop. 2.11.4. The operations $[\![\_]\!]$ and $[\_]$ are compatible with substitution: given LTSs $c_1, \ldots, c_m \in$ $\mathbf{Coalg}_\bullet(F_\bullet)$ and "behaviors" $z_1, \ldots, z_m \in Z$, we have*

$$
\begin{aligned}
[\![\, t[t_i/x_i] \,]\!]\,(c_1,\ldots,c_m) &\cong [\![t]\!]\,\big(\,[\![t_1]\!](\vec{c}),\;\ldots,\;[\![t_n]\!](\vec{c})\,\big)\;; \\
[\,t[t_i/x_i]\,]\,(z_1,\ldots,z_m) &= [t]\,\big(\,[t_1](\vec{z}),\;\ldots,\;[t_n](\vec{z})\,\big)\;.
\end{aligned}
\qquad \square
$$

We present our main result on compositionality. It relates the outer and inner interpretations. The latter arose from the former via finality (Def. 4.1); the following result follows straightforward from finality, too.

Theorem 4.3 (Compositionality). *Given a Σ-term $x_1, \ldots, x_m \vdash t$ and LTSs $c_1, \ldots, c_m$ that belong to $\mathbf{Coalg}_\bullet(F_\bullet)$, we have the following diagram commute.*

$$
\begin{array}{ccc}
X_1 \times \cdots \times X_m & \xrightarrow{\mathsf{beh}(c_1) \times \cdots \times \mathsf{beh}(c_m)} & Z \times \cdots \times Z \\
{\scriptstyle (\theta_t)_{X_1,\ldots,X_m}}\downarrow & & \downarrow{\scriptstyle [t]} \\
(\!|t|\!)(X_1,\ldots,X_m) & \xrightarrow[\mathsf{beh}([\![t]\!](c_1,\ldots,c_m))]{} & Z
\end{array}
$$

Here $(\theta_t)_{X_1,\ldots,X_m}$ *is the "bookkeeping" function from (8); Z is the carrier of the final coalgebra (§3.1). Putting it equationally: for any state* $x_i \in X_i$ *of each LTS we have*

$$\mathsf{beh}(\,[\![t]\!](c_1,\ldots,c_m)\,)\,(\theta_t(x_1,\ldots,x_m)) \;=\; [t]\,\big(\mathsf{beh}(c_1)(x_1),\ldots,\,\mathsf{beh}(c_m)(x_m)\big)\ .$$

That is: the behavior of a composed system $[\![t]\!](\vec{c})$ *can be computed from the behaviors* $\mathsf{beh}(c_i)$ *of its constituent parts, using the inner operator* $[t]$. □

Let $\hat{\zeta} : \Sigma^* Z \to Z$ be the (Eilenberg-Moore) Σ^*-algebra on Z induced by a GSOS specification $(\Sigma, \mathcal{R})$, due to [22, Cor. 7.2]. Here Σ^* is the free monad induced by the signature Σ. The following result—claiming that our framework is indeed an extension of (the GSOS fragment of) [22]—holds because our ξ_t in §3, when suitably restricted, coincides with the categorical GSOS in [22].

Theorem 4.4. *Let* $x_1,\ldots,x_m \vdash t$ *be a* Σ*-term; let* κ_t *be the corresponding coprojection* $Z^m \hookrightarrow \Sigma^* Z$*. For these, the diagram on the right commutes.* □

$$\begin{array}{ccc} \Sigma^* Z & \xrightarrow{\hat{\zeta}} & Z \\ {\scriptstyle \kappa_t}\uparrow & \nearrow_{[t]} & \\ Z^m & & \end{array}$$

5 Conclusions and Future Work

We have extended our previous work [14] so that any process operator specified by GSOS rules can now be interpreted as a component connector that combines LTSs as components. This outer interpretation gives rise to a canonical inner interpretation that coincides with what is derived by the bialgebraic modeling of SOS [22].

Our framework is categorical, hence comes with great potential generality. This includes application to systems other than LTS—like bialgebraic modeling applied to weighted systems [17]—which we wish to pursue. In particular we believe our generic construction of $\mathcal{R}$-state spaces will carry over.

Regular languages and automata seem to be the first computer science example of the microcosm principle. Our current framework fails to include it. One difficulty is: the outer operators (specified in GSOS-like rules) are naturally defined on *NFAs*, while the inner operators is on regular languages that form the final *DFA*. Use of coalgebraic techniques such as trace semantics [13] is being investigated.

The notion of TLG and operations on them (§2) indicates strong relevance of rewriting logic [19] and the theory of *generalized operads/combinatorial species/clones* recently pursued by many authors, including [10]. We are especially interested in the latter, but not only because of TLGs. Roughly we can call their theory *universal algebra in varying contexts*. Here a "context" can be a monoidal one (where variables are not to be deleted, duplicated or swapped), a symmetric monoidal one (where swapping is allowed), a Cartesian one (where all three are allowed), and so on. In fact such a context can be thought of as algebraic structure itself; hence the theory may be also called *universal (algebra in algebra)*. The microcosm principle then offers a degenerate example of such, where we have the same structure on the two levels.

References

1. Aceto, L., Fokkink, W., Verhoef, C.: Structural operational semantics. In: Bergstra, J., Ponse, A., Smolka, S. (eds.) Handbook of Process Algebra, pp. 197–292. Elsevier, Amsterdam (2001)
2. Baez, J.C., Dolan, J.: Higher dimensional algebra III: n-categories and the algebra of opetopes. Adv. Math. 135, 145–206 (1998)
3. Baier, C., Blechmann, T., Klein, J., Klüppelholz, S.: A uniform framework for modeling and verifying components and connectors. In: Field, J., Vasconcelos, V.T. (eds.) COORDINATION 2009. LNCS, vol. 5521, pp. 247–267. Springer, Heidelberg (2009)
4. Baier, C., Sirjani, M., Arbab, F., Rutten, J.J.M.M.: Modeling component connectors in reo by constraint automata. Science of Comput. Progr. 61(2), 75–113 (2006)
5. Bliudze, S., Krob, D.: Modelling of complex systems: Systems as dataflow machines. Fundam. Inform. 91(2), 251–274 (2009)
6. Bloom, B., Istrail, S., Meyer, A.R.: Bisimulation can't be traced. Journ. ACM 42(1), 232–268 (1995)
7. Bonsangue, M.M., Clarke, D., Silva, A.: Automata for context-dependent connectors. In: Field, J., Vasconcelos, V.T. (eds.) COORDINATION 2009. LNCS, vol. 5521, pp. 184–203. Springer, Heidelberg (2009)
8. Bruni, R., Lanese, I., Montanari, U.: A basic algebra of stateless connectors. Theor. Comp. Sci. 366(1-2), 98–120 (2006)
9. Clarke, E., Grumberg, O., Peled, D.: Model Checking. MIT Press, Cambridge (1999)
10. Curien, P.L.: Operads, clones, and distributive laws, preprint, available online (2008)
11. Hasuo, I.: Tracing Anonymity with Coalgebras. Ph.D. thesis, Radboud Univ. Nijmegen (2008)
12. Hasuo, I.: The microcosm principle and compositionality of GSOS-based component calculi. Extended version with proofs, available online (May 2011)
13. Hasuo, I., Jacobs, B., Sokolova, A.: Generic trace semantics via coinduction. Logical Methods in Comp. Sci. 3(4:11) (2007)
14. Hasuo, I., Jacobs, B., Sokolova, A.: The microcosm principle and concurrency in coalgebra. In: Amadio, R.M. (ed.) FOSSACS 2008. LNCS, vol. 4962, pp. 246–260. Springer, Heidelberg (2008)
15. Jacobs, B.: Introduction to coalgebra. Towards mathematics of states and observations, Draft of a book (2005), `http://www.cs.ru.nl/B.Jacobs/PAPERS`
16. Klin, B.: Bialgebraic methods and modal logic in structural operational semantics. Inf. & Comp. 207(2), 237–257 (2009)
17. Klin, B.: Structural operational semantics for weighted transition systems. In: Palsberg, J. (ed.) Semantics and Algebraic Specification. LNCS, vol. 5700, pp. 121–139. Springer, Heidelberg (2009)
18. Lenisa, M., Power, J., Watanabe, H.: Category theory for operational semantics. Theor. Comp. Sci. 327(1-2), 135–154 (2004)
19. Martí-Oliet, N., Meseguer, J.: Rewriting logic: roadmap and bibliography. Theor. Comp. Sci. 285(2), 121–154 (2002)
20. Plotkin, G.D.: A structural approach to operational semantics, report DAIMI FN-19, Aarhus Univ. (1981)
21. Silva, A., Bonchi, F., Bonsangue, M.M., Rutten, J.J.M.M.: Quantitative kleene coalgebras. Inf. & Comp. 209(5), 822–849 (2011)
22. Turi, D., Plotkin, G.: Towards a mathematical operational semantics. In: Logic in Computer Science, pp. 280–291. IEEE, Computer Science Press (1997)

Bases as Coalgebras

Bart Jacobs

Institute for Computing and Information Sciences (iCIS),
Radboud University Nijmegen, The Netherlands
www.cs.ru.nl/B.Jacobs

Abstract. The free algebra adjunction, between the category of algebras of a monad and the underlying category, induces a comonad on the category of algebras. The coalgebras of this comonad are the topic of study in this paper (following earlier work). It is illustrated how such coalgebras-on-algebras can be understood as bases, decomposing each element x into primitives elements from which x can be reconstructed via the operations of the algebra. This holds in particular for the free vector space monad, but also for other monads. For instance, continuous dcpos or stably continuous frames, where each element is the join of the elements way below it, can be described as such coalgebras. Further, it is shown how these coalgebras-on-algebras give rise to a comonoid structure for copy and delete, and thus to diagonalisation of endomaps like in linear algebra.

1 Introduction

In general, algebras are used for composition and coalgebras for decomposition. An algebra $a\colon T(X) \to X$, for a functor or a monad T, can be used to produce elements in X from ingredients structured by T. Conversely, a coalgebra $c\colon X \to T(X)$ allows one to decompose an element in X into its ingredients with structure according to T. This is the fundamental difference between algebraic and coalgebraic data structures. In this paper we apply this view to the special situation where one has a coalgebra of a comonad on top of an algebra of a monad, where the comonad is canonically induced by the monad, namely as arising from the free algebra adjunction, see (1) below. Here it is proposed that such coalgebras can be seen as bases. In particular, it will be shown that the concept of basis in linear algebra gives rise to such a coalgebra $X \to \mathcal{M}(X)$ for the multiset monad $\mathcal{M}$; this coalgebra decomposes an element x of a vector space X into a formal sum $\sum_i x_i e_i \in \mathcal{M}(X)$ given by its coefficients x_i for a Hamel basis (e_i), see Theorem 1 for more details.

Other examples arise in an order-theoretic setting, formalised via the notion of monad of Kock-Zölberlein type (where $T(\eta_X) \leq \eta_{TX}$, see [14,7]). We describe how they fit in the present setting (with continuous dcpos as coalgebras), and add a new result (Theorem 4) about algebras-on-coalgebras-on-algebras, see Section 4. This builds on rather old (little noticed) work of the author [10].

A. Corradini, B. Klin, and C. Cîrstea (Eds.): CALCO 2011, LNCS 6859, pp. 237–252, 2011.

In recent work [5] in the categorical foundations of quantum mechanics it is shown that orthonormal bases in finite-dimensional Hilbert spaces are equivalent to comonoids structures (in fact, Frobenius algebras). These comonoids are used for copying and deleting elements. In Section 5 it is shown how bases as coalgebras (capturing bases-as-decomposition) also give rise to such comonoids (capturing bases-as-copier-and-deleter). These comonoids can be used to formulate in general terms what it means for an endomap to be diagonalisable. This is illustrated for the Pauli functions.

2 Comonads on Categories of Algebras

In this section we investigate the situation of a monad and the induced comonad on its category of algebras. We shall see that coalgebras of this comonad capture the notion of basis, in a very general sense. This will be illustrated later in several situations see in particular Subsection 3.2.

For an arbitrary monad $T\colon \mathbf{A} \to \mathbf{A}$, with unit η and multiplication μ, there is a category $\mathit{Alg}(T)$ or (Eilenberg-Moore) algebras, together with a left adjoint F (for free algebra functor) to the forgetful functor $U\colon \mathit{Alg}(T) \to \mathbf{A}$. This adjunction $\mathit{Alg}(T) \leftrightarrows \mathbf{A}$ induces a comonad on the category $\mathit{Alg}(T)$, which we shall write as $\overline{T} = FU$ in:

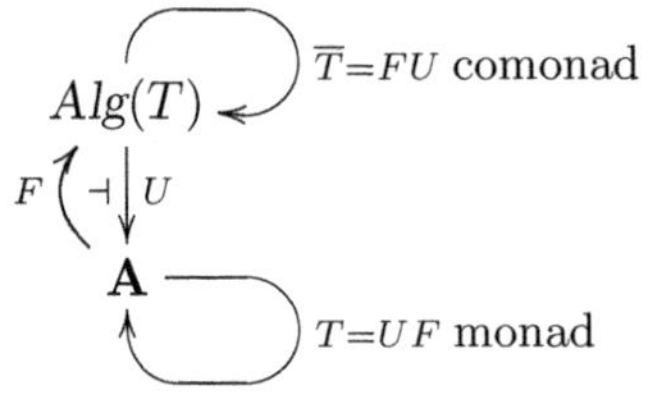

(1)

For an algebra $(TX \xrightarrow{a} X) \in \mathit{Alg}(T)$ there are counit $\varepsilon\colon \overline{T} \Rightarrow \mathrm{id}$ and comultiplication $\delta\colon \overline{T} \Rightarrow \overline{\overline{T}}$ maps in $\mathit{Alg}(T)$ given by:

$$\begin{pmatrix} TX \\ \downarrow a \\ X \end{pmatrix} \xleftarrow{\;\varepsilon = a\;} \begin{pmatrix} T^2X \\ \downarrow \mu_X \\ TX \end{pmatrix} \xrightarrow{\;\delta = T(\eta_X)\;} \begin{pmatrix} T^3X \\ \downarrow \mu_{TX} \\ T^2X \end{pmatrix} \qquad (2)$$

Definition 1. *Consider a monad $T\colon \mathbf{A} \to \mathbf{A}$ together with the induced comonad $\overline{T}\colon \mathit{Alg}(T) \to \mathit{Alg}(T)$ as in (1). A* basis *for a T-algebra $(TX \xrightarrow{a} X) \in \mathit{Alg}(T)$ is a $\overline{T}$-coalgebra on this algebra, given by a map of algebras b of the form:*

$$\begin{pmatrix} TX \\ \downarrow a \\ X \end{pmatrix} \xrightarrow{\;b\;} \overline{T}\begin{pmatrix} TX \\ \downarrow a \\ X \end{pmatrix} = FU\begin{pmatrix} TX \\ \downarrow a \\ X \end{pmatrix} = \begin{pmatrix} T^2X \\ \downarrow \mu_X \\ TX \end{pmatrix}$$

Thus, a basis b is a map $X \xrightarrow{b} TX$ in $\mathbf{A}$ satisfying $b \circ a = \mu_X \circ T(b)$ and $a \circ b = \mathrm{id}$ and $T(\eta_X) \circ b = T(b) \circ b$ in:

$$\begin{array}{ccc} T(X) & \xrightarrow{T(b)} & T^2(X) \\ {\scriptstyle a}\downarrow & & \downarrow{\scriptstyle \mu_X} \\ X & \xrightarrow[b]{} & T(X) \end{array} \qquad \begin{array}{ccc} X & \xrightarrow{b} & T(X) \\ & \searrow\!\!\!\searrow & \downarrow{\scriptstyle \varepsilon = a} \\ & & X \end{array} \qquad \begin{array}{ccc} T(X) & \xrightarrow{T(b)} & T^2(X) \\ {\scriptstyle b}\uparrow & & \uparrow{\scriptstyle \delta = T(\eta_X)} \\ X & \xrightarrow[b]{} & T(X) \end{array}$$

As we shall see a basis as described above may be understood as providing a decomposition of each element x into a collection $b(x)$ of basic elements that together form x. The actual basic elements $X_b \rightarrowtail X$ involved can be obtained as the indecomposable ones, via the following equaliser in the underlying category.

$$X_b \overset{e}{\rightarrowtail} X \underset{\eta}{\overset{b}{\rightrightarrows}} TX \tag{3}$$

One can then ask in which cases the map of algebras $T(X_b) \to X$, induced by the equaliser $e\colon X_b \to U(TX \to X)$, is an isomorphism. This is (almost always) the case for monads on **Sets**, see Proposition 1 below. But first we observe that free algebras always carry a basis.

Lemma 1. *Free algebras have a canonical basis: each* $FX = \big(T^2X \xrightarrow{\mu} TX\big) \in$ *Alg*(T) *carries a* $\overline{T}$*-coalgebra, namely given by* $T(\eta_X)$*. This gives a situation:*

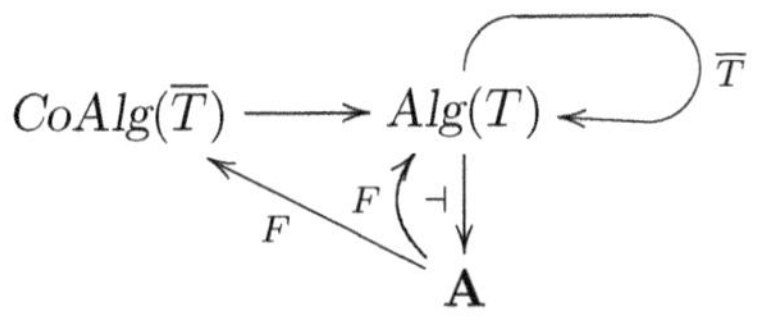

Proof. It is easy to check that $T(\eta_X)$ is a morphism in *Alg*(T) and a $\overline{T}$-coalgebra:

$$F(X) = \begin{pmatrix} T^2X \\ \downarrow{\scriptstyle \mu_X} \\ TX \end{pmatrix} \xrightarrow{\quad T(\eta_X) \quad} \begin{pmatrix} T^3X \\ \downarrow{\scriptstyle \mu_{TX}} \\ T^2X \end{pmatrix} = \overline{T}(FX). \qquad \square$$

The object X_b of basic element, as in (3), in the situation of this lemma is the original set X in case the monad T satisfies the so-called *equaliser requirement* [16], which says precisely that $\eta_X\colon X \to TX$ is the equaliser of $T(\eta_X), \eta_{TX}\colon TX \rightrightarrows T^2X$.

The comonad $\overline{T}\colon$ *Alg*$(T) \to$ *Alg*(T) from (1) gives rise to a category of coalgebras *CoAlg*$(\overline{T}) \to$ *Alg*(T), where this forgetful functor has a right adjoint, which maps an algebra $TY \to Y$ to the diagonal coalgebra $\delta\colon \mu_Y \to \mu_{TY}$ as in (2). Thus we obtain a monad on the category *CoAlg*$(\overline{T})$, written as $\overline{\overline{T}}$. On a basis $c\colon a \to \overline{T}(a)$, for an algebra $a\colon TX \to X$, there is a unit $\eta_c = c\colon c \to \delta$ and multiplication $\mu_c = T(c)\colon \delta \to \delta$ in *CoAlg*$(\overline{T})$.

By iterating this construction one obtains alternating monads and comonads. Such iterations are studied for instance in [3,10,14,18]. In special cases it is known that the iterations stop after a number of cycles. This happens after 2 iterations for monads on sets, as we shall see next, and after 3 iterations for Kock-Zölberlein monads in Section 4.

3 Set-Theoretic Examples

It turns out that for monads on the category **Sets** only free algebras have bases. This result goes back to [3]. We repeat it in the present context, with a sketch of proof. Subsequently we describe the situation for the powerset monad (from [10]) and the free vector space monad.

Proposition 1. *For a monad T on* **Sets**, *if an algebra $TX \xrightarrow{a} X$ has a basis $X \xrightarrow{b} TX$ with non-empty equaliser $X_b \rightarrowtail X \rightrightarrows TX$ as in (3), then the induced map $T(X_b) \to X$ is an isomorphism of algebras and coalgebras. In particular, in the set-theoretic case any algebra with a non-empty basis is free.*

Proof. Let's consider the equaliser $X_b \rightarrowtail X$ of $b, \eta \colon X \rightrightarrows T(X)$ from (3) in **Sets**. It is a so-called coreflexive equaliser, because there is a map $TX \to X$, namely the algebra a, satisfying $a \circ b = \mathrm{id} = a \circ \eta$. It is well-known—see *e.g.* [15, Lemma 6.5] or the dual result in [4, Volume I, Example 2.10.3.a]—that if $X_b \neq \emptyset$ such coreflexive equalisers in **Sets** are split, and thus absolute. The latter means that they are preserved under any functor application. In particular, by applying T we obtain a new equaliser in **Sets**, of the form:

$$\begin{array}{ccccc} T(X_b) & \xrightarrow{T(e)} & T(X) & \underset{T(\eta)=\delta}{\overset{T(b)}{\rightrightarrows}} & T^2(X) \\ {\scriptstyle b'}\uparrow & \nearrow_{b} & & & \\ X & & & & \end{array} \qquad (4)$$

The resulting b' is the inverse to the adjoint transpose $a \circ T(e) \colon T(X_b) \to X$, since:

- $a \circ T(e) \circ b' = a \circ b = \mathrm{id}$;
- the other equation follows because $T(e)$ is equaliser, and thus mono:

$$\begin{array}{rcll} T(e) \circ b' \circ a \circ T(e) & = & b \circ a \circ T(c) & \\ & = & \mu \circ T(b) \circ T(e) & \text{see Definition 1} \\ & = & \mu \circ T(\eta) \circ T(e) & \text{since } e \text{ is equaliser} \\ & = & T(e) \; = \; T(e) \circ \mathrm{id}. & \end{array}$$

Hence the homomorphism of algebras $a \circ T(e)$, from $F(X_b) = \mu_{X_b}$ to a is an isomorphism. In particular, $b' \colon X \to T(X_b)$ in (4) is a map of algebras, as inverse of an isomorphism of algebras. It is not hard to see that it is also an isomorphism between the coalgebras $b \colon X \to T(X)$ and $T(\eta) \colon T(X_b) \to T^2(X_b)$, as in Lemma 1. □

3.1 Complete Lattices

Consider the powerset monad $\mathcal{P}$ on **Sets**, with the category $\mathbf{CL} = Alg(\mathcal{P})$ of complete lattices and join-preserving maps as its category of algebras. The induced comonad $\overline{\mathcal{P}}\colon \mathbf{CL} \to \mathbf{CL}$ as in (1) sends a complete lattice $(L, \leq)$ to the lattice $(\mathcal{P}(L), \subseteq)$ of subsets, ignoring the original order $\leq$. The counit $\varepsilon\colon \overline{\mathcal{P}}(L) \to L$ sends a subset $U \in \mathcal{P}(L)$ to its join $\varepsilon(U) = \bigvee U$; the comultiplication $\delta\colon \overline{\mathcal{P}}(L) \to \overline{\mathcal{P}}^2(L)$ sends $U \in \mathcal{P}(L)$ to the subset of singletons $\delta(U) = \{\{x\} \mid x \in U\}$.

An (Eilenberg-Moore) coalgebra of the comonad $\overline{\mathcal{P}}$ on **CL** is a map $b\colon L \to \overline{\mathcal{P}}(L)$ in **CL** satisfying $\varepsilon \circ b = \mathrm{id}$ and $\delta \circ b = \overline{\mathcal{P}}(b) \circ b$. More concretely, this says that $\bigvee b(x) = x$ and $\{\{y\} \mid y \in b(x)\} = \{b(y) \mid y \in b(x)\}$. It is then shown in [10] that a complete lattice L carries such a coalgebra structure b if and only if L is *atomic*, where $b(x) = \{a \in L \mid a \text{ is an atom with } a \leq x\}$. Thus, such a coalgebra of the comonad $\overline{\mathcal{P}}$, if it exists, is uniquely determined and gives a decomposition of lattice elements into the atoms below it. The atoms in the lattice thus form a basis.

(The complete lattice L is atomic when each element is the join of the atoms below it. And an atom $a \in L$ is a non-zero element with no non-zero elements below it, satisfying: $a \leq \bigvee U$ implies $a \leq x$ for some $x \in U$.)

The equaliser (3) for the basic elements in this situation, for an atomic complete lattice L, is the set of atoms:

$$X_b = \{x \in L \mid \{x\} = b(x)\} = \{x \in L \mid x \text{ is an atom}\}.$$

If $X_b \neq \emptyset$, the induced map $\mathcal{P}(X_b) \to L$ is an isomorphism, by Lemma 1.

3.2 Vector Spaces

For a semiring S one can define the multiset monad $\mathcal{M}_S$ on **Sets** by $\mathcal{M}_S(X) = \{\varphi\colon X \to S \mid \mathrm{supp}(\varphi) \text{ is finite}\}$. Such an element φ can be identified with a formal finite sum $\sum_i s_i x_i$ with multiplicities $s_i \in S$ for elements $x_i \in X$. The category of algebras $Alg(\mathcal{M}_S)$ of the multiset monad $\mathcal{M}_S$ is the category of $\mathbf{Mod}_S$ of modules over S: commutative monoids with S-scalar multiplication, see *e.g.* [6] for more information. The induced comonad $\overline{\mathcal{M}_S}\colon \mathbf{Mod}_S \to \mathbf{Mod}_S$ from (1) sends such a module $X = (X, +, 0, \bullet)$ to the free module $\mathcal{M}_S(X)$ of finite multisets (formal sums) on the underlying set X, ignoring the existing module structure on X. The counit and comultiplication are given by:

$$\begin{array}{ccccc} X & \xleftarrow{\quad\varepsilon\quad} & \mathcal{M}_S(X) & \xrightarrow{\quad\delta\quad} & \mathcal{M}_S^2(X) \\ \left(\sum_j s_j \bullet x_j\right) & \longleftarrow\!\!\shortmid & \left(\sum_j s_j x_j\right) & \longmapsto & \left(\sum_j s_j (1 x_j)\right). \end{array} \qquad (5)$$

The formal sum (multiset) in the middle is mapped by the counit ε to an actual sum in X, namely to its interpretation. The comultiplication δ maps this formal sum to a multiset of multisets, with the inner multisets given by singletons $1x_j = \eta(x_j)$.

The following is a novel observation, motivating the view of coalgebras on algebras as bases.

Theorem 1. *Let X be a vector space, say over $S = \mathbb{R}$ or $S = \mathbb{C}$. Coalgebras $X \to \overline{\mathcal{M}_S}(X)$ correspond to (Hamel) bases on X.*

Proof. Suppose we have a basis $B \subseteq X$. Then we can define a coalgebra $b\colon X \to \overline{\mathcal{M}_S}(X)$ via (finite) formal sums $b(x) = \sum_j s_j a_j$, where $s_j \in S$ is the j-th coefficient of x wrt $a_j \in B \subseteq X$. By construction we have $\varepsilon \circ b = \mathrm{id}$. The equation $\delta \circ b = \mathcal{M}_S(b) \circ b$ holds because $b(a) = 1a$, for basic elements $a \in B$.

Conversely, given a coalgebra $b\colon X \to \overline{\mathcal{M}_S}(X)$ take $X_b = \{a \in X \mid b(a) = 1a\}$ as in (3). Any finite subset of elements of X_b is linearly independent: if $\sum_j s_j \bullet a_j = 0$, for finitely many $a_j \in X_b$, then in $\mathcal{M}_S(X)$,

$$0 \;=\; b(0) \;=\; b(\textstyle\sum_j s_j \bullet a_j) \;=\; \sum_j s_j b(a_j) \;=\; \sum_j s_j(1a_j) \;=\; \sum_j s_j a_j.$$

Hence $s_j = 0$, for each j. Next, since $\delta \circ b = \mathcal{M}_S(b) \circ b$, each a_j in $b(x) = \sum_j s_j a_j$ satisfies $b(a_j) = 1a_j$, so that $a_j \in X_b$. Because $\varepsilon \circ b = \mathrm{id}$, each element $x \in X$ can be expressed as sum of such basic elements. □

A basis for complete lattices in Subsection 3.1, if it exists, is uniquely determined. In the context of vector spaces bases are unique up to isomorphism.

4 Order-Theoretic Examples

Assume **C** is a poset-enriched category. This means that all homsets $\mathbf{C}(X, Y)$ are posets, and that pre- and post-composition are monotone. In this context maps $f\colon X \to Y$ and $g\colon Y \to X$ in opposite direction form an adjunction $f \dashv g$ (or Galois connection) if there inequalities $\mathrm{id}_X \leq g \circ f$ and $f \circ g \leq \mathrm{id}_Y$, corresponding to unit and counit of the adjunction. In such a situation the adjoints f, g determine each other.

A monad $T = (T, \eta, \mu)$ on such a poset-enriched category **C** is said to be of *Kock-Zölberlein type* or just a *Kock-Zölberlein monad* if $T\colon \mathbf{C}(X, Y) \to \mathbf{C}(TX, TY)$ is monotone and $T(\eta_X) \leq \eta_{TX}$ holds in the homset $\mathbf{C}\big(T(X), T^2(X)\big)$. This notion is introduced in [14] in proper 2-categorical form. Here we shall use the special 'poset' instance—like in [7] where the dual form occurs. The following result goes back to [14]; for convenience we include the proof.

Theorem 2. *Let T be a Kock-Zölberlein monad on a poset-enriched category* **C**. *For a map $a\colon T(X) \to X$ in* **C** *the following statements are equivalent.*

1. *$a\colon T(X) \to X$ is an (Eilenberg-Moore) algebra of the monad T;*
2. *$a\colon T(X) \to X$ is a left-adjoint-left-inverse of the unit $\eta\colon X \to T(X)$; this means that $a \dashv \eta_X$ is a reflection.*

Proof. First assume $a\colon T(X) \to X$ is an algebra, *i.e.* satisfies $a \circ \eta = \mathrm{id}$ and $a \circ \mu = a \circ T(a)$. It suffices to prove $\mathrm{id} \leq \eta \circ a$, corresponding to the unit of the reflection, since the equation $a \circ \eta = \mathrm{id}$ is the counit (isomorphism). This is easy, by naturality: $\eta \circ a = T(a) \circ \eta \geq T(a) \circ T(\eta) = \mathrm{id}$.

In the other direction, assume $a\colon T(X) \to X$ is left-adjoint-left-inverse of the unit $\eta\colon X \to T(X)$, so that $a \circ \eta = \mathrm{id}$ and $\mathrm{id} \leq \eta \circ a$. We have to prove $a \circ \mu = a \circ T(a)$. In one direction, we have:

$$\mu \leq T(a), \tag{6}$$

since $\mu \leq \mu \circ T(\eta \circ a) = T(a)$, and thus $a \circ \mu \leq a \circ T(a)$. For the reverse inequality we use:

$$\begin{aligned} a \circ T(a) = a \circ T(a) \circ T(\mathrm{id}) &= a \circ T(a) \circ T(\mu) \circ T(\eta) \\ &\leq a \circ T(a) \circ T(\mu) \circ \eta && \text{since } T(\eta) \leq \eta \\ &= a \circ \eta \circ a \circ \mu && \text{by naturality} \\ &= a \circ \mu. \end{aligned}$$

□

In a next step we consider the induced comonad $\overline{T}$ on the category $Alg(T)$ of algebra of a Kock-Zöberlein monad T. A first, trivial but important, observation is that the category $Alg(T)$ is also poset enriched. It is not hard to see that the comonad $\overline{T}$ is also of Kock-Zöberlein type, in the sense that for each algebra $(TX \xrightarrow{a} X)$ we have:

$$\varepsilon_{\overline{T}(a)} = \mu \leq T(a) = \overline{T}(\varepsilon_a)$$

by (6). Thus one may expect a result similar to Theorem 2 for coalgebras of this comonad $\overline{T}$. It is formulated in [14, Thm. 4.2] (and attributed to the present author). We repeat the poset version in the current context.

Theorem 3. *Let T be a Kock-Zölberlein monad on a poset-enriched category $\mathbf{C}$, with induced comonad $\overline{T}$ on the category of algebras $Alg(T)$. Assume an algebra $a\colon T(X) \to X$. For a map $c\colon X \to T(X)$, forming a map of algebras in,*

$$\begin{pmatrix} TX \\ \downarrow a \\ X \end{pmatrix} \xrightarrow{\quad c \quad} \overline{T}\begin{pmatrix} TX \\ \downarrow a \\ X \end{pmatrix} = \begin{pmatrix} T^2X \\ \downarrow \mu_X \\ TX \end{pmatrix} \tag{7}$$

the following statements are equivalent.

1. *$c\colon a \to \overline{T}(a)$ is an (Eilenberg-Moore) coalgebra of the comonad $\overline{T}$;*
2. *$c\colon a \to \overline{T}(a)$ is a left-adjoint-right-inverse of the counit $a\colon \overline{T}(a) \to a$; this means that $c \dashv a$ is a coreflection.*

Proof. Assume c is a $\overline{T}$-coalgebra, *i.e.* $c \circ a = \mu \circ T(c)$, $a \circ c = \mathrm{id}$ and $T(\eta) \circ c = T(c) \circ c$. We have to prove $c \circ a \leq \mathrm{id}$, which is obtained in:

$$c \circ a = \mu \circ T(c) \overset{(6)}{\leq} T(a) \circ T(c) = \mathrm{id}.$$

Conversely, assume a coreflection $c \dashv a$, so that $a \circ c = \mathrm{id}$ and $c \circ a \leq \mathrm{id}$. We have to prove $T(\eta) \circ c = T(c) \circ c$. In one direction we have $T(c) \leq T(\eta \circ a) \circ T(c) = T(\eta)$, and thus $T(c) \circ c \leq T(\eta) \circ c$. In the other direction, we use:

$$\begin{array}{rcll} T(c) \circ c \;=\; T^2(\mathrm{id}) \circ T(c) \circ c & = & T^2(a \circ \eta) \circ T(c) \circ c & \\ & \leq & T^2(a) \circ T(\eta) \circ T(c) \circ c & \text{since } T(\eta) \leq \eta \\ & = & T^2(a) \circ T^2(c) \circ T(\eta) \circ c & \text{by naturality} \\ & = & T(\eta) \circ c. & \end{array}$$

□

As mentioned, one can iterate the $\overline{(-)}$ construction. Below we show that for Kock-Zölberlein monads the iteration stops after 3 steps. First we need another characterisation. The proof is as before.

Lemma 2. *Let T be a Kock-Zölberlein monad on a poset-enriched category* $\mathbf{C}$, *giving rise to comonad $\overline{T}$ on Alg(T) and monad $\overline{\overline{T}}$ on CoAlg($\overline{T}$). Assume:*

- *an algebra $a\colon T(X) \to X$ in Alg(T);*
- *a coalgebra $c\colon X \to T(X)$ on a in CoAlg($\overline{T}$);*
- *an algebra $b\colon T(X) \to X$ on c in Alg($\overline{\overline{T}}$), where:*
  - *$b \circ c = \mathrm{id}$ and $b \circ T(b) = b \circ T(a)$, since b is a $\overline{\overline{T}}$-algebra;*
  - *$a \circ T(b) = b \circ \mu$, since b is a map of algebras $a \to \overline{T}(a) = \mu$;*
  - *$c \circ b = T(b) \circ T(\eta)$, since b is a map of algebras $\delta = \overline{\overline{c}} \to c$.*

The following statements are then equivalent.

1. *$b\colon \overline{\overline{T}}(c) \to c$ is an algebra of the monad $\overline{\overline{T}}$;*
2. *$b\colon \overline{\overline{T}}(c) \to c$ is a left-adjoint-left-inverse of the unit $c\colon c \to \overline{\overline{T}}(c)$.* □

The next result shows how such series of adjunctions can arise.

Lemma 3. *Assume an algebra $a\colon T(X) \to X$ of a Kock-Zöberlein monad. The free algebra $T(X)$ then carries multiple (co)reflections (algebras and coalgebras) in a situation:*

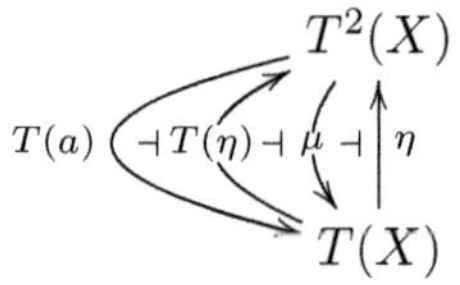

(8)

This yields a functor $T\colon$ Alg(T) $\to$ Alg($\overline{\overline{T}}$) between categories of algebras.

Proof. We check all (co)reflections from right to left.

- In the first case the counit is the identity since $\mu \circ \eta = \mathrm{id}$; because $T(\eta) \leq \eta$ for a Kock-Zöberlein monad, we get a unit $\eta \circ \mu = T(\mu) \circ \eta \geq T(\mu) \circ T(\eta) = \mathrm{id}$. (This follows already from Theorem 2.)

– In the next case we have a coreflection $T(\eta) \dashv \mu$ since the unit is the identity $\mu \circ T(\eta) = \mathrm{id}$, and: $T(\eta) \circ \mu = \mu \circ T^2(\eta) \leq \mu \circ T(\eta) = \mathrm{id}$.
– Finally one gets a reflection $T(a) \dashv T(\eta)$ from the reflection $a \dashv \eta$ from Theorem 2: $T(a) \circ T(\eta) = T(a \circ \eta) = \mathrm{id}$ and $T(\eta) \circ T(a) = T(\eta \circ a) \geq T(\mathrm{id}) = \mathrm{id}$. □

This lemma describes the only form that such structures can have. This is the main (new) result of this section.

Theorem 4. *If we have a reflection-coreflection-reflection chain $b \dashv c \dashv a \dashv \eta_X$ on an object X, like in Lemma 2, then X is a free algebra.*

Thus: for a Kock-Zöberlein monad T, the functor $T\colon \mathrm{Alg}(T) \to \mathrm{Alg}(\overline{\overline{T}})$ is an equivalence of categories.

Proof. Assume $b \dashv c \dashv a \dashv \eta_X$ on X, and consider the equaliser (3) in:

$$\begin{array}{ccccc} X_c & \overset{e}{\rightarrowtail} & X & \underset{\eta}{\overset{c}{\rightrightarrows}} & T(X) \\ {\scriptstyle k}\uparrow & & \nearrow{\scriptstyle b} & & \\ & T(X) & & & \\ X & \nearrow{\scriptstyle \eta} & & & \end{array} \tag{9}$$

We use the letter 'k' because the elements in X_c will turn out to be compact elements, in the examples later on. The first thing we note is:

$$k \circ e \;=\; \mathrm{id}_{X_c}. \tag{10}$$

This follows since e is a mono, and:

$$\begin{array}{rcll} e \circ k \circ e & = & b \circ \eta \circ e & \text{by construction of } k \\ & = & b \circ c \circ e & \text{since } e \text{ is equaliser} \\ & = & e & \text{since } b \text{ is a } \overline{\overline{T}}\text{-algebra and } c \text{ is unit.} \end{array}$$

Next we observe that the object X_c carries a T-algebra structure a_c inherited from $a\colon T(X) \to X$, as in:

$$a_c \overset{\mathrm{def}}{=} \Big(T(X_c) \xrightarrow{T(e)} T(X) \xrightarrow{a} X \xrightarrow{k} X_c\Big)$$

It is an algebra indeed, since:

$$a_c \circ \eta = k \circ a \circ T(e) \circ \eta = k \circ a \circ \eta \circ e = k \circ e \overset{(10)}{=} \mathrm{id}.$$

The other algebra equation is left to the reader.

Next we show that the transpose $a \circ T(e)\colon T(X_c) \to X$ of the equaliser $e\colon X_c \rightarrowtail X$ is an isomorphism of algebras $\mu_{X_c} \cong a$. The inverse is $T(k) \circ c\colon X \to T(X) \to T(X_c)$, since:

$$\begin{array}{lll}
\multicolumn{2}{l}{\big(a \circ T(e)\big) \circ \big(T(k) \circ c\big)} & \\
= & a \circ T(b \circ \eta) \circ c & \text{by (9)} \\
= & a \circ \mu \circ T(\eta) \circ c & \text{see the assumptions in Lemma 2} \\
= & a \circ c & \\
= & \mathrm{id} & \text{see Theorem 3} \\
\multicolumn{2}{l}{\big(T(k) \circ c\big) \circ \big(a \circ T(e)\big)} & \\
= & T(k) \circ \mu \circ T(c) \circ T(e) & \text{since } c \text{ is a map } a \to \overline{T}(a) = \mu \\
= & T(k) \circ \mu \circ T(\eta) \circ T(e) & \text{since } e \text{ is equaliser of } c \text{ and } \eta \\
= & T(k) \circ T(e) & \\
= & \mathrm{id} & \text{by (10).}
\end{array}$$

We continue to check that the assumed chain of adjunctions $b \dashv c \dashv a \dashv \eta_X$ is related to the chain $T(a_c) \dashv T(\eta) \dashv \mu \dashv \eta$ in (8) via these isomorphisms. In particular we still need to check that the following two square commute.

$$\begin{array}{ccc}
T^2(X_c) & \xrightarrow[\cong]{T(a \circ T(e))} & T(X) \\
{\scriptstyle T(\eta)}\uparrow & & \uparrow{\scriptstyle c} \\
T(X_c) & \xrightarrow[a \circ T(e)]{\cong} & X
\end{array}
\qquad\qquad
\begin{array}{ccc}
T^2(X_c) & \xrightarrow[\cong]{T(a \circ T(e))} & T(X) \\
{\scriptstyle T(a_c)}\downarrow & & \downarrow{\scriptstyle b} \\
T(X_c) & \xrightarrow[a \circ T(e)]{\cong} & X
\end{array}$$

These square commute since:

$$\begin{array}{rcll}
T(a \circ T(e)) \circ T(\eta) & = & T(a) \circ T(\eta) \circ T(e) & \text{by naturality} \\
& = & T(e) & \\
& = & c \circ a \circ T(e) & \text{see Theorem 3} \\
a \circ T(e) \circ T(a_c) & = & a \circ T(e) \circ T(k \circ a \circ T(e)) & \\
& = & a \circ T(b \circ \eta) \circ T(a \circ T(e)) & \text{by (9)} \\
& = & b \circ \mu \circ T(\eta) \circ T(a \circ T(e)) & \text{see in Lemma 2} \\
& = & b \circ T(a \circ T(e)). &
\end{array}$$

We still have to check that the functor $T\colon \mathit{Alg}(T) \to \mathit{Alg}(\overline{\overline{T}})$ is an equivalence. In the reverse direction, given a coalgebra $c\colon \overline{\overline{T}}(b) \to b$ on X, we take the induced algebra $T(X_c) \to X_c$ on the equaliser (9). Then $T(X_c) \xrightarrow{\cong} X$ is an isomorphism of $\overline{\overline{T}}$-algebras, as we have seen.

For the isomorphism in the other direction, assume we start from an algebra $a\colon T(X) \to X$, obtain the $\overline{\overline{T}}$-algebra $T(a)$ described in the chain $T(a) \dashv T(\eta) \dashv \mu \dashv \eta$ in (8), and then form the equaliser (9); it now looks as follows.

$$X \overset{\eta_X}{\rightarrowtail} T(X) \underset{\eta}{\overset{T(\eta)}{\rightrightarrows}} T^2(X)$$

This is the equaliser requirement [16], which holds since X carries an algebra structure. Clearly, $\eta \circ \eta = T(\eta) \circ \eta$ by naturality. And if a map $f: Y \to T(X)$ satisfies $\eta \circ f = T(\eta) \circ f$, then f factors through $\eta: X \to T(X)$ via $f' = a \circ f$, since

$$\eta \circ f' = \eta \circ a \circ f = T(a) \circ \eta \circ f = T(a) \circ T(\eta) \circ f = f.$$

This f' is unique with this property, since if $g: Y \to X$ also satisfies $\eta \circ g = f$, then $f' = a \circ f = a \circ \eta \circ g = g$. □

In the remainder of this section we review some examples.

4.1 Dcpos over Posets

The main example from [10] involves the ideal monad *Idl* on the category **PoSets** of partially ordered sets with monotone functions between them. In the light of Theorems 2 and 3 we briefly review the essentials.

For a poset $X = (X, \leq)$ let *Idl*(X) be the set of directed downsets in X, ordered by inclusion. This *Idl* is in fact a monad on **PoSets** with unit $X \to$ *Idl*(X) given by principal downset $x \mapsto \downarrow x$ and multiplication $\mathit{Idl}^2(X) \to \mathit{Idl}(X)$ by union. This monad is of Kock-Zölberlein type since for $U \in \mathit{Idl}(X)$ we have:

$$\begin{aligned} \mathit{Idl}(\Downarrow)(U) &= \Downarrow\{\downarrow x \mid x \in U\} = \{V \in \mathit{Idl}(X) \mid \exists x \in U. V \subseteq \downarrow x\} \\ &\subseteq \{V \in \mathit{Idl}(X) \mid V \subseteq U\} \qquad \text{since } U \text{ is a downset} \\ &= \downarrow U. \end{aligned}$$

Applying Theorem 2 to the ideal monad yields the (folklore) equivalence of the following points.

1. X is a directed complete partial order (dcpo): each directed subset $U \subseteq X$ has a join $\bigvee U$ in X;
2. The unit $\downarrow : X \to \mathit{Idl}(X)$ has a left adjoint—which is the join;
3. X carries a (necessarily unique) algebra structure $\mathit{Idl}(X) \to X$, which is also the join.

Additionally, algebra maps are precisely the continuous functions. Thus we may use as category **Dcpo** = *Alg*(*Idl*).

The monad *Idl* on **PoSets** induces a comonad on **Dcpo**, written $\overline{\mathit{Idl}}$, with counit $\varepsilon = \bigvee: \mathit{Idl}(X) \to X$ and comultiplication $\delta = \mathit{Idl}(\Downarrow): \mathit{Idl}(X) \to \mathit{Idl}^2(X)$, so that $\delta(U) = \Downarrow\{\downarrow x \mid x \in U\}$. In order to characterise coalgebras of this comonad $\overline{\mathit{Idl}}$ we need the following. In a dcpo X, the way below relation $\ll$ is defined as: for $x, y \in X$,

$$x \ll y \iff \text{for each directed } U \subseteq X, \text{ if } y \leq \bigvee U \text{ then } \exists z \in U. x \leq z.$$

A *continuous poset* is then a dcpo in which for each element $x \in X$ the set $\twoheaddownarrow x = \{y \in X \mid y \ll x\}$ is directed and has x as join. These elements way-below x may be seen as a (local) basis.

The following equivalence formed the basis for [14, Thm. 4.2] (of which Theorem 3 is a special case). The equivalence of points (1) and (2) is known from the literature, see e.g. [11, VII, Proposition 2.1], [9, Proposition 2.3], or [8, Theorem I-1.10]. The equivalence of points (2) and (3) is given by Theorem 3.

For a dcpo X, the following statements are equivalent.

1. X is a continuous poset;
2. The counit $\bigvee \colon \mathit{Idl}(X) \to X$ of the comonad $\overline{\mathit{Idl}}$ on **Dcpo** has a left adjoint (in **Dcpo**); it is $x \mapsto \twoheaddownarrow x$.
3. X carries a (necessarily unique) $\overline{\mathit{Idl}}$-coalgebra structure $X \to \mathit{Idl}(X)$, which is also $\twoheaddownarrow(-)$.

Theorem 4 says that another iteration $\overline{\overline{\mathit{Idl}}}$ yields nothing new.

4.2 Frames over Semi-lattices

For a poset X, the set $\mathit{Dwn}(X) = \{U \subseteq X \mid U \text{ is downclosed}\}$ of downsets of X is a frame (or complete Heyting algebra, or locale), see [11]. If the poset X has finite meets $\top, \wedge$, then the downset map $\downarrow \colon X \to \mathit{Dwn}(X)$ preserves meets: $\downarrow\top = X$ and $\downarrow(x \wedge y) = \downarrow x \cap \downarrow y$. Hence it is a morphism in the category **MSL** of meet semi-lattices. It is not hard to see that Dwn is a monad on **MSL** that is of Kock-Zöberlein type. For a (meet) semi-lattice $X = (X, \top, \wedge)$ the following are equivalent.

1. X is a frame: X has arbitrary joins and its finite meets distribute over these joins: $x \wedge \left(\bigvee_i y_i\right) = \bigvee_i (x \wedge y_i)$;
2. The unit $\downarrow \colon X \to \mathit{Dwn}(X)$ has a left adjoint in **MSL**—which is the join;
3. X carries a (necessarily unique) algebra structure $\mathit{Dwn}(X) \to X$ in **MSL**, which is also the join.

Moreover, the algebra maps are precisely the frame maps, preserving arbitrary joins and finite meets; thus $\mathbf{Frm} = \mathit{Alg}(\mathit{Dwn})$.

In a next step, for a frame X, the following statements are equivalent.

1. X is a stably continuous frame, *i.e.* a frame that is continuous as a dcpo, in which $\top \ll \top$, and also $x \ll y$ and $x \ll z$ implies $x \ll y \wedge z$;
2. The counit $\bigvee \colon \mathit{Dwn}(X) \to X$ of the comonad $\overline{\mathit{Dwn}}$ on **Frm** has a left adjoint in **Frm**; it is $x \mapsto \twoheaddownarrow x$.
3. X carries a (necessarily unique) $\overline{\mathit{Dwn}}$-coalgebra structure $X \to \mathit{Dwn}(X)$, which is also $\twoheaddownarrow(-)$. □

One can show that coalgebra homomorphisms are the *proper* frame homomorphisms (from [2]) that preserve $\ll$. We recall from [11, VII, 4.5] that for a sober topological space X, its opens $\Omega(X)$ form a continuous lattice iff X is a locally compact space. Further, the stably continuous frames are precisely the retracts of frames of the form $\mathit{Dwn}(X)$, for X a meet semi-lattice—here via the coreflection $\twoheaddownarrow \dashv \bigvee$.

5 Comonoids from Bases

A recent insight, see [5], is that orthonormal bases in finite-dimensional Hilbert spaces can be described via so-called Frobenius algebras. In general, such an algebra consists of an object carrying both a monoid and a comonoid structure that interact appropriately. In the self-dual category of Hilbert spaces, it suffices to have either a monoid or a comonoid, since the dual is induced by the dagger / adjoint transpose $(-)^{\dagger}$. In this section we show that the kind of coalgebras (on algebras) considered in this paper also give rise to comonoids, assuming that the category of algebras has monoidal (tensor) structure.

In a (symmetric) monoidal category $\mathbf{A}$ a comonoid is the dual of a monoid, given by maps $I \stackrel{u}{\leftarrow} X \stackrel{d}{\rightarrow} X \otimes X$ satisfying the duals of the monoid equations. Such comonoids are used for copying and deletion, in linear and quantum logic. If $\otimes$ is cartesian product $\times$, each object carries a unique comonoid structure $1 \stackrel{!}{\leftarrow} X \stackrel{\Delta}{\rightarrow} X \times X$. The no-cloning theorem in quantum mechanics says that copying arbitrary states is impossible. But copying wrt a basis is allowed, see [5,17].

If a monad T on a symmetric monoidal category $\mathbf{A}$ is a commutative (aka. symmetric monoidal) monad, and the category $Alg(T)$ has enough coequalisers, then it is also symmetric monoidal, and the free functor $F\colon \mathbf{A} \to Alg(T)$ preserves this monoidal structure. This classical result goes back to [13,12]. We shall use it for the special case where the monoidal structure on the base category $\mathbf{A}$ is cartesian.

Proposition 2. *In the setting described above, assume the category of algebras $Alg(T)$ is symmetric monoidal, for a monad T on a cartesian category $\mathbf{A}$. Each $\overline{T}$-coalgebra / basis $b\colon X \to F(X)$, say on algebra $a\colon T(X) \to X$, gives rise to a commutative comonoid in $Alg(T)$ by:*

$$\begin{aligned} d_b &= \Big(X \xrightarrow{b} FX \xrightarrow{T(\Delta)} F(X \times X) \xrightarrow[\cong]{\xi^{-1}} FX \otimes FX \xrightarrow{a \otimes a} X \otimes X\Big) \\ u_b &= \Big(X \xrightarrow{b} F(X) \xrightarrow{T(!)} F(1) = I\Big), \end{aligned} \qquad (11)$$

where we use the underlying comonoid structure $1 \stackrel{!}{\leftarrow} X \stackrel{\Delta}{\rightarrow} X \times X$ on X in the underlying category $\mathbf{A}$.

Proof. It is not hard to see that these d_b and u_b are maps of algebras. For instance,

$$\mu_1 \circ T(u_b) = \mu_1 \circ T^2(!) \circ T(b) = T(!) \circ \mu_X \circ T(b) = T(!) \circ b \circ a = u_b \circ a.$$

The verification of the comonoid properties involves lengthy calculations, which are basically straightforward. We just show that u is neutral element for d, using the equations from Definition 1.

$$
\begin{aligned}
&(u_b \otimes \mathrm{id}) \circ d_d \\
&\quad = (T(!) \otimes \mathrm{id}) \circ (b \otimes \mathrm{id}) \circ (a \otimes a) \circ \xi^{-1} \circ T(\Delta) \circ b \\
&\quad = (T(!) \otimes \mathrm{id}) \circ (\mu \otimes \mathrm{id}) \circ (T(b) \otimes a) \circ \xi^{-1} \circ T(\Delta) \circ b \\
&\quad = (T(!) \otimes a) \circ (\mu \otimes T(a)) \circ (T(b) \otimes T(b)) \circ \xi^{-1} \circ T(\Delta) \circ b \\
&\quad = (T(!) \otimes a) \circ (\mu \otimes \mu) \circ \xi^{-1} \circ T(b \otimes b) \circ T(\Delta) \circ b \\
&\quad = (T(!) \otimes a) \circ (\mu \otimes \mu) \circ \xi^{-1} \circ T(\Delta) \circ T(b) \circ b \\
&\quad = (T(!) \otimes a) \circ (\mu \otimes \mu) \circ \xi^{-1} \circ T(\Delta) \circ T(\eta) \circ b \\
&\quad = (T(!) \otimes a) \circ (\mu \otimes \mu) \circ \xi^{-1} \circ T(\eta \times \eta) \circ T(\Delta) \circ b \\
&\quad = (T(!) \otimes a) \circ (\mu \otimes \mu) \circ (T(\eta) \times T(\eta)) \circ \xi^{-1} \circ T(\Delta) \circ b \\
&\quad = (T(!) \otimes a) \circ \xi^{-1} \circ T(\Delta) \circ b \\
&\quad = (\mathrm{id} \otimes a) \circ (T(!) \otimes \mathrm{id}) \circ \xi^{-1} \circ T(\Delta) \circ b \\
&\quad = (\mathrm{id} \otimes a) \circ \xi^{-1} \circ T(! \times \mathrm{id}) \circ T(\Delta) \circ b \\
&\quad = (\mathrm{id} \otimes a) \circ \xi^{-1} \circ T(\lambda^{-1}) \circ b \qquad \text{where } \lambda \colon 1 \times X \xrightarrow{\cong} X \\
&\quad = (\mathrm{id} \otimes a) \circ \lambda^{-1} \circ b \qquad \text{since } \xi \text{ is monoidal, where } \lambda \colon I \otimes X \xrightarrow{\cong} X \\
&\quad = \lambda^{-1} \circ a \circ b \\
&\quad = \lambda^{-1} \colon X \xrightarrow{\cong} I \otimes X.
\end{aligned}
$$

□

Example 1. To make the comonoid construction (11) more concrete, let V be a vector space, say over the complex numbers $\mathbb{C}$, with a basis, described as a coalgebra $b\colon V \to \mathcal{M}_{\mathbb{C}}(V)$ like in Theorem 1, with basic elements (e_j), satisfying $b(e_j) = 1e_j$. The counit $u_b = \mathcal{M}_{\mathbb{C}}(!) \circ b\colon V \to \mathbb{C}$ from (11) is:

$$v \longmapsto \textstyle\sum_j v_j e_j \longmapsto \sum_j v_j.$$

Similarly, the comultiplication $d_b\colon V \to V \otimes V$ as in (11) is the composite:

$$v \longmapsto \textstyle\sum_j v_j e_j \longmapsto \sum_j v_j(e_j \otimes e_j),$$

like in [5]. (For Hilbert spaces one uses orthonormal bases instead of Hamel bases; the counit u of the comonoid then exists only in the finite-dimensional case. The comultiplication d seems more relevant, see also below, and may thus also be studied on its own, like in [1], without finiteness restriction.)

In general, given a comonoid $I \stackrel{u}{\leftarrow} X \stackrel{d}{\rightarrow} X \otimes X$, an endomap $f\colon X \to X$ may be called *diagonalisable*—wrt. this comonoid, or actually, comultiplication d—if there is a "map of eigenvalues" $v\colon X \to I$ such that f equals the composite:

$$X \xrightarrow{\ d\ } X \otimes X \xrightarrow{v \otimes \mathrm{id}} I \otimes X \xrightarrow[\cong]{\ \lambda\ } X. \tag{12}$$

In the special case where the comonoid comes from a coalgebra (basis) $b\colon X \to T(X)$, like in (11), an endomap of algebras $f\colon X \to X$, say on $a\colon T(X) \to X$, is diagonalisable if there is a map of algebras $v\colon X \to I = T(1)$ such that f is:

$$X \xrightarrow{b} T(X) \xrightarrow{T(\langle v, \mathrm{id}\rangle)} T(T(1) \times X) \xrightarrow{T(\mathsf{st})} T^2(1 \times X) \xrightarrow{\cong} T^2(X) \xrightarrow{\mu} T(X) \xrightarrow{a} X,$$

where st is a strength map of the form $T(X) \times Y \to T(X \times Y)$, which exists because the monad T is assumed to be commutative.

We illustrate what this means for Pauli matrices.

Example 2. We consider the set $\mathbb{C}^2$ as vector space over $\mathbb{C}$, and thus as algebra of the (commutative) multiset monad $\mathcal{M}_{\mathbb{C}}\colon \mathbf{Sets} \to \mathbf{Sets}$ via the map $\mathcal{M}_{\mathbb{C}}(\mathbb{C}^2) \xrightarrow{a} \mathbb{C}^2$ that sends a formal sum $s_1(z_1, w_1) + \cdots + s_n(z_n, w_n)$ of pairs in $\mathbb{C}^2$ to the pair of sums $(s_1 \cdot z_1 + \cdots + s_n \cdot z_n, s_1 \cdot w_1 + \cdots + s_n \cdot w_n) \in \mathbb{C}^2$.

The familiar Pauli spin functions $\sigma_{\mathsf{x}}, \sigma_{\mathsf{y}}, \sigma_{\mathsf{z}}\colon \mathbb{C}^2 \to \mathbb{C}^2$ are given by:

$$\sigma_{\mathsf{x}}(z, w) = (w, z) \qquad \sigma_{\mathsf{y}}(z, w) = (-iw, iz) \qquad \sigma_{\mathsf{z}}(z, w) = (z, -w).$$

We concentrate on σ_{x}; it satisfies $\sigma_{\mathsf{x}}(1,1) = (1,1)$ and $\sigma_{\mathsf{x}}(1,-1) = -(1,-1)$. These eigenvectors $(1,1)$ and $(1,-1)$ are organised in a basis $b_{\mathsf{x}}\colon \mathbb{C}^2 \to \mathcal{M}_{\mathbb{C}}(\mathbb{C}^2)$, as in Definition 1, via the following formal sum.

$$b_{\mathsf{x}}(z, w) \;=\; \tfrac{z+w}{2}(1,1) + \tfrac{z-w}{2}(1,-1).$$

It expresses an arbitrary element of $\mathbb{C}^2$ in terms of this basis of eigenvectors. It is not hard to see that b_{x} is a $\overline{\mathcal{M}_{\mathbb{C}}}$-coalgebra; for instance:

$$\big(a \circ b_{\mathsf{x}}\big)(z, w) = a\big(\tfrac{z+w}{2}(1,1) + \tfrac{z-w}{2}(1,-1)\big) = \big(\tfrac{z+w}{2} + \tfrac{z-w}{2}, \tfrac{z+w}{2} - \tfrac{z-w}{2}\big) = (z, w).$$

The comonoid structure $\mathbb{C} \xleftarrow{u_{\mathsf{x}}} \mathbb{C}^2 \xrightarrow{d_{\mathsf{x}}} \mathbb{C}^2 \otimes \mathbb{C}^2$ induced by b_{x} as in (11) is given by $u_{\mathsf{x}}(z, w) = z$ and $d_{\mathsf{x}}(z, w) = \frac{z+w}{2}\big((1,1) \otimes (1,1)\big) + \frac{z-w}{2}\big((1,-1) \otimes (1,-1)\big)$. The eigenvalue map $v_{\mathsf{x}}\colon \mathbb{C}^2 \to \mathbb{C}$ is given by $v_{\mathsf{x}}(z, w) = w$. The eigenvalues $1, -1$ appear by application to the basic elements: $v_{\mathsf{x}}(1,1) = 1$ and $v_{\mathsf{x}}(1,-1) = -1$. Further, the Pauli function σ_{x} is diagonalised as in (12) via these $d_{\mathsf{x}}, v_{\mathsf{x}}$, since:

$$\begin{aligned}
&\big(\lambda \circ (v_{\mathsf{x}} \otimes \mathrm{id}) \circ d_{\mathsf{x}}\big)(z, w) \\
&= \big(\lambda \circ (v_{\mathsf{x}} \otimes \mathrm{id})\big)\Big(\tfrac{z+w}{2}\big((1,1) \otimes (1,1)\big) + \tfrac{z-w}{2}\big((1,-1) \otimes (1,-1)\big)\Big) \\
&= \lambda\Big(\tfrac{z+w}{2}\big(1 \otimes (1,1)\big) + \tfrac{z-w}{2}\big(-1 \otimes (1,-1)\big)\Big) \\
&= \tfrac{z+w}{2}(1,1) - \tfrac{z-w}{2}(1,-1) \\
&= (w, z) \;=\; \sigma_{\mathsf{x}}(z, w).
\end{aligned}$$

In a similar way one defines for the other Pauli functions σ_{y} and σ_{z}:

$$\begin{aligned}
b_{\mathsf{y}}(z, w) &= \tfrac{iz+w}{2}(-i, 1) + \tfrac{iz-w}{2}(i, 1) \qquad & v_{\mathsf{y}}(z, w) &= iz \\
b_{\mathsf{z}}(z, w) &= z(1,0) - w(0,1) \qquad & v_{\mathsf{z}}(z, w) &= z - w.
\end{aligned}$$

References

1. Abramsky, S., Heunen, C.: H*-algebras and nonunital Frobenius algebras: first steps in infinite-dimensional categorical quantum mechanics (2010), arxiv.org/abs/1011.6123
2. Banaschewski, B., Brümmer, G.: Stably continuous frames. Math. Proc. Cambridge Phil. Soc. 114, 7–19 (1988)
3. Barr, M.: Coalgebras in a category of algebras. In: Category Theory, Homology Theory and their Applications I. Lect. Notes Math., vol. 86, pp. 1–12. Springer, Berlin (1969)
4. Borceux, F.: Handbook of Categorical Algebra. Encyclopedia of Mathematics, vol. 50, 51, 52. Cambridge Univ. Press, Cambridge (1994)
5. Coecke, B., Pavlović, D., Vicary, J.: A new description of orthogonal bases. Math. Struct. in Comp. Sci. (to appear), arxiv.org/abs/0810.0812
6. Coumans, D., Jacobs, B.: Scalars, monads and categories (2010), http://arxiv.org/abs/1003.0585
7. Escardó, M.: Properly injective spaces and function spaces. Topology and its Applications 98(1-2), 75–120 (1999)
8. Gierz, G., Hofmann, K.H., Keimel, K., Lawson, J.D., Mislove, M., Scott, D.: Continuous Lattices and Domains. Encyclopedia of Mathematics, vol. 93. Cambridge Univ. Press, Cambridge (2003)
9. Hoffmann, R.-E.: Continuous posets and adjoint sequences. Semigroup Forum 18, 173–188 (1979)
10. Jacobs, B.: Coalgebras and approximation. In: Nerode, A., Matiyasevich, Y.V. (eds.) Logical Foundations of Computer Science. LNCS, vol. 813, pp. 173–183. Springer, Berlin (1994)
11. Johnstone, P.T. (ed.): Stone Spaces. Cambridge Studies in Advanced Mathematics, vol. 3. Cambridge Univ. Press, Cambridge (1982)
12. Kock, A.: Bilinearity and cartesian closed monads. Math. Scand. 29, 161–174 (1971)
13. Kock, A.: Closed categories generated by commutative monads. Journ. Austr. Math. Soc. XII, 405–424 (1971)
14. Kock, A.: Monads for which structures are adjoint to units. Journ. of Pure & Appl. Algebra 104, 41–59 (1995)
15. Mesablishvili, B.: Monads of effective descent type and comonadicity. Theory and Applications of Categories 16(1), 1–45 (2006)
16. Moggi, E.: Partial morphisms in categories of effective objects. Inf. & Comp. 76(2/3), 250–277 (1988)
17. Nielsen, M.A., Chuang, I.L.: Quantum Computation and Quantum Information. Cambridge Univ. Press, Cambridge (2000)
18. Rosebrugh, R., Wood, R.J.: Constructive complete distributivity II. Math. Proc. Cambridge Phil. Soc. 10, 245–249 (1991)

A Coalgebraic Approach to Supervisory Control of Partially Observed Mealy Automata

Jun Kohjina[1], Toshimitsu Ushio[1], and Yoshiki Kinoshita[2]

[1] Graduate School of Engineering Science, Osaka University, Japan
[2] National Institute of Advanced Industrial Science and Technology, Japan

Abstract. Supervisory control is a logical control method of discrete event systems introduced by Ramadge and Wonham. We propose a novel coalgebraic formulation of a supervisory control problem and design a controller called supervisor satisfying a given specification under partial observations. In this paper, plants, specifications, and supervisors are modeled by Mealy automata, automata, and Moore automata, respectively. We define a composition of a supervisor and a plant coinductively, which is called a supervisory composition, to represent a behavior of the controlled plant. We formulate a supervisory control problem using the supervisory composition. We define two relations: a partial bisimulation relation and a modified normal relation. We show that these relations are related to the controllability/observability and the modified normality which are the key notions in the supervisory control theory.

Keywords: discrete event systems, supervisory control, coalgebra.

1 Introduction

Supervisory control is a logical control method of discrete event systems introduced by Ramadge and Wonham [1,7]. It gives a method to restrict, or control, a plant so that it behaves exactly as specified. Restriction is done by disabling some of the events in the plant.

Ramadge and Wonham formulated the plant as an automaton M whose input alphabet is the set A of events of the plant. It is controlled by a Moore automaton S, called a supervisor, whose input alphabet is the same as the input alphabet of M, and whose output alphabet is a subset of the input plant alphabet of M. Output symbols of S disables previously determined transitions of M at the next step. In this way, the automaton M combined with S is defined. Ramadge and Wonham studied the *supervisory control problem*: given M and a *prefix closed* sublanguage L of the language accepted by M, find S such that the language generated by the combination of M with S exactly agrees with L. Moreover, the output symbol of S must be taken from the subset A_c of A specified in advance. The element of A_c and $A_u = A \setminus A_c$ are called *controllable* and *uncontrollable* events, respectively.

To solve the supervisory control problem, Ramadge and Wonham defined a notion of *controllable languages*, showed that the supervisory control problem

A. Corradini, B. Klin, and C. Cîrstea (Eds.): CALCO 2011, LNCS 6859, pp. 253–267, 2011.

has a solution if and only if L is controllable, and gave a construction of the supervisor [7].

Observability of events were introduced in [2,6]. The subset A_o of the *observable events* is considered. A supervisor is *under partial observations* if A_o is a proper subset of A. The notion of observability is extended to specification languages, and it was shown that there exists a supervisor satisfying the specification if and only if the specification language is controllable and observable.

The supervisory control problem for Mealy automata is recently studied by Ushio and Takai [10]. Here, the plant is formulated as a Mealy automaton with input alphabet A and output alphabet B, and the supervisor is a function from B^* to $\mathcal{P}(A_c)$. Ushio and Takai introduced the notion of P-observability of specification languages, and showed that there exists a supervisor satisfying the specification if and only if the specification is controllable and P-observable [10]. They also gave an explicit construction for the supervisor.

Rutten [8] gave a coalgebraic framework to the notions in supervisory control problem. In particular, he introduced the coalgebraic notion of controllable relations corresponding to the controllable language in non-coalgebraic setting, and proposed an algorithm to compute the supremal controllable sublanguage using partial bisimulations. Komenda and van Schuppen further extended Rutten's work to the case of partial observations [4] and modular control [5].

We focus on extending Ushio and Takai's work in the light of coalgebraic techniques developed by Rutten, Komenda and van Schuppen. So, we model a plant by a *partially observed Mealy automaton* whose input alphabet (event set) is partitioned into a controllable and an uncontrollable event set, and whose output alphabet (event set) is partitioned into an observable and an unobservable event set. A control specification is given by a nonempty prefix-closed language over the input event set of the plant. So, the control specification is represented by an automaton whose input event set is the same as the plant. A supervisor is a Moore automaton where its input event set is the observed output event set of the plant and its output event set is the power set of the controllable event set. Each element of an output event is a controllable event disabled by the supervisor so that some event sequences of the plant are prevented from occurring.

Mealy automata and Moore automata coexist in our framework principally by a historical reason. In this paper, we formulate a plant by a Mealy automaton, but a supervisor tended to be formulated as a Moore automaton [7]. We could have formulated supervisors as Mealy automata, but the Moore automaton treatment makes it easier to compare with the previous results. Incidentally, the specification language was often formulated as a prefix closed language, but the set of all prefix closed languages is the base set of a final $(1+-)^A$-coalgebra, and $(1+-)^A$-coalgebras are nothing but finite automata, a special case of Mealy automata. This is why we formulate the specification as a finite automaton, rather than its accepting language.

In order to formulate a supervisory control problem, we introduced a notion of *supervisory composition* for a plant and a supervisor by coinductive definition. It is a generalization of Rutten's supervised product [8]. We define two relations: a partial bisimulation and a modified normal relation. It is shown that the partial bisimulations characterize the controllability and observability. This result makes it easier to compare the current work and the earlier work can be seen easily, because controllability and observability are the two key notions in the supervisory control theory. We introduce *modified normality*, which is a coalgebraic notion of normality for Mealy automata.

The rest of this paper is organized as follows: Section 2 reviews mathematical notations and final coalgebra for three type functors. In Section 3, we formulate the supervisory control using coalgebra, and show a necessary and sufficient condition for the existence of a supervisor satisfying a specification. In Section 4, we introduce the modified normality, and propose an algorithm to compute the supremal modified normal and controllable sublanguage. Finally, Section 5 concludes the paper.

2 Preliminaries

In this section, we give mathematical notations and final coalgebras for three functors $(1 + B \times -)^A$, $B \times (-)^A$ and $(1 + -)^A$. After setting up the notations, we review the explicit construction of these final coalgebras.

2.1 Notations

Let X and Y be sets. A function from X to $1 + Y$ is called a partial function from X to Y, where $1 = \{\bot\}$ and $+$ is the disjoint union; if $f(x) = \bot$ holds, f is considered to be undefined at x. We write $f : X \rightharpoonup Y$ for a partial function f from X to Y. The composition $g \circ f : X \rightharpoonup Z$ of $f : X \rightharpoonup Y$ and $g : Y \rightharpoonup Z$ is defined by

$$g \circ f(x) = \begin{cases} g(f(x)) & \text{if } f(x) \neq \bot, \\ \bot & \text{otherwise.} \end{cases}$$

We also use $\circ$ for composition of total functions. For a partial function $f : X \rightharpoonup Y$, $\mathrm{dom}(f)$ is defined by

$$\mathrm{dom}(f) = \{x \in X \mid f(x) \neq \bot\}.$$

For a set A, A^* is the Kleene closure of A, that is, the set of all finite sequences over A, and let $\varepsilon \in A^*$ be the empty sequence. The monoid $(A^*, \varepsilon, \cdot)$ is the free monoid over A, where $u \cdot v$ denotes the concatenation of u and v. We write $u \preceq w$, if u is a prefix of w, that is, there exists $v \in A^*$ such that $u \cdot v = w$. The length of a sequence w is denoted by $|w|$.

A language $L \subseteq A^*$ is said to be prefix-closed if the following condition holds:

$$\forall u, w \in A^* : (u \preceq w) \wedge (w \in L) \Rightarrow u \in L.$$

2.2 Mealy Automata as $(1 + B \times -)^A$-coalgebras

A (partial) Mealy automaton is a $(1+B \times -)^A$-coalgebra (see [9], for instance). By construction of the functor $(1 + B \times -)^A$, we know there exists a final $(1 + B \times -)^A$-coalgebra, and there is an explicit construction for it [3]. We shall use the construction later, so we give its overview in this section.

Let $(X, \xi : X \to (1 + B \times X)^A)$ be a Mealy automaton, where X is a set of states, A is a set of input events, B is a set of output events, and ξ is a transition function.

For the simplicity, we introduce the following notations:

$$x \xrightarrow{a|-} \text{ iff } \xi(x)(a) \neq \bot,$$

$$x \xrightarrow{a|b} x' \text{ iff } \xi(x)(a) = \langle b, x' \rangle \, .$$

We extend these notations to strings in A^* and B^* by induction.

We now give a final $(1 + B \times -)^A$-coalgebra $(\mathcal{M}, m)$. Its state set

$$\mathcal{M} = \left\{ M : A^* \rightharpoonup B^* \;\middle|\; \begin{array}{r} M \text{ is length-preserving and prefix-preserving,} \\ \mathrm{dom}(M) \neq \emptyset. \end{array} \right\}$$

consists of prefix-preserving and length-preserving functions. The structure map $m : \mathcal{M} \to (1 + B \times \mathcal{M})^A$ is defined by is defined as follows:

$$m(M)(a) = \begin{cases} \langle M(a), M_a \rangle & \text{if } a \in \mathrm{dom}(M), \\ \bot & \text{otherwise,} \end{cases}$$

where a-derivative $M_a \in \mathcal{M}$ of $M \in \mathcal{M}$ is defined by

$$M_a(w) = \mathsf{tail} \circ M(a \cdot w)$$

and $\mathsf{tail} : B^* \rightharpoonup B^*$ is defined by

$$\mathsf{tail}(\varepsilon) = \bot, \;\; \mathsf{tail}(b \cdot w) = w \text{ if } b \in B \text{ and } w \in B^*.$$

To complete the definition of $\mathcal{M}$ and m, we further define *length-preserving* and *prefix-preserving.*

- A partial function $f : A^* \rightharpoonup B^*$ is said to be prefix-preserving if $\mathrm{dom}(f)$ is prefix-closed and, for all $u, w \in \mathrm{dom}(f)$, the following condition holds:

$$u \preceq w \Rightarrow f(u) \preceq f(w).$$

- A partial function $f : A^* \rightharpoonup B^*$ is said to be length-preserving if the following condition holds:

$$\forall w \in \mathrm{dom}(f), |w| = |f(w)|.$$

Proposition 1 ([3]). *$(\mathcal{M}, m)$ is the final $(1 + B \times -)^A$-coalgebraD*

2.3 Moore Automata as $B \times (-)^A$-coalgebras

A Moore automaton is a $B \times (-)^A$-coalgebra [8]. A final $B \times (-)^A$-coalgebra $(\mathcal{S}, \langle o, t \rangle)$ constructed as follows. The state set $\mathcal{S}$ is defined by

$$\mathcal{S} = (B^*)^{A^*} = \{S : A^* \to B^*\}$$

The structure map $\langle o, t \rangle : \mathcal{S} \to B \times \mathcal{S}^A$ is defined by

$$o(S) = S(\varepsilon), \; t(S)(a) = S_a,$$

where the a-derivative $S_a \in \mathcal{S}$ of $S \in \mathcal{S}$ is defined by

$$S_a(w) = S(a \cdot w).$$

The functions $o : \mathcal{S} \to B$ and $t : \mathcal{S} \to \mathcal{S}^A$ are an output function and a transition function, respectively.

Proposition 2 ([8]). *$(\mathcal{S}, \langle o, t \rangle)$ is the final $B \times (-)^A$-coalgebra.*

2.4 Finite Automata as $(1 + -)^A$-coalgebras

A finite automaton is a $(1 + -)^A$-coalgebra. This is a special case($B = 1$) of $(1 + B \times -)^A$-coalgebra. A final $(1 + -)^A$-coalgebra $(\mathcal{L}, l)$ is constructed as follows. The state set $\mathcal{L}$ is defined by

$$\mathcal{L} = \{L \subseteq A^* \mid L \text{ is prefix-closed}, L \neq \emptyset\}.$$

The structure map $l : \mathcal{L} \to (1 + \mathcal{L})^A$ is defined by

$$l(L)(a) = \begin{cases} L_a & \text{if } a \in L, \\ \perp & \text{otherwise}, \end{cases}$$

where L_a is an a-derivative of the language L defined by

$$L_a = \{w \in A^* \mid a \cdot w \in L\}.$$

We extend the definition of the derivative to strings in A^*.

Proposition 3. *$(\mathcal{L}, l)$ is the final $(1 + -)^A$-coalgebraD*

For the simplicity, we introduce the following notations:

$$x \xrightarrow{a} x' \text{ iff } \xi(x)(a) = x',$$
$$x \not\xrightarrow{a} \text{ iff } \xi(x)(a) = \perp,$$
$$x \to \text{ iff } \exists a \in A, \xi(x)(a) \neq \perp.$$

We extend these notations to strings in A^* by induction. For all $M \in \mathcal{M}$, $\text{dom}(M)$ is a prefix-closed language. The following condition holds:

$$\text{dom}(M) \xrightarrow{a} \text{dom}(M') \text{ iff } M \xrightarrow{a|-} M'.$$

Definition 1 (simulation relation). *Let* $(X, \xi : X \to (1+X)^A)$ *and* $(Y, \eta : Y \to (1+Y)^A)$ *be* $(1+-)^A$*-coalgebras. A binary relation* $R \subseteq X \times Y$ *is said to be a simulation relation from* X *to* Y *if, for all* $x \in X$ *and* $y \in Y$*, the following condition holds:*

$$\forall a \in A : x \; R \; y \wedge x \xrightarrow{a} x' \Rightarrow y \xrightarrow{a} y' \wedge x' \; R \; y'$$

The simulation relation closed under the union. Thus, there exists the greatest simulation relation denoted by $\sqsubseteq$. $\sqsubseteq$ is the union of all simulation relations from X to Y. The simulation relation is symmetric and transitive, that is, it is a preorder. We shall later use the following lemma.

Lemma 1. *For all* $K, L \in \mathcal{L}$,

$$K \sqsubseteq L \Rightarrow K \subseteq L.$$

Lemma 2. *Let* (X, ξ) *and* (Y, η) *be* $(1+-)^A$*-coalgebras and* $R \subseteq X \times Y$ *be a binary relation. Then, the following conditions are equivalent.*

1. *R is a $(1+-)^A$-bisimulation relation.*
2. *For all $a \in A$, $x \in X$, and $y \in Y$, the following two conditions hold:*
   (a) $\forall x' \in X : x \; R \; y \wedge x \xrightarrow{a} x' \Rightarrow y \xrightarrow{a} y' \wedge x' \; R \; y'$,
   (b) $\forall y' \in Y : x \; R \; y \wedge y \xrightarrow{a} y' \Rightarrow x \xrightarrow{a} x' \wedge x' \; R \; y'$.

3 Supervisory Control of Mealy Automata

In this section, we give a solution to the supervisory control problem in the setting of Mealy automata.

3.1 Coalgebraic Formulation of Supervisory Control

Let A be a set and A_c be its subset; our intension is that A is the set of *input events* which occurs in the plant. We partition A as $A = A_c + A_u$, where A_c is intended to be the set of *controllable events*, and A_u be that of *uncontrollable events* [7]. Let B be a set of *output events* and B_o be its subset of *observable output events.*

Let M be a Mealy automaton whose set of output symbols is B, and S be a Moore automaton whose set of input symbols is B_o whose set of output symbols is the powerset of A_c; so, S is a $(\mathcal{P}(A_c) \times (-)^{B_o})$-coalgebra. The Mealy automaton M models the *plant* where input events in A occur sequentially, M changes its state according to those input events, causing output events in B. Aside from M, there is a *supervisor* S which takes as input only *observable* output events caused by A, and reacts by designating a set of input events, for which are forced to be disabled in M. The designated set of disabled input events remain unchanged until S designates another set by transition.

The Mealy automaton M induces a partial map from A^* to B^*. Its behavior is restricts by specification of a subset of A^*. The subset is specified not arbitrary subset of A^*, but by means of prefix closed language.

We consider an automaton $(\mathcal{S} \times \mathcal{M}, \mathsf{spv})$ which represents so-called a closed loop system. Its set of inputs is A. So, the automaton is a $(1+-)^A$-coalgebra. Its set of states is $\mathcal{S} \times \mathcal{M}$, where $\mathcal{M}$ is the set of prefix- and length-preserving partial functions from A^* to B^*, and $\mathcal{S}$ is the state set of the final $(\mathcal{P}(A_c) \times (-)^{B_o})$-coalgebra $(\mathcal{S}, \langle o, t\rangle)$. So, the state set $\mathcal{S}$ is defined by

$$\mathcal{S} = (\mathcal{P}(A_c))^{(B_o)^*} = \{S : (B_o)^* \to \mathcal{P}(A_c)\}.$$

The structure map $\mathsf{spv} : \mathcal{S} \times \mathcal{M} \to (1 + \mathcal{S} \times \mathcal{M})^A$ is defined as follows:

$$\mathsf{spv}\,\langle S, M\rangle\,(a) = \begin{cases} \langle S_{(P \circ M)(a)}, M_a\rangle & \text{if } M \xrightarrow{a|-} M_a \wedge a \notin o(S), \\ \bot & \text{otherwise,} \end{cases}$$

where the projection $P : B^* \to (B_o)^*$ is inductively defined by

$$P(\varepsilon) = \varepsilon, \;\; P(w \cdot b) = \begin{cases} P(w) \cdot b & \text{if } b \in B_o, \\ P(w) & \text{otherwise.} \end{cases}$$

Since $(\mathcal{L}, l)$ is a final $(1+-)^A$-coalgebra, there exists a unique $(1+-)^A$-coalgebra homomorphism

$$/ : (\mathcal{S} \times \mathcal{M}, \mathsf{spv}) \to (\mathcal{L}, l).$$

In other words, there exists a unique function $/ : \mathcal{S} \times \mathcal{M} \to \mathcal{L}$ which makes the following diagram commute:

$$\begin{array}{ccc} \mathcal{S} \times \mathcal{M} & \overset{/}{\dashrightarrow} & \mathcal{L} \\ {\scriptstyle \mathsf{spv}}\downarrow & & \downarrow {\scriptstyle l}\ \text{final} \\ (1 + \mathcal{S} \times \mathcal{M})^A & \underset{(\mathrm{id}_1 + /)^A}{\dashrightarrow} & (1 + \mathcal{L})^A \end{array}$$

If S is a Moore automaton in $\mathcal{S}$ and M is a Mealy automaton in $\mathcal{M}$, we write S/M for $/\,\langle S, M\rangle$, the image of $\langle S, M\rangle$ under $/$. Any event a in $o(S)$ is considered to be disabled by the supervisor S. If $M(a)$ is an observable event, the supervisor S transits to the next state $S_{M(a)}$. Otherwise it stays at the same state.

The supervisory composition is an extension of the supervised product defined by Rutten [8], which is a composition of a plant and a supervisor which generates specifications. In the case of $B = A$, $M = \mathrm{id}\,|_L$ and $S(w) = A_c \setminus \{a \in A_c \mid \exists u \in A^* : (K \xrightarrow{u \cdot a}) \wedge (P(u) = w)\}$, S/M corresponds to the supervised product $K/^O_U L$ introduced by Komenda and van Schuppen [4]. We formulate the supervisory control problem as follows. Given

- a set $A = A_c + A_u$ of input events (and its partition to A_c and A_u),
- a set B of output events and its subset B_o of observable output events,
- a Mealy automaton M whose set of inputs is A and whose set of outputs is B and
- a prefix closed language L, called the *specification* of the problem,

construct a Moore automaton S whose set of inputs is B_o and whose set of outputs is $\mathcal{P}(A_c)$, so that $S/M = L$.

3.2 Solution to the Supervisory Control Problem for Mealy Automata

Let (X, ξ) be a $(1 + -)^A$-coalgebra and $x_0 \in X$ be an initial state. The binary relation $\approx_{M_0}^{x_0}$ on X is defined by

$$\approx_{M_0}^{x_0} = \left\{ \langle x, x' \rangle \;\middle|\; \exists w, w' \in A^*, x_0 \xrightarrow{w} x, x_0 \xrightarrow{w'} x', P \circ M_0(w) = P \circ M_0(w') \right\}.$$

Intuition behind this definition may be described as follows: $x \approx_{M_0}^{x_0} x'$ if and only if there are execution paths p from x_0 to x and p' from x_0 to x' in M_0 which traces the same observable output events. The binary relation $\approx_{M_0}^{x_0}$ is symmetric. However, $\approx_{M_0}^{x_0}$ does not satisfy neither reflexivity nor transitivity in general. Reflexivity holds if all states are reachable from x_0.

Definition 2 (**(M_0, x_0)-partial bisimulation relation**). *Let (X, ξ) and (Y, η) be $(1+-)^A$-coalgebras. A binary relation $R \subseteq X \times Y$ is said to to be an (M_0, x_0)-partial bisimulation relation for $x_0 \in X$ and $M_0 \in \mathcal{M}$, if, for all $x \in X$ and $y \in Y$, the following conditions hold:*

1. $\forall a \in A, \forall x' \in X, \exists y' \in Y : x \, R \, y \wedge x \xrightarrow{a} x' \Rightarrow y \xrightarrow{a} y' \wedge x' \, R \, y'$;
2. $\forall a \in A_u, \forall y' \in Y, \exists x' \in X : x \, R \, y \wedge y \xrightarrow{a} y' \Rightarrow x \xrightarrow{a} x' \wedge x' \, R \, y'$;
3. $\forall a \in A_c, \forall y' \in Y, \exists x' \in X$:
   $x \, R \, y \wedge y \xrightarrow{a} y' \wedge (\exists q \in X, \ (x \approx_{M_0}^{x_0} q) \wedge (q \xrightarrow{a})) \Rightarrow x \xrightarrow{a} x' \wedge x' \, R \, y'$.

The condition 3 is equivalent to the following condition.
3'. $\forall a \in A_c, \forall y' \in Y, \exists x' \in X$:
$x \, R \, y \wedge y \xrightarrow{a} y' \Rightarrow (x \xrightarrow{a} x' \wedge x' \, R \, y') \vee (\forall q \in X, \ x \approx_{M_0}^{x_0} q \Rightarrow q \not\xrightarrow{a})$.
The partial bisimulation defined in Definition 2 is an extension of the partial bisimulation defined in [4].

Theorem 1 below gives a solution to the supervisory control problem. It turns out that the solution does not always exist, but we give a necessary and sufficient condition for the existence with the construction of the supervisor when it exists.

Theorem 1. *Given a specification $K_0 \in \mathcal{L}$ for a plant $M_0 \in \mathcal{M}$, the following two conditions are equivalent:*

1. $\exists S \in \mathcal{S}, \ S/M_0 = K_0$.
2. *There exists an (M_0, x_0)-partial bisimulation relation $R \subseteq \mathcal{L} \times \mathcal{L}$ such that $K_0 \, R \, \mathrm{dom}(M_0)$.*

Proof. ($1 \Rightarrow 2$) Let $S_0 \in \mathcal{S}$ be a supervisor such that $S_0/M_0 = K_0$. We consider a binary relation

$$R = \{ \langle K, \mathrm{dom}(M) \rangle \mid K \in \mathcal{L}, M \in \mathcal{M}, \exists S \in \mathcal{S}, S/M = K \}.$$

Since $\mathrm{dom}(M)$ is prefix-closed and nonempty, $\mathrm{dom}(M)$ is an element of $\mathcal{L}$. It suffices to show that R is an (M_0, x_0)-partial bisimulation relation.

To prove it, we consider $K \in \mathcal{L}$ and $M \in \mathcal{M}$ such that $K \, R \, \mathrm{dom}(M)$. Then, there exists a supervisor $S \in \mathcal{S}$ such that $S/M = K$. We check the three conditions for (M_0, x_0)-partial bisimulations below.

1. Consider $a \in A$ such that $K \xrightarrow{a} K_a$. $S/M = K$ implies $S/M \xrightarrow{a} S_{P \circ M(a)}/M_a$ and $S_{P \circ M(a)}/M_a = K_a$. Thus, $K_a \; R \; \mathrm{dom}(M_a)$ holds. By the definition of $/$, $M \xrightarrow{a|-} M_a$, that is, $\mathrm{dom}(M) \xrightarrow{a} \mathrm{dom}(M_a)$ holds.
2. Consider $a \in A_u$ such that $\mathrm{dom}(M) \xrightarrow{a} \mathrm{dom}(M_a)$. $\mathrm{dom}(M) \xrightarrow{a} \mathrm{dom}(M_a)$ implies $M \xrightarrow{a|-} M_a$. Since it is an element of A_u, a is not an element of $o(S)$. By the definition of $/$, $S/M \xrightarrow{a} S_{P \circ M(a)}/M_a$ holds. $S/M = K$ implies $K \xrightarrow{a} K_a$ and $K_a = S_{P \circ M(a)}/M_a$, so $K_a \; R \; \mathrm{dom}(M_a)$ holds.
3. Consider $a \in A_c$ such that $\mathrm{dom}(M) \xrightarrow{a} \mathrm{dom}(M_a)$ and $(K \approx_{M_0}^{K_0} L) \wedge (L \xrightarrow{a})$ and $L \in \mathcal{L}$. $S/M = K$ implies $((S/M) \approx_{M_0}^{K_0} L) \wedge (L \xrightarrow{a})$. Because $(S/M) \approx_{M_0}^{K_0} L$, there exist two sequences $w, w' \in A^*$ such that $K_0 \xrightarrow{w} (S/M)$, $K_0 \xrightarrow{w'} L$, and $P \circ M_0(w) = P \circ M_0(w')$. Since $S_0/M_0 = K_0$, we have

$$S/M = (S_0)_{P \circ M_0(w)}/(M_0)_w \text{ and } L = (S_0)_{P \circ M_0(w')}/(M_0)_{w'}.$$

The assumption $L \xrightarrow{a}$ implies $(S_0)_{P \circ M_0(w')}/(M_0)_{w'} \xrightarrow{a}$. Thus, we have

$$a \notin o\left((S_0)_{P \circ M_0(w')}\right) = o\left((S_0)_{P \circ M_0(w)}\right) = o(S).$$

The assumption $\mathrm{dom}(M) \xrightarrow{a} \mathrm{dom}(M_a)$ implies $M \xrightarrow{a|-} M_a$. By the definition of $/$, we have $S/M \xrightarrow{a} S_{P \circ M(a)}/M_a$. $S/M = K$ implies $K \xrightarrow{a} K_a$. Since $K_a = S_{P \circ M(a)}/M_a$, $K_a \; R \; \mathrm{dom}(M_a)$ holds.

($2 \Rightarrow 1$) Let $\simeq$ be an (M_0, K_0)-partial bisimulation relation such that $K_0 \simeq \mathrm{dom}(M_0)$. For $w \in B_o^*$, we construct the supervisor $\hat{S}$ as follows.

$$\hat{S}(w) = A_c \setminus \{a \in A_c \mid \exists u \in A^* : (K_0 \xrightarrow{u \cdot a}) \wedge (P \circ M_0(u) = w)\}.$$

We show $\hat{S}/M_0 = K_0$ by coinduction. Consider a binary relation $R \subseteq \mathcal{L} \times \mathcal{L}$

$$R = \left\{ \langle S/M, K \rangle \;\middle|\; \exists w \in A^* : \hat{S}/M_0 \xrightarrow{w} S/M, K_0 \xrightarrow{w} K, K \simeq \mathrm{dom}(M) \right\}.$$

Since $(\hat{S}/M_0) \; R \; K_0$ holds, it is sufficient to prove that R is a bisimulation relation. We assume that $(S/M) \; R \; K$ for S/M and $K \in \mathcal{L}$. Then, there exists a string $w \in A^*$ such that $K_0 \xrightarrow{w} K$ and $\hat{S}/M_0 \xrightarrow{w} S/M$.

- Take $a \in A$ such that $S/M \xrightarrow{a} S_{P \circ M(a)}/M_a$. By definition of $/$, $M \xrightarrow{a|-} M_a$, which means $\mathrm{dom}(M) \xrightarrow{a} \mathrm{dom}(M_a)$.
  - If $a \in A_u$, $K \sim \mathrm{dom}(M)$ implies $(K \xrightarrow{a} K_a) \wedge (K_a \simeq \mathrm{dom}(M_a))$. We have $\hat{S}/M_0 \xrightarrow{w \cdot a} S_{P \circ M(a)}/M_a$ and $K_0 \xrightarrow{w \cdot a} K_a$. Thus, $(S_{P \circ M(a)}/M_a) \; R \; K_a$ holds.
  - If $a \in A_c$, $(K \simeq \mathrm{dom}(M)) \wedge ((S/M) \approx_{M_0}^{K_0} K)$, therefore $(K \xrightarrow{a} K') \wedge (K' \simeq \mathrm{dom}(M'))$. We have $\hat{S}/M_0 \xrightarrow{w \cdot a} S_{P \circ M(a)}/M_a$ and $K_0 \xrightarrow{w \cdot a} K_a$. Thus, $(S_{P \circ M(a)}/M_a) \; R \; K_a$ holds.

– Let $a \in A$ and $K \xrightarrow{a} K_a$. $K \simeq \mathrm{dom}(M)$ implies $(\mathrm{dom}(M) \xrightarrow{a} \mathrm{dom}(M_a)) \wedge (K_a \simeq \mathrm{dom}(M_a))$. Then, $M \xrightarrow{a|-} M'$ holds. $(K_0 \xrightarrow{w} K) \wedge (K \xrightarrow{a} K_a)$ implies $K_0 \xrightarrow{w \cdot a} K_a$. By the definition of $\hat{S}$, we have

$$a \notin \hat{S}\big(P \circ M_0(w)\big) = \hat{S}_{P \circ M_0(w)}(\varepsilon) = o(S).$$

By the definition of $/$, we have $S/M \xrightarrow{a} S_{P \circ M(a)}/M_a$, which implies $\hat{S}/M_0 \xrightarrow{w \cdot a} S_{P \circ M(a)}/M_a$. Thus, $(S_{P \circ M(a)}/M_a)\ R\ K_a$ holds.

By Lemma 2, R is a bisimulation relation. □

4 Modified Normal Relations

It turned out that, for a given specification, there does not always exist a supervisor. So, we deal with the existence problem of the largest sublanguage of the specification with a supervisor which satisfies it. In this section, we introduce the notion of modified normal relations and show that they are closed under arbitrary union. So, there is always the largest modified normal relation. Moreover, there is a supervisor which satisfies it, so it is the solution of our problem.

In this section, we fix a Mealy automaton M and notations such as $x \xrightarrow{a}$ are used for M.

4.1 Modified Normality

Definition 3 (controllable and modified normal relation). *Let (X, ξ) and (Y, η) be $(1 + -)^A$-coalgebras. A binary relation $R \subseteq X \times Y$ is said to be a* controllable and modified normal relation *with respect to $x_0 \in X$ and M if the following conditions hold:*

1. *y simulates x, namely, $\forall a \in A$, $\forall x, x' \in X$, $\forall y \in Y$, $x\ R\ y \wedge x \xrightarrow{a} x' \Rightarrow \exists y' \in Y, y \xrightarrow{a} y' \wedge x'\ R\ y'$;*
2. *x simulates y on A_u, namely, $\forall a \in A_u$, $\forall x, \in X$, $\forall y, y' \in Y$, $x\ R\ y \wedge y \xrightarrow{a} y' \Rightarrow \exists x' \in X, x \xrightarrow{a} x' \wedge x'\ R\ y'$;*
3. *x simulates y on A_c under some condition, namely, $\forall a \in A_c$, $\forall x, \in X$, $\forall y, y' \in Y$, $x\ R\ y \wedge y \xrightarrow{a} y' \wedge (\exists q \in X, (q \rightarrow) \wedge x \approx_M^{x_0} q) \Rightarrow x \xrightarrow{a} x' \wedge x'\ R\ y'$.*

It can be shown that controllable and modified normal relations between fixed sets are closed under union. So, there is the maximum such; we write $\bowtie_{X,Y}$ for it. We often omit the subscripts X and Y when they are obvious from the context.

Theorem 2. *Let (X, ξ) and (Y, η) be $(1 + -)^A$-coalgebras. A controllable and modified normal relation from X to Y is an (M, x_0)-partial bisimulation relation.*

Proof. Obvious from the definition of controllable and modified normal relations. □

This theorem indicates that if there exists a controllable and modified normal relation then there exists a supervisor satisfying the specification.

Definition 4 (modified normal relation). *Let (X, ξ) and (Y, η) be $(1+-)^A$-coalgebras. A binary relation $R \subseteq X \times Y$ is a* modified normal relation *with respect to $x_0 \in X$ and M if the following conditions hold:*

1. *y simulates x; namely, $\forall a \in A, \forall x, x' \in X, \forall y \in Y, x \, R \, y \wedge x \xrightarrow{a} x' \Rightarrow \exists y \in Y, y \xrightarrow{a} y' \wedge x' \, R \, y'$;*
2. *x simulates y under some condition; namely, $\forall a \in A, \forall x \in X, \forall y, y' \in Y$, $x \, R \, y \wedge y \xrightarrow{a} y' \wedge (\exists q \in X, (q \rightarrow) \wedge x \approx_M^{x_0} q) \Rightarrow \exists x \in X, x \xrightarrow{a} x' \wedge x' \, R \, y'$.*

Proposition 4. *The following conditions are equivalent.*

1. *A binary relation R is a controllable and modified normal relation.*
2. *A binary relation R satisfies following two conditions: R is a control relation defined in [8], and R is a modified normal relation.*

Definition 5 (modified normal language). *A language K with the same alphabet as L is said to be* modified normal *with respect to L and $\tilde{P} \circ M$ if the following condition holds:*

$$\forall w \in L, \ \forall w' \in K, \ \tilde{P} \circ M(w) = \tilde{P} \circ M(w') \Rightarrow w \in K,$$

where $\tilde{P} = P \circ \mathsf{init}$ and $\mathsf{init}(\varepsilon) = \bot$, $\mathsf{init}(w \cdot b) = w$ if $w \in B^*$ and $b \in B$.

Languages which are normal and observable play a key role in supervisory control under partial observations in [1,2,6]. It is noted that a modified normal language is a generalization of a normal language, which is defined by

$$\forall w \in K, \forall w' \in L, P \circ M(w) = P \circ M(w') \Rightarrow w' \in K,$$

and is not always observable, that is,

$$\forall w, w' \in K, \forall a \in A_c, wa \in K, w'a \in L, P \circ M(w) = P \circ M(w') \Rightarrow w'a \in K.$$

So, we introduce the modified normal languages. Since modified normal languages are closed under union, there always exists the supremal, or maximal, modified normal sublanguage of a given language K.

Theorem 3. *The following two conditions for a sublanguage K_0 of L_0 are equivalent:*

1. *K_0 is a modified normal language with respect to L_0 and $\tilde{P} \circ M$.*
2. *There exists a modified normal relation $R \subseteq \mathcal{L} \times \mathcal{L}$ such that $K_0 \, R \, L_0$.*

Proof. (1) $\Rightarrow$ (2) Let $R = \{\langle (K_0)_w, (L_0)_w \rangle \mid w \in K_0\}$. Since $K_0 \, R \, L_0$ holds, it suffices to show that R is a modified normal relation.

1. Take $a \in A$ such that $K \, R \, L \wedge K \xrightarrow{a} K_a$. $K \, R \, L$ implies that there exists a string $w \in K_0$ such that $K = (K_0)_w$ and $L = (L_0)_w$. $K \xrightarrow{a} K_a$ implies $w \cdot a \in K_0$. Since $K_0 \subseteq L_0$, we have $w \cdot a \in L_0$, which implies $L \xrightarrow{a} L_a$D Recall $K_a = (K_0)_{w \cdot a}$ and $L_a = (L_0)_{w \cdot a}$. Thus, $K_a \, R \, L_a$ holds.

2. Take $a \in A$ such that $K\ R\ L \wedge L \xrightarrow{a} L_a \wedge (\exists K' \in \mathcal{L}, \exists a' \in A, (K' \xrightarrow{a'}) \wedge K \approx_M^{K_0} K')$. $K\ R\ L$ implies that there exists a string $w \in K_0$ such that $K = (K_0)_w$ and $L = (L_0)_w$. $L \xrightarrow{a} L_a$ implies $w \cdot a \in L_0$. Since there exists a language $K' \in \mathcal{L}$ such that $(K' \rightarrow) \wedge K \approx_M^{K_0} K'$, there exists a string $w' \in K_0$ such that $K_0 \xrightarrow{w'} K' \wedge P \circ M(w) = P \circ M(w')$. Since there exists $a' \in A$ such that $K_0 \xrightarrow{w' \cdot a'}$, we have $w' \cdot a' \in K_0$. Then, $\tilde{P} \circ M(w \cdot a) = \tilde{P} \circ M(w' \cdot a')$. Since K_0 is modified normal, we have $w \cdot a \in K_0$, that is, $K_a\ R\ L_a$.

(2) $\Rightarrow$ (1)
Let R be a modified normal relation such that $K_0\ R\ L_0$. Then,

$$\forall w \in L_0 : \exists w' \in K_0, \tilde{P} \circ M(w) = \tilde{P} \circ M(w') \Rightarrow w \in K_0. \qquad (1)$$

We prove (1) by induction on construction of w. If $w = \varepsilon$, then, there exists no string w'eith $w' \neq \varepsilon$ such that $\tilde{P} \circ M(w) = \tilde{P} \circ M(w')$. Therefore, (1) holds for εD We consider $w \in L_0$ such that (1) holds. We consider any $w \cdot a \in L_0$, $u \in K_0$ and $\tilde{P} \circ M(w \cdot a) = \tilde{P} \circ M(u)$. Since $\tilde{P} \circ M(w \cdot a) = \tilde{P} \circ M(u)$, we have

$$\exists w' \preceq u, \tilde{P} \circ M(w) = \tilde{P} \circ M(w').$$

By the induction hypothesis, we have $w \in K_0$. Let $K = (K_0)_w$ and $L = (L_0)_w$. Since $K_0\ R\ L_0$ holds, we have $K\ R\ L$ by the definition of the modified normal relation. We can set $u = u' \cdot a'$ for some $u' \in K_0$ and $a' \in A$ since $w \cdot a \neq \varepsilon$. Let $K' = (K_0)_{u'}$. Since $K \approx_M^{K_0} K' \wedge K' \xrightarrow{a'}$ holds, we have $K \xrightarrow{a} K_a$ by the definition of the modified normal relation. Therefore, $w \cdot a \in K_0$ holds. □

Moreover, Rutten showed following theorem in [8].

Theorem 4. *The following two conditions for a sublanguage K_0 of L_0 are equivalent:*

1. *K_0 is a controllable language with respect to L_0, that is, $K_0 A_u \cap L_0 \subseteq K_0$, where $K_0 A_u$ is a concatenation of two languages K_0 and L_0.*
2. *There exists a control relation $R \subseteq \mathcal{L} \times \mathcal{L}$ such that $K_0\ R\ L_0$.*

Corollary 1. *The following two conditions are equivalent:*

1. *K_0 is a controllable and modified normal language.*
2. *There exists a controllable and modified normal relation $R \subseteq \mathcal{L} \times \mathcal{L}$ such that $K_0\ R\ L_0$.*

4.2 Supremal Controllable and Modified Normal Sublanguage

Let (X, ξ) and (Y, η) be $(1 + -)^A$-coalgebras, and $x_0 \in X$ and $y_0 \in Y$ be initial states. Let R be a binary relation between X and Y. We define an operator $\Phi_R : \mathcal{P}(R) \rightarrow \mathcal{P}(R)$ as follows:

$$\Phi_R(H) = \left\{ \langle x, y \rangle \in H \;\middle|\; \begin{array}{l} \forall a \in A_u : y \xrightarrow{a} y' \Rightarrow x \xrightarrow{a} x' \wedge x'\ H\ y' \text{ and} \\ \forall a \in A_c : \begin{array}{r} y \xrightarrow{a} y' \wedge (\exists q \in X, (q \rightarrow) \wedge x \approx_M^{x_0} q) \\ \Rightarrow x \xrightarrow{a} x' \wedge x'\ H\ y'. \end{array} \end{array} \right\}.$$

Proposition 5. *Φ_R is monotone, i.e., $R_1 \subseteq R_2$ implies $\Phi_R(R_1) \subseteq \Phi_R(R_2)$.*

Proof. Let $\langle x, y\rangle$ be an element of $\Phi_R(R_1)$. By the definition of Φ_R, the element $\langle x, y\rangle$ satisfies following conditions:

- $\langle x, y\rangle \in R_1 \subseteq R_2$;
- $\forall a \in A_u : y \xrightarrow{a} y' \Rightarrow x \xrightarrow{a} x' \wedge \langle x', y'\rangle \in R_1 \subseteq R_2$;
- $\forall a \in A_c : y \xrightarrow{a} y' \wedge (\exists q \in X, (q \rightarrow) \wedge x \approx_M^{x_0} q) \Rightarrow x \xrightarrow{a} x' \wedge \langle x', y'\rangle \in R_1 \subseteq R_2$.

Therefore, $\langle x', y'\rangle$ is an element of $\Phi_R(R_2)$. □

Knaster-Tarski fixpoint theorem for the complete lattice $\mathcal{P}(R)$ states that Φ_R has its largest fixpoint. We denote it by $\tilde{R}$.

Consider a binary relation

$$R_0 = \{\langle x, y\rangle \mid \exists w \in A^* : x_0 \xrightarrow{w} x \text{ and } y_0 \xrightarrow{w} y\}$$

and construct the greatest fixpoint of Φ_{R_0} denoted by $\tilde{R}_0$. We define the $(1+-)^A$-coalgebra $(\tilde{R}_0, \alpha)$ as follows:

$$\alpha\,\langle x, y\rangle\,(a) = \begin{cases} \langle x', y'\rangle & \text{if } x \xrightarrow{a} x', y \xrightarrow{a} y' \text{ and } x' \tilde{R}_0\, y', \\ \bot & \text{otherwise.} \end{cases}$$

Theorem 5. *The following statement holds:*

1. *For any state v in a $(1+-)^A$-coalgebra (V, α_V): if $v \sqsubseteq x$ and $v \bowtie y$, then $x\, \tilde{R}_0\, y$ and $v \sqsubseteq \langle x, y\rangle$.*
2. *If $x\, \tilde{R}_0\, y$, then $\langle x, y\rangle \sqsubseteq x$ and $\langle x, y\rangle \bowtie y$.*

Proof. 1. We consider a state v in a $(1+-)^A$-coalgebra (V, α_V) with $v \sqsubseteq x$ and $v \bowtie y$. Let $H \subseteq V \times Y$ be a simulation relation such that $v\, H\, x$. Let $Q \subseteq V \times T$ be a controllable and modified normal relation such that $v\, Q\, y$. Define

$$P = \{\langle p, q\rangle \in R \mid \exists z \in V, z\, H\, p \text{ and } z\, Q\, q\}.$$

First, we show $x\, \tilde{R}_0\, y$. Since $x\, P\, y$ and $P \subseteq R$, it is sufficient to prove $P \subseteq \Phi_{R_0}(P)$. For $\langle p, q\rangle \in P$, there exists $z \in V$ such that $z\, H\, p$ and $z\, Q\, q$. We consider any $a \in A_u$ and $q \xrightarrow{a} q'$. Since Q is a controllable relation, we have $z \xrightarrow{a} z'$ and $z'\, Q\, q'$. Since H is a simulation relation, we have $p \xrightarrow{a} p'$ and $z'\, H\, p'$. Thus, we have $\langle p', q'\rangle \in P$. We consider any $a \in A_c$ and $q' \in X$ such that $q \xrightarrow{a} q'$ and $(\exists u \in V, (u \rightarrow) \wedge x \approx_M^{v_0} u)$. Since Q is a modified normal relation, we have $z \xrightarrow{a} z'$ and $z'\, Q\, p'$. We have $\langle p', q'\rangle \in P$ in a similar way. Therefore, we have $\langle p, q\rangle \in \Phi_{R_0}(P)$. Next, we show $v \sqsubseteq \langle x, y\rangle$. Define

$$P' = \{\langle z, \langle p, q\rangle\rangle \mid z \in V, p\, \tilde{R}_0\, q, z\, H\, p \text{ and } z\, Q\, q\}.$$

We show that P' is a simulation relation. We consider the case that $z \xrightarrow{a} z'$. Since H is a simulation relation and Q is a modified normal relation, we have $p \xrightarrow{a} p'$, $z'\, H\, p'$, $q \xrightarrow{a} q'$ and $z'\, Q\, q'$. Since $\tilde{R}_0 = \Phi_{R_0}(\tilde{R}_0)$, we have $p'\, \tilde{R}_0\, q'$. Thus, we have $\langle p, q\rangle \xrightarrow{a} \langle p', q'\rangle$ and $z'\, P'\, \langle p', q'\rangle$.

2. We assume that $x \tilde{R_0} y$. First we show that $\langle x, y\rangle \sqsubseteq x$. By the definition of $(\tilde{R_0}, \alpha)$,

$$\{\langle\langle p, q\rangle, p\rangle \mid p \tilde{R_0} q\}$$

is a simulation relation obviously. Next we show that $\langle x, y\rangle \bowtie y$. Define

$$H = \{\langle\langle p, q\rangle, q\rangle \mid p \tilde{R_0} q\}.$$

We show that H is a controllable and modified normal relation between $\tilde{R_0}$ and Y. For a pair $\langle p, q\rangle H q$,

(a) If $\langle p, q\rangle \xrightarrow{a} \langle p', q'\rangle$ for $a \in A$, then $q \xrightarrow{a} q'$ and $\langle p', q'\rangle H q'$ by the definition of $(\tilde{R_0}, \alpha)$.

(b) If $q \xrightarrow{a} q'$ for $a \in A_u$, then $p \xrightarrow{a} p'$ and $p' \tilde{R_0} q'$, since $\langle p, q\rangle \in \tilde{R_0} = \Phi_{R_0}(\tilde{R_0})$. Thus, we have $\langle p, q\rangle \xrightarrow{a} \langle p', q'\rangle$ and $\langle p', q'\rangle H q'$.

(c) We assume $q \xrightarrow{a} q'$ and $(\exists \langle u, v\rangle \in \tilde{R_0}, (\langle u, v\rangle \rightarrow) \wedge \langle p, q\rangle \approx_M^{\langle x_0, y_0\rangle} \langle u, v\rangle)$ for some $a \in A_c$. $\langle u, v\rangle \rightarrow$ and $\langle p, q\rangle \approx_M^{\langle x_0, y_0\rangle} \langle u, v\rangle$ implies $u \rightarrow$ and $p \approx_M^{x_0} u$. Since $\langle p, q\rangle \in \tilde{R_0} = \Phi_{R_0}(\tilde{R_0})$, we have $p \xrightarrow{a} p'$ and $p' R q'$. Thus, we have $\langle p, q\rangle \xrightarrow{a} \langle p', q'\rangle$ and $\langle p', q'\rangle H q'$. □

Corollary 2. *Let K and L be two prefix closed languages. We assume that there exists a prefix closed language K' such that $K' \sqsubseteq K$ and $K' \bowtie L$. The greatest fixpoint $\tilde{R_0}$ satisfies that $K \tilde{R_0} L$ and* beh $\langle K, L\rangle$ *is the supremal controllable and modified normal sublanguage, where* beh : $(\tilde{R_0}, \alpha) \rightarrow (\mathcal{L}, l)$ *is a unique homomorphism.*

Proof. We take $(\mathcal{L}, l, K)$ as (X, ξ, x_0), $(\mathcal{L}, l, L)$ as (Y, η, y_0) and $(\mathcal{L}, l, K')$ as (V, α, v). By the first part of Theorem 5, if $K' \subseteq K$ and $K' \bowtie L$, then $K \tilde{R_0} L$ and $K' \subseteq$ beh $\langle K, L\rangle$. From the assumption, there exists a prefix closed language K' such that $K' \sqsubseteq K$ and $K' \bowtie L$. Therefore, $\langle K, L\rangle$ is an element of $\tilde{R_0}$. By the second part of Theorem 5, beh $\langle K, L\rangle \subseteq K$ and beh $\langle K, L\rangle \bowtie L$. □

Corollary 2 shows that the language corresponding to the largest fixpoint $\tilde{R_0}$ is the supremal controllable and modified normal language if there exists a nonempty controllable and modified normal language.

5 Conclusion

This paper studied a supervisory control problem of Mealy automata under partial observations. We formulated the supervisory control problem in a coalgebraic framework, and show a necessary and a sufficient condition for the existence of a supervisor satisfying a specification. We introduced the notion of modified normality and propose an algorithm to compute the supremal modified normal and controllable sublanguage. It is future work to investigate the relationship between the (M_0, x_0)-partial bisimulation relation and the controllable and modified normal relation. It is also future work to extend the coalgebraic formulation of a supervisor to modular and decentralized control.

Acknowledgments. This work was supported in part by a Grant-in-Aid for Scientific Research (C), No. 21560464, from the Ministry of Education, Culture, Sports Science, and Technology of Japan. The third author acknowledges the partial support by DEOS (Dependable Embedded Operating Systems) programme conducted in CREST scheme of JST. The authors thank anonymous reviewers for their fruitful comments to improve the quality of the paper.

References

1. Cassandras, C., Lafortune, S.: Introduction to Discrete Event Systems. Springer, Heidelberg (2008)
2. Cieslak, R., Desclaux, C., Fawaz, A., Varaiya, P.: Supervisory control of discrete-event processes with partial observations. IEEE Transactions on Automatic Control 33(3), 249–260 (1988)
3. Hansen, H.: Coalgebraic Modelling: Applications in Automata theory and Modal logic. Ph.D. thesis (2009)
4. Komenda, J., van Schuppen, J.: Control of discrete-event systems with partial observations using coalgebra and coinduction. Discrete Event Dynamic Systems 15(3), 257–315 (2005)
5. Komenda, J., van Schuppen, J.: Modular control of discrete-event systems with coalgebra. IEEE Transactions on Automatic Control 53(2), 447–460 (2008)
6. Lin, F., Wonham, W.: On observability of discrete-event systems. Information Sciences 44(3), 173–198 (1988)
7. Ramadge, P., Wonham, W.: Supervisory control of a class of discrete event processes. SIAM J. Control & Optimiz. 25(1), 206–230 (1987)
8. Rutten, J.: Coalgebra, concurrency and control. CWI. Software Engineering (SEN) (R 9921), 1–31 (1999)
9. Rutten, J.J.: Universal coalgebra: a theory of systems. Theoretical Computer Science 249(1), 3–80 (2000)
10. Ushio, T., Takai, S.: Supervisory control of discrete event systems modeled by Mealy automata with nondeterministic output functions. In: American Control Conference, pp. 4260–4265 (2009)

Coalgebraic Semantics for Derivations in Logic Programming

Ekaterina Komendantskaya[1,*] and John Power[2,**]

[1] Department of Computing, University of Dundee, UK
[2] Department of Computer Science, University of Bath, UK

Abstract. Every variable-free logic program induces a $P_f P_f$-coalgebra on the set of atomic formulae in the program. The coalgebra p sends an atomic formula A to the set of the sets of atomic formulae in the antecedent of each clause for which A is the head. In an earlier paper, we identified a variable-free logic program with a $P_f P_f$-coalgebra on Set and showed that, if $C(P_f P_f)$ is the cofree comonad on $P_f P_f$, then given a logic program P qua $P_f P_f$-coalgebra, the corresponding $C(P_f P_f)$-coalgebra structure describes the parallel and-or derivation trees of P. In this paper, we extend that analysis to arbitrary logic programs. That requires a subtle analysis of lax natural transformations between *Poset*-valued functors on a Lawvere theory, of locally ordered endofunctors and comonads on locally ordered categories, and of coalgebras, oplax maps of coalgebras, and the relationships between such for locally ordered endofunctors and the cofree comonads on them.

Keywords: Logic programming, SLD-resolution, Coalgebra, Lawvere theories, Lax natural transformations, Oplax maps of coalgebras.

1 Introduction

In the standard formulations of logic programming, such as in Lloyd's book [18], a first-order logic program P consists of a finite set of clauses of the form

$$A \leftarrow A_1, \ldots, A_n$$

where A and the A_i's are atomic formulae, typically containing free variables, and where $A_1, \ldots, A_n$ is understood to mean the conjunction of the A_i's: note that n may be 0.

SLD-resolution, which is a central algorithm for logic programming, takes a goal G, typically written as

$$\leftarrow B_1, \ldots, B_n$$

* The work was supported by the Engineering and Physical Sciences Research Council, UK; Postdoctoral Fellow research grant EP/F044046/2.

** John Power would like to acknowledge the support of SICSA for an extended visit to Dundee funding this research.

A. Corradini, B. Klin, and C. Cîrstea (Eds.): CALCO 2011, LNCS 6859, pp. 268–282, 2011.

where the list of B_i's is again understood to mean a conjunction of atomic formulae, typically containing free variables, and constructs a proof for an instantiation of G from substitution instances of the clauses in P [18]. The algorithm uses Horn-clause logic, with variable substitution determined universally to make the first atom in G agree with the head of a clause in P, then proceeding inductively.

A running example of a logic program in this paper is as follows.

Example 1. Let ListNat denote the logic program

```
          nat(0) ←
       nat(s(x)) ← nat(x)
       list(nil) ←
list(cons x y) ← nat(x), list(y)
```

The program involves variables `x` and `y`, function symbols `0`, `s`, `nil` and `cons`, and predicate symbols `nat` and `list`, with the choice of notation designed to make the intended meaning of the program clear.

Logic programs resemble, and indeed induce, transition systems or rewrite systems, hence coalgebras. That fact has been used to study their operational semantics, e.g., [4,6]. In [15], we developed the idea for variable-free logic programs. Given a set of atoms At, and a variable-free logic program P built over At, one can construct a $P_f P_f$-coalgebra structure on At, where P_f is the finite powerset functor: each atom is the head of finitely many clauses in P, and the body of each of those clauses contains finitely many atoms. Our main result was that if $C(P_f P_f)$ is the cofree comonad on $P_f P_f$, then, given a logic program P qua $P_f P_f$-coalgebra, the corresponding $C(P_f P_f)$-coalgebra structure characterises the parallel and-or derivation trees of P: see Section 2 for a definition and for more detail.

Modulo a concern about recursion, which can be addressed by extending from finiteness to countability, one can construct a variable-free logic program from an arbitrary logic program by taking all ground instances of all clauses in the original logic program. The resulting variable-free logic program is of equivalent power to the original one, but one has factored out all the analysis of substitution that appears in SLD-resolution. So, in order to model the substitution in the SLD-resolution algorithm, in this paper, we extend our coalgebraic analysis of logic programming from variable-free logic programs to arbitrary logic programs. In particular, we study the relationship between coalgebras for an extension of $P_f P_f$ and the coalgebras for the comonad induced by it.

There have been several category theoretic models of logic programs and computations, and several of them have involved the characterisation of the first-order language underlying a logic program as a *Lawvere theory*, e.g., [2,4,5,14], and that of most general unifiers (mgu's) as *equalisers*, e.g., [3] or as *pullbacks*, e.g., [5,2]. We duly adopt those ideas here, see Section 3.

Given a signature Σ of function symbols, let $\mathcal{L}_\Sigma$ denote the Lawvere theory generated by Σ. Given a logic program P with function symbols in Σ, we would like to consider the functor category $[\mathcal{L}_\Sigma^{op}, Set]$, extending the set At of atoms in a variable-free logic program to the functor from $\mathcal{L}_\Sigma^{op}$ to Set sending a natural number n to the set $At(n)$ of atomic formulae with at most n variables generated by the predicate symbols in P. One can extend any endofunctor H on Set to the endofunctor $[\mathcal{L}_\Sigma^{op}, H]$ on $[\mathcal{L}_\Sigma^{op}, Set]$ that sends $F : \mathcal{L}_\Sigma^{op} \to Set$ to the composite HF. So we would then like to model P by the putative $[\mathcal{L}_\Sigma^{op}, P_f P_f]$-coalgebra $p : At \longrightarrow P_f P_f At$ that, at n, takes an atomic formula $A(x_1, \ldots, x_n)$ with at most n variables, considers all substitutions of clauses in P whose head agrees with $A(x_1, \ldots, x_n)$, and gives the set of sets of atomic formulae in antecedents, mimicking the construction for variable-free logic programs. Unfortunately, that does not work.

Consider the logic program ListNat of Example 1. There is a map in $\mathcal{L}_\Sigma$ of the form $0 \to 1$ that models the nullary function symbol 0. So, naturality of the map $p : At \longrightarrow P_f P_f At$ in $[\mathcal{L}_\Sigma^{op}, Set]$ would yield commutativity of the diagram

$$\begin{array}{ccc} At(1) & \longrightarrow & P_f P_f At(1) \\ \big\downarrow & & \big\downarrow \\ At(0) & \longrightarrow & P_f P_f At(0) \end{array}$$

There being no clause of the form `nat(x)` $\leftarrow$ in ListNat, commutativity of the diagram would in turn imply that there cannot be a clause in ListNat of the form `nat(0)` $\leftarrow$ either, but in fact there is one!

In order to model examples such as ListNat, we need to relax the naturality condition on p: if naturality could be relaxed to a subset condition, so that, in general,

$$\begin{array}{ccc} At(m) & \longrightarrow & P_f P_f At(m) \\ \big\downarrow & \geq & \big\downarrow \\ At(n) & \longrightarrow & P_f P_f At(n) \end{array}$$

need not commute, but rather the composite via $P_f P_f At(m)$ need only yield a subset of that via $At(n)$, it would be possible for $p_1(\mathtt{nat(x)})$ to be the empty set while $p_0(\mathtt{nat(0)})$ is non-empty in the ListNat example above.

In order to express such a lax naturality condition, we need to extend Set to $Poset$ and we need to extend P_f from Set to $Poset$. The category $Lax(\mathcal{L}_\Sigma^{op}, Poset)$ of strict functors and lax natural transformations is not complete, so the usual construction of a cofree comonad on an endofunctor no longer works directly. On the other hand, $Poset$ is finitely cocomplete as a locally ordered category, so we can adopt the subtle work of [13] on categories of lax natural transformations, which is what we do: see Section 4.

A mild problem arises in regard to the finiteness of the outer occurrence of P_f in $P_f P_f$. The problem is that substitution can generate infinitely many instances of clauses with the same head. For instance, suppose one extends ListNat with a clause of the form $A \leftarrow$ `nat(x)` with no occurrences of x in A. Substitution yields the clause $A \leftarrow$ `nat(s`$^{\mathtt{n}}$`(0))`, for every natural number n, giving rise to a countable set of clauses with head A. We need to allow for possibilities such as this as the infiniteness arises even from a finite signature. So we extend from $P_f P_f$ to $P_c P_f$, where P_c extends the countable powerset functor.

Those are the key technical difficulties that we address in the paper. Note that, in contrast to [15], we do not model the ordering of subgoals and repetitions. These have been modelled in relevant literature, notably in Corradini and Montanari's landmark papers [7,8], but we defer making precise the relationship with the ideas herein.

We end the paper by making a natural construction of a locally ordered endofunctor to extend $P_c P_f$ in Section 5, checking how coalgebra models our leading example, and comparing the trees we obtain with parallel and-or derivation trees.

2 Parallel and-or Derivation Trees and Coalgebra

In this section, we briefly recall from [15] the definition of the parallel and-or derivation trees generated by an arbitrary logic program, and how, in the case of variable-free logic programs, they can be seen in terms of coalgebraic structure.

Key motivating texts for the definition of parallel and-or derivation tree are [9] and [12], as explained in [15]. We freely use the usual logic programming conventions for substitution and most general unifiers, see Section 3.

Definition 1. *Let P be a logic program and let $\leftarrow A$ be an atomic goal (possibly with variables). The* parallel and-or derivation tree *for A is the possibly infinite tree T satisfying the following properties.*

- *A is the root of T.*
- *Each node in T is either an and-node or an or-node.*
- *Each or-node is given by $\bullet$.*
- *Each and-node is an atom.*
- *For every node A' occurring in T, if A' is unifiable with only one clause $B \leftarrow B_1, \ldots, B_n$ in P with mgu θ, then A' has n children given by and-nodes $B_1\theta, \ldots B_n\theta$.*
- *For every node A' occurring in T, if A' is unifiable with exactly $m > 1$ distinct clauses $C_1, \ldots, C_m$ in P via mgu's $\theta_1, \ldots, \theta_m$, then A' has exactly m children given by or-nodes, such that, for every $i \in m$, if $C_i = B^i \leftarrow B^i_1, \ldots, B^i_{n_i}$, then the ith or-node has n_i children given by and-nodes $B^i_1\theta_i, \ldots, B^i_{n_i}\theta_i$.*

We now recall the coalgebraic development of [15].

Proposition 1. *For any set At, there is a bijection between the set of variable-free logic programs over the set of atoms At and the set of $P_f P_f$-coalgebra structures on At, where P_f is the finite powerset functor on Set.*

Proposition 2. *Let $C(P_f P_f)$ denote the cofree comonad on $P_f P_f$. Then, for $p : At \longrightarrow P_f P_f(At)$, the corresponding $C(P_f P_f)$-coalgebra is given as follows: $C(P_f P_f)(At)$ is a limit of a diagram of the form*

$$\ldots \longrightarrow At \times P_f P_f(At \times P_f P_f(At)) \longrightarrow At \times P_f P_f(At) \longrightarrow At.$$

Put $At_0 = At$ and $At_{n+1} = At \times P_f P_f At_n$, and define the cone

$$p_0 = id : At \longrightarrow At (= At_0)$$
$$p_{n+1} = \langle id, P_f P_f(p_n) \circ p \rangle : At \longrightarrow At \times P_f P_f At_n (= At_{n+1})$$

Then the limiting property determines the coalgebra $\overline{p} : At \longrightarrow C(P_f P_f)(At)$.

In [15], we gave a general account of the relationship between a variable-free logic program qua $P_f P_f$-coalgebra and the parallel and-or derivation trees it generates. Here we recall a representative example.

Example 2. Consider the variable-free logic program:

```
q(b,a) ←
s(a,b) ←
  p(a) ← q(b,a), s(a,b)
q(b,a) ← s(a,b)
```

The program has three atoms, namely `q(b,a)`, `s(a,b)` and `p(a)`. So At = {`q(b,a)`, `s(a,b)`, `p(a)`}. The program can be identified with the $P_f P_f$-coalgebra structure on At given by
p(`q(b,a)`) = {{}, {`s(a,b)`}}, where {} is the empty set.
p(`s(a,b)`) = {{}}, i.e., the one element set consisting of the empty set.
p(`p(a)`) = {{`q(b,a)`,`s(a,b)`}}.

Consider the $C(P_f P_f)$-coalgebra corresponding to p. It sends `p(a)` to the parallel refutation of `p(a)` depicted on the left side of Figure 1. Note that the nodes of the tree alternate between those labelled by atoms and those labelled by bullets ($\bullet$). The set of children of each bullet represents a goal, made up of the conjunction of the atoms in the labels. An atom with multiple children is the head of multiple clauses in the program: its children represent these clauses. We use the traditional notation $\Box$ to denote {}.

Where an atom has a single $\bullet$-child, we can elide that node without losing any information; the result of applying this transformation to our example is shown on the right in Figure 1. The resulting tree is precisely the parallel and-or derivation tree for the atomic goal ← `p(a)` as in Definition 1. So the two trees express equivalent information.

In the first-order case, direct use of Definition 2 yields inconsistent derivations, as explained e.g. in [12]. So *composition (and-or parallel) trees* were introduced [12].

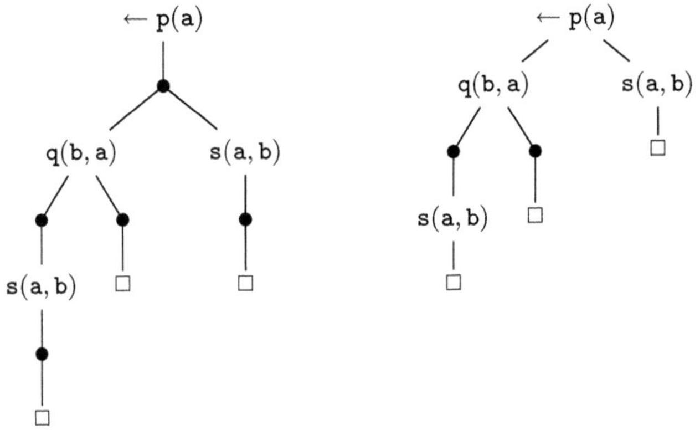

Fig. 1. The action of $\overline{p} : \mathrm{At} \longrightarrow C(P_f P_f)(\mathrm{At})$ on p(a), and the corresponding parallel and-or derivation tree

Construction of composition trees involves additional algorithms that synchronise branches created by or-nodes. Composition trees contain a special kind of *composition node* used whenever both and- and or-parallel computations are possible for one goal. Every composition node is a list of atoms in the goal. If, in a goal $G = \leftarrow B_1, \ldots B_n$, an atom B_i is unifiable with $k > 1$ clauses, then the algorithm adds k children (k composition nodes) to the node G; similarly for every atom in G that is unifiable with more than one clause. Every such composition node has the form $B_1, \ldots B_n$, and n and-parallel edges. Thus, all possible combinations of all possible or-choices at every and-parallel step are given. In this paper, we do not study composition trees directly but rather suggest an alternative.

3 Using Lawvere Theories to Model First-Order Signatures and Substitution

In this section, we start to move towards using coalgebra to model arbitrary logic programs by recalling the relationship between first-order signatures and Lawvere theories, in particular how the former give rise to the latter. Then we recall how to use that to model most general unifiers as equalisers.

A *signature* Σ consists of a set of *function symbols* $f, g, \ldots$ each equipped with a fixed *arity* given by a natural number indicating the number of arguments it is supposed to have. Nullary (0-ary) function symbols are allowed and are called *constants*. Given a countably infinite set *Var* of variables, the set $Ter(\Sigma)$ of *terms* over Σ is defined inductively:

- $x \in Ter(\Sigma)$ for every $x \in Var$.
- If f is an n-ary function symbol ($n \geq 0$) and $t_1, \ldots, t_n \in Ter(\Sigma)$, then $f(t_1, \ldots, t_n) \in Ter(\Sigma)$.

Definition 2. *Given a signature Σ and a category C with strictly associative finite products, an* interpretation *of Σ in C is an object X of C, together with, for each function symbol f of arity n, a map in C from X^n to X.*

Proposition 3. *Given a signature Σ, there exists a category $\mathcal{L}_\Sigma$ with strictly associative finite products and an interpretation $\| \ \|_\Sigma$ of Σ in $\mathcal{L}_\Sigma$, such that for any category C with strictly associative finite products, and interpretation γ of Σ in C, there exists a unique functor $g : \mathcal{L}_\Sigma \to C$ that strictly preserves finite products, such that g composed with $\| \ \|_\Sigma$ gives γ, as in the following diagram:*

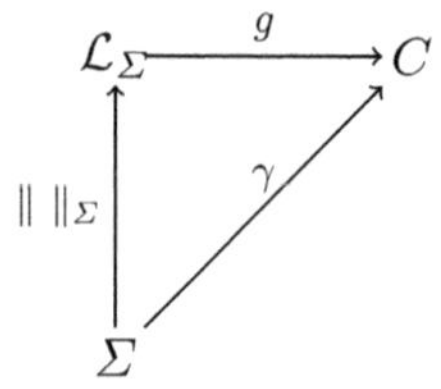

Proof. Define the set $\mathtt{ob}(\mathcal{L}_\Sigma)$ to be the set of natural numbers.

For each natural number n, let $x_1, \ldots, x_n$ be a specified list of distinct variables. Define $\mathtt{ob}(\mathcal{L}_\Sigma)(n, m)$ to be the set of m-tuples $(t_1, \ldots, t_m)$ of terms generated by the function symbols in Σ and variables $x_1, \ldots, x_n$. Define composition in $\mathcal{L}_\Sigma$ by substitution. The interpretation $\| \ \|_\Sigma$ sends an n-ary function symbol f to $f(x_1, \ldots, x_n)$.

One can readily check that these constructions satisfy the axioms for a category and for an interpretation, with $\mathcal{L}_\Sigma$ having strictly associative finite products given by the sum of natural numbers. The terminal object of $\mathcal{L}_\Sigma$ is the natural number 0. The universal property follows directly from the construction.

Definition 3. *Given a signature Σ, the category $\mathcal{L}_\Sigma$ determined by Proposition 3 is called the* Lawvere theory *generated by Σ [17].*

One can describe $\mathcal{L}_\Sigma$ without the need for a specified list of variables for each n: in a term t, a variable context is always implicit, i.e., $x_1, \ldots, x_m \vdash t$, and the variable context is considered as a binder.

In contrast to the usual practice in category theory, sorting is not modelled by using a sorted finite product theory but rather by modelling predicates for sorts such as `nat` or `list` using the structure of the category $[\mathcal{L}_\Sigma, Set]$ or, more subtly, of $Lax(\mathcal{L}_\Sigma, Poset)$, as illustrated below: Lloyd's book [18] gives a representataive account of logic programming, and although category theorists may disapprove, it is not sorted.

Example 3. Consider `ListNat`. It is naturally two-sorted, with one sort for natural numbers and one for lists. Traditionally, category theory would not use Proposition 3 but rather a two-sorted version of it: see [16]. But `ListNat` is a legitimate untyped logic program and is representative of such.

The constants `0` and `nil` are modelled by maps from 0 to 1 in $\mathcal{L}_\Sigma$, `s` is modelled by a map from 1 to 1, and `cons` is modelled by a map from 2 to 1. The term `s(0)` is therefore modelled by the map from 0 to 1 given by the composite of the maps modelling `s` and `0`; similarly for the term `s(nil)`, although the latter does not make semantic sense.

A key construct in standard accounts of SLD-resolution such as [18] is that of a most general unifier, which we now recall. It is typically expressed using distinctive notation for substitution. Note that the coalgebraic approach does not require us to model substitution by most general unifiers; it does not even require us to take syntax over Lawvere theories, as we may take it over more general categories: note the generality of Section 4 and see [14].

Definition 4. *A* substitution *is a function θ from Var to $Ter(\Sigma)$ that is the identity on all but finitely many variables. Each substitution canonically generates a function from $Ter(\Sigma)$ to itself defined inductively by the following:*

$$\theta(f(t_1,\ldots,t_n)) \equiv f(\theta(t_1),\ldots,\theta(t_n))$$

Following the usual convention in logic programming, we denote $\theta(t)$ by $t\theta$[18].

Definition 5. *Let S be a finite set of terms. A substitution θ is called a* unifier *for S if, for any pair of terms t_1 and t_2 in S, applying the substitution θ yields $t_1\theta = t_2\theta$. A unifier θ for S is called a* most general unifier *(mgu) for S if, for each unifier σ of S, there exists a substitution γ such that $\sigma = \theta\gamma$.*

The structure of $\mathcal{L}_\Sigma$ allows us to characterise most general unifiers in terms of equalisers as follows, cf [21], where they are modelled by coequalisers in the Kleisli category for a the monad T_Σ on Set induced by $\mathcal{L}_\Sigma$.

Proposition 4. *Given a signature Σ, for any pair of terms (`s`,`t`) with variables among $x_1,\ldots,x_n$, a most general unifier of `s` and `t` exists if and only if an equaliser of `s` and `t` qua maps in $\mathcal{L}_\Sigma$ exists, in which case the most general unifier is given by the equaliser.*

Example 4. A most general unifier of the terms `cons(x,nil)` and `cons(s(0),y)` of Example 3 exists and is given by the substitution σ : {`s(0)/x`, `nil/y`}.

4 Coalgebra on Categories of Lax Maps

Assume we have a signature Σ of function symbols and, for each natural number n, a specified list of variables $x_1,\ldots,x_n$. Then, given an arbitrary logic program with signature Σ, we can extend our study of the set At of atoms for a variable-free logic program in [15] by considering the functor $At : \mathcal{L}_\Sigma^{op} \to Set$ that sends a natural number n to the set of all atomic formulae with variables among $x_1,\ldots,x_n$ generated by the function symbols in Σ and the predicate symbols appearing in the logic program. A map $f : n \to m$ in $\mathcal{L}_\Sigma$ is sent to the function $At(f) : At(m) \to At(n)$ that sends an atomic formula $A(x_1,\ldots,x_m)$ to $A(f_1(x_1,\ldots,x_n)/x_1,\ldots,f_m(x_1,\ldots,x_n)/x_m)$, i.e., $At(f)$ is defined by substitution.

As explained in the Introduction, we cannot model a logic program by a natural transformation of the form $p : At \longrightarrow P_f P_f At$ as naturality breaks down even in simple examples such as ListNat. We need lax naturality. In order even to define it, we first need to extend $At : \mathcal{L}_\Sigma^{op} \to Set$ to have codomain $Poset$. That is routine, given by composing At with the inclusion of Set into $Poset$. Mildly overloading notation, we denote the composite by $At : \mathcal{L}_\Sigma^{op} \to Poset$, noting that it is trivially locally ordered.

Definition 6. *Given locally ordered functors $H, K : D \longrightarrow C$, a* lax natural transformation *from H to K is the assignment to each object d of D, of a map $\alpha_d : Hd \longrightarrow Kd$ such that for each map $f : d \longrightarrow d'$ in D, one has $(Kf)(\alpha_d) \leq (\alpha_{d'})(Hf)$.*

Locally ordered functors and lax natural transformations, with pointwise composition and pointwise ordering, form a locally ordered category we denote by $Lax(D, C)$.

As explained in the Introduction, we need to extend the endofunctor $P_c P_f$ on Set rather than extending $P_f P_f$ as, even with finitely many function symbols, substitution could give rise to countably many clauses with the same head. So we need to extend $P_c P_f$ from an endofunctor on Set to a locally ordered endofunctor on $Lax(\mathcal{L}_\Sigma^{op}, Poset)$. A natural way to do that, while retaining the role of $P_c P_f$, is first to extend $P_c P_f$ to a locally ordered endofunctor E on $Poset$, then to consider the locally ordered endofunctor $Lax(\mathcal{L}_\Sigma^{op}, E)$ on $Lax(\mathcal{L}_\Sigma^{op}, Poset)$ that sends $H : \mathcal{L}_\Sigma^{op} \to Poset$ to the composite EH.

We shall return to the question of extending to $P_c P_f$ to $Poset$, but what about the cofree comonad $C(P_c P_f)$ on $P_c P_f$?

The locally ordered category $Lax(\mathcal{L}_\Sigma^{op}, Poset)$ is neither complete nor cocomplete, so it does not follow from the usual general theory that a cofree comonad on a locally ordered endofunctor on it need exist at all, let alone be given by a limiting construct resembling that of Proposition 2. Moreover, the laxness in $Lax(\mathcal{L}_\Sigma^{op}, Poset)$ makes the category of coalgebras for an endofunctor on it problematic, as the strictness in the definition of map of coalgebras does not cohere well with the laxness in the definition of map in $Lax(\mathcal{L}_\Sigma^{op}, Poset)$.

Using techniques developed by Kelly in Section 3.3 of [13], we can negotiate these obstacles. Rather than directly considering a cofree comonad on $Lax(\mathcal{L}_\Sigma^{op}, E)$, we can extend the comonad $C(P_c P_f)$ from Set to $Lax(\mathcal{L}_\Sigma^{op}, Poset)$, mimicking our extension of $P_c P_f$. We can then use a variant of the fact that, if it exists, a cofree comonad $C(H)$ on an arbitrary endofunctor H is characterised by a canonical isomorphism of categories

$$H\text{-}coalg \simeq C(H)\text{-}Coalg$$

where $-coalg$ stands for functor coalgebras while $-Coalg$ is for Eilenberg-Moore coalgebras. Although the categories of coalgebras and strict maps are problematic in the lax setting, categories of coalgebras and oplax maps do respect the laxness of $Lax(\mathcal{L}_\Sigma^{op}, Poset)$, allowing a suitable variant. The details are as follows.

Proposition 5. *Given a locally ordered comonad G on a locally ordered category C, the data given by $Lax(D,G) : Lax(D,C) \to Lax(D,C)$ and pointwise liftings of the structural natural transformations of G yield a locally ordered comonad we also denote by $Lax(D,G)$ on $Lax(D,C)$.*

The proof of Proposition 5 is not entirely trivial as it involves a mixture of the strict structure in the definition of comonad with the lax structure in the definition of $Lax(D,C)$. Nevertheless, with attention to detail, a proof is routine, and it means that, once we have extended the comonad $C(P_cP_f)$ to $Poset$, we can further extend it axiomatically to $Lax(\mathcal{L}_\Sigma^{op}, Poset)$.

Let E be an arbitrary locally ordered endofunctor on an arbitrary locally ordered category C. Denote by $E\text{-}coalg_{oplax}$ the locally ordered category whose objects are E-coalgebras and whose maps are oplax maps of E-coalgebras, meaning that, in the square

$$\begin{array}{ccc} X & \longrightarrow & Y \\ \downarrow & \leq & \downarrow \\ EX & \longrightarrow & EY \end{array}$$

the composite via EX is less than or equal to the composite via Y, with the evident composition and locally ordered structure. Since C and E are arbitrary, one can replace C by $Lax(D,C)$ and replace E by $Lax(D,E)$, yielding the locally ordered category $Lax(D,E)\text{-}coalg_{oplax}$. The following result is also not immediate, but it again follows from routine checking. It is an instance of a general phenomenon that allows laxness to commute exactly with oplaxness but not with any other variant of laxness such as laxness itself or strictness or pseudoness.

Proposition 6. *The locally ordered category $Lax(D,E)\text{-}coalg_{oplax}$ is canonically isomorphic to $Lax(D, E\text{-}coalg_{oplax})$.*

Proposition 6 gives us an easy way to make constructions with, and check claims regarding, $Lax(D,E)$-coalgebras : it characterises such coalgebras in terms of locally ordered functors into $E\text{-}coalg_{oplax}$; the latter locally ordered category, i.e., $E\text{-}coalg_{oplax}$, is simpler to study than $Lax(D,E)\text{-}coalg_{oplax}$ as it only involves one kind of laxness rather than two.

Definition 7. *Given a locally ordered comonad G on C, the locally ordered category $G\text{-}Coalg_{oplax}$ has objects given by (strict) G-coalgebras and maps given by oplax maps of coalgebras, where maps are defined as in $E\text{-}coalg_{oplax}$.*

With care, Proposition 6 can be extended from locally ordered endofunctors to locally ordered comonads, yielding the following:

Proposition 7. *Given a locally ordered comonad G, the locally ordered category $Lax(D,G)\text{-}Coalg_{oplax}$ is canonically isomorphic to $Lax(D, G\text{-}Coalg_{oplax})$.*

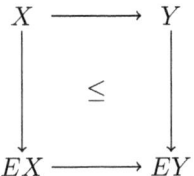

The analysis of [13], but expressed there in terms of laxness rather than oplaxness and in terms of monads rather than comonads, yields the following:

Theorem 1. *Given a locally ordered endofunctor E on a locally ordered category with finite colimits C, if $C(E)$ is the cofree comonad on E, then E-$coalg_{oplax}$ is canonically isomorphic to $C(E)$-$Coalg_{oplax}$.*

Combining Proposition 6, Proposition 7 and Theorem 1, we can conclude the following:

Theorem 2. *Given a locally ordered endofunctor E on a locally ordered category with finite colimits C, if $C(E)$ is the cofree comonad on E, then there is a canonical isomorphism*

$$Lax(D,E)\text{-}coalg_{oplax} \simeq Lax(D,C(E))\text{-}Coalg_{oplax}$$

Corollary 1. *For any locally ordered endofunctor E on Poset, if $C(E)$ is the cofree comonad on E, then there is a canonical isomorphism*

$$Lax(\mathcal{L}_{\Sigma}^{op},E)\text{-}coalg_{oplax} \simeq Lax(\mathcal{L}_{\Sigma}^{op},C(E))\text{-}Coalg_{oplax}$$

Corollary 1 provides us with the central axiomatic result we need to extend our analysis of variable-free logic programs in [15] to arbitrary logic programs. The bulk of the analysis of this section holds axiomatically, so that seems the best way in which to explain it although we have only one leading example, that determined by an extension of P_cP_f to *Poset*. In Section 5, we shall investigate such an extension.

5 Coalgebraic Semantics for Arbitrary Logic Programs

The reason we need to extend P_cP_f from *Set* to *Poset* is to allow for lax naturality, and the reason for that is to take advantage of the partial order structure of the set $P_c(X)$: we neither need nor want to change the set $P_c(X)$ itself; we just need to exploit its natural partial order structure given by subset inclusion. Nor do we want to change the nature of the relationship between a variable-free logic program P and the associated coalgebra $p : At \longrightarrow P_fP_f(At)$: as best we can, we simply want to extend that relationship by making it pointwise relative to the indexing category $\mathcal{L}_{\Sigma}^{op}$.

In order to give a locally ordered endofunctor on *Poset*, we need to extend P_cP_f from acting on a set X to acting on a partially ordered set P, respecting the partial order structure. This leads to a natural choice as follows:

Definition 8. *Define $P_f : Poset \longrightarrow Poset$ by letting $P_f(P)$ be the partial order given by the set of finite subsets of P, with $A \leq B$ if for all $a \in A$, there exists $b \in B$ for which $a \leq b$ in P, with behaviour on maps given by image. Define P_c similarly but with countability replacing finiteness.*

As *Poset* is complete and cocomplete, and as P_cP_f has a rank, a cofree comonad $C(P_cP_f)$ necessarily exists on P_cP_f. Moreover, it is given by the transfinite (just allowing for countability) extension of the construction in Proposition 2.

By the work of Section 4, the $Lax(\mathcal{L}_{\Sigma}^{op}, P_cP_f)$-coalgebra structure, i.e., the lax natural transformation, $p : At \longrightarrow P_cP_fAt$ associated with an arbitrary logic program P, evaluated at a natural number n, sends an atomic formula $A(x_1, \ldots, x_n)$ to the set of sets of antecedents in substitution instances of clauses in P for which the head of the substituted instance agrees with $A(x_1, \ldots, x_n)$. Extending Section 2, this can be expressed as a tree of the nature of the left hand tree in Figure 1, interleaving two kinds of nodes.

Comparing these trees with the definition of parallel and-or derivation tree, i.e., with Definition 1, these trees are more intrinsic: parallel and-or derivation trees have most general unifiers built into a single tree, whereas, for each natural number n, coalgebra yields trees involving at most n free variables, then models substitution by replacing them by related, extended trees. We shall illustrate with our leading example.

The two constructs are obviously related, but the coalgebraic one makes fewer identifications, SLD-resolution being modelled by a list of trees corresponding to a succession of substitutions rather than by a single tree. We would suggest that this list of trees may be worth considering as a possible refinement of the notion of parallel and-or derivation tree, lending itself to a tree-rewriting understanding of the SLD-algorithm. Providing such an account is a priority for us as future research.

Example 5. Consider ListNat as in Example 3. Suppose we start with $A(x,y)\epsilon At(2)$ given by the atomic formula $\mathtt{list(cons(x, cons(y, x)))}$. Then $p(A(x,y))$ is the element of $P_cP_fAt(2)$ expressible by the tree on the left hand side of Figure 2.

This tree agrees with the first part of the parallel and-or derivation tree for $\mathtt{list(cons(x, cons(y, x)))}$ as determined by Definition 1. But the tree here has leaves `nat(x)`, `nat(y)` and `list(x)`, whereas the parallel and-or derivation tree follows those nodes, using substitutions determined by mgu's. Moreover, those substitutions need not be consistent with each other: not only are there two ways to unify each of `nat(x)`, `nat(y)` and `list(x)`, but also there is no consistent substitution for `x` at all.

In contrast, the coalgebraic structure means any substitution, whether determined by an mgu or not, applies to the whole tree. The lax naturality means a substitution potentially yields two different trees: one given by substitution into the tree, then pruning to remove redundant branches, the other given by substitution into the root, then applying p.

For example, suppose we substitute $s(z)$ for both x and y in $\mathtt{list(cons(x, cons(y, x)))}$. This substitution is given by applying At to the map $(s, s) : 1 \longrightarrow 2$ in $\mathcal{L}_{\Sigma}$. So $At((s, s))(A(x, y))$ is an element of $At(1)$. Its image under $p_1 : At(1) \longrightarrow P_cP_fAt(1)$ is the element of $P_cP_fAt(1)$ expressible by the tree on the right hand side of Figure 2.

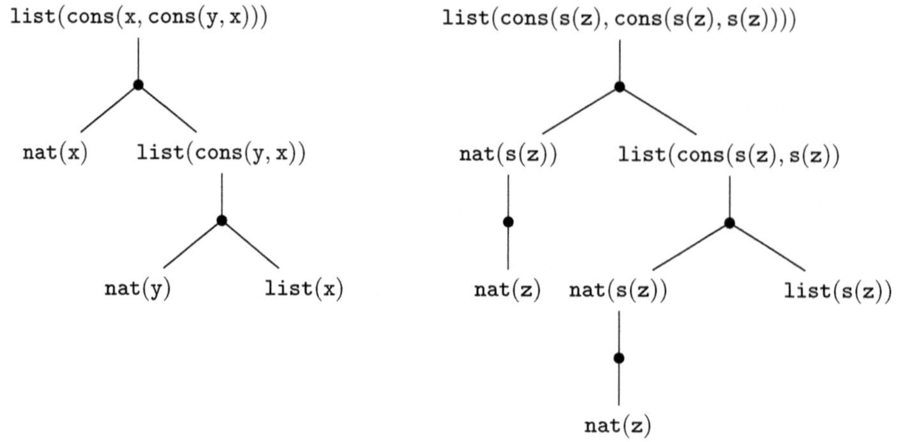

Fig. 2. The left hand tree represents $p(\texttt{list(cons(x, cons(y, x)))})$ and the right hand tree represents $pAt((s, s))(\texttt{list(cons(x, cons(y, x)))})$, i.e., $p(\texttt{list(cons(s(z), cons(s(z), s(z))))})$

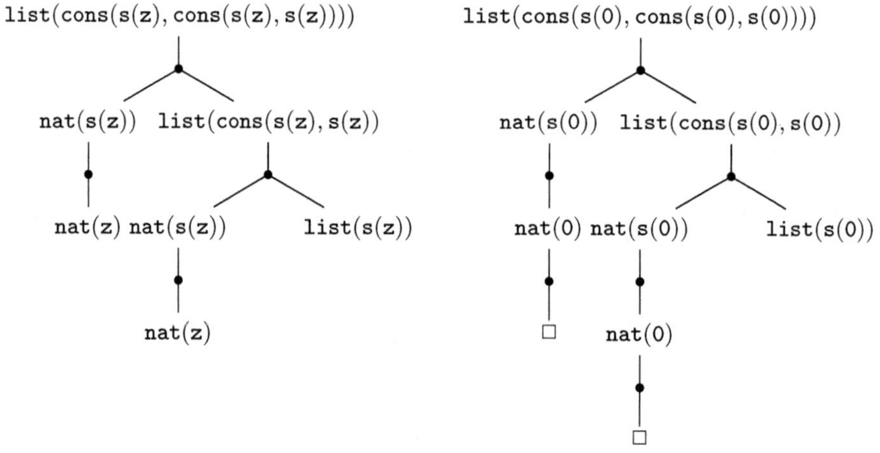

Fig. 3. On the left is the tree depicting $pAt((s, s))(\texttt{list(cons(x, cons(y, x)))})$ as also appears on the right of Figure 2, and on the right is the tree depicting $pAt(0)At((s, s))(\texttt{list(cons(x, cons(y, x)))})$

The laxness of the naturality of p is indicated by the increased length, in two places, of the second tree when compared with the first tree. Observe that, before those two places, the two trees have the same structure: that need not always be exactly the case, as substitution in a tree could involve pruning if substitution instances of two different atoms yield the same atom.

Now suppose we make the further substitution of 0 for z. This substitution is given by applying At to the map $0 : 0 \rightarrow 1$ in $\mathcal{L}_\Sigma$. In Figure 3, we depict $p_1 At((s,s))(A(x,y))$ on the left, repeating the right hand tree of Figure 2, and we depict $p_0 At(0)At((s,s))(A(x,y))$ on the right.

Two of the leaves of the latter tree are labelled by $\Box$, but one leaf, namely `list(s(0))` is not, so the tree does not yield a proof. Again, observe the laxness.

6 Conclusions and Further Work

Using sophisticated category theoretic techniques surrounding the notion of laxness, we have extended the coalgebraic analysis of variable-free logic programs in [15] to arbitrary logic programs. For variable-free logic programs, the cofree comonad on $P_f P_f$ allowed us to represent the parallel and-or derivation trees generated by a logic program. For arbitrary logic programs, the situation is more subtle, as coalgebra naturally gives rise to a list of trees determined by substitutions, whereas a parallel and-or derivation tree has all the information squeezed into one tree.

A natural question to arise in the light of this is whether the coalgebraic structure given here suggests a more subtle semantics for SLD-resolution than that given by parallel and-or derivation trees, perhaps one based upon tree-rewriting. That is one direction in which we propose to continue research.

The key fact driving our analysis has been the observation that the implication $\leftarrow$ acts at a meta-level, like a sequent rather than a logical connective. That observation extends to first-order fragments of linear logic and the Logic of Bunched Implications [10,20]. So we plan to extend the work in the paper to logic programming languages based on such logics.

The situation regarding higher-order logic programming languages such as λ-*PROLOG* [19] is more subtle. Despite their higher-order nature, such logic programming languages typically make fundamental use of sequents. So it may well be fruitful to consider modelling them in terms of coalgebra too, albeit probably on a sophisticated base category such as a category of Heyting algebras.

More generally, the results of this paper can be applied to the studies of Higher-order recursion schemes, [1].

A further direction is to investigate the operational meaning of coinductive logic programming [11,22]. That requires a modification to the algorithm of SLD-resolution we have considered in this paper. In particular, given a logic program that defines an infinite stream (similarly to our running example of `list`, but without the base case for `nil`), the interpreter for coinductive logic programs of this kind would be able to deduce a finite atom `stream(cons(x,y))` from the infinite derivations.

References

1. Adámek, J., Milius, S., Velebil, J.: Semantics of higher-order recursion schemes. In: CoRR, abs/1101.4929 (2011)
2. Amato, G., Lipton, J., McGrail, R.: On the algebraic structure of declarative programming languages. Theor. Comput. Sci. 410(46), 4626–4671 (2009)
3. Asperti, A., Martini, S.: Projections instead of variables: A category theoretic interpretation of logic programs. In: ICLP, pp. 337–352 (1989)
4. Bonchi, F., Montanari, U.: Reactive systems (semi-)saturated semantics and coalgebras on presheaves. Theor. Comput. Sci. 410(41), 4044–4066 (2009)
5. Bruni, R., Montanari, U., Rossi, F.: An interactive semantics of logic programming. TPLP 1(6), 647–690 (2001)
6. Comini, M., Levi, G., Meo, M.C.: A theory of observables for logic programs. Inf. Comput. 169(1), 23–80 (2001)
7. Corradini, A., Montanari, U.: An algebraic semantics of logic programs as structured transition systems. In: Proc. NACLP 1990. MIT Press, Cambridge (1990)
8. Corradini, A., Montanari, U.: An algebraic semantics for structured transition systems and its application to logic programs. TCS 103, 51–106 (1992)
9. Costa, V.S., Warren, D.H.D., Yang, R.: Andorra-I: A parallel prolog system that transparently exploits both and-and or-parallelism. In: PPOPP, pp. 83–93 (1991)
10. Girard, J.-Y.: Linear logic. Theor. Comput. Sci. 50, 1–102 (1987)
11. Gupta, G., Bansal, A., Min, R., Simon, L., Mallya, A.: Coinductive logic programming and its applications. In: Dahl, V., Niemelä, I. (eds.) ICLP 2007. LNCS, vol. 4670, pp. 27–44. Springer, Heidelberg (2007)
12. Gupta, G., Costa, V.S.: Optimal implementation of and-or parallel prolog. In: Conference proceedings on PARLE 1992, pp. 71–92. Elsevier North-Holland, Inc., New York (1994)
13. Kelly, G.M.: Coherence theorems for lax algebras and for distributive laws. In: Category Seminar. LNM, vol. 420, pp. 281–375 (1974)
14. Kinoshita, Y., Power, A.J.: A fibrational semantics for logic programs. In: Proceedings of the Fifth International Workshop on Extensions of Logic Programming. LNCS (LNAI), vol. 1050. Springer, Heidelberg (1996)
15. Komendantskaya, E., McCusker, G., Power, J.: Coalgebraic semantics for parallel derivation strategies in logic programming. In: Johnson, M., Pavlovic, D. (eds.) AMAST 2010. LNCS, vol. 6486, pp. 111–127. Springer, Heidelberg (2011)
16. Komendantskaya, E., Power, J.: Fibrational semantics for many-valued logic programs: Grounds for non-groundness. In: Hölldobler, S., Lutz, C., Wansing, H. (eds.) JELIA 2008. LNCS (LNAI), vol. 5293, pp. 258–271. Springer, Heidelberg (2008)
17. Lawvere, W.: Functional semantics of algebraic theories. PhD thesis, Columbia University (1963)
18. Lloyd, J.: Foundations of Logic Programming, 2nd edn. Springer, Heidelberg (1987)
19. Miller, D., Nadathur, G.: Higher-order logic programming. In: ICLP, pp. 448–462 (1986)
20. Pym, D.: The Semantics and Proof Theory of the Logic of Bunched Implications. Applied Logic Series, vol. 26. Kluwer Academic Publishers, Dordrecht (2002)
21. Rydeheard, D., Burstall, R.: Computational Category theory. Prentice Hall, Englewood Cliffs (1988)
22. Simon, L., Bansal, A., Mallya, A., Gupta, G.: Co-logic programming: Extending logic programming with coinduction. In: Arge, L., Cachin, C., Jurdziński, T., Tarlecki, A. (eds.) ICALP 2007. LNCS, vol. 4596, pp. 472–483. Springer, Heidelberg (2007)

Hybridization of Institutions

Manuel A. Martins[1], Alexandre Madeira[1,2,4], Răzvan Diaconescu[3], and Luís S. Barbosa[2]

[1] Department of Mathematics, University of Aveiro
[2] Department of Informatics, Minho University
[3] Institute of Mathematics "Simion Stoilow" of the Romanian Academy
[4] Critical Software S.A., Portugal

Abstract. Modal logics are successfully used as specification logics for reactive systems. However, they are not expressive enough to refer to individual states and reason about the local behaviour of such systems. This limitation is overcome in hybrid logics which introduce special symbols for naming states in models. Actually, hybrid logics have recently regained interest, resulting in a number of new results and techniques as well as applications to software specification.

In this context, the first contribution of this paper is an attempt to 'universalize' the hybridization idea. Following the lines of [15], where a method to modalize arbitrary institutions is presented, the paper introduces a method to hybridize logics at the same institution-independent level. The method extends arbitrary institutions with Kripke semantics (for multi-modalities with arbitrary arities) and hybrid features. This paves the ground for a general result: any encoding (expressed as *comorphism*) from an arbitrary institution to first order logic ($\mathcal{FOL}$) determines a comorphism from its hybridization to $\mathcal{FOL}$. This second contribution opens the possibility of effective tool support to specification languages based upon logics with hybrid features.

Keywords: Institution theory, hybrid logic, formal specification.

1 Introduction

Modern societies are increasingly dependent on software systems and services whose reliability is crucial for their own development, security, privacy, and quality of life. On the other hand, software is large and complex, deals with a multitude of different concerns and has to meet requirements formulated (and verified) at different abstraction levels. For the last three decades this has put forward a research agenda on mathematically sound development methods that seem to be finally emerging as a key concern for industry.

Typically, three issues in this agenda need to be rigorously addressed. The first concerns the sort of *mathematical structures* suitable to model software systems; the second focus on the *languages* in which such models can be specified and, finally, the last one addresses the *satisfaction relation* between the (semantic) mathematical structures and the (syntactic) formulation of requirements as

A. Corradini, B. Klin, and C. Cîrstea (Eds.): CALCO 2011, LNCS 6859, pp. 283–297, 2011.

sentences in the specification language. A fourth concern, which is becoming more and more relevant in practice, should be added: the fact that the working software engineer often has to capture and relate different kinds of requirements entails the need for a uniform specification framework in which different formalisms can be expressed and related. A quite canonical way to answer this challenge resorts to the notion of an institution [16,13] which, as an abstract representation of a logical system, encompasses syntax, semantics and satisfaction, and provides a formal framework for relating, comparing and combining specification logics.

Institution theory [16] is a categorical abstract model theory that arose about three decades ago within specification theory as a response to the explosion in the population of logics in use there. Its original aim was to develop as much computing science as possible in a general uniform way independently of particular logical systems. This has now been achieved to an extent even greater than originally thought, as institution theory became the most fundamental mathematical theory underlying algebraic specification theory, also being increasingly used in other area of computer science. Moreover, institution theory constitutes a major trend in the so-called 'universal logic' (in the sense envisaged by Jean-Yves Béziau) which is considered by many a true renaissance of mathematical logic.

Modal logics have been successfully used as specification languages for state transition systems, which, on their turn, are taken as basic, underlying structures in program development. From a proof theoretic point of view, such logics have interesting algorithmic proprieties, and, moreover, they can naturally be translated to first order logic. However, (non-hybrid) modal logics do not allow explicit references to specific states of the underlying transition system which, in a number of cases, is a desirable feature in a specification. For instance, such modal logics are adequate to specify systems as dynamic processes which evolve in response to events. But, on the other hand, they are not expressive enough to identify particular states in a system's evolution, neither to express (local) properties referring to one such state or a group thereof. Hybrid logic [2], on the other hand, overcomes this limitation by introducing *nominals* as references to specific states in a modal framework, taking together features from first-order logic and modal logic.

Historically, hybrid logic was introduced by Arthur Prior [22] in the 50's. Afterwards, his student Robert Bull extended the theory significantly by establishing a number of completeness results for generalizations of Prior's hybrid logic. After a period without much developments, in the 80's the Bulgarian school of logic (namely Passy, Tinchev, Gargov and Goranko) revived the interest in hybrid logic, studying, in particular, the possible roles of the binder operator [21]. More recently, Areces and Blackburn intensely expanded the theory (cf. the dedicated web page at `http://hylo.loria.fr/`), addressing, notably, the complexity of the satisfiability problem. The work of Braüner on proof theory for hybrid logic should also be mentioned [5]. His study of quantified hybrid logic is, in a sense, at the origin of the results presented in this paper. Actually, the way first order and hybrid logics are combined in quantified hybrid logic, was a first motivation for the quest for a general, institution-independent approach to the hybridization of logics which constitutes the main contribution of this paper.

In fact, the idea of introducing nominals to explicitly refer to individual states, can be applied to any logic with a Kripke semantics. Quoting [1], *"(...)Strictly speaking, not all modal logics are hybrid, but certainly any modal logics can be hybridized, and in our view many of them should be (...)"*. This principle is reflected in a recent trend of hybridization of specification formalisms and process calculi. Beyond the classical cases of hybrid versions of propositional and first order logic, hybrid accounts of intuitionistic logic [6], $\mathcal{CTL}$ [24,19], $\mathcal{LTL}$ [11], μ-calculus [23] among others, are already studied.

What is, thus, in such a context the contribution of this paper? First of all, as stated above, we put forward an institution-independent method to hybridize arbitrary logics, shedding light on the generic pattern of hybridization. In other words, we liberate the essence of hybridization from logical details that are orthogonal to the hybrid idea and that are tributary to other logics.

The hybridization process is also a mechanism for combining logics. Combination of logical system (or institutions), in which typically different roles are played by the different logics to be composed, is, in itself, a relevant research topic. The approach discussed in this paper is in line with the process of modalization of an institution, proposed in [15], in which a modal logic is combined with an arbitrary institution in a systematic way. We take a further step by replacing modal by hybrid logic and allowing multi-modalities.

The paper's second contribution is also a general result: it is shown that any encoding (expressed as 'comorphism' in the sense of [17]) from an arbitrary institution to first order logic ($\mathcal{FOL}$) determines a comorphism from its (quantifier-free) hybridization to $\mathcal{FOL}$. Moreover, the proof is constructive entailing a method to define such comorphisms. This may be regarded as a first step towards a general theory of encodings of hybrid logics into $\mathcal{FOL}$ as support for borrowing formal verification tools from $\mathcal{FOL}$-based to hybrid-based specification languages.

Outline. In order to keep exposition reasonably self-contained, Section 2 reviews basic concepts on institutions and recalls a number of examples. The paper's contributions appear on Sections 3 and 4. The former introduces the hybridization process. The latter addresses the construction of comorphisms from hybrid institutions to $\mathcal{FOL}$. Finally, Section 5 concludes and points out a number of topics for future work. Proofs of all new results presented can be found in the appendix.

2 Notation and Definitions

Institutions were defined by Goguen and Burstall in [7], the seminal paper [16] being printed after a delay of many years. Below we recall the concept of institution which formalises the intuitive notion of logical system, including syntax, semantics, and the satisfaction between them.

Definition 1 (Institution). *An* institution $(\mathrm{Sign}^{\mathcal{I}}, \mathrm{Sen}^{\mathcal{I}}, \mathrm{Mod}^{\mathcal{I}}, (\models^{\mathcal{I}}_{\Sigma})_{\Sigma\in|\mathrm{Sign}^{\mathcal{I}}|})$ *consists of*

- *a category* $\mathrm{Sign}^{\mathcal{I}}$ *whose objects are called* signatures,
- *a functor* $\mathrm{Sen}^{\mathcal{I}} : \mathrm{Sign}^{\mathcal{I}} \to \mathbf{Set}$ *giving for each signature a set whose elements are called* sentences *over that signature,*
- *a functor* $\mathrm{Mod}^{\mathcal{I}} : (\mathrm{Sign}^{\mathcal{I}})^{op} \to \mathbf{CAT}$, *giving for each signature* Σ *a category whose objects are called* Σ-models, *and whose arrows are called* Σ-(model) morphisms, *and*
- *a relation* $\models^{\mathcal{I}}_{\Sigma} \subseteq |\mathrm{Mod}^{\mathcal{I}}(\Sigma)| \times \mathrm{Sen}^{\mathcal{I}}$ *for each* $\Sigma \in |\mathrm{Sen}^{\mathcal{I}}|$, *called the* satisfaction relation,

such that for each morphism $\varphi : \Sigma \to \Sigma' \in \mathrm{Sign}^{\mathcal{I}}$, *the satisfaction condition*

$$M' \models^{\mathcal{I}}_{\Sigma'} \mathrm{Sen}^{\mathcal{I}}(\varphi)(\rho) \textit{ iff } \mathrm{Mod}^{\mathcal{I}}(\varphi)(M') \models^{\mathcal{I}}_{\Sigma} \rho \tag{1}$$

holds for each $M' \in |\mathrm{Mod}^{\mathcal{I}}(\Sigma')|$ *and* $\rho \in \mathrm{Sen}^{\mathcal{I}}(\Sigma)$.

We recall the notions of *amalgamation* and *quantification space* that are crucial for what follows. The former is intensely used in institution theory, whereas the latter was introduced rather recently in [14].

Definition 2 (Amalgamation property). *Given any functor* $\mathrm{Mod} : \mathrm{Sign}^{op} \to \mathbf{CAT}$ *a commuting square of signature morphisms*

$$\begin{array}{ccc} \Sigma & \xrightarrow{\varphi_1} & \Sigma_1 \\ {\scriptstyle\varphi_2}\downarrow & & \downarrow{\scriptstyle\theta_1} \\ \Sigma_2 & \xrightarrow[\theta_2]{} & \Sigma' \end{array} \tag{2}$$

is a weak amalgamation square for Mod *if and only if, for each* Σ_1*-model* M_1 *and a* Σ_2*-model* M_2 *such that* $\mathrm{Mod}(\varphi_1)(M_1) = \mathrm{Mod}(\varphi_2)(M_2)$, *there exists a* Σ'*-model* M' *such that* $\mathrm{Mod}(\theta_1)(M') = M_1$ *and* $\mathrm{Mod}(\theta_2)(M') = M_2$. *When* M' *is required to be unique, the square is called an* amalgamation square. *The model* M' *is called an* amalgamation of M_1 and M_2 *and when it is unique it is denoted by* $M_1 \otimes_{\varphi_1,\varphi_2} M_2$.

When Mod *is the model functor* $\mathrm{Mod}^{\mathcal{I}}$ *of an institution* $\mathcal{I}$ *we say that* $\mathcal{I}$ *has the respective amalgamation properties.*

Definition 3 (Quantification space). *For any category* Sign *a subclass of arrows* $\mathcal{D} \subseteq \mathrm{Sign}$ *is called a* quantification space *if, for any* $(\chi : \Sigma \to \Sigma') \in \mathcal{D}$ *and* $\varphi : \Sigma \to \Sigma_1$, *there is a designated pushout*

$$\begin{array}{ccc} \Sigma & \xrightarrow{\varphi} & \Sigma_1 \\ {\scriptstyle\chi}\downarrow & & \downarrow{\scriptstyle\chi(\varphi)} \\ \Sigma' & \xrightarrow[\varphi[\chi]]{} & \Sigma'_1 \end{array}$$

with $\chi(\varphi) \in \mathcal{D}$ *and such that the 'horizontal' composition of such designated pushouts is again a designated pushout, i.e. for the pushouts in the following diagram*

$$\begin{array}{ccccc} \Sigma & \xrightarrow{\varphi} & \Sigma_1 & \xrightarrow{\theta} & \Sigma_2 \\ {\scriptstyle \chi}\downarrow & & \downarrow{\scriptstyle \chi(\varphi)} & & \downarrow{\scriptstyle \chi(\varphi)(\theta)} \\ \Sigma' & \xrightarrow[\varphi[\chi]]{} & \Sigma'_1 & \xrightarrow[\theta[\chi(\varphi)]]{} & \Sigma'_2 \end{array}$$

$\varphi[\chi]; \theta[\chi(\varphi)] = (\varphi;\theta)[\chi]$ *and* $\chi(\varphi)(\theta) = \chi(\varphi;\theta)$, *and such that* $\chi(1_\Sigma) = \chi$ *and* $1_\Sigma[\chi] = 1_\Sigma$.

We say that a quantification space $\mathcal{D}$ *for* Sign *is* adequate *for a functor* Mod : $\text{Sign}^{op} \to \mathbf{CAT}$ *when the designated pushouts mentioned above are weak amalgamation squares for* Mod.

Example 1 ($\mathcal{FOL}$, $\mathcal{ALG}$, $\mathcal{EQ}$, $\mathcal{REL}$ and $\mathcal{PL}$). A well known example of an institution is $\mathcal{FOL}$ — the institution of first order logic $\mathcal{FOL}$ (see [13] for a detailed account). The *signatures* are tuples (S, F, P), where S is a set of *sort symbols*, $F = \{F_{w \to s} \mid w \in S^*, s \in S\}$ is a family of sets of *operation symbols* and $P = \{P_w \mid w \in S^*\}$ is a family of sets of *relational symbols*. A signature morphism φ is a triple of functions $(\varphi_{sort}, \varphi_{ops}, \varphi_{pred}) : (S, F, P) \to (S', F', P')$ that preserves functionalities, i.e., for any $f \in F_{s_1 \ldots s_n \to s}$, $\varphi_{ops}(f) \in F'_{\varphi_{sort}(s_1) \ldots \varphi_{sort}(s_n) \to \varphi_{sort}(s)}$ and for any $\pi \in P_{s_1 \ldots s_n}$, $\varphi_{pred}(\pi) \in P'_{\varphi_{sort}(s_1) \ldots \varphi_{sort}(s_n)}$. A (S, F, P)-*model* M is family $\{M_s \mid s \in S\}$ of sets together with: for each $f \in F_{s_1 \ldots s_n \to s}$, a function $f^M : M_{s_1} \times \cdots \times M_{s_n} \to M_s$ and for any $\pi \in P_{s_1 \ldots s_n}$ a relation $\pi^M \subseteq M_{s_1} \times \cdots \times M_{s_n}$. The (S, F, P)-model homomorphisms are S-families of functions $\{h_s : M_s \to M'_s\}_{s \in S}$, such that for any $f \in F_{s_1 \ldots s_n \to s}$, and each $m_i \in M_{s_i}$, $i = 1, \ldots, n$, $h_s(f^M(m_1, \ldots, m_n)) = f^{M'}(h_{s_1}(m_1), \ldots, h_{s_n}(m_n))$ and for each $\pi \in P_{s_1, \ldots s_n}$, if $(m_1, \ldots, m_n) \in \pi^M$ then $(h_{s_1}(m_1), \ldots, h_{s_n}(m_n)) \in \pi^{M'}$. The reduct of a (S', F', P')-model M' along φ consists of the (S, F, P)-model $M' \upharpoonright_\varphi$ such that, for each $s \in S$, $(M' \upharpoonright_\varphi)_s = M'_{\varphi_{sort}(s)}$, for each $f \in F_{s_1 \ldots s_n \to s}$ $f^{M' \upharpoonright_\varphi} = \varphi_{ops}(f)^{M'}$ and for each $\pi \in P_{s_1 \ldots s_n}$ $\pi^{M' \upharpoonright_\varphi} = \varphi_{pred}(\pi)^{M'}$. The set $\text{Sen}^{\mathcal{FOL}}((S, F, P))$ of (S, F, P)-*sentences* consists of the usual first-order (S, F, P)-formulas. A signature morphism $\varphi : (S, F, P) \to (S', F', P')$ induces a translation of sentences, $\text{Sen}^{\mathcal{FOL}}(\varphi) : \text{Sen}^{\mathcal{FOL}}((S, F, P)) \to \text{Sen}^{\mathcal{FOL}}((S', F', P'))$, that replaces symbols of (S, F, P) by the respective φ-images in (S', F', P'). More precisely, let $\varphi^{trm} : T_{(S,F)} \to T_{(S',F')}$ be defined by

$$\varphi^{trm}(f(t_1, \ldots, t_n)) = \varphi_{ops}(f)(\varphi^{trm}(t_1), \ldots, \varphi^{trm}(t_n)).$$

The translation $\text{Sen}^{\mathcal{FOL}}$ is recursively defined as follows:

- $\text{Sen}^{\mathcal{FOL}}(\varphi)(t \approx t') = \varphi^{trm}(t) \approx \varphi^{trm}(t')$;
- $\text{Sen}^{\mathcal{FOL}}(\varphi)(\pi(t_1, \ldots, t_n)) = \varphi_{pred}(\pi)(\varphi^{trm}(t_1), \ldots, \varphi^{trm}(t_n))$;
- $\text{Sen}^{\mathcal{FOL}}(\varphi)(\neg\rho) = \neg\text{Sen}^{\mathcal{FOL}}(\varphi)(\rho)$;
- $\text{Sen}^{\mathcal{FOL}}(\varphi)(\rho \odot \rho') = \text{Sen}^{\mathcal{FOL}}(\varphi)(\rho) \odot \text{Sen}^{\mathcal{FOL}}(\varphi)(\rho')$, $\odot \in \{\vee, \wedge, \to\}$;
- $\text{Sen}^{\mathcal{FOL}}(\varphi)(\forall X\, \rho) = \forall X^\varphi\, \text{Sen}^{\mathcal{FOL}}(\varphi')(\rho)$, where $X^\varphi = \{(x, \varphi_{sort}(s), (S', F', P')) \mid (x, s, (S, F, P)) \in X\}$, and φ' canonically

extends φ by mapping each $(x, s, (S, F, P))$ to $(x, \varphi_{sort}(s), (S', F', P'))$. Note that we are considered a variable for (S, F, P) as a triple $(x, s, (S, F, P))$ where x is the name of the variable, s its sort, and (S, F, P) its signature (see [14]).

Finally, the *satisfaction* relation is the usual *Tarskian satisfaction relation.* We just present the case of quantifiers as an illustration:

- $\mathcal{M} \models^{\mathcal{FOL}}_{(S,F,P)} \forall X\, \rho$ iff, $\mathcal{M}' \models^{\mathcal{FOL}}_{(S,F \uplus X,P)} \rho$ for each expansion M' of M along the signature morphism $(S, F, P) \hookrightarrow (S, F \uplus X, P)$;
- $\mathcal{M} \models^{\mathcal{FOL}}_{(S,F,P)} \exists X\, \rho$ iff $\mathcal{M} \models^{\mathcal{FOL}}_{(S,F,P)} \neg\forall X\, \neg\rho$

The institution $\mathcal{ALG}$ is obtained from $\mathcal{FOL}$ by discarding the relational symbols and the corresponding interpretations in models. The institution $\mathcal{EQ}$ is defined as the sub-institution of $\mathcal{ALG}$ where the sentences are just universally quantified equations $(\forall X)\, t \approx t'$. The institution $\mathcal{REL}$ is the sub-institution of single-sorted first-order logic with signatures having only constants and relational symbols.

The institution $\mathcal{PL}$ (of propositional logic) is the fragment of $\mathcal{FOL}$ determined by signatures with empty sets of sort symbols.

3 A Method to Hybridize Arbitrary Institutions

Let us consider an institution $\mathcal{I} = (\mathrm{Sign}^{\mathcal{I}}, \mathrm{Sen}^{\mathcal{I}}, \mathrm{Mod}^{\mathcal{I}}, (\models^{\mathcal{I}}_{\Sigma})_{\Sigma \in |\mathrm{Sign}^{\mathcal{I}}|})$ with quantification space $\mathcal{D}^{\mathcal{I}} \subseteq$ Sign. This section introduces a method to enrich the expressivity of $\mathcal{I}$ with modalities and nominals, defining a suitable semantics for it. Moreover, it is shown that the outcome still defines an institution, to which we refer as the *hybrid* $\mathcal{I}$ and denote by $\mathcal{HI}$.

The Category of $\mathcal{HI}$-signatures

The category of $\mathcal{I}$-*hybrid signatures*, denoted by $\mathrm{Sign}^{\mathcal{HI}}$, is defined as the following direct (cartesian) product of categories:

$$\mathrm{Sign}^{\mathcal{HI}} = \mathrm{Sign}^{\mathcal{I}} \times \mathrm{Sign}^{\mathcal{REL}}.$$

The $\mathcal{REL}$-signatures are denoted by (Nom, Λ), where Nom is a set of constants called *nominals* and Λ is a set of relational symbols called *modalities*; Λ_n stands for the set of modalities of arity n. General category theory entails,

Proposition 1. *The projection* $\mathrm{Sign}^{\mathcal{HI}} \to \mathrm{Sign}^{\mathcal{I}}$ *lifts small co-limits.*

The existence of co-limits of signatures is one of the properties of institutions of key practical relevance for specification in-the-large (see [16]).

Corollary 1. $\mathrm{Sign}^{\mathcal{HI}}$ *has all small co-limits.*

$\mathcal{HI}$-sentences

Let us fix a quantification space $\mathcal{D}^{\mathcal{HI}}$ for $\mathrm{Sign}^{\mathcal{HI}}$ such that for each $\chi \in \mathcal{D}^{\mathcal{HI}}$ its projection $\chi|_{\mathcal{I}}$ to $\mathrm{Sign}^{\mathcal{I}}$ belongs to $\mathcal{D}^{\mathcal{I}}$. The quantification space $\mathcal{D}^{\mathcal{HI}}$ is a parameter of the hybridization process. Whenever $\mathcal{D}^{\mathcal{HI}}$ consists of identities we say the hybridization is *quantifier-free*. Note that a quantifier-free hybridization does not necessarily mean the absence of 'local' quantification, i.e. placed at the level of base institution $\mathcal{I}$.

Let $\Delta = (\Sigma, \mathrm{Nom}, \Lambda)$. The set of sentences $\mathrm{Sen}^{\mathcal{HI}}(\Delta)$ is the least set such that

- $\mathrm{Nom} \subseteq \mathrm{Sen}^{\mathcal{HI}}(\Delta)$;
- $\mathrm{Sen}^{\mathcal{I}}(\Sigma) \subseteq \mathrm{Sen}^{\mathcal{HI}}(\Delta)$;
- $\rho \odot \rho' \in \mathrm{Sen}^{\mathcal{HI}}(\Delta)$ for any $\rho, \rho' \in \mathrm{Sen}^{\mathcal{HI}}(\Delta)$ and any $\odot \in \{\vee, \wedge, \rightarrow\}$;
- $\neg\rho \in \mathrm{Sen}^{\mathcal{HI}}(\Delta)$, for any $\rho \in \mathrm{Sen}^{\mathcal{HI}}(\Delta)$;
- $@_i\rho \in \mathrm{Sen}^{\mathcal{HI}}(\Delta)$ for any $\rho \in \mathrm{Sen}(\Sigma)$ and $i \in \mathrm{Nom}$;
- $[\lambda](\rho_1, \ldots, \rho_n) \in \mathrm{Sen}^{\mathcal{HI}}(\Delta)$, for any $\lambda \in \Lambda_{n+1}, \rho_i \in \mathrm{Sen}^{\mathcal{HI}}(\Delta), i \in \{1, \ldots, n\}$;
- $(\forall\chi)\rho \in \mathrm{Sen}^{\mathcal{HI}}(\Delta)$, for any $\rho \in \mathrm{Sen}^{\mathcal{HI}}(\Delta')$ and $\chi : \ \Delta \to \Delta' \in \mathcal{D}^{\mathcal{HI}}$;
- $(\exists\chi)\rho \in \mathrm{Sen}^{\mathcal{HI}}(\Delta)$, for any $\rho \in \mathrm{Sen}^{\mathcal{HI}}(\Delta')$ and $\chi : \ \Delta \to \Delta' \in \mathcal{D}^{\mathcal{HI}}$.

Translations of $\mathcal{HI}$-sentences

Let $\varphi = (\varphi_{\mathrm{Sig}}, \varphi_{\mathrm{Nom}}, \varphi_{\mathrm{MS}}) : \ (\Sigma, \mathrm{Nom}, \Lambda) \to (\Sigma', \mathrm{Nom}', \Lambda')$ be a morphims of $\mathcal{HI}$-signatures.

The translation $\mathrm{Sen}^{\mathcal{HI}}(\varphi)$ is defined as follows:

- $\mathrm{Sen}^{\mathcal{HI}}(\varphi)(\rho) = \mathrm{Sen}^{\mathcal{I}}(\varphi_{\mathrm{Sig}})(\rho)$ for any $\rho \in \mathrm{Sen}^{\mathcal{I}}(\Sigma)$;
- $\mathrm{Sen}^{\mathcal{HI}}(\varphi)(i) = \varphi_{\mathrm{Nom}}(i)$;
- $\mathrm{Sen}^{\mathcal{HI}}(\varphi)(\neg\rho) = \neg\mathrm{Sen}^{\mathcal{HI}}(\varphi)(\rho)$;
- $\mathrm{Sen}^{\mathcal{HI}}(\varphi)(\rho \odot \rho') = \mathrm{Sen}^{\mathcal{HI}}(\varphi)(\rho) \odot \mathrm{Sen}^{\mathcal{HI}}(\varphi)(\rho')$, $\odot \in \{\vee, \wedge, \rightarrow\}$;
- $\mathrm{Sen}^{\mathcal{HI}}(\varphi)(@_i\rho) = @_{\varphi_{\mathrm{Nom}}(i)}\mathrm{Sen}^{\mathcal{HI}}(\rho)$;
- $\mathrm{Sen}^{\mathcal{HI}}(\varphi)([\lambda](\rho_1, \ldots, \rho_n)) = [\varphi_{\mathrm{MS}}(\lambda)](\mathrm{Sen}^{\mathcal{HI}}(\rho_1), \ldots, \mathrm{Sen}^{\mathcal{HI}}(\rho_n))$;
- $\mathrm{Sen}^{\mathcal{HI}}(\varphi)((\forall\chi)\rho) = (\forall\chi(\varphi))\mathrm{Sen}^{\mathcal{HI}}(\varphi[\chi])(\rho)$;
- $\mathrm{Sen}^{\mathcal{HI}}(\varphi)((\exists\chi)\rho) = (\exists\chi(\varphi))\mathrm{Sen}^{\mathcal{HI}}(\varphi[\chi])(\rho)$.

Proposition 2. $\mathrm{Sen}^{\mathcal{HI}}$ *is a functor* $\mathrm{Sign}^{\mathcal{HI}} \to \mathbf{Set}$.

$\mathcal{HI}$-models

The $(\Sigma, \mathrm{Nom}, \Lambda)$-*models* are pairs $\mathcal{M} = (M, R)$ where

- R is a (Nom, Λ)-model in $\mathcal{REL}$;
- M is a function $|R| \to |\mathrm{Mod}^{\mathcal{I}}(\Sigma)|$.

The carrier set $|R|$ forms *the set of the states of* $\mathcal{M}$; $\{n^R \mid n \in \mathrm{Nom}\}$ represents the interpretations of the *nominals* Nom, whereas relations $\{\lambda^R \mid \lambda \in \Lambda_n, n \in \omega\}$ represent the interpretation of the *modalities* Λ. We denote $M(s)$ simply by M_s.

A $(\Sigma, \mathrm{Nom}, \Lambda)$-*model homomorphism* $h : \ (M, R) \to (M', R')$ consists of a pair aggregating

- a (Nom, Λ)-model homomorphism in $\mathcal{REL}$, $h_{st} : R \to R'$; i.e., a function $h_{st} : |R| \to |R'|$ such that for $i \in \text{Nom}$, $i^{R'} = h_{st}(i^R)$; and for any $s_1, \ldots, s_n \in |R|$, and $\lambda \in \Lambda_n$, $(s_1, \ldots, s_n) \in \lambda^R$, $(h_{st}(s_1), \ldots, h_{st}(s_n)) \in \lambda^{R'}$.
- a natural transformation $h_{mod} : M \Rightarrow M' \circ h_{st}$; note that h_{mod} is a $|R|$-indexed family of Σ-model homomorphisms $\{h^s_{mod} : M_s \to M'_{h_{st}(s)}\}_{s \in |R|}$.

The composition of hybrid model homomorphisms is defined canonically as

$$h; h' = (h_{st}; h'_{st}, h_{mod}; (h'_{mod} \circ h_{st})).$$

Fact 1. *Let Δ be any hybrid signature over an institution $\mathcal{I}$. Then Δ-models together with their homomorphisms constitute a category.*

Reducts of $\mathcal{HI}$-models

Let $\Delta = (\Sigma, \text{Nom}, \Lambda)$ and $\Delta' = (\Sigma', \text{Nom}', \Lambda')$ be two hybrid signatures, $\varphi = (\varphi_{\text{Sig}}, \varphi_{\text{Nom}}, \varphi_{\text{MS}})$ a morphism between Δ and Δ' and (M', R') a Δ'-model. The **reduct** of (M', R') along φ, denoted by $\text{Mod}^{\mathcal{HI}}(\varphi)(M', R')$, is the Δ-model (M, R) such that

- $|R| = |R'|$;
- for any $n \in \text{Nom}, n^R = \varphi_{\text{Nom}}(n)^{R'}$;
- for any $\lambda \in \Lambda$, $\lambda^R = \varphi_{\text{MS}}(\lambda)^{R'}$;
- for any $s \in |R|$, $M_s = \text{Mod}^{\mathcal{I}}(\varphi_{\text{Sig}})(M'_s)$.

Theorem 1. *A pushout square of $\mathcal{HI}$-signature morphisms is a (weak) amalgamation square (for $\text{Mod}^{\mathcal{HI}}$) if the underlying square of signature morphisms in $\mathcal{I}$ is a (weak) amalgamation square.*

Corollary 2. *$\mathcal{D}^{\mathcal{HI}}$ is adequate for $\text{Mod}^{\mathcal{HI}}$.*

The Satisfaction Relation

For any $(\Sigma, \text{Nom}, \Lambda)$-model (M, R) and for any $s \in |R|$:

- $(M, R) \models^s \rho$ iff $M_s \models^{\mathcal{I}} \rho$; when $\rho \in \text{Sen}^{\mathcal{I}}(\Sigma)$,
- $(M, R) \models^s i$ iff $i^R = s$; when $i \in \text{Nom}$,
- $(M, R) \models^s \neg\rho$ iff $(M, R) \not\models^s \rho$,
- $(M, R) \models^s \rho \vee \rho'$ iff $(M, R) \models^s \rho$ or $(M, R) \models^s \rho'$,
- $(M, R) \models^s \rho \wedge \rho'$ iff $(M, R) \models^s \rho$ and $(M, R) \models^s \rho'$,
- $(M, R) \models^s \rho \to \rho'$ iff $(M, R) \models^s \rho$ implies that $(M, R) \models^s \rho'$,
- $(M, R) \models^s [\lambda](\rho_1, \ldots \rho_n)$ iff $(M, R) \models^{s_i} \rho_i$ for $1 \leq i \leq n$, $\lambda \in \Lambda_{n+1}$ and any $(s, s_1, \ldots, s_n) \in R_\lambda$,
- $(M, R) \models^s @_j \rho$ iff $(M, R) \models^{j^R} \rho$,
- $(M, R) \models^s (\forall\chi)\rho$ iff $(M', R') \models^s \rho$ for any (M', R') such that $\text{Mod}^{\mathcal{HI}}(\chi)(M', R') = (M, R)$, and
- $(M, R) \models^s (\exists\chi)\rho$ iff $(M', R') \models^s \rho$ for some (M', R') such that $\text{Mod}^{\mathcal{HI}}(\chi)(M', R') = (M, R)$.

We write $(M, R) \models \rho$ iff $(M, R) \models^s \rho$ for any $s \in |R|$.

The Satisfaction Condition

Theorem 2. *Let* $\Delta = (\Sigma, \mathrm{Nom}, \Lambda)$ *and* $\Delta' = (\Sigma', \mathrm{Nom}', \Lambda')$ *be two* $\mathcal{HI}$-*signatures and* $\varphi : \Delta \to \Delta'$ *a morphism of signatures. For any* $\rho \in \mathrm{Sen}^{\mathcal{HI}}(\Delta)$, $(M', R') \in |\mathrm{Mod}^{\mathcal{HI}}(\Delta')|$, *and* $s \in |R|$

$$\mathrm{Mod}^{\mathcal{HI}}(\varphi)(M', R') \models^s \rho \text{ iff } (M', R') \models^s \mathrm{Sen}^{\mathcal{HI}}(\varphi)(\rho). \tag{3}$$

Proof. The proof is by induction on the structure of ρ.

Corollary 3 (The Satisfaction Condition). $(\mathrm{Sign}^{\mathcal{HI}}, \mathrm{Sen}^{\mathcal{HI}}, \mathrm{Mod}^{\mathcal{HI}}, \models^{\mathcal{HI}})$ *is an institution.*

Example 2 ($\mathcal{HPL}$). Let $\mathcal{APL}$ be the sub-institution of $\mathcal{PL}$ whose sentences are the propositional symbols. Applying the hybridization method described above to $\mathcal{APL}$ and fixing $\Lambda_2 = \{\Box\}$ and $\Lambda_n = \emptyset$ for each $n \neq 2$, we obtain the institution of the "standard" *hybrid propositional logic* (without state quantifiers): the category of signatures is $\mathrm{Sign}^{\mathcal{HPL}} = \mathbf{Set} \times \mathbf{Set}$ with objects denoted by (P, Nom) and morphisms by $(\varphi_{\mathrm{Sig}}, \varphi_{\mathrm{Nom}})$; sentences are the usual hybrid propositional formulas, i.e., modal formulas closed by boolean connectives, $\Box$, and by the operator $@_i, i \in \mathrm{Nom}$; models consists of pairs $\mathcal{P} = (M, R)$ where R consists of a carrier set, interpretations $i^R \in S$ for each $i \in \mathrm{Nom}$, and a binary relation $\Box^R \subseteq |R| \times |R|$, and for each $s \in |R|$, M_s is a propositional model, i.e., a function $M_s : P \to \{\top, \bot\}$. The quantification space $\mathcal{D}^{\mathcal{HPL}}$ is the trivial one, consisting of the identities, which means this process is a quantifier-free hybridization. The satisfaction relation is defined as above on top of the propositional satisfaction relation, i.e., $\mathcal{P} \models^s p$ iff $M_s(p) = \top$.

A challenging issue concerns finding suitable quantification spaces to capture other versions of hybrid propositional logic. For instance, it would be interesting, along the hybridization process, to capture the quantifiers $\mathbf{A}$ and $\mathbf{E}$, where $\mathbf{A}\rho$ (respectively, $\mathbf{E}\rho$) means that "*ρ is true in all the states of the model*" (respectively, "*ρ is true in some state of the model*") [1]. This can be achieved by taking as a quantification space the extensions of signatures with nominal symbols; for instance one may express $\mathcal{P} \models \mathbf{E}\rho$ by $\mathcal{P} \models^s (\exists i)@_i\rho$.

Example 3 ($\mathcal{HFOL}$, $\mathcal{HEQ}$). The application of the hybridization method to $\mathcal{FOL}$ taking as a quantification space signature extensions both with $\mathcal{FOL}$ variables and variables over nominals, captures the state-variables quantification of *first-order hybrid logic* of [4].

Binding "state variables" to the point of evaluation highly increase the expressive power of a hybrid logic, which is enabled through the binder operator $\downarrow$ (e.g. [2,4]). This may be achieved by taking i-expansions $\chi : (\Sigma, \mathrm{Nom}, \Lambda) \hookrightarrow (\Sigma, \mathrm{Nom} \uplus \{i\}, \Lambda)$ as a quantification space and including, when defining satisfaction, the condition

– $\mathcal{P} \models^s (\downarrow \chi)\rho$ iff for any χ-expansion $\mathcal{P}'$ of $\mathcal{P}$ such that $i^R = s$, we have $\mathcal{P}' \models^s \rho$.

As a final example, let us mention the hybridization of $\mathcal{EQ}$ with the trivial quantification space. The resulting hybrid equational institution provides a suitable setting for specifying evolving systems in which each state is endowed with a specific algebra [20].

4 $\mathcal{FOL}$ as a Support to Hybrid Specification

This section studies the existence of encodings of hybrid institutions into $\mathcal{FOL}$. The relevance of such encodings is to provide proof theoretic support to hybrid specifications. In particular, we show that any encoding of the base institution $\mathcal{I}$ to $\mathcal{FOL}$ may be lifted to an encoding of the quantifier-free hybrid institution $\mathcal{HI}$ to $\mathcal{FOL}$. Our approach to logic encodings relies upon the concept of comorphism, recalled below from the literature (e.g. [17]).

Definition 4 (Comorphisms). *Given institutions* $\mathcal{I} = (\mathrm{Sign}, \mathrm{Sen}, \mathrm{Mod}, \models)$ *and* $\mathcal{I}' = (\mathrm{Sign}', \mathrm{Sen}', \mathrm{Mod}', \models')$ *a* comorphism $(\Phi, \alpha, \beta) : \mathcal{I} \to \mathcal{I}'$ *consists of*

1. *a functor* $\Phi : \mathrm{Sign} \to \mathrm{Sign}'$,
2. *a natural transformation* $\alpha : \mathrm{Sen} \Rightarrow \Phi; \mathrm{Sen}'$, *and*
3. *a natural transformation* $\beta : \Phi^{op}; \mathrm{Mod}' \Rightarrow \mathrm{Mod}$

such that the following satisfaction condition *holds*

$$M' \models'_{\Phi(\Sigma)} \alpha_\Sigma(e) \;\; \textit{iff} \;\; \beta_\Sigma(M') \models_\Sigma e$$

for each signature $\Sigma \in |\mathrm{Sign}|$, $\Phi(\Sigma)$*-model* M'*, and* Σ*-sentence* e.

The comorphism is conservative *whenever, for each* Σ*-model* M *in* $\mathcal{I}$*, there exists a* $\Phi(\Sigma)$*-model* M' *in* $\mathcal{I}'$ *such that* $M = \beta_\Sigma(M')$.

The following is a consequence of conservativity, with the important proof theoretic implication that can be proved properties in the source institution by using the proof system of the target institution in a sound and complete way.

Fact 2. *For any set* $\Gamma \subseteq \mathrm{Sen}(\Sigma)$ *and sentence* $\rho \in \mathrm{Sen}(\Sigma)$,

$$\Gamma \models_\Sigma \rho \;\; \textit{iff} \;\; \alpha_\Sigma(\Gamma) \models'_{\Phi(\Sigma)} \alpha_\Sigma(\rho).$$

Example 4. One may legitimately wonder about the existence of a canonical embedding of the base institution $\mathcal{I}$ into its hybridization $\mathcal{HI}$ in the form of a comorphism $(\Phi, \alpha, \beta) : \mathcal{I} \to \mathcal{HI}$. The answer is as follows:

- $\Phi(\Sigma) = (\Sigma, \{i\}, \emptyset)$,
- $\alpha_\Sigma(\rho) = @_i \rho$, and
- $\beta_\Sigma(M, R) = M_{i^R}$.

It is easy to show that this is a conservative comorphism.

Thus, let $\mathcal{HI}$ be the quantifier-free hybridization of institution $\mathcal{I}$. Given any comorphism $\mathcal{I} \xrightarrow{(\Phi,\alpha,\beta)} \mathcal{FOL}$ we define a comorphism $\mathcal{HI} \xrightarrow{(\Phi',\alpha',\beta')} \mathcal{FOL}$ by

Translation of Signatures:

$\Phi'(\Sigma, \text{Nom}, \Lambda) = (S_\Sigma + \{\text{ST}\}, \overline{F_\Sigma} + \overline{\text{Nom}}, \overline{P_\Sigma} + \overline{\Lambda})$ where

- $\Phi(\Sigma) = (S_\Sigma, P_\Sigma, P_\Sigma)$ (a $\mathcal{FOL}$ signature);
- $\overline{F_\Sigma} = \begin{cases} (\overline{F_\Sigma})_{\text{ST}w \to s} = (F_\Sigma)_{w \to s} & \text{for any } s \in S_\Sigma, w \in S_\Sigma^* \\ \emptyset, & \text{for the other cases} \end{cases};$
- $\overline{P_\Sigma} = \begin{cases} (\overline{P_\Sigma})_{\text{ST}w} = (P_\Sigma)_w & \text{for any } w \in S_\Sigma^*; \\ \emptyset, & \text{for the other cases} \end{cases}$
- $\overline{\text{Nom}} = \{i : \ \to \text{ST} \mid i \in \text{Nom}\}$
- $\overline{\Lambda} = \{\lambda : \ \text{ST}^n \mid \lambda \in \Lambda_n\}$.

Translation of Models:

$\beta'_{(\Sigma, \text{Nom}, \Lambda)}(M) = (M', R)$ where

- R is the reduct $M \restriction_{(\{\text{ST}\}, \overline{\text{Nom}}, \overline{\Lambda})}$, and
- $M' : \text{ST}^R \to |\text{Mod}^{\mathcal{I}}(\Sigma)|$ is defined for each $s \in S$ by $M'_s = \beta_\Sigma(M_s)$ where M_s is the $\Phi(\Sigma) = (S_\Sigma, P_\Sigma, P_\Sigma)$-model defined by
  - for each sort $\in S_\Sigma$, $\text{sort}^{M_s} = \text{sort}^M$;
  - for each $f \in F_\Sigma$, $f^{M_s}(m) = f^M(s, m)$;
  - for each $\pi \in P_\Sigma$, $m \in \pi^{M_s}$ iff $(s, m) \in \pi^M$.

Translation of Sentences:

$\alpha'_{(\Sigma, \text{Nom}, \Lambda)}(\rho) = \forall x\, \alpha'^x_{(\Sigma, \text{Nom}, \Lambda)}(\rho)$, where
$\alpha'^x_{(\Sigma, \text{Nom}, \Lambda)} : \text{Sen}^{\mathcal{HI}}(\Sigma, \text{Nom}, \Lambda) \to \text{Sen}^{\mathcal{FOL}}(\Phi'(\Sigma, \text{Nom}, \Lambda) \cup \{x\})$ with x a constant of sort ST, is defined by

- for each $\rho \in \text{Sen}^{\mathcal{I}}(\Sigma)$, $\alpha'^x(\rho) = \alpha^x(\alpha_\Sigma(\rho))$ where $\alpha^x_{(\Sigma, \text{Nom}, \Lambda)} :$ $\text{Sen}^{\mathcal{FOL}}(\Phi(\Sigma)) \to \text{Sen}^{\mathcal{FOL}}(\Phi'(\Sigma, \text{Nom}, \Lambda) \cup \{x\})$ is defined by
  - $\alpha^x(t \approx t') = \alpha^x(t) \approx \alpha^x(t')$ where $\alpha^x(f(t_1, \ldots, t_n))$ $= f(x, \alpha^x(t_1), \ldots, \alpha^x(t_n))$;
  - $\alpha^x(\pi(t)) = \pi(x, \alpha^x(t))$;
  - $\alpha^x(\rho_1 \odot \rho_2) = \alpha^x(\rho_1) \odot \alpha^x(\rho_2)$, $\odot \in \{\vee, \wedge, \to\}$;
  - $\alpha^x(\neg\rho) = \neg\alpha^x(\rho)$;
  - $\alpha^x(\forall y\, \rho) = \forall y\, \alpha^x(\rho)$;
- $\alpha'^x(i) = i \approx x$, $i \in \text{Nom}$;
- $\alpha'^x(@_i \rho) = \alpha'^i(\rho)$;
- $\alpha'^x([\lambda](\rho_1, \ldots, \rho_n)) = \forall y_1, \ldots, y_n\, (\lambda(x, y_1, \ldots, y_n) \to \bigwedge_{1 \leq i \leq n} \alpha'^{y_i}(\rho_i))$;
- $\alpha'^x(\rho_1 \odot \rho_2) = \alpha'^x(\rho_1) \odot \alpha'^x(\rho_2)$, $\odot \in \{\vee, \wedge, \to\}$;

Lemma 1. *For any $\Delta \in |\text{Sign}^{\mathcal{HI}}|$, $\rho \in \text{Sen}^{\mathcal{FOL}}(\Phi(\Sigma))$, $M' \in \text{Mod}^{\mathcal{FOL}}(\Phi'(\Delta))$ and $s \in S$,*

$$M'_s \models_{\Phi(\Sigma)} \rho \ \textit{ iff } \ M'^s \models_{\Phi'(\Delta)+x} \alpha^x(\rho), \tag{4}$$

where M'^s denotes the expansion of M' to $\Phi'(\Delta) + x$ defined by $x^{M'^s} = s$.

Theorem 3. *For any* $\Delta \in |\mathrm{Sign}^{\mathcal{HI}}|$, $\rho \in \mathrm{Sen}^{\mathcal{HI}}(\Delta)$ $M' \in \mathrm{Mod}^{\mathcal{FOL}}(\Phi'(\Delta))$ *and* $s \in S$,

$$\beta'_\Delta(M')(\models^{\mathcal{HI}}_\Delta)^s \rho \text{ iff } M'^s \models^{\mathcal{FOL}}_{\Phi'(\Delta)+x} \alpha'^x_\Delta(\rho), \tag{5}$$

where M'^s *denotes the expansion of* M' *to* $\Phi'(\Delta) + x$ *defined by* $x^{M'^s} = s$.

Proof. The proof is by induction on the structure of ρ.

Corollary 4 (Satisfaction condition for (Φ', α', β')). *(Φ', α', β') is comorphism $\mathcal{HI} \to \mathcal{FOL}$, i.e. for any $\Delta \in |\mathrm{Sign}^{\mathcal{HI}}|$, $\rho \in \mathrm{Sen}^{\mathcal{HI}}(\Delta)$ and $M' \in \mathrm{Mod}^{\mathcal{FOL}}(\Phi'(\Delta))$,*

$$\beta'_\Delta(M') \models^{\mathcal{HI}}_\Delta \rho \text{ iff } M' \models^{\mathcal{FOL}}_{\Phi'(\Delta)} \alpha'_\Delta(\rho).$$

Example 5 ($\mathcal{HEQ}2\mathcal{FOL}$). A simple, but useful example of the construction proposed above arises by its application to the embedding of $\mathcal{EQL}$ into $\mathcal{FOL}$, entailing a comorphism $\mathcal{HEQL} \to \mathcal{FOL}$.

5 Conclusions and Further Work

The paper's contribution is twofold: first it defines a method to hybridize arbitrary institutions; then it is shown that a comorphism from an arbitrary institution to $\mathcal{FOL}$ gives rise to another comorphism from its (quantifier-free) hybridization to $\mathcal{FOL}$.

Beyond their intrinsic theoretical interest, the application of these results seems promising. On the one hand, hybridization of logics is achieved, by this method, in a systematic way which applies to a broad class of logics. On the other, our second result paves the way to effective tool support to reasoning about hybrid specifications, by resorting to $\mathcal{FOL}$-oriented verification tools.

This work also opens a number of interesting research directions. We discuss below the set of main topics in our agenda.

Remark 1. An aspect of our method, which increases the complexity of hybridizing arbitrary institutions, is the need for "desconstructing" the base institution. For instance, in order to hybridize $\mathcal{FOL}$, we have to take in the role of a base institution its sub-institution of atomic formulas (without quantifiers and boolean connectives). The same happens in the hybridization of propositional logic (see Ex. 2). In order to overcome this situation, it is necessary to find a way to proscribe the overloading of connectives at the base and hybrid levels. The problem may be solved by resorting to the (abstract) notion of boolean connective (cf. [13, Chap. 3]). For instance, suppose that the institution $\mathcal{I}$ has semantical negation, i.e., that for any $\rho \in \mathrm{Sen}^{\mathcal{I}}(\Sigma)$ there is a $\rho' \in \mathrm{Sen}^{\mathcal{I}}(\Sigma)$ such that for any $\mathcal{M} \in |\mathrm{Mod}^{\mathcal{I}}(\Sigma)|$, $\mathcal{M} \models^{\mathcal{I}}_\Sigma \rho'$ iff it is false that $\mathcal{M} \models^{\mathcal{I}}_\Sigma \rho$. Then, in order to avoid the connective negation, we may replace, in the definition of the hybrid sentences, negation introduction by

$$\text{If } \rho \in \mathrm{Sen}^{\mathcal{HI}}(\Sigma, \mathrm{Nom}, \Lambda) \setminus \mathrm{Sen}^{\mathcal{I}}(\Sigma), \text{ then } \neg\rho \in \mathrm{Sen}^{\mathcal{HI}}(\Sigma, \mathrm{Nom}, \Lambda),$$

and similarly for the other boolean connectives. This seems to be enough to obtain the $\mathcal{HFOL}$ from $\mathcal{FOL}$.

Hybridization of modal logics is a more challenging question: *how to introduce nominals into institutions that already have Kripke semantics?* For instance it is known that $\mathcal{CTL}$ defines an institution (cf. [8]) and that there are hybrid extensions of this logic currently being studied (cf. [19,24]). Actually, this sort of hybridization falls out of the scope of the method discussed in this paper. Its application to $\mathcal{CTL}$ leads to a kind of "graph of graphs", raising the question of how such a *double modalization* can be avoided. Certainly, there are tricky technical aspects to overcome. However, the hybridization of a (concrete) institution with Kripke semantics, i.e., the introduction of nominals and a satisfaction operator on a institution whose models are already of the form $(S, (M_s)_{s\in S})$ seems to be an easy task. Hence, an answer to this problem resorts to the decomposition of the hybridization process into two steps: a *modalization* followed by a *hybridization*. The former, may be defined as in [15] just making a straightforward generalization to sets of modal symbols Λ. The latter is then applied to the resulting institution.

Remark 2 (Calculus for hybrid institutions). Comparing the calculus of [2] for hybrid propositional logic with that of [4] for hybrid first-order logic, a common structure pops out: they "share" rules involving sentences with nominals and satisfaction operators (i.e., formulas with "hybrid nature") and have specific rules to reason about "atomic sentences" that come from the base institution. Hence, it makes sense to think about the development of a general proof calculus for hybrid institutions built on top of the calculus equipping the base institution in the style of [3,10].

Remark 3 (Modal symbols quantification). Another interesting point to explore is the power of quantification over modal symbols, for instance by considering in the quantification space inclusions of the form $\Lambda \hookrightarrow \Lambda + \lambda$. Using this quantification it seems possible to express general properties about the state space of a model. For instance, $\mathcal{P} \models^s (\forall\lambda)p \to [\lambda]p$ means that if p holds in s then it is invariant in all the model and $\mathcal{P} \models^s (\forall\lambda)p \to [\lambda]q$ says that if p holds on s then q holds in another state of the model.

Remark 4 (New case studies). There are many interesting hybrid institutions that may be obtained by application of the method proposed in this paper. Particularly interesting case studies are the derivation of both *intuitionistic hybrid logic* [9,6] and *many-valued hybrid logic* [18] from their respective bases.

Remark 5 (On encoding hybridizations to $\mathcal{FOL}$). An important property of logic encodings, which guarantees the sound and complete borrowing of formal reasoning from the target into the source of the encoding, is that they keep unchanged the consequence relation of the encoded logic (see Fact 2). In the case of the encoding $\mathcal{HI} \to \mathcal{FOL}$ defined as a comorphism in Section 4 this would have followed immediately if $(\Phi', \alpha', \beta') : \mathcal{HI} \to \mathcal{FOL}$ were conservative which in its turn, should be a natural consequence of the conservativeness of $(\Phi, \alpha, \beta) : \mathcal{I} \to \mathcal{FOL}$. Unfortunately this latter step does not work in general, unless the approach is extended by considering also a 'rigid' part for signatures and models as in [15].

Our current encoding of hybridizations to $\mathcal{FOL}$ is limited in its applicability by the fact that it applies only to encodings of the base institution that can be expressed as plain comorphisms to $\mathcal{FOL}$. This means that our current result may be in reality applied to hybridizations of various fragments of $\mathcal{FOL}$ but not to any of the myriad of specification logics that are encoded into $\mathcal{FOL}$ by the so-called 'theoroidal comorphisms' [17]. We plan to extend our encoding result to this more general situation, thus widely enlarging the $\mathcal{FOL}$-oriented formal reasoning support for hybridized logics.

We also plan to extend our encoding result to quantified hybridizations.

Remark 6 (Model theory for hybridized institutions). A deeper development of the model theory of a specification formalism always results into a better understanding of its specification power. Our general hybridization method opens the door for a general institution-independent approach to the model theory of hybrid(ized) logics by using techniques from [13]. We believe that the end result of such investigation would make yet another point in favour of the hybrid variants of modal logics, as they are expected to exhibit better model theoretic properties than their non-hybrid variants.

In particular we will consider extending the method of ultraproducts of [15] from modalized to hybridized institutions, to investigate a general method of diagrams and the existence of initial semantics for hybridized institutions. The latter has a special specification theoretic significance: it would give foundational support for classical algebraic specification style with hybrid(ized) logics. The method of diagrams, which is a very common model theoretic property of logics and a technique that pervades a lot of model theoretic results (see [12,13] for its institution-independent rendering), unfortunately fails on modal logics. However because of the special "hybrid features" we expect it to hold in some form in hybrid(ized) logics.

Acknowledgments. This research was partially supported by FCT (the Portuguese Foundation for Science and Technology) under contract `PTDC/EIA-CCO/108302/2008` (the MONDRIAN project) and CIDMA at University of Aveiro. M. Martins was further supported by the project *Nociones de Completud*, reference FFI2009-09345 (Spain), and A. Madeira was also supported by `SFRH/BDE/33650/2009`, a PhD grant jointly supported by FCT and *Critical Software S.A., Portugal.*

References

1. Areces, C., Blackburn, P., Delany, S.R.: Bringing them all together. Journal of Logic and Computation 11, 657–669 (2001)
2. Blackburn, P.: Representation, reasoning, and relational structures: a hybrid logic manifesto. Logic Journal of IGPL 8(3), 339–365 (2000)
3. Borzyszkowski, T.: Logical systems for structured specifications. Theoretical Computer Science 286(2), 197–245 (2002)
4. Braüner, T.: Natural deduction for first-order hybrid logic. Journal of Logic, Language and Information 14, 173 (2005)

5. Braüner, T.: Hybrid Logic and its Proof-Theory. Applied Logic Series, vol. 37. Springer, Heidelberg (2011)
6. Braüner, T., de Paiva, V.: Intuitionistic hybrid logic. J. Applied Logic 4(3), 231–255 (2006)
7. Burstall, R., Goguen, J.: The semantics of Clear, a specification language. In: Bjorner, D. (ed.) 1979 Copenhagen Winter School on Abstract Software Specification. LNCS, vol. 86, pp. 292–332. Springer, Heidelberg (1980)
8. Cengarle, M.V.: The temporal logic institution. Technical report, Universitat Munchen, Insitut fur informatik (1998)
9. Chadha, R., Macedonio, D., Sassone, V.: A hybrid intuitionistic logic: Semantics and decidability. Journal of Logic and Computation 16, 2006 (2005)
10. Codescu, M., Găină, D.: Birkhoff completeness in institutions. Logica Universalis 2(2), 277–309 (2008)
11. Demri, S., Lazic, R., Nowak, D.: On the freeze quantifier in constraint ltl: decidability and complexity. Technical Report LSV-05-03, Laboratoire Specification et Verification (2005)
12. Diaconescu, R.: Elementary diagrams in institutions. Journal of Logic and Computation 14(5), 651–674 (2004)
13. Diaconescu, R.: Institution-independent Model Theory. Birkhäuser, Basel (2008)
14. Diaconescu, R.: Quasi-boolean encodings and conditionals in algebraic specification. Journal of Logic and Algebraic Programming 79(2), 174–188 (2010)
15. Diaconescu, R., Stefaneas, P.S.: Ultraproducts and possible worlds semantics in institutions. Theor. Comput. Sci. 379(1-2), 210–230 (2007)
16. Goguen, J., Burstall, R.: Institutions: Abstract model theory for specification and programming. Journal of the Association for Computing Machinery 39(1), 95–146 (1992)
17. Goguen, J., Roşu, G.: Institution morphisms. Formal Aspects of Computing 13, 274–307 (2002)
18. Hansen, J., Bolander, T., Braüner, T.: Many-valued hybrid logic. In: Areces, C., Goldblatt, R. (eds.) Advances in Modal Logic, pp. 111–132. College Publications (2008)
19. Kara, A., Lange, M., Schwentick, T., Weber, V.: On the hybrid extension of CTL and CTL+. In: CoRR, abs/0906.2541 (2009)
20. Martins, M., Madeira, A., Barbosa, L.: Reasoning about complex requirements in a uniform setting. Technical report, DI-CCTC-2-1-2011 (2011)
21. Passy, S., Tinchev, T.: An essay in combinatory dynamic logic. Inf. Comput. 93, 263–332 (1991)
22. Prior, A.N.: Past, Present and Future. Oxford University Press, Oxford (1967)
23. Sattler, U., Vardi, M.Y.: The hybrid μ-calculus. In: Goré, R., Leitsch, A., Nipkow, T. (eds.) IJCAR 2001. LNCS (LNAI), vol. 2083, pp. 76–91. Springer, Heidelberg (2001)
24. Weber, V.: On the complexity of branching-time logics. In: Grädel, E., Kahle, R. (eds.) CSL 2009. LNCS, vol. 5771, pp. 530–545. Springer, Heidelberg (2009)

Linearly-Used State in Models of Call-by-Value

Rasmus Ejlers Møgelberg[1,*] and Sam Staton[2,**]

[1] IT University of Copenhagen, Denmark
[2] Computer Laboratory, University of Cambridge, UK

Abstract. We investigate the phenomenon that *every monad is a linear state monad*. We do this by studying a fully-complete state-passing translation from an impure call-by-value language to a new linear type theory: the enriched call-by-value calculus. The results are not specific to store, but can be applied to any computational effect expressible using algebraic operations, even to effects that are not usually thought of as stateful. There is a bijective correspondence between generic effects in the source language and state access operations in the enriched call-by-value calculus.

From the perspective of categorical models, the enriched call-by-value calculus suggests a refinement of the traditional Kleisli models of effectful call-by-value languages. The new models can be understood as enriched adjunctions.

1 Introduction

Computational effects such as store effects, input/output and control effects are usually associated with the imperative style of programming, and functional programming languages exhibiting such behaviour are thought of as "impure". However, computational effects can be encapsulated within a purely functional language by the use of monads [12]. The central idea behind this is to distinguish between a type of values such as (nat), and a type of computations $T(\text{nat})$ that may return a value of type nat but can also do other things along the way. Imperative behaviour can then be encoded using *generic effects* in the sense of Plotkin and Power [19]. For example, one can add global store by adding a pair of terms, $\texttt{assign}_l \colon \text{Val} \to T(1)$ and $\texttt{deref}_l \colon T(\text{Val})$, for each cell l in the store, or one can add nondeterminism by adding a constant $\texttt{random}\colon T(1+1)$ computing a random boolean. Computational effects that can be described using generic effects are called *algebraic* and these account for a wide range of effects with the notable exception of control effects such as continuations.

It is striking that there is no notion of state in the theory of monads. After all, imperative behaviour is often about changing or branching on the state of the machine. The notion of state is most naturally associated with certain effects

* Research supported by the Danish Agency for Science, Technology and Innovation.
** Research supported by EPSRC Fellowship EP/E042414/1 and ANR Projet CHOCO.

A. Corradini, B. Klin, and C. Cîrstea (Eds.): CALCO 2011, LNCS 6859, pp. 298–313, 2011.

like store effects, but in this paper we shall see that all algebraic effects can be viewed in this way.[1]

The notion of state plays an important role in operational semantics and Hoare logic. Indeed, Plotkin and Power [15] have suggested that configurations, i.e. pairs $\langle M, s\rangle$ consisting of a term M to be evaluated and the current state s of the machine, might form a basis for defining general operational semantics.

In this paper we show how the theory of algebraic effects can be formulated by taking the notion of state as primitive, rather than the notion of monad. On the syntactic side we introduce the *enriched call-by-value calculus* (ECBV). We show that a special state type in ECBV gives rise to a language that is equivalent to a fine-grained monadic call-by-value calculus (FGCBV) [11]. On the semantic side we introduce a notion of enriched call-by-value model which generalises monad models.

Central to our treatment of state is the idea of linear usage: computations cannot copy the state and save it for later, nor can they discard the state and insert a new one instead. This special status of the state was already noted by Strachey [25] and Scott [24] and has been developed by O'Hearn and Reynolds [13].

Linear usage of state can be expressed syntactically using a *linear state passing style*, in which a stateful computation of type $A \to B$ is considered as a linear map of type $!A \otimes \underline{S} \multimap !B \otimes \underline{S}$. The type $\underline{S}$ of states must be used linearly, but A and B can be used arbitrarily. These type constructions form the basis of ECBV, which can be considered as a kind of non-commutative linear logic that is expressive enough to describe the linear usage of state.

Earlier metalanguages for effects, such as the monadic metalanguage [12], call-by-push-value [10], and the enriched effect calculus [2] have an explicit monadic type constructor. There is no monadic type constructor in ECBV: there is a state type $\underline{S}$ instead. Still, in this fragment one can express all algebraic notions of effects, even the ones that we are not used to thinking of as "state-like", using what we call *state access operations*. For example, the generic effects $\texttt{assign}_l$, $\texttt{deref}_l$, and $\texttt{random}$ correspond to the following state access operations:

$$\texttt{write}_l\colon !\mathrm{Val} \otimes \underline{S} \multimap \underline{S}, \quad \texttt{read}_l\colon \underline{S} \multimap !\mathrm{Val} \otimes \underline{S}, \quad \texttt{random}\colon \underline{S} \multimap !(1+1) \otimes \underline{S}\,. \quad (1)$$

The equivalence of FGCBV and ECBV is proved for extensions of the two calculi along any algebraic effect theory. The generalisation is formulated using a notion of effect theory [16] which captures notions of algebraic effects.

The categorical models of ECBV provide a new general notion of model for call-by-value languages. In brief, an enriched model consists of two categories **V** and **C** such that **V** has products and distributive coproducts, and **C** is enriched in **V** with copowers and coproducts. The objects of **V** interpret ordinary "value" types, and the objects of **C** interpret "computation" types (such as the state type $\underline{S}$) which must be used linearly. This class of models encompasses all Kleisli categories (which have been axiomatised as Freyd categories) and many

[1] For this reason we use the terminology "store effects" for the specific (memory access operations) and reserve "state" for the general notion.

Eilenberg-Moore categories (which provide a natural notion of model for call-by-push-value and the enriched effect calculus).

Our approach to proving the equivalence of FGCBV and ECBV is semantic. We show that the traditional models of effectful call-by-value languages, using monads and Kleisli constructions, form a coreflective subcategory of the models of ECBV. The state-passing translation is the unit of the coreflection. Our main semantic result provides a bijective correspondence between comodel structures on a state object and model structures on the induced linear state monad. This extends Plotkin and Power's correspondence between algebraic operations and generic effects [19] with a third component: state access operations.

The enriched call-by-value calculus is a fragment of the enriched effect calculus (EEC, [2]). In Section 8 we show that EEC is a conservative extension of ECBV. This shows that the linear state monad translation from FGCBV into EEC is fully complete: any term of translated type in the target language corresponds to a unique term in the source language. This result indicates that EEC is a promising calculus for reasoning about linear usage of effects. The related paper [3] shows how the linear-use continuation passing translation arises from a natural dual model construction on models of EEC. In fact, from the point of view of EEC the two translations are surprisingly similar: the linearly used state translation is essentially dual to the linearly used continuations translation.

2 Source Calculus: Fine-Grained Call-by-Value

Our source language is a call-by-value language equipped with an equational theory to be thought of as generated by some operational semantics, as in [14]. We use a variant of fine-grained call-by-value [11], because the explicit separation of judgements into value and producer judgements fits well with a similar division in the target language.

We use α to range over type constants. The types are given by the grammar

$$\sigma ::= \alpha \mid 1 \mid \sigma \times \sigma \mid 0 \mid \sigma + \sigma \mid \sigma \rightharpoonup \sigma\,.$$

The fine-grained call-by-value calculus (FGCBV) has two typing judgements, one for values and one for producers. These are written $\Gamma \vdash^{v} V : \sigma$ and $\Gamma \vdash^{p} M : \sigma$. The latter should be thought of as typing computations which produce values in the type judged but may also perform side-effects along the way. In both judgements the variables of the contexts are to be considered as placeholders for values. The function space $\rightharpoonup$ is a call-by-value one, which takes a value and produces a computation. In fact this language is equivalent to Moggi's monadic λ_c: the type construction $(1 \rightharpoonup (-))$ is a monad. Typing rules along with equality rules are given in Figure 1.

We can define derived case constructs on producer terms:

$$\begin{aligned}&\texttt{case}^{p}\ M\ \texttt{of}\ (\texttt{in}_1(x_1).N_1; \texttt{in}_2(x_2).N_2)\\ &\qquad \stackrel{\text{def}}{=}\ M\,\texttt{to}\,x.\,(\texttt{case}\ x\ \texttt{of}\ (\texttt{in}_1(x_1).\lambda w\!:\!1.\,N_1; \texttt{in}_2(x_2).\lambda w\!:\!1.\,N_2))(\star)\\ &\texttt{image}^{p}(M)\ \stackrel{\text{def}}{=}\ M\,\texttt{to}\,x.\,(\texttt{image}(x))\,\star\end{aligned}$$

$$\frac{}{\Gamma, x\colon \sigma, \Gamma' \vdash^v x\colon\sigma} \qquad \frac{}{\Gamma \vdash^v \star\colon 1} \qquad \frac{\Gamma \vdash^v V\colon \sigma_1 \times \sigma_2}{\Gamma \vdash^v \pi_i(V)\colon \sigma_i} \qquad \frac{\Gamma \vdash^v V\colon \sigma_i}{\Gamma \vdash^v \mathtt{in}_i(V)\colon \sigma_1 + \sigma_2}$$

$$\frac{\Gamma \vdash^v V_1\colon \sigma_1 \quad \Gamma \vdash^v V_2\colon \sigma_2}{\Gamma \vdash^v \langle V_1, V_2\rangle\colon \sigma_1 \times \sigma_2} \qquad \frac{\Gamma \vdash^v V\colon \sigma_1 + \sigma_2 \quad \Gamma, x_i\colon \sigma_i \vdash^v W_i\colon \tau \quad (i=1,2)}{\Gamma \vdash^v \mathtt{case}\ V\ \mathtt{of}\ (\mathtt{in}_1(x_1).W_1; \mathtt{in}_2(x_2).W_2)\colon \tau}$$

$$\frac{\Gamma \vdash^v V\colon 0}{\Gamma \vdash^v \mathtt{image}(V)\colon \sigma} \qquad \frac{\Gamma \vdash^v V\colon \sigma}{\Gamma \vdash^p \mathtt{return}\, V\colon \sigma} \qquad \frac{\Gamma \vdash^p M\colon \sigma \quad \Gamma, x\colon \sigma \vdash^p N\colon \tau}{\Gamma \vdash^p M\,\mathtt{to}\,x.\,N\colon \tau}$$

$$\frac{\Gamma, x\colon \sigma \vdash^p N\colon \tau}{\Gamma \vdash^v \lambda x\colon\sigma.\, N\colon \sigma \rightharpoonup \tau} \qquad \frac{\Gamma \vdash^v V\colon \sigma \rightharpoonup \tau \quad \Gamma \vdash^v W\colon \sigma}{\Gamma \vdash^p V\,W\colon \tau}$$

$$M = \star \qquad \pi_i(\langle V_1, V_2\rangle) = V_i \qquad \langle \pi_1(V), \pi_2(V)\rangle = V \qquad \mathtt{image}(V) = W[V/x]$$

$$\mathtt{case}\ \mathtt{in}_i(V)\ \mathtt{of}\ (\mathtt{in}_1(x_1).W_1; \mathtt{in}_2(x_2).W_2) = W_i[V/x_i] \qquad \lambda x\colon\sigma.\, M(V) = M[V/x]$$

$$\mathtt{case}\ V\ \mathtt{of}\ (\mathtt{in}_1(x_1).W[\mathtt{in}_1(x_1)/x]; \mathtt{in}_2(x_2).W[\mathtt{in}_2(x_2)/x]) = W[V/x] \qquad \lambda x\colon\sigma.\,(V\,x) = V$$

$$M\,\mathtt{to}\,x.\,\mathtt{return}\,x = M \qquad \mathtt{return}\,V\,\mathtt{to}\,x.\,N = N[V/x]$$

$$(M\,\mathtt{to}\,x.\,N)\,\mathtt{to}\,y.\,P = M\,\mathtt{to}\,x.\,(N\,\mathtt{to}\,y.\,P)$$

Fig. 1. Fine-grained call-by-value. (Equality rules subject to usual conventions.)

where z, w are fresh variables. These constructions have derived typing rules

$$\frac{\Gamma \vdash^p M\colon \sigma_1 + \sigma_2 \quad \Gamma, x_i\colon \sigma_i \vdash^p N_i\colon \tau\ (i=1,2)}{\Gamma \vdash^p \mathtt{case}^p\ M\ \mathtt{of}\ (\mathtt{in}_1(x_1).N_1; \mathtt{in}_2(x_2).N_2)\colon \tau} \qquad \frac{\Gamma \vdash^p M\colon 0}{\Gamma \vdash^p \mathtt{image}^p(M)\colon A} \tag{2}$$

FGCBV is a skeleton on which one can add specific effects. We will make this precise in Section 2.1, but we begin with some examples. In the case of global store, given by some set of cells Loc holding values of some type Val, we add the following *generic effects* [19] to FGCBV: for each cell $l \in$ Loc, we add producer term constants $\mathtt{deref}_l$ and $\mathtt{assign}_l$ with typing judgements $\Gamma \vdash^p \mathtt{deref}_l\colon \mathrm{Val}$ and $\Gamma \vdash^p \mathtt{assign}_l(V)\colon 1$ if $\Gamma \vdash^v V\colon \mathrm{Val}$. We add to the theory of equality in Figure 1 the seven equations for global store proposed by Plotkin and Power [18], for example the two equations

$$\mathtt{deref}_l\,\mathtt{to}\,x.\,\mathtt{assign}_l(x) = \mathtt{return}\,(\star) \tag{3}$$

$$\mathtt{assign}_l(V)\,\mathtt{to}\,x.\,\mathtt{assign}_l(W) = \mathtt{assign}_l(W) \tag{4}$$

which state that reading a cell and then writing the same value is the same as doing nothing, and that the effect of two writes equals that of the second.

In the case of non-determinism, the generic effect is "random" with typing judgement $\Gamma \vdash^p \mathtt{random}\colon 1 + 1$. The equations are perhaps most easily described using the *algebraic operation* corresponding to random, defined as

$M \,\mathtt{or}\, N \stackrel{\text{def}}{=} \mathtt{case}^p\ \mathtt{random}\ \mathtt{of}\ (\mathtt{in}_1(x).M; \mathtt{in}_2(x).N)$. The derived typing rule says that $\Gamma \vdash^p M \,\mathtt{or}\, N : \sigma$ if $\Gamma \vdash^p M : \sigma$ and $\Gamma \vdash^p N : \sigma$. There are three equations: associativity, commutativity and idempotency of "or".

2.1 Effect Theories

We do not want to allow arbitrary extensions of FGCBV. In this section we define *effect theories*, which are particularly well-behaved extensions that include the examples above, of global store and nondeterminism. Effect theories are important from the semantic point of view because they are a kind of presentation for enriched algebraic theories, as will be clarified in Section 6.

Plotkin and Pretnar have also defined a notion of effect theory [17, §3]. Their effect theories can be accommodated in our model. The main difference is in the presentation: we use generic effects rather than algebraic operations.

By a value signature we shall simply mean a signature for a many-sorted algebraic theory in the usual sense. This means a set of type constants ranged over by α, β, and a set of term constants f with a given arity $f\colon (\alpha_1, \ldots, \alpha_n) \to \beta$, where the α_i, β range over type constants. We can extend FGCBV along a value signature by adding the type constants and the typing rule

$$\frac{\Gamma \vdash^v t_i : \alpha_i \; (i = 1, \ldots, n)}{\Gamma \vdash^v f(t_1, \ldots, t_n) : \beta} \tag{5}$$

for every term constant $f\colon (\alpha_1, \ldots, \alpha_n) \to \beta$ in the signature. A value theory is a value signature with a set of equations, i.e. pairs of terms typable in the same context $\Gamma \vdash^v V = W : \beta$, where V, W are formed only using variable introduction and the rule (5).

An *effect signature* consists of a value theory and a set of effect constants each with an assigned arity $e : \bar{\beta}; \bar{\alpha}_1 + \ldots + \bar{\alpha}_n$ consisting of a list of type constants and a formal sum of lists of type constants. FGCBV can be extended along an effect signature by adding, for every $e : \bar{\beta}; \bar{\alpha}_1 + \ldots + \bar{\alpha}_n$ a typing judgement

$$\frac{\Gamma \vdash^v \bar{V} : \bar{\beta}}{\Gamma \vdash^p e(\bar{V}) : \bar{\alpha}_1 + \ldots + \bar{\alpha}_n} \tag{6}$$

The hypothesis is to be understood as a vector of typing judgements, and in the conclusion, the vectors $\bar{\alpha}_i$ should be interpreted as the product of the types in the vector.

For example, the theory for global store has one value type constant Val, and for each location $l \in \mathrm{Loc}$ a pair of effect constants $(\mathtt{deref}_l \colon 1; \mathrm{Val})$ and $(\mathtt{assign}_l \colon \mathrm{Val}; 1)$. In this case term constants in the value theory can be used to add basic operations manipulating values in Val. In the case of nondeterminism, the effect constant `random` has arity $1; 1 + 1$.

An effect theory comprises an effect signature and a set of equations. The equations are pairs of producer terms-in-context $\Gamma \vdash^p M = N : \bar{\alpha}_1 + \ldots + \bar{\alpha}_n$ of a restricted kind. We impose the following restrictions: firstly, Γ must consist of variables with type constants, i.e., of the form $x\colon \alpha$. Secondly, the terms M and N must be built from a first order fragment of *effect terms*. This first order fragment consists of the first nine rules of Figure 1, the derived rules for sum types in producer terms (2) and the rules (5) and (6).

We write FGCBV_E for FGCBV augmented with an effect theory E.

3 Target Calculus: Enriched Call-by-Value

The target language for the linear state translation is a new calculus called the *enriched call-by-value calculus* (ECBV), that we now introduce. It is a fragment of the enriched effect calculus (EEC), which was introduced by Egger et al. [2] as a calculus for reasoning about linear usage in computational effects. The types of ECBV can be understood as a fragment of linear logic that is expressive enough to describe the linear state monad, $\underline{\mathsf{S}} \multimap\, !(-) \otimes \underline{\mathsf{S}}$. We will not dwell on the connection with linear logic here.

The enriched call-by-value calculus has two collections of types: value types and computation types. We use $\alpha, \beta, \ldots$ to range over a set of *value type constants*, and $\underline{\alpha}, \underline{\beta}, \ldots$ to range over a disjoint set of *computation type constants*. We then use $\mathsf{A}, \mathsf{B}, \ldots$ to range over value types, and $\underline{\mathsf{A}}, \underline{\mathsf{B}}, \ldots$ to range over computation types, which are specified by the grammar below:

$$\mathsf{A} ::= \alpha \mid 1 \mid \mathsf{A} \times \mathsf{B} \mid 0 \mid \mathsf{A} + \mathsf{B} \mid \underline{\mathsf{A}} \multimap \underline{\mathsf{B}}$$
$$\underline{\mathsf{A}} ::= \underline{\alpha} \mid \underline{0} \mid \underline{\mathsf{A}} \oplus \underline{\mathsf{B}} \mid\, !\mathsf{A} \otimes \underline{\mathsf{B}}\ .$$

Note that the construction $!\mathsf{A} \otimes \underline{\mathsf{B}}$ is indivisible: the strings $!\mathsf{A}$ and $\underline{\mathsf{A}} \otimes \underline{\mathsf{B}}$ are not well-formed types. Note also that unlike EEC [2] there is no inclusion of computation types into value types. Moreover, there are no type constructors corresponding to F or U as known from CBPV [10].

The enriched call-by-value calculus has two basic typing judgements, written

$$\Gamma \mid - \vdash t\colon \mathsf{B} \qquad \text{and} \qquad \Gamma \mid z\colon \underline{\mathsf{A}} \vdash t\colon \underline{\mathsf{B}} \tag{7}$$

In the first judgement, B is a value type, and in the second judgement, both $\underline{\mathsf{A}}$ and $\underline{\mathsf{B}}$ need to be computation types. The second judgement should be thought of as a judgement of linearity in the variable $z\colon \underline{\mathsf{A}}$. The typing rules are given in Figure 2. In the figure, Γ is an assignment of value types to variables, and Δ is an assignment of a computation type to a single variable, as in (7). The equality theory includes α, β and η rules and is exactly as for EEC [2, Sec. 3].

We can talk about type isomorphisms in ECBV in the usual way. For value types, an isomorphism $\mathsf{A} \cong \mathsf{B}$ is given by two judgements, $x\colon \mathsf{A} \mid - \vdash t\colon \mathsf{B}$ and $y\colon \mathsf{B} \mid - \vdash u\colon \mathsf{A}$, such that $u[t/y] = x$, $t[u/x] = y$. For computation types, $\underline{\mathsf{A}} \cong \underline{\mathsf{B}}$ is witnessed by closed terms of type $\underline{\mathsf{A}} \multimap \underline{\mathsf{B}}$, $\underline{\mathsf{B}} \multimap \underline{\mathsf{A}}$ composing in both directions to identities. We note the following type isomorphisms, inherited from EEC:

$$\underline{\mathsf{A}} \cong\, !1 \otimes \underline{\mathsf{A}} \qquad !\mathsf{A} \otimes (!\mathsf{B} \otimes \underline{\mathsf{C}}) \cong\, !(\mathsf{A} \times \mathsf{B}) \otimes \underline{\mathsf{C}} \tag{8}$$
$$\underline{0} \cong\, !0 \otimes \underline{\mathsf{B}} \qquad (!\mathsf{A} \otimes \underline{\mathsf{C}}) \oplus (!\mathsf{B} \otimes \underline{\mathsf{C}}) \cong\, !(\mathsf{A} + \mathsf{B}) \otimes \underline{\mathsf{C}} \tag{9}$$

$$\frac{}{\Gamma, x\colon\mathsf{A} \mid - \vdash x\colon \mathsf{A}} \qquad \frac{}{\Gamma \mid z\colon\underline{\mathsf{A}} \vdash z\colon \underline{\mathsf{A}}} \qquad \frac{}{\Gamma \mid - \vdash \star\colon 1}$$

$$\frac{\Gamma \mid - \vdash t\colon \mathsf{A} \quad \Gamma \mid - \vdash u\colon \mathsf{B}}{\Gamma \mid - \vdash \langle t, u\rangle\colon \mathsf{A}\times\mathsf{B}} \qquad \frac{\Gamma \mid - \vdash t\colon \mathsf{A}_1\times\mathsf{A}_2}{\Gamma \mid - \vdash \pi_i(t)\colon \mathsf{A}_i}$$

$$\frac{\Gamma \mid - \vdash t\colon \mathsf{A}_i}{\Gamma \mid - \vdash \mathrm{in}_i(t)\colon \mathsf{A}_1 + \mathsf{A}_2} \qquad \frac{\Gamma \mid \Delta \vdash t\colon \underline{\mathsf{A}}_i}{\Gamma \mid \Delta \vdash \underline{\mathrm{in}}_i(t)\colon \underline{\mathsf{A}}_1 \oplus \underline{\mathsf{A}}_2}$$

$$\frac{\Gamma \mid - \vdash t\colon 0}{\Gamma \mid - \vdash \mathrm{image}(t)\colon \mathsf{A}} \qquad \frac{\Gamma \mid - \vdash s\colon \mathsf{A}_1 + \mathsf{A}_2 \quad \Gamma, x_i\colon\mathsf{A}_i \mid - \vdash t_i\colon \mathsf{C}\ (i=1,2)}{\Gamma \mid - \vdash \mathrm{case}\, s\, \mathrm{of}\, (\mathrm{in}_1(x_1).\, t_1; \mathrm{in}_2(x_2).\, t_2)\colon \mathsf{C}}$$

$$\frac{\Gamma \mid \Delta \vdash t\colon \underline{0}}{\Gamma \mid \Delta \vdash \underline{\mathrm{image}}(t)\colon \underline{\mathsf{A}}} \qquad \frac{\Gamma \mid \Delta \vdash s\colon \underline{\mathsf{A}}_1 \oplus \underline{\mathsf{A}}_2 \quad \Gamma \mid x_i\colon\underline{\mathsf{A}}_i \vdash t_i\colon \underline{\mathsf{C}}\ (i=1,2)}{\Gamma \mid \Delta \vdash \underline{\mathrm{case}}\, s\, \mathrm{of}\, (\underline{\mathrm{in}}_1(x_1).\, t_1; \underline{\mathrm{in}}_2(x_2).\, t_2)\colon \underline{\mathsf{C}}}$$

$$\frac{\Gamma \mid z\colon\underline{\mathsf{A}} \vdash t\colon \underline{\mathsf{B}}}{\Gamma \mid - \vdash \underline{\lambda} z\colon\underline{\mathsf{A}}.\, t\colon \underline{\mathsf{A}} \multimap \underline{\mathsf{B}}} \qquad \frac{\Gamma \mid - \vdash s\colon \underline{\mathsf{A}} \multimap \underline{\mathsf{B}} \quad \Gamma \mid \Delta \vdash t\colon \underline{\mathsf{A}}}{\Gamma \mid \Delta \vdash s[t]\colon \underline{\mathsf{B}}}$$

$$\frac{\Gamma \mid - \vdash t\colon \mathsf{A} \quad \Gamma \mid \Delta \vdash u\colon \underline{\mathsf{B}}}{\Gamma \mid \Delta \vdash\, !t \otimes u\colon !\mathsf{A} \otimes \underline{\mathsf{B}}} \qquad \frac{\Gamma \mid \Delta \vdash s\colon !\mathsf{A}\otimes\underline{\mathsf{B}} \quad \Gamma, x\colon\mathsf{A} \mid z\colon\underline{\mathsf{B}} \vdash t\colon \underline{\mathsf{C}}}{\Gamma \mid \Delta \vdash \mathrm{let}\ !x \otimes z\ \mathrm{be}\ s\ \mathrm{in}\ t\colon \underline{\mathsf{C}}}$$

Fig. 2. Typing rules for the enriched call-by-value calculus

Given an effect signature E (Sec. 2.1), we add effects to ECBV as follows. We assume that there is a distinguished computation type constant $\underline{\mathrm{S}}$, called the state type. For each effect constant $e : \bar{\beta}; \bar{\alpha}_1 + \ldots + \bar{\alpha}_n$, we add a closed term, called a *state access operation*:

$$e\colon !\bar{\beta} \otimes \underline{\mathrm{S}} \multimap\, !(\bar{\alpha}_1 + \cdots + \bar{\alpha}_n) \otimes \underline{\mathrm{S}} \tag{10}$$

In order to add the equations from an effect theory to ECBV, we need to give interpretations to effect terms. In Section 4 we are going to translate all of FGCBV into ECBV, so we postpone this to there.

We write $\mathrm{ECBV}_E^{\underline{\mathrm{S}}}$ for the enriched effect calculus extended over the effect theory E as described above.

Examples 1. The effect theories of global store and of non-determinism will give rise to the state access operations $\texttt{read}_l$, $\texttt{write}_l$, and $\texttt{random}$ in (1).

Note that $\texttt{read}_l\colon \underline{\mathrm{S}} \multimap\, !\mathrm{Val} \otimes \underline{\mathrm{S}}$ returns a state that must be used linearly and a result value of the read operation that can be used arbitrarily. One of the equations (3) for global store requires $\texttt{write}_l(\texttt{read}_l(s)) = s$; another one (4) says $\texttt{write}_l(!v \otimes (\texttt{write}_l(!w \otimes s))) = \texttt{write}_l(!v \otimes s)$.

4 The State-Passing Translation

We now describe the state-passing translation from FGCBV to ECBV. We translate FGCBV$_E$ types σ to ECBV$_E^{\underline{S}}$ value types σ°:

$$\alpha^\circ = \alpha \qquad (\sigma \times \tau)^\circ = \sigma^\circ \times \tau^\circ \qquad 1^\circ = 1$$
$$(\sigma \rightharpoonup \tau)^\circ = !(\sigma^\circ) \otimes \underline{S} \multimap !(\tau^\circ) \otimes \underline{S} \qquad (\sigma + \tau)^\circ = \sigma^\circ + \tau^\circ \qquad 0^\circ = 0$$

The translation takes value type judgements $\Gamma \vdash^v V : \sigma$ to ECBV judgements $\Gamma^\circ \,|\, - \vdash V^\circ : \sigma^\circ$ and it takes producer judgements $\Gamma \vdash^p M : \sigma$ to ECBV judgements $\Gamma^\circ \,|\, s : \underline{S} \vdash M^\circ : !(\sigma^\circ) \otimes \underline{S}$, as follows.

$$x^\circ = x \qquad \star^\circ = \star \qquad \langle V, W\rangle^\circ = \langle V^\circ, W^\circ\rangle$$
$$(\pi_i(V))^\circ = \pi_i(V^\circ) \quad (\texttt{image}(V))^\circ = \mathrm{image}(V^\circ) \quad (\texttt{in}_i(V))^\circ = \mathrm{in}_i(V^\circ)$$
$$(\lambda x : \sigma.\, N)^\circ = \underline{\lambda} z : !\sigma^\circ \otimes \underline{S}.\ \mathrm{let}\ !x \otimes s\ \mathrm{be}\ z\ \mathrm{in}\ N^\circ \qquad (V\,W)^\circ = V^\circ[!(W^\circ) \otimes s]$$
$$(M\,\texttt{to}\,x.\,N)^\circ = \mathrm{let}\ !x \otimes s\ \mathrm{be}\ M^\circ\ \mathrm{in}\ N^\circ \qquad (\texttt{return}\,V)^\circ = !(V^\circ) \otimes s$$
$$(\texttt{case}\ V\ \texttt{of}\ (\texttt{in}_1(x_1).W_1; \texttt{in}_2(x_2).W_2))^\circ = \mathrm{case}\ V^\circ\ \mathrm{of}\ (\mathrm{in}_1(x_1).W_1^\circ; \mathrm{in}_2(x_2).W_2^\circ)$$

We translate generic effects to state operations: $(e(\bar{V}))^\circ = e(!\langle V_1^\circ, \ldots, V_m^\circ\rangle \otimes s)$. We are now in a position to add the equations of an effect theory to ECBV. For each equation $\Gamma \vdash^p M = N : \bar{\alpha}_1 + \ldots + \bar{\alpha}_n$ in the effect theory, we add the equation $\Gamma^\circ \,|\, s : \underline{S} \vdash M^\circ = N^\circ : !(\bar{\alpha}_1^\circ + \ldots + \bar{\alpha}_n^\circ) \otimes \underline{S}$ to ECBV$_E^{\underline{S}}$.

Theorem 2 (Soundness). *If $V = W$ then $V^\circ = W^\circ$; if $M = N$ then $M^\circ = N^\circ$.*

Theorem 3 (Fullness on types). *Let A be a value type of ECBV formed using no other computation type constants than $\underline{S}$. Then there exists a FGCBV type σ such that $\sigma^\circ \cong \mathsf{A}$.*

Proof. By induction on the structure of types. The interesting case $\underline{\mathsf{A}} \multimap \underline{\mathsf{B}}$ uses the fact that any computation type not using any $\underline{\alpha}$ other than $\underline{S}$ is isomorphic to one of the form $!\mathsf{C} \otimes \underline{S}$, which follows from the isomorphisms (8) – (9). □

We now state our main syntactic result.

Theorem 4 (Full completeness). *Suppose $\Gamma \vdash^v V, W : \sigma$ and $\Gamma \vdash^p M, N : \sigma$.*

1. *If $V^\circ = W^\circ$ then $V = W$. If $M^\circ = N^\circ$ then $M = N$.*
2. *For any $\Gamma^\circ \,|\, - \vdash t : \sigma^\circ$ there exists a term $\Gamma \vdash^v V : \sigma$ such that $t = V^\circ$.*
3. *For any $\Gamma^\circ \,|\, s : \underline{S} \vdash t : !(\sigma^\circ) \otimes \underline{S}$ there exists $\Gamma \vdash^p M : \sigma$ such that $t = M^\circ$.*

Theorem 4 can be proved syntactically as follows. Consider first the fragment of ECBV with no other computation type constants than $\underline{S}$, and only the value type constants of FGCBV$_E$. This fragment is equivalent to a variant of ECBV where the only computation types are the ones of the form $!\mathsf{A} \otimes \underline{S}$ with corresponding variants of the typing rules for $!\mathsf{A} \otimes \underline{\mathsf{B}}$. The translation $(-)^\circ$ gives a bijection from FGCBV$_E$ types to value types of ECBV$_E^{\underline{S}}$, and one can define an inverse to this translation. Further type constants can be added to ECBV$_E^{\underline{S}}$ without changing the result; this can be proved via a normalization theorem for ECBV$_E^{\underline{S}}$ which follows the one for EEC (to appear in [4]).

In Section 7.1 we sketch a semantic proof of Theorems 3 and 4.

5 Categorical Models

By studying categorical models, we are able to give a canonical, universal status to the two calculi that we have considered so far, and also to the state-passing translation. In Section 7, the full completeness of the state-passing translation will be explained as an equivalence of free categories.

5.1 Monad Models of the Fine-Grained Call-by-Value Calculus

Terminology. Recall that a distributive category is a category with finite products and coproducts, such that the canonical morphisms $((A \times B) + (A \times C)) \to (A \times (B + C))$ and $0 \to A \times 0$ are isomorphisms.

Definition 5. *A* monad model of FGCBV *(or simply a* monad model*) is a distributive category* $\mathbf{V}$ *with a strong monad* T *and Kleisli-exponentials (that is, exponentials of the form* $(A \to T(B))$*).*

A semantics for FGCBV is given in a monad-model in a standard way. For instance, $[\![\sigma \rightharpoonup \tau]\!] = ([\![\sigma]\!] \to T([\![\tau]\!]))$. A value type judgement $\Gamma \vdash^v V : \sigma$ is taken to a morphism $[\![\Gamma]\!] \to [\![\sigma]\!]$, and a producer type judgement $\Gamma \vdash^p M : \sigma$ is taken to a morphism $[\![\Gamma]\!] \to T([\![\sigma]\!])$. This defines a sound and complete notion of model for FGCBV (e.g. [11]). In particular, the types and terms of FGCBV form a syntactic model, which is initial (with respect to an appropriate notion of morphism).

5.2 Enriched Call-by-Value Models

The categorical notion of model for ECBV involves basic concepts from enriched category theory [9]. Let us recall some rudiments. Following [7,6], we begin with actions of categories. Let $\mathbf{V}$ be a category with finite products. Recall that an action of $\mathbf{V}$ on a category $\mathbf{C}$ is a functor $\cdot \colon \mathbf{V} \times \mathbf{C} \to \mathbf{C}$ together with coherent natural unit and associativity isomorphisms, $(1 \cdot \underline{A}) \cong \underline{A}$ and $((A \times B) \cdot \underline{C}) \cong (A \cdot (B \cdot \underline{C}))$. (We underline objects of $\mathbf{C}$ to distinguish them from objects of $\mathbf{V}$.) An *enrichment of a category* $\mathbf{C}$ *in* $\mathbf{V}$ *with copowers* is determined by an action of $\mathbf{V}$ on $\mathbf{C}$ such that each functor $(- \cdot \underline{A}) \colon \mathbf{V} \to \mathbf{C}$ has a right adjoint, $\mathbf{C}(\underline{A}, -) \colon \mathbf{C} \to \mathbf{V}$. Then $A \cdot \underline{B}$ is called a copower, and $\mathbf{C}(\underline{A}, \underline{B})$ is called enrichment. Recall also that a *power* is a right adjoint to $(A \cdot -) \colon \mathbf{C} \to \mathbf{C}$ (we will need this in Section 6).

If $\mathbf{C}$ is enriched in $\mathbf{V}$ with copowers, and $\mathbf{C}$ has finite coproducts, then the coproducts in $\mathbf{C}$ are *enriched* if each functor $(A \cdot -) \colon \mathbf{C} \to \mathbf{C}$ preserves them.

Definition 6. *An* enriched call-by-value model *(or simply* enriched model*) is given by a distributive category* $\mathbf{V}$ *and a category* $\mathbf{C}$ *enriched in* $\mathbf{V}$ *with copowers and enriched finite coproducts. A model of* $\mathrm{ECBV}^{\underline{S}}$ *is given by an enriched call-by-value model together with a chosen object* $\underline{S}$ *of* $\mathbf{C}$*.*

A semantics for ECBV in an enriched model is given similarly to the semantics of EEC [2]. For each value type A, an object $[\![\mathsf{A}]\!]$ of $\mathbf{V}$ is given, and for each computation type $\underline{\mathsf{A}}$, an object $[\![\underline{\mathsf{A}}]\!]$ of $\mathbf{C}$ is given. The product and sum types are interpreted as products and coproducts in $\mathbf{V}$ and $\mathbf{C}$. We let $[\![!\mathsf{A} \otimes \underline{\mathsf{B}}]\!] = ([\![\mathsf{A}]\!] \cdot [\![\underline{\mathsf{B}}]\!])$, and $[\![\underline{\mathsf{A}} \multimap \underline{\mathsf{B}}]\!] = \mathbf{C}([\![\underline{\mathsf{A}}]\!], [\![\underline{\mathsf{B}}]\!])$. The specified object $\underline{S}$ in an ECBV$^{\underline{\mathsf{S}}}$ model interprets $\underline{\mathsf{S}}$. A judgement $\Gamma \mid - \vdash t \colon \mathsf{A}$ is interpreted as a morphism $[\![\Gamma]\!] \to [\![\mathsf{A}]\!]$ in $\mathbf{V}$, and a judgement $\Gamma \mid \Delta \vdash t \colon \underline{\mathsf{A}}$ is interpreted as a morphism $[\![\Gamma]\!] \cdot [\![\Delta]\!] \to [\![\underline{\mathsf{A}}]\!]$ in $\mathbf{C}$. The types and terms of ECBV form a syntactic model which is initial with respect to an appropriate notion of morphism.

5.3 From Enriched Models to Monad Models and Back

Given a monad model $(\mathbf{V}, T)$ there is a monoidal action $\mathbf{V} \times \mathbf{Kl}(T) \to \mathbf{Kl}(T)$ on the Kleisli category defined on objects as the product functor and defined on morphisms using the strength of T. The Kleisli category $\mathbf{Kl}(T)$ is $\mathbf{V}$-enriched because $\mathbf{V}$ has Kleisli exponentials.

Proposition 7. *If* $(\mathbf{V}, T)$ *is a monad model (in the sense of Definition 5) then* $(\mathbf{V}, \mathbf{Kl}(T), 1)$ *is an* ECBV$^{\underline{\mathsf{S}}}$ *model (in the sense of Definition 6).*

On the other hand, if $(\mathbf{V}, \mathbf{C})$ is an enriched model, we will say that an adjunction $F \dashv U \colon \mathbf{C} \to \mathbf{V}$ is *enriched* if there is a natural coherent isomorphism $F(A \times B) \cong A \cdot F(B)$. When $\mathbf{V}$ is cartesian closed, this is equivalent to the usual definition, i.e. a natural isomorphism $\mathbf{C}(F(-), =) \cong \mathbf{V}(-, U(=))$ (see e.g. [8]).

The choice of $\underline{S}$ in ECBV$^{\underline{\mathsf{S}}}$ models gives an enriched adjunction, since $(- \cdot \underline{S})$ is left adjoint to $\mathbf{C}(\underline{S}, -) \colon \mathbf{C} \to \mathbf{V}$. The following proposition (first noted noted for EEC [3], though it does not appear explicitly there) shows that every enriched adjunction arises in this way:

Proposition 8 ([3]). *Let* $(\mathbf{V}, \mathbf{C})$ *be an enriched model. If* $F \dashv U \colon \mathbf{C} \to \mathbf{V}$ *is an enriched adjunction then it is naturally isomorphic to the enriched adjunction induced by* $F(1)$.

So we can equivalently consider ECBV$^{\underline{\mathsf{S}}}$ models as enriched adjunctions.

Given an enriched adjunction, the corresponding monad gives a monad model. In particular:

Proposition 9. *If* $(\mathbf{V}, \mathbf{C}, \underline{S})$ *is an* ECBV$^{\underline{\mathsf{S}}}$ *model then* $(\mathbf{V}, \mathbf{C}(\underline{S}, - \cdot \underline{S}))$ *is a monad model.*

If we start with a monad model, take the corresponding ECBV$^{\underline{\mathsf{S}}}$ model (via Prop. 7) and then go back (via Prop. 9), we get a monad model that is equivalent to the one that we started with. This is simply because $\mathbf{Kl}(T)(1, A \times 1) \cong T(A)$. We have the slogan: *Every monad is a linear state monad.*

We shall prove later that this connection between monad models and enriched models can be understood as a coreflection.

5.4 Remark: Closed Freyd Categories

Closed Freyd categories [20] are an alternative way of presenting monad models. Freyd categories are usually defined using premonoidal categories [23], but we will use the following equivalent definition using actions (following [10, App. B]). A distributive closed Freyd category [22] can be described as an enriched model $(\mathbf{V}, \mathbf{C})$ together with an identity on objects functor $J \colon \mathbf{V} \to \mathbf{C}$ that preserves the action (i.e. $J(\mathsf{A} \times \mathsf{B}) = \mathsf{A} \cdot J(\mathsf{B})$).

If $(\mathbf{V}, T)$ is a monad model the inclusion $\mathbf{V} \to \mathbf{Kl}(T)$ is a distributive closed Freyd category, and every distributive closed Freyd category arises in this way. By removing the requirement that $\mathbf{V}$ and $\mathbf{C}$ have the same objects, we discover the more general class of enriched models.

6 Models and Comodels of Effect Theories

We define what it means for monad models and enriched models to model an effect theory (in the sense of Sec. 2.1).

Models of Value Theories. Let $\mathbf{V}$ be a distributive category. An interpretation of a value signature in $\mathbf{V}$ is given by interpretations of the type constants α as objects $[\![\alpha]\!]$ of $\mathbf{V}$, and interpretations of term constants $f \colon \bar{\alpha} \to \beta$ as morphisms $[\![f]\!] \colon [\![\bar{\alpha}]\!] \to [\![\beta]\!]$. This is extended to interpret a term in context $\Gamma \vdash^{v} V \colon \beta$ as a morphism $[\![V]\!] \colon [\![\Gamma]\!] \to [\![\beta]\!]$. An interpretation of a value theory is an interpretation of the signature such that $[\![V]\!] = [\![W]\!]$ for each equation $\Gamma \vdash^{v} V = W \colon \beta$.

Interpreting Effect Theories in Monad Models. An interpretation of an effect theory E in a monad model $(\mathbf{V}, T)$ is an interpretation of the value theory in $\mathbf{V}$ and an interpretation of each effect constant $e \colon \bar{\beta}; \bar{\alpha}_1 + \cdots + \bar{\alpha}_n$ in E as a Kleisli map $[\![e]\!] \colon [\![\bar{\beta}]\!] \to T([\![\bar{\alpha}_1]\!] + \cdots + [\![\bar{\alpha}_n]\!])$, satisfying the equations of the theory.

Effects and Enriched Models. In enriched models, according to (10), every effect constant should be interpreted as a morphism $[\![e]\!] \colon [\![\bar{\beta}]\!] \cdot \underline{S} \to ([\![\bar{\alpha}_1]\!] + \cdots + [\![\bar{\alpha}_n]\!]) \cdot \underline{S}$ in $\mathbf{C}$. We can relate these to the Kleisli maps of the monad model, extending the bijective correspondence between algebraic operations and generic effects [19]:

Proposition 10. *Let* $(\mathbf{V}, \mathbf{C})$ *be an enriched model and consider* $\underline{S}$ *in* $\mathbf{C}$*. The following sets are in natural bijection:*

1. State access operations*: morphisms* $A \cdot \underline{S} \to B \cdot \underline{S}$ *in* $\mathbf{C}$*.*
2. Generic effects *for the induced monad: morphisms* $A \to \mathbf{C}(\underline{S}, B \cdot \underline{S})$ *in* $\mathbf{V}$*.*
3. Algebraic operations *in* $\mathbf{C}$*: families of morphisms* $\mathbf{C}(\underline{S}, \underline{X})^{B} \to \mathbf{C}(\underline{S}, \underline{X})^{A}$ *natural in* $\underline{X}$ *in* $\mathbf{C}$*.*

(Although we do not assume that $\mathbf{V}$ is cartesian closed, the exponentials mentioned in Item 3 always exist.)

To explain the status of the special object $\underline{S}$ we provide a general notion of model for effect theories.

Interpretations of Effect Theories in General. For a moment, let $\mathbf{V}$ be a distributive category, and let $\mathbf{A}$ be a category enriched in $\mathbf{V}$ with *powers*. Consider an effect signature E and an interpretation of the value theory in $\mathbf{V}$. A *model* of E in $\mathbf{A}$ consists of an object A of $\mathbf{A}$ together with, for each effect constant $e\colon \bar{\beta}; \bar{\alpha}_1 + \cdots + \bar{\alpha}_n$ in E, a morphism $[\![e]\!]\colon A^{[\![\bar{\alpha}_1]\!]+\cdots+[\![\bar{\alpha}_n]\!]} \to A^{[\![\bar{\beta}]\!]}$ in $\mathbf{A}$.

To describe when a model satisfies equations in an effect theory, we need to give a semantics to effect terms. (Recall that an effect term is a first order term of FGCBV.) In any model A of an effect signature, we interpret an effect term typing judgement $\Gamma \vdash^p M : \tau$ as a morphism $[\![M]\!]\colon A^{[\![\tau]\!]} \to A^{[\![\Gamma]\!]}$ in $\mathbf{A}$, by induction on the structure of typing derivations. For instance, consider the $\mathtt{case}^p$ rule in (2). Given interpretations $[\![M]\!]\colon A^{[\![\sigma_1+\sigma_2]\!]} \to A^{[\![\Gamma]\!]}$ and $[\![N_i]\!]\colon A^{[\![\tau]\!]} \to A^{[\![\Gamma,\sigma_i]\!]}$ $(i = 1, 2)$, we define $[\![\mathtt{case}^p\ M\ \mathtt{of}\ (\mathtt{in}_1(x_1).N_1; \mathtt{in}_2(x_2).N_2)]\!]$ to be the composite

$$A^{[\![\tau]\!]} \xrightarrow{([\![N_1]\!],[\![N_2]\!])} A^{[\![\Gamma,\sigma_1]\!]} \times A^{[\![\Gamma,\sigma_2]\!]} \cong A^{([\![\sigma_1+\sigma_2]\!])\times[\![\Gamma]\!]} \xrightarrow{[\![M]\!]^{[\![\Gamma]\!]}} A^{[\![\Gamma]\!]\times[\![\Gamma]\!]} \xrightarrow{A^{\Delta}} A^{[\![\Gamma]\!]}.$$

As another example, $[\![\mathtt{return}\,V]\!] = A^{[\![V]\!]}$. A model of an effect theory in $\mathbf{A}$ is a model of the effect signature such that every effect equation $\Gamma \vdash M = N : \tau$ in the theory is satisfied, i.e. $[\![M]\!] = [\![N]\!]$.

In an enriched model $(\mathbf{V}, \mathbf{C})$, we have a category $\mathbf{C}$ enriched in $\mathbf{V}$ with *co*powers. This means that $\mathbf{C}^{\mathrm{op}}$ is enriched in $\mathbf{V}$ with powers. A *comodel* in $\mathbf{C}$ is a model in $\mathbf{C}^{\mathrm{op}}$.

Definition 11. *An* $\mathrm{ECBV}^{\underline{\mathrm{S}}}_{E}$ model *is an* $\mathrm{ECBV}^{\underline{\mathrm{S}}}$ *model with a given* E*-comodel structure on* $\underline{S}$.

Proposition 12. *Let* $(\mathbf{V}, \mathbf{C})$ *be an enriched model and consider* $\underline{S}$ *in* $\mathbf{C}$. *The following data are equivalent.*

1. *An* E*-comodel structure for the object* $\underline{S}$.
2. *An* E*-model structure for the induced monad model.*
3. *For each effect constant* $e\colon \bar{\beta}; \bar{\alpha}_1 + \cdots + \bar{\alpha}_n$ *a family of morphisms*

$$\textstyle\prod_i \mathbf{C}(\underline{S}, \underline{X})^{[\![\bar{\alpha}_i]\!]} \to \mathbf{C}(\underline{S}, \underline{X})^{[\![\bar{\beta}]\!]} \qquad \textit{natural in } \underline{X}$$

equipping each $\mathbf{C}(\underline{S}, \underline{X})$ *with the structure of a model of* E.

Example 13. Let Val, Loc be sets of values and locations respectively, and let $\underline{S}$ be the set of functions (Loc $\to$ Val). The category **Set** is enriched in itself with copowers given by products, and indeed $\underline{S}$ is a comodel for the theory for global store in the enriched model (**Set**, **Set**). The induced monad on **Set** is $((-)\times\underline{S})^{\underline{S}}$. Power and Shkaravska [21] showed that $\underline{S}$ is the final comodel of global store.

7 Categories of Models and Full Completeness

We sketch how the constructions of Propositions 7 and 9 extend to define adjoint 2-functors between a 2-category of monad models and a 2-category of enriched models. We sketch how to use these results to prove full completeness of the linear state monad translation.

Let **ENR** be the 2-category whose objects are $\text{ECBV}^{\underline{\mathsf{S}}}$ models $(\mathbf{V}, \mathbf{C}, \underline{S})$. A 1-cell $(\mathbf{V}, \mathbf{C}, \underline{S}) \to (\mathbf{V}', \mathbf{C}', \underline{S}')$ is a pair of functors $F \colon \mathbf{V} \to \mathbf{V}'$, $G \colon \mathbf{C} \to \mathbf{C}'$ together with an isomorphism $G\underline{S} \cong \underline{S}'$ and a natural isomorphism $G(A \cdot \underline{B}) \cong (F(A)) \cdot (G(\underline{B}))$ whose mate is an isomorphism $F(\mathbf{C}(\underline{B}, \underline{C})) \cong \mathbf{C}'(G\underline{B}, G\underline{C})$, and such that F preserves products and coproducts and G preserves coproducts (up to isomorphism). The 2-cells are natural coherent isomorphisms.

Let **MND** be the 2-category whose objects are monad models $(\mathbf{V}, T)$. A 1-cell $(\mathbf{V}, T) \to (\mathbf{V}', T')$ is a functor $F \colon \mathbf{V} \to \mathbf{V}'$ together with a natural isomorphism $\phi \colon T'F \cong FT$ making (F, ϕ) a monad morphism [26], and such that F preserves products and coproducts, strengths and Kleisli exponentials. The 2-cells are natural coherent isomorphisms.

These definitions can be extended to 2-categories $\mathbf{ENR}_E$, $\mathbf{MND}_E$ whose objects are models of effect theories in the sense of Section 6 and whose 1-cells are required to preserve the interpretations of the theories.

Theorem 14. *The constructions of Propositions 7 and 9 extend to a 2-adjunction whose unit is an isomorphism:* $\mathbf{Kleisli} \dashv \mathbf{StateMnd} \colon \mathbf{ENR}_E \to \mathbf{MND}_E$.

The 2-adjunction is a restriction of a well known 2-adjunction between the category of monads and the category of adjunctions.

7.1 Full Completeness

We now provide a semantic argument to explain Theorems 3 and 4. Since the 2-functor $\mathbf{Kleisli} \colon \mathbf{MND}_E \to \mathbf{ENR}_E$ is a left adjoint, it preserves free constructions up to equivalence. In particular it takes the syntactic monad model $(\mathcal{V}_{\text{FGCBV}}, 1 \rightharpoonup (-))$, built from the syntax of FGCBV, to the syntactic enriched model $(\mathcal{V}_{\text{ECBV}}, \mathcal{C}_{\text{ECBV}})$, built from the syntax of $\text{ECBV}^{\underline{\mathsf{S}}}$ with exactly one computation type constant, $\underline{\mathsf{S}}$. In consequence, the morphism of monad models that describes the state-passing translation of Section 4,

$$(\mathcal{V}_{\text{FGCBV}}, 1 \rightharpoonup (-)) \longrightarrow (\mathcal{V}_{\text{ECBV}}, \underline{\mathsf{S}} \multimap (!(-) \otimes \underline{\mathsf{S}}))$$

is equivalent to the unit of the 2-adjunction $\mathbf{Kleisli} \dashv \mathbf{StateMnd}$, and thus it is an equivalence of categories. In other words, it is essentially surjective and full and faithful, providing a categorical proof of Theorems 3 and 4 respectively (under the assumption that ECBV has exactly one computation type constant).

8 The Enriched Effect Calculus

The enriched effect calculus (EEC) of Egger et al. [2] extends the enriched call-by-value calculus that we introduced in Section 3 with some type constructions:

$$\begin{aligned} \mathsf{A} &::= \ldots \mid \mathsf{A} \to \mathsf{B} \mid \underline{\alpha} \mid \underline{0} \mid \underline{\mathsf{A}} \oplus \underline{\mathsf{B}} \mid !\mathsf{A} \otimes \underline{\mathsf{B}} \mid !\mathsf{A} \\ \underline{\mathsf{A}} &::= \cdots \mid 1 \mid \underline{\mathsf{A}} \times \underline{\mathsf{B}} \mid \mathsf{A} \to \underline{\mathsf{B}} \mid !\mathsf{A}\ . \end{aligned}$$

The additional types have been used to describe other aspects of effectful computation, such as the traditional monadic call-by-name and call-by-value interpretations, and continuation-passing. The additional types of EEC do not affect the full completeness of the linear state-passing translation (Thm. 4), for the following reason. In Proposition 16 we show that every model of ECBV embeds in a model of EEC; conservativity of EEC over ECBV then follows from a strong normalisation result for EEC [4]. Thus the linear state-passing translation of Section 4 can be understood as a fully complete translation into EEC.

Definition 15 ([2]). *A model of EEC* $(\mathbf{V}, \mathbf{C}, F, U)$ *is given by a cartesian closed category* $\mathbf{V}$ *with coproducts, a* $\mathbf{V}$*-enriched category* $\mathbf{C}$ *with products and coproducts and powers and copowers, and an enriched adjunction* $F \dashv U \colon \mathbf{C} \to \mathbf{V}$.

We refer to [2] for the term calculus and interpretation of EEC into EEC models. Here, for brevity, we work directly with models.

Clearly every EEC model is an enriched model of ECBV in the sense of Definition 6. Conversely:

Proposition 16. *Every enriched model of ECBV embeds in an EEC model.*

Proof (sketch). Consider an enriched model $(\mathbf{V}, \mathbf{C})$ (in the sense of Def. 6).

For any category $\mathbf{A}$ with finite coproducts, let $\mathrm{FP}(\mathbf{A}^{\mathrm{op}}, \mathbf{Set})$ be the category of finite product preserving functors $\mathbf{A}^{\mathrm{op}} \to \mathbf{Set}$ and natural transformations between them. This category is the cocompletion of $\mathbf{A}$ as a category with finite coproducts (e.g. [5], [23], [9, Thms 5.86, 6.11]).

We will show that $(\mathrm{FP}(\mathbf{V}^{\mathrm{op}}, \mathbf{Set}), \mathrm{FP}(\mathbf{C}^{\mathrm{op}}, \mathbf{Set}))$ is an EEC model, and that $(\mathbf{V}, \mathbf{C})$ embeds in it as an enriched model. For general reasons, $\mathrm{FP}(\mathbf{V}^{\mathrm{op}}, \mathbf{Set})$ and $\mathrm{FP}(\mathbf{C}^{\mathrm{op}}, \mathbf{Set})$ have products and coproducts, and the Yoneda embeddings $(\mathbf{V} \hookrightarrow \mathrm{FP}(\mathbf{V}^{\mathrm{op}}, \mathbf{Set}), \mathbf{C} \hookrightarrow \mathrm{FP}(\mathbf{C}^{\mathrm{op}}, \mathbf{Set}))$ preserve them. Since $\mathbf{V}$ is distributive, $\mathrm{FP}(\mathbf{V}^{\mathrm{op}}, \mathbf{Set})$ is cartesian closed (see [5]).

We now show that $\mathrm{FP}(\mathbf{C}^{\mathrm{op}}, \mathbf{Set})$ is enriched in $\mathrm{FP}(\mathbf{V}^{\mathrm{op}}, \mathbf{Set})$ with powers and copowers. Recall the construction of Day [1], which induces a monoidal biclosed structure on $\hat{\mathbf{A}}$ $(= [\mathbf{A}^{\mathrm{op}}, \mathbf{Set}])$ for every monoidal structure on any category $\mathbf{A}$. We develop this in two ways. First, the monoidal action of $\mathbf{V}$ on $\mathbf{C}$ induces a monoidal action $\hat{\mathbf{V}} \times \hat{\mathbf{C}} \to \hat{\mathbf{C}}$ which has right adjoints in both arguments. Secondly, the monoidal action of $\hat{\mathbf{V}}$ on $\hat{\mathbf{C}}$ restricts to a monoidal action $\mathrm{FP}(\mathbf{V}^{\mathrm{op}}, \mathbf{Set}) \times \mathrm{FP}(\mathbf{C}^{\mathrm{op}}, \mathbf{Set}) \to \mathrm{FP}(\mathbf{C}^{\mathrm{op}}, \mathbf{Set})$ and the right adjoints restrict too. (This second observation relies on the fact that $\mathbf{V}$ is considered with the cartesian monoidal structure.) Thus $\mathrm{FP}(\mathbf{C}^{\mathrm{op}}, \mathbf{Set})$ is enriched in $\mathrm{FP}(\mathbf{V}^{\mathrm{op}}, \mathbf{Set})$ with copowers and powers.

Finally, the enriched adjunction $F \dashv U \colon \mathrm{FP}(\mathbf{C}^{\mathrm{op}}, \mathbf{Set}) \to \mathrm{FP}(\mathbf{V}^{\mathrm{op}}, \mathbf{Set})$ can be induced by any choice of object of $\mathrm{FP}(\mathbf{C}^{\mathrm{op}}, \mathbf{Set})$, by Proposition 8. □

The construction described in this proof is inspired by the following situation. Let $\mathbf{Set}_{\mathrm{f}}$ be the category of finite sets, and let $\mathbb{T}$ be a Lawvere theory. Then $(\mathbf{Set}_{\mathrm{f}}, \mathbb{T}^{\mathrm{op}})$ is almost an enriched model, except that the category $\mathbb{T}^{\mathrm{op}}$ is not $\mathbf{Set}_{\mathrm{f}}$-enriched in general. Our construction, applied to $(\mathbf{Set}_{\mathrm{f}}, \mathbb{T}^{\mathrm{op}})$, yields the basic motivating example of an EEC model: FP($\mathbf{Set}_f^{op}$, **Set**) is the category of sets (since $\mathbf{Set}_f$ is the free category with finite coproducts on one generator) and FP($\mathbb{T}$, **Set**) is the category of algebras of the theory.

Acknowledgements. We thank Alex Simpson for help and encouragement. Also thanks to Lars Birkedal, Jeff Egger, Masahito Hasegawa, Shin-ya Katsumata and Paul Levy for helpful discussions.

References

1. Day, B.: On closed categories of functors. Lect. Notes Math., vol. 137, pp. 1–38. Springer, Heidelberg (1970)
2. Egger, J., Møgelberg, R., Simpson, A.: Enriching an effect calculus with linear types. In: Grädel, E., Kahle, R. (eds.) CSL 2009. LNCS, vol. 5771, pp. 240–254. Springer, Heidelberg (2009)
3. Egger, J., Møgelberg, R., Simpson, A.: Linearly-used continuations in the enriched effect calculus. In: Ong, L. (ed.) FOSSACS 2010. LNCS, vol. 6014, pp. 18–32. Springer, Heidelberg (2010)
4. Egger, J., Møgelberg, R., Simpson, A.: The enriched effect calculus (2011) (in preparation)
5. Fiore, M.P.: Enrichment and representation theorems for categories of domains and continuous functions (March 1996) (unpublished manuscript)
6. Gordon, R., Power, A.: Enrichment through variation. J. Pure Appl. Algebra 120, 167–185 (1997)
7. Janelidze, G., Kelly, G.: A note on actions of a monoidal category. Theory Appl. of Categ. 9(4), 61–91 (2001)
8. Kelly, G.M.: Adjunction for enriched categories. Lect. Notes Math., vol. 106, pp. 166–177. Springer, Heidelberg (1969)
9. Kelly, G.M.: Basic Concepts of Enriched Category Theory. Cambridge University Press, Cambridge (1982)
10. Levy, P.B.: Call By Push Value. Kluwer, Dordrecht (2003)
11. Levy, P., Power, J., Thielecke, H.: Modelling environments in call-by-value programming languages. Inform. and Comput. 185 (2003)
12. Moggi, E.: Notions of computation and monads. Inform. and Comput. 93, 55–92 (1991)
13. O'Hearn, P.W., Reynolds, J.C.: From Algol to polymorphic linear lambda-calculus. J. ACM 47 (2000)
14. Plotkin, G.: Call-by-name, call-by-value, and the λ-calculus. Theoret. Comp. Sci. 1, 125–159 (1975)
15. Plotkin, G., Power, J.: Tensors of comodels and models for operational semantics. In: Proc. MFPS XXIV. Electr. Notes Theor. Comput. Sci., vol. 218, pp. 295–311. Elsevier, Amsterdam (2008)
16. Plotkin, G., Pretnar, M.: A logic for algebraic effects. In: Proc. LICS 2008. IEEE Press, Los Alamitos (2008)
17. Plotkin, G., Pretnar, M.: Handlers of algebraic effects. In: Castagna, G. (ed.) ESOP 2009. LNCS, vol. 5502, pp. 80–94. Springer, Heidelberg (2009)
18. Plotkin, G.D., Power, J.: Notions of computation determine monads. In: Nielsen, M., Engberg, U. (eds.) FOSSACS 2002. LNCS, vol. 2303, p. 342. Springer, Heidelberg (2002)

19. Plotkin, G.D., Power, J.: Algebraic operations and generic effects. Appl. Categ. Structures 11(1), 69–94 (2003)
20. Power, J., Thielecke, H.: Closed freyd- and κ-categories. In: Wiedermann, J., Van Emde Boas, P., Nielsen, M. (eds.) ICALP 1999. LNCS, vol. 1644, p. 625. Springer, Heidelberg (1999)
21. Power, A.J., Shkaravska, O.: From comodels to coalgebras: State and arrays. In: Proc. CMCS 2004. Electr. Notes Theor. Comput. Sci., vol. 106, pp. 297–314. Elsevier, Amsterdam (2004)
22. Power, J.: Generic models for computational effects. Theoret. Comput. Sci. 364(2), 254–269 (2006)
23. Power, J., Robinson, E.: Premonoidal categories and notions of computation. Math. Structures Comput. Sci. 7(5), 453–468 (1997)
24. Scott, D.: Mathematical concepts in programming language semantics. In: Proceedings of the Spring Joint Computer Conference, pp. 225–234. ACM, New York (1972)
25. Strachey, C.: The varieties of programming language. In: Proc. International Computing Symposium, pp. 222–233. Cini Foundation, Venice (1972); Also Tech. Monograph PRG-10, Univ. Oxford (1973)
26. Street, R.: The formal theory of monads. J. Pure Appl. Algebra 2(2), 149–168 (1972)

Proving Safety Properties of Rewrite Theories

Camilo Rocha and José Meseguer

University of Illinois at Urbana-Champaign
{hrochan2,meseguer}@cs.illinois.edu

Abstract. Rewrite theories are a general and expressive formalism for specifying concurrent systems in which states are axiomatized by equations and transitions among states are axiomatized by rewrite rules. We present a deductive approach for verifying *safety properties* of rewrite theories in which all formal temporal reasoning about concurrent transitions is ultimately reduced to purely equational inductive reasoning. Narrowing modulo axioms is extensively used in our inference system to further simplify the equational proof obligations to which all proofs of safety formulas are ultimately reduced. In this way, existing equational reasoning techniques and tools can be directly applied to verify safety properties of concurrent systems. We report on the implementation of this deductive system in the Maude Invariant Analyzer tool, which provides a substantial degree of automation and can automatically discharge many proof obligations without user intervention.

1 Introduction

Safety properties of concurrent systems are among the most important properties to verify and have received extensive attention in many formal approaches, both algorithmic and deductive. Algorithmic approaches such as model checking are quite attractive because they are automatic. However, they often assume a finite-state system and are not directly applicable to infinite state systems.

This paper is part of a broader effort to develop *generic* methods to reason about temporal logic properties of concurrent systems. It advances such an effort by developing deductive methods and tools for proving two key safety properties, namely, *stability* and *invariance*, plus their combination by strengthening techniques. By "generic" we mean verification methods not tied to a specific programming language. By contrast, the UNITY logic [2] and deductive methods developed by Manna and Pnueli [11], are tailored to verify safety properties of concurrent programs in specific imperative languages.

Of course, any such generic approach requires a *logical framework* general enough to encompass many different models and languages. In our case we use the rewriting logic framework [12], which has been shown to express very naturally many different models of concurrent computation and many concurrent languages. In rewriting logic, a concurrent system such as, for example, a network protocol or an entire concurrent programming language such as Java, is specified as a *rewrite theory* $\mathcal{R} = (\Sigma, E, R)$, with (Σ, E) an equational theory specifying

A. Corradini, B. Klin, and C. Cîrstea (Eds.): CALCO 2011, LNCS 6859, pp. 314–328, 2011.

the system's *states* as elements of the *initial algebra* $\mathcal{T}_{\Sigma/E}$, and R a collection of (non-equational) rules specifying the system's *concurrent transitions* in the *initial reachability model* $\mathcal{T}_{\mathcal{R}}$ [1]. This precisely means that the states of such an initial model are the elements of $\mathcal{T}_{\Sigma/E}$ and that its one-step transitions between such states are *provable* rewrite steps by the rules R.

Safety properties are a special type of *inductive* properties. That is, they do not hold for just any model of the given rewrite theory $\mathcal{R}$ but for $\mathcal{T}_{\mathcal{R}}$. Therefore, given any safety property φ we are interested in the model-theoretic satisfaction relation $\mathcal{T}_{\mathcal{R}} \models \varphi$, which we approximate deductively by an inductive inference relation $\mathcal{R} \Vdash \varphi$ which we prove always implies $\mathcal{T}_{\mathcal{R}} \models \varphi$. The inference rules transform pairs of the form $\mathcal{R} \Vdash \varphi$ into other pairs $\mathcal{R}' \Vdash \varphi'$ in such a way that: (i) all temporal logic formulas eventually disappear and are replaced by *purely* equational *conditional* formulas of the form $(\forall X)\ t = u$ **if** C, and (ii) the rewrite theory $\mathcal{R} = (\Sigma, E, R)$ is eventually replaced by the equational theory (Σ, E). Our methods focus on automatically discharging as many such conditional formulas as possible by procedures which either: (i) *narrow* the condition C, (ii) show C to be *unfeasible*, or (iii) prove $t = u$ *assuming* condition C. Proofs of all results presented here are included in [16].

We report on the implementation of the above-mentioned inductive inference system in the Invariant Analyzer tool (InvA), which provides a substantial degree of mechanization and can automatically discharge many proof obligations without user intervention. InvA uses functionality from Maude and its Church-Rosser Checker tool to discharge as many proof obligations as possible. It then returns the remaining proof obligations for the user to interactively discharge them by using, for instance, Maude's Inductive Theorem Prover. Throughout we use Lamport's infinite-state "bakery" protocol as a running example in the syntax of Maude [4]; other examples and the InvA tool are available from `http://camilorocha.info`.

2 Preliminaries

We assume an *order sorted signature* $\Sigma = (S, \leq, F)$ with finite poset of sorts $(S, \leq)$ and a finite set of function symbols F. We also assume that: (i) each connected component of a sort $s \in S$ in the poset ordering has a top sort, denoted by $[s]$, and (ii) for each operator declaration $f \in F_{s_1 \ldots s_n, s}$ there is also a declaration $f \in F_{[s_1]\ldots[s_n],[s]}$. We let $X = \{X_s\}_{s \in S}$ be an S-sorted family of disjoint sets of variables with each X_s countably infinite. The set of terms of sort s is denoted by $T_\Sigma(X)_s$ and the set of ground terms of sort s is denoted by $T_{\Sigma,s}$, which we assume nonempty for each s. $\mathcal{T}_\Sigma(X)$ and $\mathcal{T}_\Sigma$ denote the respective term algebras. The set of variables of a term t is written $vars(t)$ and is extended to sets of terms in the natural way. A *substitution* θ is a sorted map from a finite subset $dom(\theta) \subseteq X$ to $T_\Sigma(X)$ and extends homomorphically in the natural way; $ran(\theta)$ denotes the set of variables introduced by θ and $t\theta$ the application of θ to a term t. Substitution $\theta_1\theta_2$ is the composition of substitutions θ_1 and θ_2. A substitution θ is called *ground* if $ran(\theta) = \emptyset$.

A Σ-*equation* is a sentence $(\forall X)\ t = u$ **if** C, where $t = u$ is a Σ-*equality* with $t, u \in T_\Sigma(X)_s$ for some sort $s \in S$ and the *condition* C is a finite conjunction of Σ-equalities. An *equational theory* is a pair (Σ, E) with order-sorted signature Σ and finite set of Σ-equations E. For φ a Σ-equation, $(\Sigma, E) \vdash \varphi$ iff φ can be proved from (Σ, E) by the deduction rules in [13] iff φ is valid in all models of (Σ, E). An equational theory (Σ, E) induces the congruence relation $=_E$ on $T_\Sigma(X)$ defined for any $t, u \in T_\Sigma(X)$ by $t =_E u$ iff $(\Sigma, E) \vdash (\forall X)\, t = u$. Σ-algebras $\mathcal{T}_{\Sigma/E}(X)$ and $\mathcal{T}_{\Sigma/E}$ denote the quotient algebras induced by $=_E$ over the algebras $\mathcal{T}_\Sigma(X)$ and $\mathcal{T}_\Sigma$; we call $\mathcal{T}_{\Sigma/E}$ the *initial algebra* of (Σ, E). An E-*unifier* for a Σ-equality $t = u$ is a substitution θ such that $t\theta =_E u\theta$. A *complete* set of E-unifiers for a Σ-equality $t = u$ is written $\mathrm{CSU}_E(t = u)$, and it is called *finitary* if it contains a finite number of E-unifiers. We let $\mathrm{GU}_E(t = u)$ denote the set of ground E-unifiers of a Σ-equality $t = u$.

A Σ-*rule* is a sentence $(\forall X)\ t \to u$ **if** C, where $t \to u$ is a Σ-*sequent* with $t, u \in T_\Sigma(X)_s$ for some sort $s \in S$ and the *condition* C is a finite conjunction of Σ-equations. A *rewrite theory* is a tuple $\mathcal{R} = (\Sigma, E, R)$ with equational theory $\mathcal{E}_\mathcal{R} = (\Sigma, E)$ and a finite set of Σ-rules R. A *topmost rewrite theory* is a rewrite theory $\mathcal{R} = (\Sigma, E, R)$, where for some top sort $\mathfrak{s} = [\mathfrak{s}]$, terms $l, r \in T_\Sigma(X)_\mathfrak{s}$ in each $(\forall X)\ l \to r$ **if** $C \in R$, $l \notin X$, and no operator in Σ has $\mathfrak{s}$ as argument sort. For $\mathcal{R} = (\Sigma, E, R)$ and φ a Σ-rule, $\mathcal{R} \vdash \varphi$ iff φ can be obtained from $\mathcal{R}$ by the deduction rules in [1] iff φ is valid in all models of $\mathcal{R}$. For φ a Σ-equation, $\mathcal{R} \vdash \varphi$ iff $\mathcal{E}_\mathcal{R} \vdash \varphi$. A rewrite theory $\mathcal{R} = (\Sigma, E, R)$ induces the rewrite relation $\to_\mathcal{R}$ on $T_{\Sigma/E}(X)$ defined for every $t, u \in T_\Sigma(X)$ by $[t]_E \to_\mathcal{R} [u]_E$ iff there is a *one-step* rewrite proof $\mathcal{R} \vdash (\forall X)\ t \to u$. We let $\mathcal{R} \vdash (\forall X)\ t \to u$ and $\mathcal{R} \vdash (\forall X)\ t \overset{*}{\to} u$ respectively denote a one-step rewrite proof and an arbitrary length (but finite) rewrite proof in $\mathcal{R}$ from t to u. $\mathcal{T}_\mathcal{R} = (\mathcal{T}_{\Sigma/E}, \overset{*}{\to}_\mathcal{R})$ is the *initial reachability model* of $\mathcal{R} = (\Sigma, E, R)$ [1]. A Σ-sentence $(\forall Y)\, \varphi$ is an *inductive consequence* of $\mathcal{R}$ iff $\mathcal{R} \Vdash (\forall Y)\, \varphi$ iff $(\forall \theta : Y \longrightarrow T_\Sigma)\, \mathcal{R} \vdash \varphi\theta$ iff $\mathcal{T}_\mathcal{R} \models \varphi$.

State Predicates. A set of *state predicates* Π for $\mathcal{R} = (\Sigma, E, R)$ can be equationally-defined by an equational theory $\mathcal{E}_\Pi = (\Sigma_\Pi, E \uplus E_\Pi)$. Signature Σ_Π contains Σ, two sorts Bool $\leq$ [Bool] with constants $\top$ and $\bot$ of sort Bool, predicate symbols $p : \mathfrak{s} \longrightarrow$ [Bool] for each $p \in \Pi$, and optionally some auxiliary function symbols. Equations in E_Π define the predicate symbols in Σ_Π and auxiliary function symbols, if any; they protect[1] (Σ, E) and the equational theory specifying sort Bool, constants $\top$ and $\bot$, and the Boolean operations. It is easy to define a state predicate $p \in \Pi$ as a Boolean combination of other already-defined state predicates $\{p_1, \ldots, p_n\}$ in Σ_Π, so that the choice of focusing on atomic state predicates is mainly to simplify the exposition but does not limit the general applicability of our results. For the safety properties treated in this paper, *only* the positive case is needed to define a predicate's p semantics. The reason why p has typing $p : \mathfrak{s} \longrightarrow$ [Bool] instead of $p : \mathfrak{s} \longrightarrow$ Bool, is to allow partial definitions of p with equations that only define the *positive* case by

[1] A theory inclusion $(\Sigma, E) \subseteq (\Sigma', E')$ is *protecting* iff the unique Σ-homomorphism $\mathcal{T}_{\Sigma/E} \longrightarrow \mathcal{T}_{\Sigma'/E'}|_\Sigma$ to the Σ-reduct of the initial algebra $\mathcal{T}_{\Sigma'/E'}$ is an isomorphism.

equations $(\forall Y)\ p(t) = \top$ **if** C, and either leave the *negative* case implicit or may only define some negative cases with equations $(\forall Y)\ p(t') = \bot$ **if** C' without necessarily covering all the cases.

LTL Semantics. For $p \in \Pi$ and $[t]_E \in T_{\Sigma/E,\mathfrak{s}}$, the *semantics of* p in $\mathcal{T}_{\mathcal{R}}$ is defined by $\mathcal{E}_\Pi$ as follows: we say that $p([t]_E)$ *holds* in $\mathcal{T}_{\mathcal{R}}$ iff $\mathcal{E}_\Pi \vdash p(t) = \top$. This defines the Kripke structure $\mathcal{K}_{\mathcal{R}}^{\Pi} = (T_{\Sigma/E,\mathfrak{s}}, \rightarrow_{\mathcal{R}}, L_\Pi)$ with labeling function L_Π such that, for each $[t]_E \in T_{\Sigma/E,\mathfrak{s}}$, $p \in L_\Pi([t]_E)$ iff $p([t]_E)$ holds in $\mathcal{T}_{\mathcal{R}}$. Then, all of LTL can be interpreted in $\mathcal{K}_{\mathcal{R}}^{\Pi}$ in the standard way [3], including its first-order version. This remark is used in what follows to view some of our results as inference rules for reasoning about LTL properties of $\mathcal{K}_{\mathcal{R}}^{\Pi}$.

Executability Conditions. We assume that the set of equations of a rewrite theory $\mathcal{R}$ can be decomposed into a disjoint union $E \uplus A$, with A a collection of axioms (such as associativity, and/or commutativity, and/or identity) for which there exists a *matching algorithm modulo* A producing a finite number of A-matching substitutions, or failing otherwise. The second condition is that the equations E can be oriented into a set of *ground sort-decreasing*, *ground confluent*, and *ground terminating* rules $\overrightarrow{E}$ modulo A. We let $[\mathrm{can}_{\Sigma,E/A}(t)]_A \in T_{\Sigma/A,s}$ denote the *E-canonical form* of $[t]_A$. The rules R in $\mathcal{R}$ are assumed to be *ground coherent* relative to the equations E modulo A [20].

Free Constructors. For $\mathcal{R} = (\Sigma, E \uplus A, R)$ we say that $\Omega \subseteq \Sigma$ is a signature of *free constructors* modulo A if for each sort s in Σ and $t \in T_{\Sigma,s}$ there is $u \in T_{\Omega,s}$ satisfying $t =_{E \uplus A} u$ and for any $v \in T_{\Omega,s}$ $\mathrm{can}_{\Sigma,E/A}(v) =_A v$. Since $\mathcal{R}$ is ground coherent, the requirement $l \in T_\Omega(X)$ for each $(\forall X)\ l \rightarrow r$ **if** $C \in R$ is natural.

3 Ground Stability

For $\mathcal{R} = (\Sigma, E, R)$ a topmost rewrite theory, let p be a state predicate on the set of states $T_{\Sigma/E,\mathfrak{s}}$ of $\mathcal{T}_{\mathcal{R}}$. The property p being *(ground) stable* for $\mathcal{R}$ is the safety property $\mathcal{T}_{\mathcal{R}} \models p \Rightarrow \Box p$. That is, if p holds in a state $[t]_E \in T_{\Sigma/E,\mathfrak{s}}$ of $\mathcal{T}_{\mathcal{R}}$, then p holds in any state $[t']_E$ such that $[t]_E \xrightarrow{*}_{\mathcal{R}} [t']_E$. The concept of ground stability for $\mathcal{R}$ is intimately related to the notion of $\mathcal{T}_{\Sigma/E}$ being closed under $\rightarrow_{\mathcal{R}}$, namely, $\mathcal{R}$ being ground p-stable exactly means that the subset of $T_{\Sigma/E,\mathfrak{s}}$ satisfying p is closed under $\rightarrow_{\mathcal{R}}$.

Definition 1. *Let $\mathcal{R} = (\Sigma, E, R)$ be a topmost rewrite theory and let Π be a set of state predicates for $\mathcal{R}$ equationally defined in $\mathcal{E}_\Pi = (\Sigma_\Pi, E \uplus E_\Pi)$. For $p \in \Pi$, $\mathcal{R}$ is called* ground p-stable *under $R_0 \subseteq R$ iff, for each $t, u \in T_{\Sigma,\mathfrak{s}}$, $\mathcal{E}_\Pi \vdash p(t) = \top$ and $(\Sigma, E, R_0) \vdash t \xrightarrow{*} u$ imply $\mathcal{E}_\Pi \vdash p(u) = \top$. $\mathcal{R}$ is* ground p-stable, *written $\mathcal{R} \Vdash p \Rightarrow \Box p$, iff $\mathcal{R}$ is ground p-stable under R.*

For a topmost rewrite theory $\mathcal{R} = (\Sigma, E, R)$, the reachability condition in the definition of ground stability can be reduced to a simpler 1-step rewrite condition, obtaining an equivalent notion that avoids arbitrary depth proof-search.

$$\frac{\mathcal{R} \Vdash p \Rightarrow \bigcirc p}{\mathcal{R} \Vdash p \Rightarrow \Box p}\ \text{G-ST}$$

$$\frac{\mathcal{E}_\Pi \Vdash \bigwedge_{\substack{((\forall X)\ l \to r \textbf{ if } C) \in R \\ (\theta, w, D) \in \Theta(l)}} (\forall ran(\theta))\ p(r\theta) = \top \textbf{ if } C\theta \wedge D\theta \wedge w\theta = \top}{\mathcal{R} \Vdash p \Rightarrow \bigcirc p}\ \text{NR1}$$

Fig. 1. Ground p-stability for $\mathcal{R} = (\Sigma, E, R)$, with Θ as in Theorem 1

Lemma 1. *Let $\mathcal{R}$, $\mathcal{E}_\Pi$, p, and R_0 be as in Definition 1. Then $\mathcal{R}$ is ground p-stable under R_0 iff, for each $t, u \in T_{\Sigma,\mathfrak{s}}$, $\mathcal{E}_\Pi \vdash p(t) = \top$ and $(\Sigma, E, R_0) \vdash t \to u$ imply $\mathcal{E}_\Pi \vdash p(u) = \top$.*

In the notation of Linear Time Temporal Logic (LTL), Lemma 1 justifies the soundness of the inference rule G-ST in Figure 1 that shows how to reason about p-stability in $\mathcal{K}_\mathcal{R}^\Pi$. Symbol "$\bigcirc$" corresponds to the next operator and symbol "$\Rightarrow$" to strong implication in LTL (see [10] for details). So, for $\mathcal{K}_\mathcal{R}^\Pi \models p \Rightarrow \Box p$ to hold, it is enough to show that $\mathcal{K}_\mathcal{R}^\Pi \models p \Rightarrow \bigcirc p$ holds. Lemma 1 shows that the converse of inference rule G-ST is also sound.

The next question to ask is how to reduce the verification of the simpler condition $p \Rightarrow \bigcirc p$ to inductive equational reasoning. We use the idea of (one-step) *narrowing with equations modulo axioms* [9], a sound and complete method for ground stability analysis, to reduce the inductive reachability problem of p-stability for $\mathcal{T}_\mathcal{R}$ to equational inductive properties of $\mathcal{T}_{\mathcal{E}_\mathcal{R}}$.

Under the executability assumptions, $\mathcal{R}$ has a disjoint union $E \uplus A$ of equations, with A a collection of structural axioms on some function symbols in Σ such as associativity, commutativity, identity, etc., and E a set of ground sort-decreasing, ground confluent, ground terminating, and ground coherent (w.r.t. R) equations modulo A. For a combination of free and associative and/or commutative and/or identity axioms, except for symbols f that are associative but not commutative, a finitary A-unification algorithm exists. Instead, in general there is no finitary $E \uplus A$-unification algorithm, but for $\Omega \subseteq \Sigma$ a signature of free equational constructors modulo A and a Ω-equality $t = u$, the ground instances of $\text{CSU}_A(t = u)$ exactly characterize the set $\text{GU}_{E \uplus A}(t = u)$.

Lemma 2. *Let $\mathcal{E} = (\Sigma, E \uplus A)$ be an order-sorted equational theory with finitary A-unification algorithm, and with $\Omega \subseteq \Sigma$ a signature of free constructors modulo A. Then, for any Ω-equality $t = u$, $\alpha \in \text{GU}_{E \uplus A}(t = u)$ iff there exists $\theta \in \text{CSU}_A(t = u)$ and a ground substitution $\gamma : vars(\theta) \longrightarrow T_\Omega$ such that $\theta\gamma =_{E \uplus A} \alpha$.*

In order to show the ground p-stability of $\mathcal{R} = (\Sigma, E \uplus A, R)$ we need to prove for each rule $(\forall X)\ l \to r$ **if** $C \in R$ that if $p(l) = \top$ and C hold, then $p(r) = \top$ holds. The key observation here is that, since by assumption $l \in T_{\Omega,\mathfrak{s}}(X)$, if all left hand-sides $p(v)$ of equations $(\forall Y)\ p(v) = w$ **if** $D \in E_\Pi$ defining the state predicate $p \in \Pi$ are *Ω-patterns* (i.e., $v \in T_\Omega(X)$), then we can compute $\mathrm{CSU}_A(l = v)$ and obtain substitutions θ which, by Lemma 2, exactly characterize any ground $E \uplus A$-unifier in $\mathrm{GU}_{E\uplus A}(l = v)$. Each substitution $\theta \in \mathrm{CSU}_A(l = v)$ is such that $p(l\theta) = \top$, or at least $p(l\theta)$ *could* be equal to $\top$, and thus we are left with the task of inductively proving $p(r\theta) = \top$ under the assumptions $C\theta$, $D\theta$, and $w\theta = \top$. In this way, the inductive reachability problem of p-stability for $\mathcal{T}_\mathcal{R}$ is recast as simpler equational inductive properties of $\mathcal{T}_{\Sigma/E\uplus A}$: $\mathcal{T}_\mathcal{R}$ is ground p-stable iff $\mathcal{T}_{\Sigma/E\uplus A}$ satisfies these inductive properties. Theorem 1 proves sound and complete narrowing inference rule NR1 in Figure 1.

Theorem 1. *Let $\mathcal{R} = (\Sigma, E \uplus A, R)$ be a topmost rewrite theory with signature $\Omega \subseteq \Sigma$ of free constructors modulo A and with a finitary A-unification algorithm, and let $\mathcal{E}_\Pi = (\Sigma_\Pi, E \uplus A \uplus E_\Pi)$ be an equational definition of Π for $\mathcal{R}$. Let $p \in \Pi$ and, without loss of generality, assume that the equations $E^p_\Pi \subseteq E_\Pi$ defining $p \in \Pi$ are all conditional, have Ω-patterns as left-hand sides, and have no variable in common with the rewrite rules R. Then, $\mathcal{R}$ is ground p-stable under $(\forall Y)\ l \to r$* **if** *C iff*

$$\mathcal{E}_\Pi \Vdash \bigwedge_{(\theta,w,D)\in\Theta(l)} (\forall ran(\theta))\ p(r\theta) = \top \ \textbf{if}\ C\theta \wedge D\theta \wedge w\theta = \top,$$

where $\Theta(l) = \{(\theta, w, D) \mid ((\forall Z)\ p(v) = w\ \textbf{if}\ D) \in E^p_\Pi \wedge \theta \in \mathrm{CSU}_A(l = v)\}$.

Observe that obtaining a complete set of unifiers in the definition of $\Theta(l)$ in Theorem 1 only involves Σ-terms and not Σ_Π-terms. This is useful in practice because the generation of proof obligations from $\Theta(l)$ does not depend on the state predicates defined in $\mathcal{E}_\Pi$ and therefore is not affected by their equational definitions, no matter how involved these definitions may be. Also observe that $\Theta(l)$ is finite for each $(\forall Y)\ l \to r$ **if** $C \in R$ because the complete set of A-unifiers is finite. Therefore, the set of proof obligations is finite because of the finiteness assumptions on E and R. As a final remark, observe that when w is $\bot$ in an equation $(\forall Z)\ p(v) = w$ **if** $D \in E^p_\Pi$, each proof obligation $(\forall ran(\theta))\ p(r\theta) = \top$ **if** $C\theta \wedge D\theta \wedge w\theta = \top$ can be *soundly* ignored, because $\mathcal{E}_\Pi$ protects sort Bool and therefore $w\theta = \bot\theta = \bot \neq \top$.

Example 1. Consider a version of Lamport's bakery protocol, slightly adapted from [6], in which processes achieve mutual exclusion by the usual method common in bakeries: there is a number dispenser and customers are served in sequential order according to the ticket that they hold. The system is specified in Maude as a topmost rewrite theory *BAKERY* with top sort *State*:

```
fmod BAKERY-SYNTAX is
 pr NAT .
 sorts ProcIdle ProcWait Proc ProcIdleSet ProcWaitSet ProcSet State .
 subsorts ProcIdle < ProcIdleSet .  subsorts ProcWait < ProcWaitSet .
 subsorts ProcIdle ProcWait < Proc < ProcSet .
 subsorts ProcIdleSet < ProcWaitSet < ProcSet .
 op idle : -> ProcIdle [ctor] .       op wait : Nat -> ProcWait [ctor] .
 op crit : Nat -> Proc [ctor] .       op none : -> ProcIdleSet [ctor] .
```

```
 op __ : ProcIdleSet ProcIdleSet -> ProcIdleSet [ctor assoc comm id: none] .
 op __ : ProcWaitSet ProcWaitSet -> ProcWaitSet [ditto] .
 op __ : ProcSet ProcSet -> ProcSet [ditto] .
 op _:_[_] : Nat Nat ProcSet -> State [ctor] .
endfm

mod BAKERY is
 pr BAKERY-SYNTAX .
 var Ps : ProcSet .  vars N M : Nat .
 rl [get]   : N : M [idle Ps] => s(N) : M [wait(N) Ps] .
 rl [serve] : N : M [wait(M) Ps] => N : M [crit(M) Ps] .
 rl [leave] : N : M [crit(M) Ps] => N : s(M) [idle Ps] .
endm
```

The equations in *BAKERY* are *all* structural axioms, namely, associativity, commutativity, and identity for sets of processes. Since there are no equations besides the structural axioms, *BAKERY*'s signature is trivially a signature of free constructors modulo *BAKERY*'s structural axioms. A ground term "$n : m\ [ps]$" of sort *State* describes the state in which n is the natural number of the next available ticket, m is the natural number of the next ticket to be served, and ps is the set of customers currently in the bakery.

We are interested in the set of state predicates $\Pi = \{\textit{bounded-tickets}\}$, expressing that tickets among customers are all bounded from above. State predicate *bounded-tickets* is equationally defined with auxiliary functions *sb*, *tkts*, and *tb* (shorthands, respectively, for subbag, tickets, and tickets below). Module *BAKERY-PROPS* defines sort *NatBag* for bags (or multisets) of natural numbers; *mtbag* denotes the empty bag and bag union is juxtaposition modulo associativity and commutativity, with identity *mtbag*.

```
fmod BAKERY-PROPS is
 pr BAKERY-SYNTAX .                            pr BOOL-OPS .
 sort NatBag .                                 subsort Nat < NatBag .
 op mtbag : -> NatBag .
 op __ : NatBag NatBag -> NatBag [assoc comm id: mtbag] .
 op bounded-tickets : State -> [Bool] .
 op tb : Nat -> NatBag .                       op tkts : ProcSet -> NatBag .
 op sb : NatBag NatBag -> [Bool] .
 var Is : ProcIdleSet .  var Ps : ProcSet .  vars N M : Nat .  vars NB NB' : NatBag .
 eq [1] : bounded-tickets(N : M [Ps]) = sb(tkts(Ps),tb(N)) .
 eq [a.1] : sb(NB,NB NB') = ⊤ .               eq [a.2] : sb(N NB,N NB') = sb(NB,NB') .
ceq [a.3] : sb(N NB,NB') = ⊥ if in?(N,NB') = ⊤.
 eq [b.1] : tkts(Is) = mtbag .                 eq [b.2] : tkts(idle Ps) = tkts(Ps) .
 eq [b.3] : tkts(wait(N) Ps) = N tkts(Ps) .    eq [b.4] : tkts(crit(N) Ps) = N tkts(Ps) .
 eq [c.1] : tb(0) = mtbag .                    eq [c.2] : tb(s(N)) = N tb(N) .
endfm
```

By Lemma 1 and Theorem 1, *BAKERY* is ground *bounded-tickets*-stable if the following sentences are inductive theorems of $\mathcal{E}_{BAKERY\text{-}PROPS}$:

$(\forall x_1, x_2 : Nat; x_3 : ProcSet)$

$$\textit{bounded-tickets}(s(x_1) : x_2[\textit{wait}(x_1)]) = \top \textbf{ if } \textit{sb}(\textit{tkts}(\textit{idle}), \textit{tb}(x_1)) = \top, \tag{1}$$

$$\textit{bounded-tickets}(s(x_1) : x_2[x_3\ \textit{wait}(x_1)]) = \top \textbf{ if } \textit{sb}(\textit{tkts}(\textit{idle}\, x_3), \textit{tb}(x_1)) = \top, \tag{2}$$

$$\textit{bounded-tickets}(x_1 : x_2[\textit{crit}(x_2)]) = \top \textbf{ if } \textit{sb}(\textit{tkts}(\textit{wait}(x_2)), \textit{tb}(x_1)) = \top, \tag{3}$$

$$\textit{bounded-tickets}(x_1 : x_2[\textit{crit}(x_2)\, x_3]) = \top \textbf{ if } \textit{sb}(\textit{tkts}(\textit{wait}(x_2)\, x_3), \textit{tb}(x_1)) = \top, \tag{4}$$

$$\textit{bounded-tickets}(x_1 : s(x_2)[\textit{idle}]) = \top \textbf{ if } \textit{sb}(\textit{tkts}(\textit{crit}(x_2)), \textit{tb}(x_1)) = \top, \tag{5}$$

$$\textit{bounded-tickets}(x_1 : s(x_2)[\textit{idle}\, x_3]) = \top \textbf{ if } \textit{sb}(\textit{tkts}(\textit{crit}(x_2)\, x_3), \textit{tb}(x_1)) = \top. \tag{6}$$

Sentences (1) and (2), (3) and (4), and (5) and (6) are obtained from equation *1* and rules *get*, *serve*, and *leave*, respectively. Sentences (1) and (5) have trivial consequents that can be automatically discharged by equational rewriting. Sentences (2)–(4) follow automatically by assuming the conditions (see Section 6 for a brief explanation of this technique). Sentence (6) can be discharged by Maude's ITP [8] with minor user interaction.

4 Ground Invariance

Invariants are the most important safety properties. For $\mathcal{R} = (\Sigma, E, R)$ a topmost rewrite theory and $p, I \in \Pi$, the property p being *ground invariant* for $\mathcal{R}$ from the set of initial states I is the safety property $\mathcal{T}_{\mathcal{R}} \models I \Rightarrow \Box p$. That is, in $\mathcal{T}_{\mathcal{R}}$ whenever I holds for a state $[t]_E \in T_{\Sigma/E,\mathfrak{s}}$, then p holds in any state $[t']_E \in T_{\Sigma/E,\mathfrak{s}}$ such that $[t]_E \xrightarrow{*}_{\mathcal{R}} [t']_E$. Since the set of initial states is defined in $\mathcal{E}_\Pi$ by a state predicate $I \in \Pi$, an equational definition of I can of course capture an infinite set of initial states.

Definition 2. *Let $\mathcal{R} = (\Sigma, E, R)$ be a topmost rewrite theory and let Π be a set of state predicates for $\mathcal{R}$ defined by $\mathcal{E}_\Pi = (\Sigma_\Pi, E \uplus E_\Pi)$. For $p, I \in \Pi$, $\mathcal{R}$ is called* ground p-invariant *from I under $R_0 \subseteq R$ iff, for each $t, u \in T_{\Sigma,\mathfrak{s}}$, $\mathcal{E}_\Pi \vdash I(t) = \top$ and $(\Sigma, E, R_0) \vdash t \xrightarrow{*} u$ imply $\mathcal{E}_\Pi \vdash p(u) = \top$. $\mathcal{R}$ is* ground p-invariant *from I, written $\mathcal{R} \Vdash I \Rightarrow \Box p$, iff $\mathcal{R}$ is ground p-invariant from I under R.*

Ground p-invariance is intimately related to ground p-stability: if every initial state defined by a predicate I satisfies p and the topmost rewrite theory $\mathcal{R}$ is p-stable, then $\mathcal{R}$ is p-invariant from I. The converse does not necessarily hold, because even if $\mathcal{R}$ is ground p-invariant from I, the set of states of $\mathcal{T}_{\mathcal{R}}$ satisfying p need not be closed under $\rightarrow_{\mathcal{R}}$. However, if $\mathcal{R}$ is ground p-invariant the set of states satisfying p over-approximates the set of states reachable from some initial state.

Lemma 3. *Let $\mathcal{R}$, Π, $\mathcal{E}_\Pi$, p, and I be as in Definition 2. Then, $\mathcal{R}$ is ground p-invariant from I under $R_0 \subseteq R$ if (i) $\mathcal{E}_\Pi \Vdash (\forall x : \mathfrak{s})\ p(x) = \top$ **if** $I(x) = \top$ and (ii) $\mathcal{R}$ is ground p-stable under R_0.*

For a topmost rewrite theory $\mathcal{R} = (\Sigma, E, R)$ and state predicates $p, q \in \Pi$, we write $q \Rightarrow p$ as a shorthand for $(\forall x : \mathfrak{s})\ p(x) = \top$ **if** $q(x) = \top$, and let $[\![p]\!]^{\mathcal{E}_\Pi} = \{[t]_E \in T_{\Sigma/E,\mathfrak{s}} \mid \mathcal{E}_\Pi \vdash p(t) = \top\}$ (or simply $[\![p]\!]$). Condition 1 in Lemma 3 states that every initial state specified by I must satisfy property p. That is, for Π and $\mathcal{E}_\Pi$ defined as in Lemma 3, $\mathcal{E}_\Pi \Vdash I \Rightarrow p$ holds iff $[\![I]\!] \subseteq [\![p]\!]$. Observe that this condition does not depend on the dynamics of $\mathcal{T}_{\mathcal{R}}$, but only on its set of states $T_{\Sigma/E,\mathfrak{s}}$. Conditions (i) and (ii) in Lemma 3 are used in the literature to define the notion of *inductive invariant*, i.e., of a predicate holding in the set of initial states and mantained true by every transition.

In LTL terms, Lemma 3 proves the soundness of inference rule G-Inv in Figure 2 for proving p invariant from I in $\mathcal{K}^\Pi_{\mathcal{R}}$. The only remaining question is how to prove $I \Rightarrow p$. Theorem 2 answers this question, by giving a necessary and sufficient condition for proving statements of the form $q \Rightarrow p$, with $p, q \in \Pi$ state predicates. It also proves the soundness of inference rule C$\Rightarrow$ in Figure 2.

$$\frac{\mathcal{R} \Vdash I \Rightarrow p \qquad \mathcal{R} \Vdash p \Rightarrow \Box p}{\mathcal{R} \Vdash I \Rightarrow \Box p} \text{ G-INV}$$

$$\frac{\mathcal{E}_\Pi \Vdash \bigwedge_{((\forall Y)\ q(v)=w \textbf{ if } C) \in E^q_\Pi} (\forall Y)\ p(v) = \top \textbf{ if } C \wedge w = \top}{\mathcal{R} \Vdash q \Rightarrow p} \text{ C}\Rightarrow$$

Fig. 2. Ground p-invariance from I for $\mathcal{R} = (\Sigma, E, R)$, with E^q_Π as in Theorem 2

Theorem 2. *Let $\mathcal{R}$, Π, $\mathcal{E}_\Pi$, and p be as in Definition 2, and let $q \in \Pi$. Without loss of generality, assume that the equations $E^q_\Pi \subseteq E_\Pi$ defining $q \in \Pi$ are all conditional. Then $[\![q]\!] \subseteq [\![p]\!]$ iff*

$$\mathcal{E}_\Pi \Vdash \bigwedge_{((\forall Y)\ q(v)=w \textbf{ if } C) \in E^q_\Pi} (\forall Y)\ p(v) = \top \textbf{ if } C \wedge w = \top.$$

Example 2. Recall Example 1 from Section 3. Here we are interested in state predicates $\Pi = \{$*bounded-tickets*, *init*$\}$ for *BAKERY*, with *bounded-tickets* as defined in *BAKERY-PROPS*. State predicate *init* defines the set of initial states for $\mathcal{T}_{BAKERY}$ in module *BAKERY-PROPS-EXT1*, which protects *BAKERY-PROPS*.

```
fmod BAKERY-PROPS-EXT1 is
 protecting BAKERY-PROPS .
 op init : State -> [Bool] .
 var Is : ProcIdleSet .
 eq [2] : init(0 : 0 [Is]) = ⊤ .
endfm
```

An initial state for $\mathcal{T}_{BAKERY}$ is any state in which the next available ticket and the ticket to be served have value zero, and customers in the "bakery" are all in state *idle*. Observe that no constraint is imposed on the initial number of customers. We want to prove *BAKERY* ground *bounded-tickets*-invariant for *init*. By Lemma 3, it is sufficient to prove: (i) inductively in $\mathcal{E}_{BAKERY\text{-}PROPS\text{-}EXT1}$ sentence $(\forall x : \mathfrak{s})$ *bounded-tickets*$(x) = \top$ **if** *init*$(x) = \top$ and (ii) *BAKERY* is ground *bounded-tickets*-stable. Condition (i) is proved in Example 1 and, by Theorem 2, condition (i) holds iff Sentence (7) is an inductive theorem of $\mathcal{E}_{BAKERY\text{-}PROPS\text{-}EXT1}$:

$$(\forall x_1 : \mathit{ProcIdleSet})\ \mathit{bounded\text{-}tickets}(0 : 0\,[x_1]) = \top. \tag{7}$$

Sentence (7) admits a simple proof by equational rewriting because $\mathit{tkts}(x_1) = \mathit{mtbag}$ and $\mathit{tb}(0) = \mathit{mtbag}$. So, we have $\mathit{BAKERY} \Vdash \mathit{init} \Rightarrow \Box \mathit{bounded\text{-}tickets}$.

5 Strengthenings for Ground Invariance

For state predicates $p, I \in \Pi$, a *strengthening* for the ground p-invariance from I of a topmost rewrite theory $\mathcal{R}$ is a state predicate $q \in \Pi$ such that $\mathcal{R}$ is ground q-invariant from I and, moreover, q can be used to prove $\mathcal{R} \Vdash I \Rightarrow \Box p$. State predicate q is often the result of gradually refining a too-weakly defined p for which $\mathcal{R}$ being ground p-invariant cannot be proven directly by Lemma 3. We present two strengthening techniques for ground invariance, prove their correctness, and illustrate their application using the running example.

Recall Lemma 3 in Section 4 stating that if $[\![I]\!] \subseteq [\![p]\!]$ and $\mathcal{R}$ is ground p-stable, then $\mathcal{R}$ is ground p-invariant from I. The first key observation for an strengthening technique is the one previously made in Section 4: a topmost rewrite theory $\mathcal{R}$ may be ground p-invariant from I and yet not be ground p-stable. For the ground p-invariance from I of $\mathcal{R}$, the only states from which p need not be falsified are precisely those $[t]_E$ reachable from any state in $[\![I]\!]$. The idea is then to strengthen p as follows: if $\mathcal{R}$ is ground q-invariant from I and every state satisfying q also satisfies p (i.e., $[\![q]\!] \subseteq [\![p]\!]$), then clearly $\mathcal{R}$ is ground p-invariant from I even if not necessarily ground p-stable, because any state in $\mathcal{T}_{\mathcal{R}}$ reachable from $[\![I]\!]$ is also in $[\![p]\!]$. Theorem 4 states that for proving $\mathcal{R} \Vdash I \Rightarrow \Box p$ assuming $\mathcal{R} \Vdash J \Rightarrow \Box q$, it is sufficient to equationally check $[\![q]\!] \subseteq [\![p]\!]$ and $[\![I]\!] \subseteq [\![J]\!]$. In LTL terms, Theorem 4 proves the soundness of inference rule STR1 in Figure 3.

Lemma 4. *Let $\mathcal{R}$, Π, $\mathcal{E}_\Pi$, and p be as in Definition 2, and let $q, J \in \Pi$. If $\mathcal{R}$ is ground q-invariant from J and $[\![q]\!] \subseteq [\![p]\!]$, then $\mathcal{R}$ is ground p-invariant for any $I \in \Pi$ such that $[\![I]\!] \subseteq [\![J]\!]$.*

The second strengthening technique follows from observing that if $\mathcal{R}$ is ground q-invariant from I, then for all states $[t]_E$ reachable from I the equivalence $[t]_E \rightarrow_{\mathcal{R}} [u]_E \iff [t]_E \rightarrow_{\mathcal{R}} [u]_E \wedge [t]_E \in [\![q]\!]$ is logically valid. The strengthening technique introduced in Theorem 3 is particularly useful in situations in which $[\![q]\!] \not\subseteq [\![p]\!]$ and also when proving $[\![q]\!] \subseteq [\![p]\!]$ may be difficult. Theorem 3 also proves the soundness of narrowing inference rule STR2 in Figure 3.

Theorem 3. *Let $\mathcal{R} = (\Sigma, E \uplus A, R)$ be a topmost rewrite theory with signature $\Omega \subseteq \Sigma$ of free constructors modulo A and with a finitary A-unification algorithm, and let $\mathcal{E}_\Pi = (\Sigma_\Pi, E \uplus A \uplus E_\Pi)$ be an equational definition of Π for $\mathcal{R}$. Let $p \in \Pi$ and, without loss of generality, assume that the equations $E_\Pi^p \subseteq E_\Pi$ defining $p \in \Pi$ are all conditional, have Ω-patterns as left-hand sides, and have no variable in common with the rules R. Then, $\mathcal{R}$ is ground p-invariant from $I \in \Pi$ under $(\forall Y)\ l \rightarrow r$ **if** C if $\mathcal{R}$ is ground q-invariant from I, $[\![I]\!] \subseteq [\![p]\!]$, and*

$$\mathcal{E}_\Pi \Vdash \bigwedge_{(\theta,w,D)\in\Theta(l)} (\forall ran(\theta))\ p(r\theta) = \top \ \textbf{if}\ C\theta \wedge D\theta \wedge w\theta = \top \wedge q(l\theta) = \top,$$

where $\Theta(l) = \{(\theta, w, D) \mid ((\forall Z)\ p(v) = w\ \textbf{if}\ D) \in E_\Pi^p \wedge \theta \in \mathrm{CSU}_A(v = l)\}$.

$$\frac{\mathcal{R} \Vdash J \Rightarrow \Box q \qquad \mathcal{R} \Vdash I \Rightarrow J \qquad \mathcal{R} \Vdash q \Rightarrow p}{\mathcal{R} \Vdash I \Rightarrow \Box p} \textsc{Str1}$$

$$\frac{\mathcal{R} \Vdash I \Rightarrow \Box q \qquad \mathcal{R} \Vdash I \Rightarrow p \qquad \mathcal{R} \Vdash q \wedge p \Rightarrow \bigcirc p}{\mathcal{R} \Vdash I \Rightarrow \Box p} \textsc{Str2}$$

$$\frac{\mathcal{E}_{\Pi} \Vdash \bigwedge_{\substack{((\forall X)\ l \to r \ \mathbf{if}\ C) \in R \\ (\theta, w, D) \in \Theta(l)}} (\forall ran(\theta))\ p(r\theta) = \top \ \mathbf{if}\ C\theta \wedge D\theta \wedge w\theta = \top \wedge q(l\theta) = \top}{\mathcal{R} \Vdash q \wedge p \Rightarrow \bigcirc p} \textsc{Nr2}$$

Fig. 3. Strengthenings for $\mathcal{R} = (\Sigma, E, R)$, with Θ as in Theorem 3

Example 3. Recall examples 1 and 2 from sections 3 and 4. We are interested in state predicates $\Pi = \{$*bounded-tickets*, *init*, *unique-tickets*, *good-state*, *mutex*$\}$ for *BAKERY*, with *bounded-tickets* and *init* as defined in examples 1 and 2, respectively. Predicate *mutex* defines a mutual exclusion property for *BAKERY*, and *unique-tickets* and *good-state* are strengthenings for *mutex*. These predicates and some auxiliary functions are defined in module *BAKERY-PROPS-EXT2*.

```
fmod BAKERY-PROPS-EXT2 is
 pr BAKERY-PROPS-EXT1 .
 ops unique-tickets good-state mutex : State -> [Bool] .
 op set : NatBag -> Bool .
 op in : Nat NatBag -> Bool .
 op = : Nat Nat -> Bool [comm] .
 var Ws : ProcWaitSet .  var Ps : ProcSet .
 vars N M M' N' : Nat .
 ...
 eq [3]   : unique-tickets(N : M [Ps]) = set(tkts(Ps)) .
 eq [4.1] : good-state(N : M [Ws]) = true .
 eq [4.2] : good-state(N : M [crit(M) Ws]) = true .
 eq [4.3] : good-state(N : M [crit(M') crit(N') Ps]) = false .
 eq [5.1] : mutex(N : M [Ws]) = true .
 eq [5.2] : mutex(N : M [crit(M') Ws]) = true .
 eq [5.3] : mutex(N : M [crit(M') crit(N') Ps]) = false .
endfm
```

The mutual exclusion property, completely defined by *mutex* for the sort Bool, holds in a state iff such state has at most one customer being served. State predicate *good-state* is a stronger version of *mutex* in which, for it to hold, the customer being served must have the least ticket number among all customers. State predicate *unique-tickets* holds whenever the tickets among the customers are all distinct. Auxiliary predicate *set*(b) holds if bag b is a set.

We prove *BAKERY* ground *mutex*-invariant for *init*, by reducing its proof to three simpler goals. Namely, to prove (i) *BAKERY* $\Vdash$ *init* $\Rightarrow \Box$*unique-tickets* by assuming *BAKERY* $\Vdash$ *init* $\Rightarrow \Box$*bounded-tickets*, (ii) *BAKERY* $\Vdash$ *init* $\Rightarrow$ $\Box$*good-state* by assuming (i), and (iii) *BAKERY* $\Vdash$ *init* $\Rightarrow \Box$*mutex* by assuming (ii). Applying Theorem 3 to (i) and (ii), and Lemma 4 to (iii) yields a total of 31 equational proof obligations, of which 29 are automatically discharged and only 2 require trivial inductive theorem proving interaction (see Section 6).

6 Maude's Invariant Analyzer

For a topmost rewrite theory $\mathcal{R}$ and of a set of state predicates Π in Maude, the InvA tool mechanizes inference rules G-ST, G-INV, STR1, STR2, NR1, and NR2. Given a ground stability or ground invariance property φ, it generates equational proof obligations such that if they hold, then $\mathcal{T}_{\mathcal{R}} \models \varphi$. It also mechanizes inference rule C$\Rightarrow$. Thanks to the availability in Maude 2.6 of unification modulo commutativity (C), associativity and commutativity (AC), and modulo these theories plus identities (U), and to the narrowing modulo infrastructure available in Full Maude 2.6, InvA can handle modules with operators declared C, CU, AC, and ACU.

Automatic Discharge of Proof Obligations. After applying rules G-ST, G-INV, STR1, STR2, NR1, and NR2 according to the user commands, the InvA tool uses rewriting and narrowing-based reasoning for automatically discharging as many of the generated equational proof obligations as possible. For $\mathcal{E} = (\Sigma, E \uplus A)$ and a conditional proof obligation $\varphi = (\forall X)\ t = u$ **if** C, the InvA tool applies a proof-search strategy such that if it succeeds, then $\mathcal{T}_{\mathcal{E}} \models \varphi$. Otherwise, the proof obligation is output to the user. Let $\overline{t}$, $\overline{u}$, $\overline{C}$ be obtained by replacing each variable $x \in X$ by a new constant $\overline{x} \in \overline{X}$, with $\Sigma \cap \overline{X} = \emptyset$. First, the strategy checks if φ holds *trivially*, i.e., if $can_{\Sigma,E/A}(t) =_A can_{\Sigma,E/A}(u)$ or there is $t_i = u_i$ in C with $can_{\Sigma,E/A}(t_i), can_{\Sigma,E/A}(u_i) \in T_\Sigma$ but $can_{\Sigma,E/A}(t_i) \neq_A can_{\Sigma,E/A}(u_i)$. Second, it checks if φ is *context-joinable* [5]: φ is context-joinable if $\overline{t}$ and $\overline{u}$ are joinable in the rewrite theory $\mathcal{R}^{\varphi}_{\mathcal{E}} = (\Sigma(\overline{X}), A, \overrightarrow{E} \cup \overrightarrow{C})$, obtained from orienting equations E as rewrite rules $\overrightarrow{E}$ and *heuristically* orienting each equality $t_i = u_i$ in C as a sequent $\overline{t_i} \to \overline{u_i}$ in $\overrightarrow{C}$. Third, it checks if the proof obligation is *unfeasible* [5]: φ is unfeasible if there is a conjunct $\overline{t_i} \to \overline{u_i}$ in $\overrightarrow{C}$ and $v, w \in T_\Sigma(X)$ such that $\mathcal{R}^{\varphi}_{\mathcal{E}} \vdash \overline{t_i} \to \overline{v} \wedge \overline{t_i} \to \overline{w}$, $CSU_A(v = w) = \emptyset$, and v and w are strongly irreducible with $\overrightarrow{E}$ modulo A. Because of the executability assumptions on $(\Sigma, E \uplus A)$, the first test of the strategy either succeeds or fails in finitely many equational rewrite steps. For the second and third tests, the strategy is not guaranteed to succeed or fail in finitely many rewrite steps because the oriented sequents $\overrightarrow{C}$ can falsify the termination assumption, and hence InvA uses a bound for the depth of the proof-search.

Tool Snapshot. We show an interaction with InvA in which, by assuming $BAKERY \Vdash init \Rightarrow \Box good\text{-}state$, we prove $BAKERY \Vdash init \Rightarrow \Box mutex$:

```
Checking BAKERY-PROPS ||- init => good-state ...
Proof obligations generated:  1
Proof obligations discharged: 1
Success!
rewrites: 4241 in 17ms cpu (21ms real) (245045 rewrites/second)

Checking BAKERY U BAKERY-PROPS ||- good-state => O good-state under
```

```
    strengthening unique-tickets ...
Proof obligations generated:  19
Proof obligations discharged: 19
Success!
rewrites: 26879 in 67ms cpu (71ms real) (395972 rewrites/second)

Checking BAKERY-PROPS ||- good-state => mutex ...
Proof obligations generated:  3
Proof obligations discharged: 3
Success!
rewrites: 9121 in 17ms cpu (18ms real) (525282 rewrites/second)
```

The InvA tool generates a total of 37 proof obligations for examples 1, 2, and 3. It automatically discharges 34 of them in approximately 200 milliseconds; the remaining 3 proof obligations can be discharged in Maude's ITP by structural induction on the sort *Nat* with the help of some simple lemmas (see [16]).

7 Related Work

A comprehensive account of the vast literature on deductive approaches for verifying invariants of concurrent systems is beyond the scope of this work. The aim here is more modest, namely, we focus on related work using deductive rewriting techniques for verifying invariants.

Rusu [17] and Rusu and Clavel [18] propose an approach for verifying invariant properties of (possibly infinite-state) concurrent systems specified by an unconditional topmost rewrite theory. Their approach consists in casting an invariance problem of the form $\mathcal{T}_\mathcal{R} \models I \Rightarrow \Box p$ as an inductive problem of an equational theory $\mathcal{M}(\mathcal{R}, I)$ as follows: $\mathcal{T}_\mathcal{R} \models I \Rightarrow \Box p$ iff $p(t) = \top$ holds in the initial algebra $\mathcal{T}_{\mathcal{M}(\mathcal{R},I)}$ for every ground term t of sort *Reachable*; a term t has sort *Reachable* in $\mathcal{T}_{\mathcal{M}(\mathcal{R},I)}$ iff t is reachable in $\mathcal{T}_\mathcal{R}$ from I. The key difference between their approach and ours is that the proof obligations generated for proving $\mathcal{T}_{\mathcal{M}(\mathcal{R},I)} \models p(t) = \top$ do not take advantage of p's equational definition, in contrast to our approach in which theorems 1 and 3 are very useful for simplifying the user's interactive theorem proving burden. Our approach can benefit from using narrowing for symbolically testing state predicates, a bounded symbolic execution technique achieved by narrowing with the rules, although more research is required for handling conditional rewrite theories.

Proof scores in the OTS/CafeOBJ method are used to prove invariant properties of concurrent systems specified by *observational transition systems* [15]. This approach has been applied for verifying safety properties of large specifications, including communication protocols. The idea is to exploit properties of Boolean operators for decomposing an invariant property into proof scores (reasonably smaller formulas), which are discharged interactively by equational rewriting. The main difference between this approach and ours is that proof scores are constructed and manipulated manually by the user, which is time-consuming in a verification process. The interesting idea of exploiting the properties of Boolean operators needs to be further studied and considered within our approach, as well as the development of more challenging case studies.

Combinations of deductive and algorithmic techniques have also been proposed for proving temporal logic properties φ of a (possibly infinite-state) concurrent system specified by a rewrite theory $\mathcal{R} = (\Sigma, E, R)$, including equational abstractions [14] and reductions with invisible transitions [7]. Both equational abstractions and invisible transition techniques reduce the verification problem of infinite-state systems to finite-state ones so that model checking methods are decidable. These two approaches, as it is also the case in our approach, require user-intervention for defining, respectively, the abstraction predicates and the invisible rules, and for discharging the inductive proof obligations resulting from the corresponding transformations. In particular, for a state predicate p, the checking algorithms based on narrowing presented in this paper can be easily extended to generate necessary and sufficient proof obligations for checking p-invisible rules $S \subseteq R$ of $\mathcal{R}$. We believe that these approaches complement each other and can be combined with our approach, resulting in a powerful and versatile framework for proving temporal properties of rewrite theories. The mechanization of these three approaches for reducing user intervention is a topic for further research.

8 Concluding Remarks

We have presented both a deductive methodology and a framework for proving ground stability and ground invariance of a (possibly infinite-state) concurrent system specified by conditional topmost rewrite theories under reasonable conditions. The proof obligations generated by our methods are equational Horn clauses. The original safety properties of the concurrent system are reduced to such equational proof obligations by the inference rules and the 1-step ground narrowing procedure. Our generic approach has been implemented in InvA, which adds new theorem proving support for verifying safety properties of infinite-state rewrite theories to the Maude environment.

Much work remains ahead. First of all, all the results presented here have a straightforward generalization to *state predicates with parameters*; that is, instead of state predicates of the form $p(s)$ with s a state, it is often very convenient to use state predicates of the form $p(s, d_1, \ldots, d_n)$, with s a state, and the $d_1, \ldots, d_n$ data parameters. All the ideas presented here can be extended to deal with predicates with data parameters. Another worthwhile direction is endowing InvA with automatic proof methods for strengthening invariants in the style of [19]. Finally, a wider range of case studies stressing the tool's capabilities, including application to rewrite theories $\mathcal{R}_{\mathcal{L}}$ that formally specify a programming languange $\mathcal{L}$, should be developed. More ambitiously, the transformational approach to safety property verification presented here should be extended to a wider set of of LTL formulas, including formulas stating liveness properties.

Acknowledgments. This work has been partially supported by NSF grants CNS 07-16638 and 09-04749, and CCF 09-05584.

References

1. Bruni, R., Meseguer, J.: Semantic foundations for generalized rewrite theories. Theoretical Computer Science 360(1-3), 386–414 (2006)
2. Chandy, K.M., Misra, J.: Parallel Program Design, A foundation. Addison Wesley, Reading (1988)
3. Clarke, E.M., Grumberg, O., Peled, D.A.: Model Checking. The MIT Press, Cambridge (1999)
4. Clavel, M., Durán, F., Eker, S., Lincoln, P., Martí-Oliet, N., Bevilacqua, V., Talcott, C.: All About Maude - A High-Performance Logical Framework, 1st edn. LNCS, vol. 4350. Springer, Heidelberg (2007)
5. Durán, F., Meseguer, J.: A church-rosser checker tool for conditional order-sorted equational maude specifications. In: Ölveczky, P.C. (ed.) WRLA 2010. LNCS, vol. 6381, pp. 69–85. Springer, Heidelberg (2010)
6. Escobar, S., Bevilacqua, V.: Symbolic model checking of infinite-state systems using narrowing. In: Baader, F. (ed.) RTA 2007. LNCS, vol. 4533, pp. 153–168. Springer, Heidelberg (2007)
7. Farzan, A., Meseguer, J.: State space reduction of rewrite theories using invisible transitions. In: Johnson, M., Vene, V. (eds.) AMAST 2006. LNCS, vol. 4019, pp. 142–157. Springer, Heidelberg (2006)
8. Hendrix, J.: Decision Procedures for Equationally Based Reasoning. PhD thesis. University of Illinois at Urbana-Champaign (April 2008)
9. Jouannaud, J.-P., Kirchner, C., Kirchner, H.: Incremental construction of unification algorithms in equational theories. In: Díaz, J. (ed.) ICALP 1983. LNCS, vol. 154, pp. 361–373. Springer, Heidelberg (1983)
10. Manna, Z., Pnueli, A.: The Temporal Logic of Reactive and Concurrent Systems. Springer, New York (1992)
11. Manna, Z., Pnueli, A.: Temporal Verification of Reactive Systems. Springer, New York (1995)
12. Meseguer, J.: Conditional rewriting logic as a unified model of concurrency. Theoretical Computer Science 96(1), 73–155 (1992)
13. Meseguer, J.: Membership algebra as a logical framework for equational specification. In: Parisi-Presicce, F. (ed.) WADT 1997. LNCS, vol. 1376, pp. 18–61. Springer, Heidelberg (1998)
14. Meseguer, J., Palomino, M., Martí-Oliet, N.: Equational abstractions. Theoretical Computer Science 403(2-3), 239–264 (2008)
15. Ogata, K., Futatsugi, K.: Proof scores in the oTS/CafeOBJ method. In: Najm, E., Nestmann, U., Stevens, P. (eds.) FMOODS 2003. LNCS, vol. 2884, pp. 170–184. Springer, Heidelberg (2003)
16. Rocha, C., Meseguer, J.: Proving safety properties of rewrite theories. Technical report. University of Illinois at Urbana-Champaign (2010), `http://hdl.handle.net/2142/17407`
17. Rusu, V.: Combining theorem proving and narrowing for rewriting-logic specifications. In: Fraser, G., Gargantini, A. (eds.) TAP 2010. LNCS, vol. 6143, pp. 135–150. Springer, Heidelberg (2010)
18. Rusu, V., Clavel, M.: Vérification d'invariants pour des systèmes spécifiés en logique de réécriture. Vingtièmes Journées Francophones des Langages Applicatifs 7.2, 317–350 (2009)
19. Tiwari, A., Rueß, H., Saïdi, H., Shankar, N.: A technique for invariant generation. In: Margaria, T., Yi, W. (eds.) TACAS 2001. LNCS, vol. 2031, pp. 113–127. Springer, Heidelberg (2001)
20. Viry, P.: Equational rules for rewriting logic. Theoretical Computer Science 285, 487–517 (2002)

Generalized Product of Coalgebraic Hybrid Logics

Katsuhiko Sano

Department of Humanistic Informatics, Kyoto University (JSPS)
Department of Computer Science, University of Leicester
katsuhiko.sano@gmail.com

Abstract. This paper proposes a modular method, called generalized product, of combining two coalgebraic hybrid logics in a parallel but non-compositional way. This is a coalgebraic generalization of a hybrid extension of product of modal logics. Our method, however, covers not only the combination of the same-type logics but also the combination of two different-type logics, e.g., graded hybrid logic and non-monotone hybrid logic. Moreover, we provide general strong completeness results for generalized products of coalgebraic hybrid logics with generic criteria.

1 Introduction and Motivation

Coalgebraic modal logic [1,2] is a logical framework that describes behavior of state-based systems, i.e., coalgebras, where modal operators are used to describe one-step behavior of the system. However, ordinary modal languages lack the ability to reason about individuals, which is a central feature of knowledge representation languages such as description logics. Coalgebraic hybrid logic [3,4,5] solves the problem by adding explicit individual names to modal syntax. This also allows to introduce non-relational (e.g. probabilistic) concepts in knowledge representations.

Schröder and Pattinson [6] proposed a framework allowing several ways of combining different-type transitions like non-deterministic and probabilistic ones. They also showed that the corresponding *heterogeneous* (coalgebraic) modal logic has a modular decision procedure. Their framework and results were recently transferred to coalgebraic hybrid logic [5].

There were several attempts (e.g. [7,8]) to extend description logics with an additional dimension representing, e.g., knowledge, time or action-dependence (cf. [9, pp.883-4]). As mentioned in [9, p.884], in a simple modal extension of the basic concept language $\mathcal{ALC}$ we can define concepts such as $\mathsf{Customer}$ as

$$\mathsf{Homo\_sapiens} \sqcap \langle \text{sometime in the past} \rangle \exists \mathsf{buys}.\mathsf{Car}$$

using a modalized concept $\langle \text{sometime in the past} \rangle \exists \mathsf{buys}.\mathsf{Car}$. In the framework of [6], however, we cannot directly express such a modalized concept, because its semantics (cf [8]) is based on the notion of *product of Kripke frames* [10]. For such reasons, a generalization of products to coalgebraic setting would be of both theoretical and practical interest; see also Examples 8 and 10 below.

Any attempt at such a generalization, however, faces difficulties of both semantic and axiomatic nature. The *semantic* difficulty consists in the lack of a suitable general

A. Corradini, B. Klin, and C. Cîrstea (Eds.): CALCO 2011, LNCS 6859, pp. 329–343, 2011.

notion of product of coalgebras (remark that 'product' is not a category-theoretic term here). While Van Benthem et. al. [11] adapted the notion of product to topological semantics of **S4**, it is not immediately clear, for example, what the right notion of product for selection function semantics of conditional logic [12] is. The *axiomatic* difficulty consists in finding generally valid interaction axioms which would allow an uniform completeness result for a large class of product models. While the Church-Rosser axiom and the commutativity axiom [10] are valid on all products of Kripke frames, Van Benthem et. al. [11] demonstrated that these two axioms are not valid on all products of topological spaces. In this sense, the Church-Rosser axiom and the commutativity axiom are not the right candidates. Let us also note here that both of these axioms are clearly of rank-2, that is, beyond the usual scope of coalgebraic axiomatizations. Moreover, even in case of Kripke semantics it is not obvious how suitable counterparts of these axioms for n-ary operators would look like.

Our paper proposes solutions to both problems. As for the *semantic* difficulty above, we propose the notion of *generalized product* of two possibly different-type coalgebras, which also covers the notion of product of Kripke frames. A categorical construction of *tensorial strength* [13] suggested to the author by Dirk Pattinson and Fredrik Dahlqvist allows to treat these products as coalgebras for products of functors. As the construction does not rely on both functors being of the same type, this opens up the possibility of modalizing concepts from probabilistic or conditional logic.

Our solution to the *axiomatic* difficulty relies on a sorted variant of *nominals*; as discussed above, such an addition turns modal logic into a weak variant of *description logic*. It turns out that this enrichment yields very simple and intuitive interaction axioms. Our nominals are *two-dimensional* with separate sorts for *horizontal* and *vertical* dimension. The idea of naming lines rather than states has already been used in author's previous studies on products of Kripke, topological and monotonic neighborhood structures [14,15]. The five interaction axioms proposed therein as a hybrid alternative to the Church-Rosser axiom and the commutativity axiom are valid both on all products of Kripke frames and all products of topologies. In the present paper, we demonstrate that an n-ary generalization of these interaction axioms (see $\mathcal{HIT}$ of Table 1) can capture the interaction between the two dimensions in an arbitrary coalgebraic setting.

Two-dimensional hybridization not only resolves the axiomatic difficulty but also provides a modular general completeness result with generic criteria. This completeness result covers any combination of two possibly different-type coalgebraic hybrid logics, e.g. hybrid modal logic for Kripke semantics, graded hybrid logic, hybrid conditional logic and other hybrid non-normal modal logic (Theorems 1, 2, and 3). We can regard our result as both a coalgebraic generalization of [14,15] and an extension of a theory of named canonical models (in our term, *Henkin-style named models*) in coalgebraic hybrid logic proposed by [4].

2 Generalized Product of Coalgebras and Predicate Lifting

Given an endofunctor T : **Sets** $\to$ **Sets**, a T-*coalgebra* is a pair (X, γ) where X is a set of *states* and $\gamma : X \to TX$ is a *transition map*.

Example 1. (i) Kripke frames for normal modal logic: a coalgebra $(X, R : X \to \mathcal{P}(X))$ for the covariant powerset functor $\mathcal{P}$ is the same thing as a Kripke frame, where $R(x)$ denotes all the states accessible from x.
(ii) Multigraph semantics for graded modal logic: an *infinite* multigraph functor $\mathcal{B}_\infty$ sends X to $(\mathbb{N} \cup \{\infty\})^X$.
(iii) Selection functions for conditional logic [12]: define $\mathcal{S}(X) := \{ f \mid f : \mathcal{P}(X) \to \mathcal{P}(X) \}$. In other words, $\mathcal{S}(X) = \mathcal{P}(X)^{\mathcal{Q}(X)}$, where $\mathcal{Q} : \mathbf{Sets} \to \mathbf{Sets}^{\mathrm{op}}$ is the contravariant powerset functor. We can regard $\mathcal{S}(X)$ as a $\mathcal{Q}(X)$-indexed Kripke frame. The subfunctor $\mathcal{S}_{\mathrm{ID}}(X)$ is defined by the condition $f(A) \subseteq A$ $(A \subseteq X)$.
(iv) Neighborhood semantics for classical modal logic: the intended functor is $\mathcal{Q} \circ \mathcal{Q}$, where $\mathcal{Q}$ is the contravariant powerset functor. Similarly, we can also consider the *monotone neighborhood* subfunctor $\mathcal{M}(X) := \{ \sigma \in \mathcal{P}\mathcal{P}(X) \mid \sigma \text{ is upward-closed} \}$.
(v) Neighborhood selection functions for classical conditional logic [12]: Define $\mathcal{S}^N (X) := \{ f \mid f : \mathcal{P}(X) \to \mathcal{P}\mathcal{P}(X) \}$, or in other words, $\mathcal{S}^N(X) = \mathcal{Q} \circ \mathcal{Q}(X)^{\mathcal{Q}(X)}$. Similarly to the previous item, we can also define the functor $\mathcal{S}^M$ giving us *monotone neighborhood selection functions*. A requirement $A \in f(A)$ $(A \subseteq X)$ gives us the subfunctors $\mathcal{S}^N_{\mathrm{ID}}$ and $\mathcal{S}^M_{\mathrm{ID}}$.

Definition 1. *Let T_1 and T_2 be endofunctors on* **Sets**. *We say that $\mathbb{C} \otimes \mathbb{D} := (C \times D, \gamma, \delta)$ is a* coalgebraic g-product frame *if $\mathbb{C} = (C, \gamma)$ is a T_1-coalgebra and $\mathbb{D} = (D, \delta)$ is a T_2-coalgebra. If $T_1 = T_2$, we call $\mathbb{C} \otimes \mathbb{D}$ a* coalgebraic product frame.

As noted by Pattinson and Dahlqvist (p.c.), we can regard $\mathbb{C} \otimes \mathbb{D}$ as a $(T_1 \times T_2)$-coalgebra $C \times D \to T_1(C \times D) \times T_2(C \times D)$. The key is to define $\gamma_h : C \times D \to T_1(C \times D)$ and $\delta_v : C \times D \to T_2(C \times D)$ in such a way that projection mappings π_1 and π_2 become, respectively, T_1- and T_2-coalgebraic homomorphisms. Let us concentrate on the horizontal dimension. Just like in case of Kripke semantics, the idea of horizontal accessibility relation is that we make the transition on the first coordinate but we do not make a transition in the second dimension [10]. Thus, γ_h should be obtained as a composition of $\gamma \times \mathrm{id}_D : C \times D \to T_1(C) \times D$ with a mapping of type $T_1(C) \times D \to T_1(C \times D)$. The latter can be defined as $\mathrm{st}_{C,D}(t, y) := T_1(\iota_y)(t)$, where $\iota_y : C \to C \times D$ sends x to (x, y). We can show that $\mathrm{st}_{C,D}$ is natural in C and D, and it is an example of the notion called *tensorial strength* [13] [1].

Given any $P \subseteq C \times D$ and $(x, y) \in C \times D$, define

$$P_{(-,y)} := \{ x' \in C \mid (x', y) \in P \}, \quad P_{(x,-)} := \{ y' \in D \mid (x, y') \in P \}.$$

It is easy to see that $P_{(-,y)} = \iota_y^{-1}[P]$ for any $P \subseteq C \times D$.

Example 2. Let $\mathbb{C} = (C, \gamma)$ and $\mathbb{D} = (D, \delta)$ be T_1- and T_2-coalgebras, respectively, $(x, y) \in C \times D$ and let us calculate $\gamma_h : C \times D \to T_1(C \times D)$ for all items in Example 1:

(i) $\gamma_h(x, y) = \{ (x', y) \mid x' \in \gamma(x) \}$.
(ii) $\gamma_h(x, y)(x', y') = \gamma(x)(x')$ (if $y = y'$); 0 (o.w.).

[1] Remark that every functor on **Sets** has a unique tensorial strength associated with it. This is pointed out by one of the reviewers.

(iii) $\gamma_h(x,y)(P) = \left\{ (x',y) \mid x' \in \gamma(x)(P_{(-,y)}) \right\}$.
(iv) $\gamma_h(x,y) = \left\{ P \subseteq C \times D \mid P_{(-,y)} \in \gamma(x) \right\}$.
(v) $\gamma_h(x,y)(P) = \left\{ P' \subseteq C \times D \mid P'_{(-,y)} \in \gamma(x)(P_{(-,y)}) \right\}$.

The notion of *predicate lifting* (cf [1,2]) gives a coalgebraic meaning to modal operators. Given any set Λ of modal operators, a Λ-*structure* consists of a functor T : **Sets** $\to$ **Sets** and a family $([\![\heartsuit]\!])_{\heartsuit \in \Lambda}$ of *predicate liftings*, i.e., $[\![\heartsuit]\!] = ([\![\heartsuit]\!]_X : \mathcal{P}(X)^n \to \mathcal{P}(TX))_{X \in \mathbf{Sets}}$ satisfying the naturality condition: $[\![\heartsuit]\!]_X \circ f^{-1} = (Tf)^{-1} \circ [\![\heartsuit]\!]_Y$ for any $f : X \to Y$. Given a Λ-structure, a T-coalgebra (X, γ), and a valuation for propositional variables, the coalgebraic semantics $[\![-]\!]$ to a (one-dimensional) coalgebraic modal language is defined as: $x \in [\![\heartsuit(\varphi_1, ..., \varphi_n)]\!]$ iff $\gamma(x) \in [\![\heartsuit]\!]_X([\![\varphi_1]\!], ..., [\![\varphi_n]\!])$. Let $(T_1, ([\![\heartsuit_1]\!])_{\heartsuit_1 \in \Lambda_1})$ be a Λ_1-structure. Given $\mathbb{C} \otimes \mathbb{D}$ and a pair (x,y), the truth condition for $\heartsuit_1(\varphi_1, ..., \varphi_n)$ is $\gamma_h(x,y) \in [\![\heartsuit_1]\!]_{C \times D}([\![\varphi_1]\!], ..., [\![\varphi_n]\!])$. A calculation of the naturality condition $[\![\heartsuit_1]\!]_C \circ \iota_y^{-1} = (T\iota_y)^{-1} \circ [\![\heartsuit_1]\!]_{C \times D}$ gives us the following simple *equivalent* condition:

$$\gamma(x) \in [\![\heartsuit_1]\!]_C([\![\varphi_1]\!]_{(-,y)}, ..., [\![\varphi_n]\!]_{(-,y)}).$$

Example 3. (i) Predicate liftings $\mathcal{P}(X) \to \mathcal{P}(\mathcal{P}(X))$ for normal modal logic are defined as $[\![\Box]\!]_X(A) := \{ B \in \mathcal{P}(X) \mid B \subseteq A \}$ and $[\![\Diamond]\!]_X(A) := \{ B \in \mathcal{P}(X) \mid B \cap A \neq \emptyset \}$.
(ii) A graded modality $\Diamond_{\geq k}$ ($k \in \omega$) counts the number of successors in Kripke semantics [16], but the naturality condition fails in Kripke semantics. A coalgebraic alternative is multigraph semantics by $\mathcal{B}_\infty$. Define $[\![\Diamond_{\geq k}]\!]_X(A) :=$ $\left\{ f : \mathcal{B}_\infty(X) \mid \sum_{y \in A} f(y) \geq k \right\}$.
(iii) Let $\Box\!\!\rightarrow$ be the conditional connective and $\Diamond\!\!\rightarrow$ the defined dual $\neg(\varphi \Box\!\!\rightarrow \neg\psi)$ of $\Box\!\!\rightarrow$. Our desired predicate liftings are $[\![\Box\!\!\rightarrow]\!]_X(A,B) := \{ f \in \mathcal{S}(X) \mid f(A) \subseteq B \}$ and $[\![\Diamond\!\!\rightarrow]\!]_X(A,B) := \{ f \in \mathcal{S}(X) \mid f(A) \cap B \neq \emptyset \}$.
(iv) The predicate lifting $\mathcal{P}(X) \to \mathcal{P}(Q \circ Q(X))$ for classical modal logic is $[\![\Box]\!]_X(A) :=$ $\{ \sigma \in \mathcal{PP}(X) \mid A \in \sigma \}$.
(v) The predicate lifting for classical conditional logic (cf. [12]) is defined as $[\![\Box\!\!\rightarrow]\!]_X(A,B) = \left\{ f \in \mathcal{S}^N(X) \mid B \in f(A) \right\}$.

3 Coalgebraic Semantics for Bi-Hybrid Language

The most fundamental semantic idea of our g-product syntax is the introduction of two *disjoint* countable sets $\mathsf{N}_1 = \{ i, j, .. \}$ and $\mathsf{N}_2 = \{ a, b, ... \}$ of *horizontal* and *vertical nominals*, which name *lines* rather than *points*. Let Λ_1 and Λ_2 be two disjoint sets of modal operators, where we assume that each $\heartsuit_l \in \Lambda_l$ has a fixed arity ($l = 1$ or 2). Given a countable set $\mathsf{P} = \{ p, q, ... \}$ of propositional variables, we define the set $\mathcal{F}(\Lambda_1; \Lambda_2)$ of all formulas inductively as:

$$\varphi ::= p \mid i \mid a \mid \neg\varphi \mid \varphi \wedge \psi \mid \heartsuit_1(\varphi_1, \ldots, \varphi_n) \mid \heartsuit_2(\varphi_1, \ldots, \varphi_m) \mid @_i\varphi \mid @_a\varphi,$$

where $i \in \mathsf{N}_1$, $a \in \mathsf{N}_2$, $p \in \mathsf{P}$, $\heartsuit_1 \in \Lambda_1$ and $\heartsuit_2 \in \Lambda_2$. We use the standard definition of the other propositional connectives: $\bot$, $\top$, $\vee$, $\to$, and $\leftrightarrow$. We say that φ is *pure* if φ

does *not* contain any propositional variable $p \in \mathsf{P}$. A *hybrid substitution* is a mapping $\sigma : \mathsf{P} \cup \mathsf{N}_1 \cup \mathsf{N}_2 \to \mathcal{F}(\Lambda_1; \Lambda_2)$ that satisfies $\sigma[\mathsf{N}_1] \subseteq \mathsf{N}_1$ and $\sigma[\mathsf{N}_2] \subseteq \mathsf{N}_2$. That is, we can substitute $p \in \mathsf{P}$ with *any* formula φ and $i \in \mathsf{N}_1$ with the same kind of nominal $j \in \mathsf{N}_1$, but we cannot substitute $i \in \mathsf{N}_1$ with $a \in \mathsf{N}_2$ or $p \in \mathsf{P}$. We also define the two sublanguages: $\mathcal{L}_1 := \mathsf{P} \cup \mathsf{N}_1 \cup \{\neg, \wedge\} \cup \Lambda_1 \cup \{ @_i \mid i \in \mathsf{N}_1 \}$ and $\mathcal{L}_2 := \mathsf{P} \cup \mathsf{N}_2 \cup \{\neg, \wedge\} \cup \Lambda_2 \cup \{ @_a \mid a \in \mathsf{N}_2 \}$. We denote by $\mathcal{F}(\Lambda_l)$ all the formulas of $\mathcal{L}_l$ ($l = 1$ or 2).

Intuitively, the denotation of $i \in \mathsf{N}_1$ is a vertical line $\{x\} \times D$ and the denotation of $a \in \mathsf{N}_2$ is a horizontal line $C \times \{y\}$ over $C \times D$. More precisely, a mapping $V : \mathsf{P} \cup \mathsf{N}_1 \cup \mathsf{N}_2 \to \mathcal{P}(C \times D)$ is a *hybrid valuation* if $\#\pi_1[V(i)] = 1$ and $\pi_2[V(i)] = D$ ($i \in \mathsf{N}_1$), and $\pi_1[V(a)] = C$ and $\#\pi_2[V(a)] = 1$ ($a \in \mathsf{N}_2$), where $\pi_1 : C \times D \to C$ and $\pi_2 : C \times D \to D$ are the projections. Let us denote the unique element of $\pi_1[V(i)]$ by i^V and the unique element of $\pi_2[V(a)]$ by a^V.

A *coalgebraic g-product model* $\mathfrak{M}$ is a pair of a coalgebraic g-product frame and a hybrid valuation on it. Given any coalgebraic g-product model $\mathfrak{M} = (C \times D, \gamma, \delta, V)$, any $(x, y) \in C \times D$, and any $\varphi \in \mathcal{F}(\Lambda_1; \Lambda_2)$, we define the satisfaction relation as follows:

$$
\begin{array}{lll}
\mathfrak{M}, (x, y) \Vdash p & \text{iff} & (x, y) \in V(p) \\
\mathfrak{M}, (x, y) \Vdash i & \text{iff} & x = i^V \\
\mathfrak{M}, (x, y) \Vdash a & \text{iff} & y = a^V \\
\mathfrak{M}, (x, y) \Vdash \neg\varphi & \text{iff} & \mathfrak{M}, (x, y) \nVdash \varphi \\
\mathfrak{M}, (x, y) \Vdash \varphi \wedge \psi & \text{iff} & \mathfrak{M}, (x, y) \Vdash \varphi \text{ and } \mathfrak{M}, (x, y) \Vdash \psi \\
\mathfrak{M}, (x, y) \Vdash \heartsuit_1(\varphi_1, \ldots, \varphi_n) & \text{iff} & \gamma(x) \in [\![\heartsuit_1]\!]_C([\![\varphi_1]\!]_{(-,y)}, \ldots, [\![\varphi_n]\!]_{(-,y)}) \\
\mathfrak{M}, (x, y) \Vdash \heartsuit_2(\varphi_1, \ldots, \varphi_m) & \text{iff} & \delta(y) \in [\![\heartsuit_2]\!]_D([\![\varphi_1]\!]_{(x,-)}, \ldots, [\![\varphi_m]\!]_{(x,-)}) \\
\mathfrak{M}, (x, y) \Vdash @_i\varphi & \text{iff} & \mathfrak{M}, (i^V, y) \Vdash \varphi \\
\mathfrak{M}, (x, y) \Vdash @_a\varphi & \text{iff} & \mathfrak{M}, (x, a^V) \Vdash \varphi,
\end{array}
$$

where $[\![\varphi]\!] := \{ (x', y') \in C \times D \mid \mathfrak{M}, (x', y') \Vdash \varphi \}$.

Remark that the behavior of $@_i\varphi$ is different from one-dimensional hybrid logic. In one-dimensional semantics, if φ holds at the state named by i, then $@_i\varphi$ holds at *all* states. In our two-dimensional semantics, $(i^V, y) \Vdash \varphi$ does not imply $[\![@_i\varphi]\!] = C \times D$ in general. Given any $(x, y) \in C \times D$, however, $(i^V, y) \Vdash \varphi$ does imply $[\![@_i\varphi]\!]_{(-,y)} = C$.

We say that a set Ψ of formulas is *valid on a coalgebraic g-product model* $\mathfrak{M}$ (notation: $\mathfrak{M} \Vdash \Psi$) if $\mathfrak{M}, (x, y) \Vdash \varphi$ for any pair (x, y) in $\mathfrak{M}$ and any $\varphi \in \Psi$. Ψ is *valid* on $\mathbb{C} \otimes \mathbb{D}$ (notation: $\mathbb{C} \otimes \mathbb{D} \Vdash \Psi$) if $(\mathbb{C} \otimes \mathbb{D}, V) \Vdash \Psi$ for any hybrid valuation V. If Ψ is a singleton $\{\psi\}$, we also use the notation such as $\mathfrak{M} \Vdash \psi$ and $\mathbb{C} \otimes \mathbb{D} \Vdash \psi$. We say that $\Psi \subseteq \mathcal{F}(\Lambda_1; \Lambda_2)$ is *satisfiable* in a coalgebraic g-product frames $\mathbb{C} \otimes \mathbb{D}$ if there exists some hybrid valuation V such that $\bigcap_{\psi \in \Psi} [\![\psi]\!] \neq \emptyset$. Ψ is *satisfiable* in a class F of coalgebraic g-product frames if there exists some $\mathbb{C} \otimes \mathbb{D} \in \mathsf{F}$ such that Ψ is satisfiable in $\mathbb{C} \otimes \mathbb{D}$. $\Psi \subseteq \mathcal{F}(\Lambda_1; \Lambda_2)$ *defines* a class F of coalgebraic g-product frames if $\mathbb{C} \otimes \mathbb{D} \in \mathsf{F}$ iff Ψ is valid on $\mathbb{C} \otimes \mathbb{D}$, for any coalgebraic g-product frame $\mathbb{C} \otimes \mathbb{D}$.

In general, we do not assume that every coalgebraic g-product model is *named*, i.e., that for every state (x, y) there exists $i \in \mathsf{N}_1$ and $a \in \mathsf{N}_2$ such that $(x, y) = (i^V, a^V)$. The importance of namedness assumption in strong completeness proofs for axiomatic extensions by pure formulas stems from the following (cf. [17, Lemma 7.22]).

Proposition 1. *Given any named* $\mathfrak{M} = (\mathbb{C} \otimes \mathbb{D}, V)$ *and any pure formula* φ, *if* $\mathfrak{M} \Vdash \varphi\sigma$ *for all hybrid substitutions* σ, *then* $\mathbb{C} \otimes \mathbb{D} \Vdash \varphi$.

Table 1. Hilbert-style Calculus $\mathcal{RAG}$ of Coalgebraic Hybrid g-Product Logic

Axioms		
$\mathcal{HIT}$	**Com@**	$@_a @_i p \leftrightarrow @_i @_a p$
	Com$\heartsuit_1$**@**$_2$	$@_a \heartsuit_1(p_1, \ldots, p_n) \leftrightarrow \heartsuit_1(@_a p_1, \ldots, @_a p_n)$
	Com$\heartsuit_2$**@**$_1$	$@_i \heartsuit_2(p_1, \ldots, p_m) \leftrightarrow \heartsuit_2(@_i p_1, \ldots, @_i p_m)$
	Red@$_1$	$@_i a \leftrightarrow a$
	Red@$_2$	$@_a i \leftrightarrow i$
$\mathcal{CHA}_1$	**K@**	$@_i(p \rightarrow q) \rightarrow (@_i p \rightarrow @_i q)$
	Sd	$\neg @_i p \leftrightarrow @_i \neg p$
	Ref	$@_i i$
	Intro	$i \wedge p \rightarrow @_i p$
	Agree	$@_i @_j p \rightarrow @_j p$
	Mob	$@_i p \rightarrow (\heartsuit_1(q_1, \ldots, q_n) \leftrightarrow \heartsuit_1(@_i p \wedge q_1, \ldots, @_i p \wedge q_n))$ $(\heartsuit_1 \in \Lambda_1)$
$\mathcal{CHA}_2$	The corresponding $\mathcal{L}_2$-axioms to $\mathcal{CHA}_1$	
$\mathcal{R}_1$	A set of one-step $\mathcal{L}_1$-axioms, i.e., a subset of $\mathrm{Prop}(\Lambda_1(\mathrm{Prop}(\mathsf{P})))$	
$\mathcal{R}_2$	A set of one-step $\mathcal{L}_2$-axioms, i.e., a subset of $\mathrm{Prop}(\Lambda_2(\mathrm{Prop}(\mathsf{P})))$	
$\mathcal{A}$	A set of extra axioms, i.e., a subset of $\mathcal{F}(\Lambda_1; \Lambda_2)$	
	Propositional Tautlogies	
Rules		
$\mathcal{CHR}_1$	**Cong**	From $\varphi_1 \leftrightarrow \psi_1, \ldots, \varphi_n \leftrightarrow \psi_n$, we may infer $\heartsuit_1(\varphi_1, \ldots, \varphi_n) \leftrightarrow \heartsuit_1(\psi_1, \ldots, \psi_n)$ $(\heartsuit_1 \in \Lambda_1)$
	Nec@	From φ, we may infer $@_i \varphi$
$\mathcal{CHR}_2$	The corresponding $\mathcal{L}_2$-rules to $\mathcal{CHA}_1$	
$\mathcal{G}_1$	A set of non-orthodox $\mathcal{L}_1$-rules	
$\mathcal{G}_2$	A set of non-orthodox $\mathcal{L}_2$-rules	
	Modus Ponens	
	Hybrid Substitutions	

Proposition 2. *$\varphi \in \mathcal{F}(\Lambda_1)$ is valid on $\mathbb{C} \otimes \mathbb{D}$* iff *$\varphi$ is valid $\mathbb{C}$. Similarly, $\varphi \in \mathcal{F}(\Lambda_2)$ is valid on $\mathbb{C} \otimes \mathbb{D}$* iff *$\varphi$ is valid $\mathbb{D}$.*

Proof. We can show these by the similar argument as in [14, Proposition 2.2]. □

4 Generalized Product of Coalgebraic Hybrid Logics

$\mathcal{HIT}$ in Table 1 are our *five interaction axioms*, which are both coalgebraic and hybrid alternative to the Church-Rosser axiom and the commutativity axiom in Kripke semantics. In this section, we first show that $\mathcal{HIT}$ are valid. Second, we introduce the notion of *one-step axiom*, which captures the one-step behavior of modal operators [1]. Finally, we present a Hilbert-style proof system for generalized g-product of coalgebraic hybrid logics, by the help of the axioms and the inference rules for coalgebraic hybrid logics already proposed by [3].

Proposition 3. *$\mathcal{HIT}$ in Table 1 are valid on all coalgebraic g-product frames.*

Proof. We only check **Com**$\heartsuit_1$**@**$_2$. Let $(C \times D, \gamma, \delta, V)$ be a coalgebraic g-product model and $(x, y) \in C \times D$. $[\![p]\!]_{(-,a^V)} = [\![@_a p]\!]_{(-,y)}$ gives us the desired equivalence. □

Definition 2 (one-step axiom and one-step soundness). *Let Z be a set. Given any set Λ of modal operators, define $\Lambda(Z) := \{ \heartsuit(z_1, \ldots, z_n) \mid \heartsuit \in \Lambda$ and $z_1, ..., z_n \in Z \}$.* Prop(Z) *denotes the set of all propositional (or Boolean) combinations of Z. Let $(T, ([\![\heartsuit]\!])_{\heartsuit \in \Lambda})$ be a Λ-structure. For any set X, any $\tau : Z \rightarrow \mathcal{P}(X)$, and any $\alpha \in$* Prop(Λ(Z)), *we define*

$\|\alpha\|_{TX,\tau} \subseteq TX$ *inductively as usual except that* $\|\heartsuit(z_1,...,z_n)\|_{TX,\tau} = [\![\heartsuit]\!]_X(\tau(z_1),\ldots,\tau(z_n))$. *A* one-step axiom *(or,* Rank-1 axiom*) over* Λ *is a formula* $\psi \in \mathrm{Prop}(\Lambda(\mathrm{Prop}(\mathsf{P})))$. *A one-step axiom* ψ *is* one-step sound *if* $\|\psi\|_{TX,\tau} = TX$, *for any* X *and any* $\tau : \mathsf{P} \to \mathcal{P}(X)$.

Since our syntax has two disjoint sets Λ_1 and Λ_2 of modal operators, we consider two kinds of one-step axioms. One-step sound axioms are sound in the following sense.

Proposition 4. *Let* ψ *be a one-step axiom over* Λ_l $(l = 1$ *or* $2)$ *and* σ *a hybrid substitution. If* ψ *is one-step sound, then* $\psi\sigma$ *is valid on all coalgebraic* g*-product frames.*

We say that an inference rule φ/ψ is *non-orthodox* if it involves a syntactic side-condition in the premise. A typical example in hybrid logic is **Name**-rule. **Name**$_1$ and **Name**$_2$ are the following non-orthodox rules of $\mathcal{L}_1$ and $\mathcal{L}_2$, respectively:

$$\frac{i \to \varphi}{\varphi}\ \mathbf{Name}_1 \qquad \frac{a \to \psi}{\psi}\ \mathbf{Name}_2$$

where i does not appear in φ and a does not appear in ψ. It is not difficult to show that **Name**$_l$ preserves the validity on all coalgebraic g-product frames ($l = 1$ or 2).

Definition 3 (coalgebraic hybrid g-product logic). *Let* $\mathcal{R}_l$ *be a set of one-step axioms over* Λ_l $(l = 1$ *or* $2)$, $\mathcal{A} \subseteq \mathcal{F}(\Lambda_1;\Lambda_2)$ *a set of formulas, and* $\mathcal{G}_l$ *be a set of non-orthodox rules in* $\mathcal{F}(\Lambda_l)$ $(l = 1$ *or* $2)$. *We define a Hilbert system* $\mathcal{RAG}$ *as follows: we write* $\vdash_{\mathcal{RAG}} \varphi$ *if* φ *belongs to the smallest set that contains all the axioms in Table 1 and is closed under all the rules in Table 1. We say that* φ *is* $\mathcal{RAG}$-derivable *from* Ψ *(written:*$\Psi \vdash_{\mathcal{RAG}} \varphi$*) if there exists* $\psi_1,...,\psi_n \in \Psi$ *such that* $\vdash_{\mathcal{RAG}} \psi_1 \wedge \cdots \wedge \psi_n \to \varphi$.

Our axioms $\mathcal{CHA}_1$ and rules $\mathcal{CHA}_2$ in Table 1 are modified from the axioms and rules for one-dimensional coalgebraic hybrid logic given in [3,4]. **Mob** (‘*make-or-break*’) is a major difference from the ordinary hybrid logic for Kripke semantics. It is a coalgebraic and n-ary generalization of **Back**-axiom: $@_i p \to \Box @_i p$ [18], since **Back** defines the non-emptiness of the neighborhood frames $\tau(x)$ over neighborhood semantics, though it is valid over Kripke semantics.

Proposition 5. *Let* $\mathcal{R}_l$ *be one-step sound and* $\mathcal{G}_l$ *preserve validity on all coalgebraic* g*-product frames* $(l = 1$ *or* $2)$. *If* φ *is a* $\mathcal{RAG}$*-theorem, then* φ *is valid on the class of all coalgebraic* g*-product frames defined by* $\mathcal{A}$.

Proof. Propositions 2, 3, 4 give us the desired validity of φ. □

Example 4. (i) One-step axioms for axiomatizing Kripke frames is $\Box\top$ and $\Box(p \wedge q) \leftrightarrow (\Box p \wedge \Box q)$. Or, if we prefer $\Diamond$ to $\Box$, the one-step axioms: $\neg\Diamond\bot$ and $\Diamond(p \vee q) \leftrightarrow (\Diamond p \vee \Diamond q)$ also give us the same logic.

(ii) The one-step axioms for graded modal logic is the following [16] (see also [19, p.676]): $\Box(p \to q) \to (\Box p \to \Box q)$, $\Diamond_{\geq k} p \to \Diamond_{\geq l} p$ $(l < k)$, $\Diamond_{\geq k} p \leftrightarrow \bigvee_{0 \leq i \leq k} \Diamond_{\geq i}(p \wedge q) \wedge \Diamond_{\geq k-i}(p \wedge \neg q)$, and $\Box(p \to q) \to (\Diamond_{\geq k} p \to \Diamond_{\geq k} q)$, where we define $\Box := \neg\Diamond_{\geq 1}\neg$.

(iii) For the minimal conditional logic, $r \boxright \top$ and $(r \boxright p \wedge q) \leftrightarrow ((r \boxright p) \wedge (r \boxright q))$, which allows us to regard $r \boxright$ as a normal modal operator. A requirement $f(A) \subseteq A$ $(A \subseteq X)$ of selection functions is captured by the one-step axiom $p \boxright p$.
(iv) While our set of one-step axioms for neighborhood frames is an empty set, $\Box(p \wedge q) \rightarrow (\Box p \wedge \Box q)$ is the one-step axiom for monotone neighborhood frames.
(v) $(r \boxright p \wedge q) \rightarrow ((r \boxright p) \wedge (r \boxright q))$ is the one-step axiom for monotone neighborhood selection functions. $p \boxright p$ corresponds to $A \in f(A)$ $(A \subseteq X)$.

5 Pure Completeness for Coalgebraic Hybrid g-Product Logic

Definition 4. *A set* Ψ *is* $\mathcal{RAG}$-consistent *if* $\Psi \nvdash_{\mathcal{RAG}} \bot$. Ψ *is* maximally $\mathcal{RAG}$-consistent *if* Ψ *is* $\mathcal{RAG}$*-consistent and it satisfies:* $\varphi \in \Psi$ *or* $\neg\varphi \in \Psi$ *for any* $\varphi \in \mathcal{F}(\Lambda_1; \Lambda_2)$.

If Ψ is maximally $\mathcal{RAG}$-consistent, it clearly contains all the $\mathcal{RAG}$-theorems.

Definition 5. *A Hilbert system* $\mathcal{RAG}$ *is* (locally) strongly complete *with respect to a class* F *of coalgebraic* g*-product frames if every* $\mathcal{RAG}$*-consistent set is satisfiable in* F.

Definition 6. *Let* Ψ *be a maximal* $\mathcal{RAG}$*-consistent set. Define* $C_\Psi = \{\, [i] \mid i \in \mathsf{N}_1 \,\}$ *and* $D_\Psi = \{\, |a| \mid a \in \mathsf{N}_2 \,\}$, *where* $[i] = \{\, j \in \mathsf{N}_1 \mid @_i j \in \Psi \,\}$ *and* $|a| = \{\, b \in \mathsf{N}_2 \mid @_a b \in \Psi \,\}$. *Define also* $V_\Psi : \mathsf{P} \cup \mathsf{N}_1 \cup \mathsf{N}_2 \rightarrow \mathcal{P}(C_\Psi \times D_\Psi)$ *by* $V_\Psi(p) = \{\, ([i], |a|) \mid @_i @_a p \in \Psi \,\}$ $(p \in \mathsf{P})$, $V_\Psi(j) = \{\, ([i], |a|) \mid @_i @_a j \in \Psi \,\}$ $(j \in \mathsf{N}_1)$ *and* $V_\Psi(b) = \{\, ([i], |a|) \mid @_i @_a b \in \Psi \,\}$ $(b \in \mathsf{N}_2)$.

V_Ψ is a *hybrid valuation* as follows: We show that $V_\Psi(j) := \left\{ [i] \mid @_j i \in_\Psi \right\} \times D_\Psi$ and that $\left\{ [i] \mid @_j i \in_\Psi \right\}$ is a singleton. We can establish the first clause, since we have $\vdash @_i @_b j \leftrightarrow @_i j$ (by **Nec@** and **Red@**$_2$) and $\vdash @_j i \leftrightarrow @_i j$. As for the second clause, it suffices to note that we have $\vdash @_i i$ and $\vdash @_i j \wedge @_j k \rightarrow @_i k$.

Definition 7. *Given a maximal* $\mathcal{RAG}$*-consistent* Ψ, *we say that* $(C_\Psi \times D_\Psi, \gamma, \delta, V_\Psi)$ *is a* coalgebraic Henkin-style g-product model *if* γ *and* δ *satisfies both the* h-coherence condition*: For any* $\heartsuit_1 \in \Lambda_1$, *any* $i \in \mathsf{N}_1$, *any* φ,

$$\gamma([i]) \in [\![\heartsuit_1]\!]_{C_\Psi}(\widehat{\varphi_1}^1, \ldots, \widehat{\varphi_n}^1) \text{ iff } @_i \heartsuit_1(\varphi_1, \ldots, \varphi_n) \in \Psi,$$

and the v-coherence condition*: For any* $\heartsuit_2 \in \Lambda_2$, *any* $a \in \mathsf{N}_2$ *and any* φ,

$$\delta(|a|) \in [\![\heartsuit_2]\!]_{D_\Psi}(\widehat{\varphi_1}^2, \ldots, \widehat{\varphi_m}^2) \text{ iff } @_a \heartsuit_2(\varphi_1, \ldots, \varphi_m) \in \Psi,$$

where $\widehat{\varphi}^1 = \left\{ [j] \in C_\Psi \mid @_j \varphi \in \Psi \right\}$ *and* $\widehat{\varphi}^2 = \{\, [b] \in D_\Psi \mid @_b \varphi \in \Psi \,\}$.

Lemma 1 (Truth Lemma). *If* $\mathfrak{M}_\Psi = (C_\Psi \times D_\Psi, \gamma, \delta, V_\Psi)$ *is a coalgebraic Henkin-style* g*-product model, then we have:*

$$\mathfrak{M}_\Psi, ([i], |a|) \Vdash \varphi \text{ iff } @_i @_a \varphi \in \Psi,$$

for any $\varphi \in \mathfrak{F}(\Lambda_1; \Lambda_2)$ *and any* $([i], |a|) \in C_\Psi \times D_\Psi$.

Proof. Define $\widehat{\varphi} := \{([i], |a|) \mid @_i @_a \varphi \in \Psi\}$. We can show the case where $\varphi \equiv @_i\psi$, $@_a\psi$ similarly to the proof of [14, Lemma 3.11, p.466]. Here, let us concentrate on the case where $\varphi \equiv \heartsuit_1(\varphi_1, .., \varphi_n)$. It is easy to show that $\widehat{\psi}_{(-,|a|)} = \widehat{@_a\psi}^1$ for any ψ. Then, $([i], |a|) \Vdash \heartsuit_1(\varphi_1, .., \varphi_n)$ iff :

$$\begin{aligned}
&\gamma([i]) \in [\![\heartsuit_1]\!]_{C_\Psi}([\![\varphi_1]\!]_{(-,|a|)}, ..., [\![\varphi_n]\!]_{(-,|a|)}) \\
\text{iff } &\gamma([i]) \in [\![\heartsuit_1]\!]_{C_\Psi}(\widehat{\varphi_1}_{(-,|a|)}, ..., \widehat{\varphi_n}_{(-,|a|)}) \quad \text{by induction hypothesis} \\
\text{iff } &\gamma([i]) \in [\![\heartsuit_1]\!]_{C_\Psi}(\widehat{@_a\varphi_1}^1, ..., \widehat{@_a\varphi_n}^1) \quad \text{by the above equation} \\
\text{iff } &@_i\heartsuit_1(@_a\varphi_1, ..., @_a\varphi_n) \in \Psi \quad \text{by the } h\text{-coherence condition.}
\end{aligned}$$

Equivalently, we obtain $@_i @_a \heartsuit_1(\varphi_1, ..., \varphi_n) \in \Psi$ by **Com**$\heartsuit_1 @_2$. □

Given a Λ-structure, now our interest consists in: how one-step axioms give us the *existence* of transition maps γ and δ on C_Ψ and D_Ψ, respectively. First, we will give a criterion for any combination of two hybrid logics from (iv) and (v) in Example 3. Second, we will provide a criterion for any combination of two hybrid logics from (i), (ii), and (iii) in Example 3. Finally, we will also give a criterion for any combination of hybrid logic from (i), (ii), and (iii), and a hybrid logic from (iv) and (v).

5.1 Generalized Product of Strongly One-Step Complete Logics

Lemma 2. *If* **Name**$_l \in \mathcal{G}_l$ *($l = 1$ or 2), the following are valid derivation rules in $\mathcal{RAG}$:*

(i) If $\vdash @_i\varphi$, then $\vdash \varphi$, where i is fresh in φ.
(ii) If $\vdash @_i \bigwedge_{1\le r\le n}(\varphi_r \leftrightarrow \psi_r)$, then $\vdash @_j(\heartsuit_1(\varphi_1, ..., \varphi_n) \leftrightarrow \heartsuit_1(\psi_1, ..., \psi_n))$, where i is fresh in $\bigwedge_{1\le r\le n}(\varphi_r \leftrightarrow \psi_r)$.
(iii) If $\vdash \theta \to (@_i \bigwedge_{1\le r\le n}(\varphi_r \leftrightarrow \psi_r))$, then $\vdash \theta \to (@_j(\heartsuit_1(\varphi_1, ..., \varphi_n) \leftrightarrow \heartsuit_1(\psi_1, ..., \psi_n)))$, where i is fresh in θ and $\bigwedge_{1\le r\le n}(\varphi_r \leftrightarrow \psi_r)$.

The corresponding $\mathcal{L}_2$-rules are also derivable in $\mathcal{RAG}$.

(ii) is an n-ary generalization of the rule called **NameCong** in [4].

Proof. (i) holds by **Intro** and propositional inferences. We can establish (ii) by (i), **Cong**, and **Nec@**. Let us show (iii). Assume that i is fresh in θ and $\bigwedge_{1\le r\le n}(\varphi_r \leftrightarrow \psi_r)$ and suppose that $\vdash \theta \to (@_i \bigwedge_{1\le r\le n}(\varphi_r \leftrightarrow \psi_r))$. The latter is equivalent to $\vdash \bigwedge_{1\le r\le n}((@_i\varphi_r \wedge \theta) \leftrightarrow (@_i\psi_r \wedge \theta))$. Here, let $k \neq i$ be a fresh nominal in θ and $\bigwedge_{1\le r\le n}(\varphi_r \leftrightarrow \psi_r)$. We deduce from **Nec@** that $\vdash @_k \bigwedge_{1\le r\le n}((@_i\varphi_r \wedge \theta) \leftrightarrow (@_i\psi_r \wedge \theta))$. By **Agree** and distributivity of @ over Boolean connectives[2], we obtain $\vdash @_i \bigwedge_{1\le r\le n}((\varphi_r \wedge @_k\theta) \leftrightarrow (\psi_r \wedge @_k\theta))$. It follows from (ii) that $\vdash @_j(\heartsuit_1(\varphi_1 \wedge @_k\theta, ..., \varphi_n \wedge @_k\theta) \leftrightarrow \heartsuit_1(\psi_1 \wedge @_k\theta, ..., \psi_n \wedge @_k\theta))$. By **Mob**, we obtain $\vdash @_j(@_k\theta \to (\heartsuit_1(\varphi_1, ..., \varphi_n) \leftrightarrow \heartsuit_1(\psi_1, ..., \psi_n)))$. Again by **Agree** and distributivity of @ over Boolean connectives, $\vdash @_k(\theta \to @_j(\heartsuit_1(\varphi_1, ..., \varphi_n) \leftrightarrow \heartsuit_1(\psi_1, ..., \psi_n)))$. Finally, (i) gives us the desired conclusion. □

[2] Since we can regard $@_i$ as a *normal* modal operator (i.e., $@_i$ preserves finite (possibly empty) conjunctions) and $@_i$ preserves the negation as in **Sd** of Table 1, we can syntactically obtain the distributivity of @ over Boolean connectives.

Definition 8. *Let Ψ be a set of formulas. We say that Ψ is* labelled *if $i \wedge a \in \Psi$ for some $(i, a) \in \mathsf{N}_1 \times \mathsf{N}_2$. Ψ is* horizontally 0-pasted *if whenever $@_i \bigwedge_{1 \le r \le n}(\varphi_r \leftrightarrow \psi_r) \in \Psi$ for all $i \in \mathsf{N}_1$, then $@_j(\heartsuit_1(\varphi_1, ..., \varphi_n) \leftrightarrow \heartsuit_1(\psi_1, ..., \psi_n)) \in \Psi$ for all $j \in \mathsf{N}_1$ ($\heartsuit_1 \in \Lambda_1$). Similarly, Ψ is* vertically 0-pasted *if whenever $@_a \bigwedge_{1 \le r \le m}(\varphi_r \leftrightarrow \psi_r) \in \Psi$ for all $a \in \mathsf{N}_2$, then $@_b(\heartsuit_2(\varphi_1, ..., \varphi_m) \leftrightarrow \heartsuit_2(\psi_1, ..., \psi_m)) \in \Psi$ for all $b \in \mathsf{N}_2$ ($\heartsuit_2 \in \Lambda_2$).*

Lemma 3. *Let* **Name**$_l \in \mathcal{G}_l$ *($l = 1$ or 2). Every $\mathcal{RAG}$-consistent formulas can be extended to a labelled, horizontally and vertically 0-pasted maximal $\mathcal{RAG}$-consistent set, by adding countably many new nominals to the language.*

Definition 9. *Let $\mathcal{R} \subseteq$ Prop(Λ(Prop(P))) be a set of one-step axioms and X a set. We say that $\Xi \subseteq$ Prop($\Lambda(\mathcal{P}(X))$) is* one-step consistent *with respect to X if the union of Ξ and $\{\alpha\tau \mid \alpha \in \mathcal{R}$ and $\tau : \mathsf{P} \to \mathcal{P}(X)\}$ is propositionally consistent. $\Xi \subseteq$ Prop($\Lambda(\mathcal{P}(X))$) is* one-step satisfiable *if $\bigcap_{\alpha \in \Xi} \|\alpha\|_{TX,\mathrm{id}} \neq \emptyset$, where* id *is the identity function on $\mathcal{P}(X)$.*

Definition 10. *$\mathcal{R}$ is* strongly one-step complete *if, for any set X, every one-step consistent set $\Xi \subseteq$ Prop($\Lambda(\mathcal{P}(X))$) is one-step satisfiable.*

Lemma 4. *Let Ψ be maximally $\mathcal{RAG}$-consistent. If Ψ is horizontally 0-pasted and $\mathcal{R}_1$ is strongly one-step complete, then there exists $\gamma : C_\Psi \to T_1(C_\Psi)$ such that it satisfies the h-coherence condition. Similarly, we have the corresponding statement about v-coherence condition.*

Theorem 1. *Let $\mathcal{G}_l = \{$*__Name__*$_l\}$ ($l = 1$ or 2). Let also $\mathcal{R}_1$ and $\mathcal{R}_2$ be strongly one-step complete and $\mathcal{A}$ a set of pure formulas. Then, $\mathcal{RAG}$ is strongly complete with respect to the class of coalgebraic g-product frames defined by $\mathcal{A}$.*

Proof. Suppose that Ψ_0 is $\mathcal{RAG}$-consistent. By Lemma 3, we can extend Ψ_0 to some Ψ such that Ψ is labelled, horizontally and vertically 0-pasted, and maximally $\mathcal{RAG}$-consistent. Since $\mathcal{R}_1$ and $\mathcal{R}_2$ are one-step complete, we can find a coalgebraic Henkin-style g-product model $\mathfrak{M}_\Psi = (C_\Psi \times D_\Psi, \gamma, \delta, V_\Psi)$ by Lemma 4. Since Ψ is labelled, let us fix some (i, a) such that $i \wedge a \in \Psi$. By **Intro**, we can establish that $@_i @_a \varphi \in \Psi$ for any $\varphi \in \Psi$. So, we derive from Truth Lemma (Lemma 1) that Ψ_0 is satisfiable in $(C_\Psi \times D_\Psi, \gamma, \delta)$. Finally, let us show that $(C_\Psi \times D_\Psi, \gamma, \delta)$ belongs to the class of coalgebraic g-product frames defined by $\mathcal{A}$. For any $\psi \in \mathcal{A}$, we have $\mathfrak{M}_\Psi \Vdash \psi\sigma$ for all hybrid substitutions σ. Therefore, $(C_\Psi \times D_\Psi, \gamma, \delta) \Vdash \mathcal{A}$ by Proposition 1. □

Example 5. Theorem 1 establishes previously unknown strong completeness results of product of any pure extensions of non-monotone hybrid logics. It also gives an alternative proof of the known pure completeness result of product of monotone hybrid logics [15, Theorem 5.11], since the axioms for monotone neighborhood frames in Example 4 (iv) are strongly one-step complete. Remark that we *explicitly defined* the horizontal and vertical neighborhood frames in the proof of [15, Theorem 5.11]. Moreover, Theorem 1 newly establishes unknown pure completeness results of both product of classical conditional hybrid logics and g-product of non-normal hybrid logic and classical conditional hybrid logic, because it is easy to show, e.g., that all the axioms in Example 4 (v) are strongly one-step complete for the functor $\mathcal{S}^M_{\mathrm{ID}}$ of Example 1 (v).

5.2 Generalized Product of Bounded Logics

In the ordinary hybrid logic for Kripke semantics, the rule called **Paste** (or **BG**) plays a crucial role in establishing a pure completeness result [18], because it captures the relational character of Kripke semantics (In fact, it *characterizes* the notion of Kripke frame within monotone neighborhood frames [20]). Schröder and Pattinson [4] defined the notion of *boundedness* to capture the relational nature of Kripke semantics, multi-graph semantics, and selection function semantics as in Example 3 (i), (ii) and (iii). Remark also that we restrict our attention to the diamond-type modal operators like $\Diamond$, $\Diamond_{\geq k}$, and $\Diamond\!\!\rightarrow$, whenever we consider the notion of boundedness.

Definition 11. *Let* $(T, (\llbracket\heartsuit\rrbracket)_{\heartsuit\in\Lambda})$ *be a* Λ*-structure.* $\heartsuit \in \Lambda$ *is* k-bounded in r-th argument *with respect to* T *if for every set* X *and every* $A_1, .., A_n \subseteq X$*:*

$$\llbracket\heartsuit\rrbracket_X(A_1, .., A_r, ..., A_n) = \bigcup_{B\subseteq A_r, \#B\leq k} \llbracket\heartsuit\rrbracket_X(A_1, .., B, ..., A_n).$$

We say that Λ *is* bounded *with respect to an endofunctor* T *if every* $\heartsuit \in \Lambda$ *is* $k_\heartsuit$*-bounded in* $r_\heartsuit$*-th argument for some* $k_\heartsuit$ *and* $r_\heartsuit$*.*

If $\heartsuit \in \Lambda$ is k-bounded in r-th argument, then $\llbracket\heartsuit\rrbracket_X$ is monotone in r-th argument.

Definition 12. *Given* $\heartsuit_1 \in \Lambda_1$, $\mathbf{Paste}\heartsuit_1(r;k)$ *is the following non-orthodox rule:*

$$\frac{\left(@_{j_1}\varphi_r \wedge \cdots \wedge @_{j_k}\varphi_r \wedge @_i\heartsuit_1(\varphi_1, ..., \bigvee_{1\leq s\leq k} j_s, ..., \varphi_n)\right) \to \psi}{@_i\heartsuit_1(\varphi_1, ..., \varphi_r, ..., \varphi_n) \to \psi}\ \mathbf{Paste}\heartsuit_1(r;k),$$

where $j_1, ..., j_k$ *are pairwise distinct fresh nominals in* ψ *and* $@_i\heartsuit_1(\varphi_1, ..., \varphi_r, ..., \varphi_n)$*. Given any* $\heartsuit_2 \in \Lambda_2$*, we also define* $\mathbf{Paste}\heartsuit_2(r;k)$ *similarly.*

Example 6. (i) $\Diamond$ is 1-bounded. The corresponding $\mathbf{Paste}\Diamond(1;1)$ is the same rule as $Paste_\Diamond$ of [18].

(ii) $\Diamond_{\geq k}$ is k-bounded. $\mathbf{Paste}\Diamond_{\geq k}(1;k)$ has the following shape with the same side-condition:

$$\frac{\left(@_{j_1}\varphi \wedge \cdots \wedge @_{j_k}\varphi \wedge @_i\Diamond_{\geq k}(j_1 \vee \cdots \vee j_k)\right) \to \psi}{@_i\Diamond_{\geq k}\varphi_r \to \psi}.$$

(iii) $\Diamond\!\!\rightarrow$ is 1-bounded in second argument. The corresponding **Paste** $\Diamond\!\!\rightarrow$ (2; 1) has the following shape with the side-condition required above:

$$\frac{\left(@_j\varphi_2 \wedge @_i(\varphi_1 \Diamond\!\!\rightarrow j)\right) \to \psi}{@_i(\varphi_1 \Diamond\!\!\rightarrow \varphi_2) \to \psi}.$$

Proposition 6. *Let* $\heartsuit_l \in \Lambda_l$ *be* k*-bounded in* r*-th argument* ($l = 1$ *or* 2). $\mathbf{Paste}\heartsuit_l(r;k)$ *preserves validity on all coalgebraic g-product frames.*

Definition 13. *Let* Λ_l *be bounded with respect to* T_l ($l = 1$ *or* 2). Ψ *is* horizontally 1-pasted *if whenever* $\heartsuit_1 \in \Lambda_1$ *is* k*-bounded in* r*-th argument and* $@_i\heartsuit_1(\varphi_1, ..., \varphi_r, ..., \varphi_n) \in \Psi$*, then* $\{ @_{j_1}\varphi_r, ..., @_{j_k}\varphi_r, @_i\heartsuit_1(\varphi_1, ..., \bigvee_{1\leq s\leq k} j_s, ..., \varphi_n)\} \subseteq \Psi$ *for some* $j_1, ..., j_k \in \mathsf{N}_1$*. Similarly, we define the notion of* vertical 1-pastedness *for* Λ_2*.*

Lemma 5. *Let Λ_l be bounded and* $\{\,\mathbf{Name}_l\,\} \cup \{\,\mathbf{Paste}\heartsuit_l \mid \heartsuit_l \in \Lambda_l\,\} \subseteq \mathcal{G}_l$ $(l = 0$ *or* $1)$. *If* $\Psi \subseteq \mathcal{F}(\Lambda_1;\Lambda_2)$ *is* $\mathcal{RAG}$*-consistent, then there exists a labelled, horizontally* 0*- and* 1*-pasted and vertically* 0*- and* 1*-pasted maximally* $\mathcal{RAG}$*-consistent set by adding countably many new nominals to the language.*

The following notion is a generalization of *strongly finitary one-step completeness* for unary modal operators [4, Definition 2.18] and (*strong, strongly finitary*) *one-step completeness* for binary modal operators [4, Definition 2.22].

Definition 14. *Let Λ be bounded. Given any set X, we define:*

$$\mathrm{Bnd}(\Lambda(\mathcal{P}(X))) := \{\,\heartsuit(A_1,...,A_{r_\heartsuit},...,A_n) \in \Lambda(\mathcal{P}(X)) \mid A_1,...,A_n \subseteq X \text{ and } \#A_{r_\heartsuit} < \omega \text{ and } \heartsuit \in \Lambda\,\}.$$

We say that $\mathcal{R}$ is strongly bounded one-step complete *if for every X, every one-step consistent subset of* $\mathrm{Prop}(\mathrm{Bnd}(\Lambda(\mathcal{P}(X))))$ *is one-step satisfiable.*

Definition 15. $\mathbf{Mon}\heartsuit_l(r)$: $\heartsuit_l(p_1,...,p,...,p_n) \to \heartsuit_l(p_1,...,p \vee q,...,p_n)$ *is a monotonicity axiom of* $\heartsuit_l \in \Lambda_l$ *in r-th argument* [3].

Lemma 6. *Let Λ_l be bounded,* $\{\,\mathbf{Mon}\heartsuit_l(r) \mid \heartsuit_l \in \Lambda_l\,\} \subseteq \mathcal{R}_l$, *and $\mathcal{R}_l$ strongly bounded one-step complete* $(l = 1$ *or* $2)$. *If a maximally* $\mathcal{RAG}$*-consistent* Ψ *is horizontally* 0*- and* 1*-pasted, then there exists* $\gamma : C_\Psi \to T_1(C_\Psi)$ *such that* γ *satisfies h-coherence condition. Similarly, we have the corresponding statement about v-coherence condition.*

Theorem 2. *Let Λ_l be bounded,* $\{\,\mathbf{Mon}\heartsuit_l(r) \mid \heartsuit_l \in \Lambda_l\,\} \subseteq \mathcal{R}_l$ *and* $\mathcal{G}_l = \{\,\mathbf{Name}_l\,\} \cup \{\,\mathbf{Paste}\heartsuit_l \mid \heartsuit_l \in \Lambda_l\,\}$ $(l = 1$ *or* $2)$. *Let also* $\mathcal{R}_1$ *and* $\mathcal{R}_2$ *be strongly bounded one-step complete and* $\mathcal{A}$ *a set of pure formulas. Then,* $\mathcal{RAG}$ *is strongly complete with respect to the class of coalgebraic g-product frames defined by* $\mathcal{A}$.

Proof. An outline of the proof is similar to the proof of Theorem 1. We, however, need Lemma 5 and Lemma 6 in the places of Lemma 3 and Lemma 4, respectively. □

Example 7. Theorem 2 establishes previously unknown strong completeness results of product of any pure extensions of graded hybrid logic and any pure extensions of hybrid conditional logic. It also reproves the known pure completeness result of product of hybrid logics for Kripke semantics [14, Theorem 3.12]. Furthermore, Theorem 2 newly establishes unknown pure completeness results of g-product of any two logics from hybridization of modal logic K, hybrid conditional logic, and graded hybrid logic.

Example 8. Let us assume that the first dimension describes the domain of individuals and the second dimension describes the *temporal* structure. Let T_1 be the infinite multiset functor with graded modalities $\Diamond_{\geq k}$, T_2 be the covariant powerset functor and $\langle \mathsf{P} \rangle$ the past-tense operator 'it was the case that'. Remark that a pure formula $\neg\Diamond_{\geq 2} i$ (*the transition multiplicity between two states is always at most one*) axiomatizes the class

[3] As noted in [4], if Λ_1 consists of unary modal operators alone and $\mathcal{R}_1$ is one-step complete (i.e., any one-step consistent $\varphi \in \mathrm{Prop}(\Lambda_1(\mathcal{P}(X)))$ is one-step satisfiable), 1-pastedness implies 0-pastedness. This implication does not hold in general, e.g. in the case of $\Box\!\!\rightarrow$. This is why we need the monotonicity axiom $\mathbf{Mon}\heartsuit_l(r)$.

of Kripke frames [4, Example 2.21]. This allows us to handle the description logic $\mathcal{ALCQ}$ with qualified number restrictions in the first dimension. By using the notation of modalized description logic [8], we may define the concept Multipara by:

$$\mathsf{Human} \sqcap \mathsf{Female} \sqcap \langle \mathsf{P} \rangle \geq 2\, \mathsf{gives\_birth.Human}.$$

The A-Box $\langle \mathsf{P} \rangle(\mathsf{Multipara}(\mathsf{MARY}))$ allows to infer

$$\langle \mathsf{P} \rangle(\exists \mathsf{gives\_birth.Human}(\mathsf{MARY})).$$

In our notation, we can rewrite this inference as follows:

$$\langle \mathsf{P} \rangle @_{\mathsf{MARY}}(\mathsf{Human} \wedge \mathsf{Female} \wedge \langle \mathsf{P} \rangle \Diamond_{\geq 2} \mathsf{Human}) \rightarrow \langle \mathsf{P} \rangle @_{\mathsf{MARY}} \Diamond \mathsf{Human}.$$

where we regard Human and Female as propositional variables. The interaction axioms, $\langle \mathsf{P} \rangle \langle \mathsf{P} \rangle p \rightarrow \langle \mathsf{P} \rangle p$ and one-step axioms for graded modal logic allow us to derive this formula in our intended system. $\langle \mathsf{P} \rangle \langle \mathsf{P} \rangle p \rightarrow \langle \mathsf{P} \rangle p$ goes beyond rank-1, but a pure formula $\langle \mathsf{P} \rangle \langle \mathsf{P} \rangle i \rightarrow \langle \mathsf{P} \rangle i$ and $\mathbf{Paste}\Diamond_{\geq 1}(1; 1)$ allow us to derive $\langle \mathsf{P} \rangle \langle \mathsf{P} \rangle p \rightarrow \langle \mathsf{P} \rangle p$.

5.3 Generalized Product of Bounded and Non-bounded Logics

Theorem 3. *Let Λ_1 be bounded and $\{\mathbf{Mon}\heartsuit_1(r) \mid \heartsuit_1 \in \Lambda_1\} \subseteq \mathcal{R}_1$. Suppose $\mathcal{G}_2 = \{\mathbf{Name}_2\}$ and $\mathcal{G}_1 = \{\mathbf{Name}_1\} \cup \{\mathbf{Paste}\heartsuit_1 \mid \heartsuit_1 \in \Lambda_1\}$. Let also $\mathcal{R}_1$ be strongly bounded one-step complete and $\mathcal{R}_2$ strongly one-step complete. Given any pure axioms $\mathcal{A}$, $\mathcal{RAG}$ is strongly complete with respect to the class of coalgebraic g-product frames defined by $\mathcal{A}$.*

Proof. An essential difference from the proof of Theorem 2 is that, given any $\mathcal{RAG}$-consistent set Ψ_0, we need to construct $\Psi \supseteq \Psi_0$ such that Ψ is a labelled, horizontally 0- and 1-pasted but vertically 0-pasted MCS. However, this is easy to establish. □

Example 9. This theorem newly establishes a pure completeness of g-product of any hybrid logic for (i), (ii) or (iii) of Example 3 and any hybrid logic for (iv) or (v) of Example 3. In particular, it gives us a pure completeness of g-product of graded hybrid logic and non-monotone hybrid logic.

Example 10. Let us use the following interpretation: the first dimension is for the deontic reasoning with the monotone neighborhood functor $\mathcal{M}$ and the second dimension is for the temporal structure as in Example 8. Then, the inference "Gate A must be closed at 9 pm; Gate B must be opened at 11 pm; 11 pm is later than 9 pm; therefore, at 11 pm, it is obligatory that Gate A is closed and Gate B was opened." can be formalized as $([\mathsf{Must}] @_{\mathsf{9pm}} p \wedge [\mathsf{Must}] @_{\mathsf{11pm}} q \wedge @_{\mathsf{11pm}} \langle \mathsf{P} \rangle \mathsf{9pm}) \rightarrow @_{\mathsf{11pm}} [\mathsf{Must}](p \wedge \langle \mathsf{P} \rangle q)$. The one-step axiom $[\mathsf{Must}] p_1 \wedge [\mathsf{Must}] p_2 \rightarrow [\mathsf{Must}](p_1 \wedge p_2)$ makes this inference valid.

Remark that our result provides a complete axiomatization of the logic of irreflexive and transitive temporal orders by means of pure formulas $\neg @_i \langle \mathsf{P} \rangle i$ and $\langle \mathsf{P} \rangle \langle \mathsf{P} \rangle i \rightarrow \langle \mathsf{P} \rangle i$, respectively. As for the first dimension, we may freely add $\neg [\mathsf{Must}] \bot$, since it is both pure and one-step.

5.4 Adding Downarrow Binders

In the same sprit as did in [20,4], we can also include the *downarrow binders* $\downarrow i$ and/or $\downarrow a$ in our syntax, which binds a nominal i (or a) to the first (or, second, respectively) argument of the current state. Given any $(C \times D, \gamma, \delta, V)$, we can define

$$(C \times D, \gamma, \delta, V), (x, y) \Vdash \downarrow i.\varphi \text{ iff } (C \times D, \gamma, \delta, V[i \mapsto x](x, y) \Vdash \varphi,$$

where $V[i \mapsto x]$ is same as V except that it sends i to $\{x\} \times C$. We can also give the similar clause to $\downarrow a.\varphi$. A technique of *local definability* allows us to capture this semantics by the axiom $\mathbf{DA}_1$: $@_j(\downarrow i.\varphi \leftrightarrow \varphi[i/j])$, where $\varphi[i/j]$ is the result of replacing all free instances of i by j in φ. We can also consider the corresponding axiom $\mathbf{DA}_2$ for $\downarrow a.\varphi$. By the similar argument to [18, Theorem 5], we can immediately transfer Theorems 1, 2, and 3 to the syntax extended with the downarrow binders $\downarrow i$ and/or $\downarrow a$.

6 Conclusions and Future Work

The notion of g-product defined in this paper naturally generalizes products of Kripke frames [10] and products of topologies [11] to coalgebraic setting. As discussed in the paragraph following Definition 1, the definition is based on a natural categorical concept. Hence, as Example 2 above shows, for any specific functor(s) one can immediately calculate concrete definition of horizontal and vertical transitions. Moreover, the definition does not rely on both functors being of the same type. Future work in this direction can involve a coalgebraic generalization of the notion of *dependent product of Kripke frames* proposed by the author in [14].

Our main logical results are modular pure completeness theorems with generic criteria for g-products of coalgebraic hybrid logics. The five interaction axioms proposed first by the author in [14,15] turned out to capture the relationship between horizontal and vertical dimensions in the full coalgebraic generality. Moreover, as already mentioned in Examples 8 and 10 above, the combination of pure axioms and non-orthodox rules allows to cover a large class of non-rank-1 axioms, overcoming traditional restrictions of coalgebraic axiomatics (at least in the *bounded* case). In this stage, we do not know if we can transfer *global* strong completeness of [4] to our setting. The study of sequent calculus for generalized products is also the subject of future work.

On the practical side, we believe Examples 8 and 10 show the potential for future applications. As mentioned in the Introduction, there exists a modular framework [6] allowing a compositional way of combining transitions of different type which has recently been extended to hybrid setting [5,4]. That compositional framework and our product framework are entirely independent of each other and, e.g., our Example 8 is beyond the scope of [6,5,4]. It should be stressed, however, that both methods seem wholly compatible and it would be most interesting to investigate the potential of concrete languages arising from their combination[4].

[4] This paper was written during the author's visit to University of Leicester. I would like to thank Alexander Kurz, Dirk Pattinson, Fredrik Dalhqvist and Tomoyuki Suzuki for the stimulating discussions and comments. In particular, I wish to thank Tadeusz Litak for his very kind assistance in writing up this paper and also for his warm hospitality at Leicester. Finally, I would like to thank three reviewers for their helpful comments. However, all errors are mine.

References

1. Schröder, L.: A finite model construction for coalgebraic modal logic. The Journal of Logic and Algebraic Programming 73, 97–110 (2007)
2. Schröder, L., Pattinson, D.: Rank-1 modal logics are coalgebraic. Journal of Logic and Computation 30(5), 1113–1147 (2008)
3. Myers, R., Pattinson, D., Schröder, L.: Coalgebraic hybrid logic. In: de Alfaro, L. (ed.) FOSSACS 2009. LNCS, vol. 5504, pp. 137–151. Springer, Heidelberg (2009)
4. Schröder, L., Pattinson, D.: Named models in coalgebraic hybrid logic. In: Marion, J.Y., Schwentick, T. (eds.) Proceedings of STACS 2010. Leibniz International Proceedings in Informatics, pp. 645–656 (2010)
5. Myers, R., Pattinson, D.: Hybrid logic with the difference modality for generalisations of graphs. Journal of Applied Logic 8(4), 441–458 (2010)
6. Schröder, L., Pattinson, D.: Modular algorithms for heterogeneous modal logics. In: Arge, L., Cachin, C., Jurdziński, T., Tarlecki, A. (eds.) ICALP 2007. LNCS, vol. 4596, pp. 459–471. Springer, Heidelberg (2007)
7. Schild, K.: Combining terminological logics with tense logic. In: Damas, L.M.M., Filgueiras, M. (eds.) EPIA 1993. LNCS, vol. 727, pp. 105–120. Springer, Heidelberg (1993)
8. Wolter, F., Zakharyaschev, M.: Modal description logics: modalizing roles. Fundamenta Informaticae 39, 411–438 (1999)
9. Kurucz, A.: Combining modal logcs. In: Blackburn, P., van Benthem, J., Wolter, F. (eds.) Handbook of Modal Logic, pp. 869–924. Elsevier, Amsterdam (2007)
10. Gabbay, D.M., Shehtman, V.B.: Products of modal logics, part 1. Logic Journal of IGPL 6(1), 73–146 (1998)
11. van Benthem, J., Bezhanishvili, G., ten Cate, B., Sarenac, D.: Multimodal logics of products of topologies. Studia Logica 84(3), 369–392 (2006)
12. Chellas, B.: Basic conditional logic. Journal of Philosophcal Logic 4(2), 133–153 (1975)
13. Kock, A.: Strong functors and monoidal monads. Archiv der Mathematik 23, 114–120 (1972)
14. Sano, K.: Axiomatizing hybrid products: How can we reason many-dimensionally in hybrid logic? Journal of Applied Logic 8(4), 459–474 (2010)
15. Sano, K.: Axiomatizing hybrid products of monotone neighbourhood frames. In: Accepted for the Publication in HyLo 2010 Post-Proceedings (2010), `http://www.geocities.jp/k2sn/pro_hytop.pdf`
16. Fine, K.: In so many possible worlds. Notre Dame Journal of Formal Logic 13(4), 516–520 (1972)
17. Blackburn, P., de Rijke, M., Venema, Y.: Modal Logic. Cambridge Tracts in Theoretical Computer Science. Cambridge University Press, Cambridge (2001)
18. Blackburn, P., ten Cate, B.: Pure extensions, proof rules, and hybrid axiomatics. Studia Logica 84(3), 277–322 (2006)
19. Schröder, L., Pattinson, D.: Strong completeness of coalgebraic modal logics. In: Albers, S., Marion, J.Y. (eds.) Proceedings of STACS 2009, vol. 3, pp. 673–684. Schloss Dagstuhl-Leibniz-Zentrum fuer Informatik (2009)
20. ten Cate, B., Litak, T.: Topological perspective on the hybrid proof rules. Electronic Notes in Theoretical Computer Science 174(6), 79–94 (2007)

Distributive-Law Semantics for Cellular Automata and Agent-Based Models

Baltasar Trancón y Widemann and Michael Hauhs

Ecological Modelling, University of Bayreuth, Germany
Baltasar.Trancon@uni-bayreuth.de

Abstract. We present an effort to give formal semantics to the popular but theoretically rather unreflected scientific modelling paradigm of agent- or individual-based models. To this end, we give a generic formalization of two-dimensional cellular automata with flexible topology as the abstract basis of such models. The semantic approach of structural operational semantics a la Turi and Plotkin [7], based on bialgebras and distributive laws, leads in this case to a natural separation of the concerns of spatial structure, temporal behavior and local interaction. We give a generic distributive law for local behavior of automata and prove the equivalence to a more traditional, array-based formalization.

1 Introduction

In empirical sciences, computers have been perceived traditionally as mechanical extensions of the established modelling methods, allowing the numerical solution of analytically intractable equations. The power of computers to display complex interactive behavior has been harnessed rather lately, in particular in the form of so-called individual-based or agent-based models (ABMs). These models and their mathematical abstraction, cellular automata (CAs), are widely used in empirical areas of natural and social sciences. The epistemological and methodological implications of these models remain unexplored, not least because appropriate semantic theories are not available to modelling practitioners.

Formal theories of interactive behavior have been developed in theoretical computer science. Categorial notions, especially coalgebra and modal logic, have proven fruitful [5,6]. Here, we make results of this recent development from [7,1] accessible for modelling. To this end we show that distributive laws and their bialgebras can be used as a common semantic theory behind cellular models. In particular, we will define a large class of CAs by a parametric class of distributive laws of spatial syntax over temporal behavior.

The principle of CAs is to cover space with identical copies of simple automata. All of these share a common state space, often but not necessarily finite, and local transition rule, but may be in individually different states at any time. Global behavior emerges through the interaction of neighboring automata, with the prestate of neighbors, or some observation of it, serving as input to each automaton. The patterns of global behavior, both stationary and mobile and both periodic and asymptotic, are the principal properties of interest when studying CAs.

A. Corradini, B. Klin, and C. Cîrstea (Eds.): CALCO 2011, LNCS 6859, pp. 344–358, 2011.

The basic CA design has a number of degrees of freedom: the global shape of the world, the local shape of neighborhoods, the local state space and transition rule, the initial and boundary conditions, and the amount of observable information about neighbor states. Here, we demonstrate how structural operational semantics (SOS) in the bialgebraic style of [7] serve to structure and separate these concerns. We also demonstrate that the coalgebraic part of bialgebras leverages the powers of coinductive reasoning about the behavior of CAs, in terms of equivalence, bisimulation or corecursive definition, in a natural way. Our main result Theorem 4 is a first and promising example of such a coinductive argument. The implications of our approach for the implementation of CAs and ABMs in a functional programming language on one hand, and for the epistemological and methodological assessment of the relation of ABMs to empirical phenomena on the other hand, shall be addressed in two forthcoming companion articles.

An appendix [9] with omitted proofs and advanced examples is available as an electronic supplement to this article.

2 Theory of Bialgebras

This section introduces the basic concepts of bialgebraic semantics. The notation follows [3] in many, but not all aspects.

Definition 1 (Algebra). *Let Σ be an endofunctor. A pair (X, x) where X is an object and $x : \Sigma X \to X$ is called a Σ-**algebra**. Let (W, w) and (X, x) be two Σ-algebras. A morphism $f : W \to X$ is called a Σ-**algebra homomorphism** iff $f \circ w = x \circ \Sigma f$. We write $f : w \to x$ in this case. A Σ-algebra (A, α) such that there is a unique homomorphism $x? : \alpha \to x$ into every other Σ-algebra is called **initial**.*

Definition 2 (Coalgebra). *Let B be an endofunctor. A pair (Y, y) where Y is an object and $y : Y \to BY$ is called a B-**coalgebra**. Let (Y, y) and (Z, z) be two B-coalgebras. A morphism $f : Y \to Z$ is called a B-**coalgebra homomorphism** iff $Bf \circ y = z \circ f$. We write $f : y \to z$ in this case. A B-coalgebra (Ω, ω) such that there is a unique homomorphism $y! : y \to \omega$ from every other B-coalgebra (Y, y) is called **final**.*

Definition 3 (Distributive Law). *Let Σ, B be two endofunctors. A natural transformation $\lambda : \Sigma B \Rightarrow B\Sigma$ is called a **distributive law** of Σ over B, or just Σ/B-distributive law.*

Lemma 1. *Let λ be a Σ/B-distributive law. Then*

1. *for each Σ-algebra (X, f) there is a Σ-algebra $(BX, Bf \circ \lambda_X)$,*
2. *for each B-coalgebra (Y, g) there is a B-coalgebra $(\Sigma Y, \lambda_Y \circ \Sigma g)$.*

Definition 4 (Lifting). *Let λ be a Σ/B-distributive law. Let (X, f) be a Σ-algebra and (Y, g) be a B-coalgebra. We call*

1. $B^\lambda f = Bf \circ \lambda_X$ *the* ***behavioral lifting*** *of* f,
2. $\Sigma_\lambda g = \lambda_Y \circ \Sigma g$ *the* ***syntactic lifting*** *of* g.

Definition 5 (Bialgebra). *Let* λ *be a* Σ/B*-distributive law. A triple* (X, f, g) *is called a* λ*-**bialgebra** iff*

1. (X, f) *is a* Σ*-algebra,*
2. (X, g) *is a* B*-coalgebra, and*
3. *any of the following three, equivalent properties holds:*
   (a) $f : \Sigma_\lambda g \to g$,
   (b) $g : f \to B^\lambda f$, *or*
   (c) *the following diagram commutes.*

$$\begin{array}{ccccc} \Sigma X & \xrightarrow{f} & X & \xrightarrow{g} & BX \\ {\scriptstyle \Sigma g}\downarrow & & & & \uparrow{\scriptstyle Bf} \\ \Sigma BX & & \xrightarrow[\lambda_X]{\qquad\qquad} & & B\Sigma X \end{array}$$

A morphism that is a homomorphism on both the algebra and the coalgebra part of two λ*-bialgebras is called a* λ*-**bialgebra homomorphism**. Initial and final bialgebras are defined in the obvious way.*

The following results are trivial when the free monad of Σ and the cofree comonad of B are considered instead [7]. But we consider the bare functors precisely because we are interested in behavior with *bounded* spatio-temporal context. Without reference to the additional structure, an explicit proof is in order; cf. appendix.

Lemma 2. *Let* λ *be a* Σ/B*-distributive law. Each initial* Σ*-algebra extends uniquely to an initial* λ*-bialgebra.*

Lemma 3. *Dually, let* λ *be a* Σ/B*-distributive law. Each final* B*-coalgebra extends uniquely to a final* λ*-bialgebra.*

Theorem 1. *Let* λ *be a* Σ/B*-distributive law. For each initial* Σ*-algebra* (A, α) *and final* B*-coalgebra* (Ω, ω) *there is a unique morphism* $h : A \to \Omega$ *that is a* λ*-bialgebra homomorphism.*

Proof. Combines the previous lemmata. Each initial Σ-algebra (A, α) extends to a unique initial λ-bialgebra $\big(A, \alpha, (B^\lambda\alpha)?\big)$. Each final B-coalgebra (Ω, ω) extends to a unique λ-bialgebra $\big(\Omega, (\Sigma_\lambda\omega)!, \omega\big)$. Then there is a λ-bialgebra homomorphism $h : A \to \Omega$ that is unique by both initiality and finality; cf. appendix. □

Definition 6 (Bialgebraic Semantics). *The homomorphism induced by the preceding theorem is called the **bialgebraic semantics** of the distributive law* λ*. It is unique up to the unique isomorphisms between initial* Σ*-algebras or final* B*-coalgebras.*

In this section, we have reviewed the basic notions of bialgebraic semantics, most notably the unique homomorphisms that arise from initial algebras and final coalgebras, and the lifting operations that arise from a distributive law.

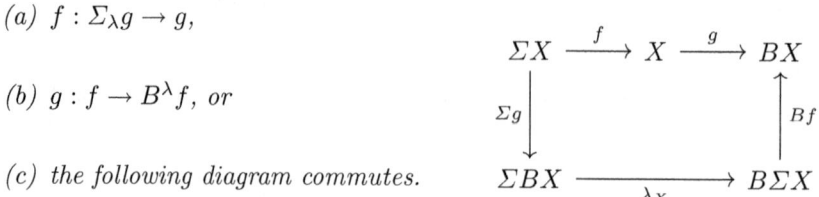

3 Syntax

The global state of a CA is composed of local states arranged in a discrete n-dimensional structure, usually a regular grid, but possibly with a nontrivial topology. We restrict our considerations to the most common case, $n = 2$, although the approach should be generalizable in a straightforward way. We allow for the syntactic description of the regular grid by a lifting operation *singleton* for local states, and binary combining operations *beside* and *above* for global states. As a simple, but instructive case of nontrivial topology, we add unary operations *vwrap* and *hwrap* for connecting opposite edges of a grid. The resulting syntax functor Σ_L, for some local state space L, is shown in Table 1. We use the graphic notation both for elements of $\Sigma_L X$ and for elements of initial Σ_L-algebras. To avoid unneccessary parentheses, it is implied that $^{\leftrightarrow}$ and $^{\updownarrow}$ bind stronger than $|$, which binds stronger than $/$.

Real-world implementations of CAs and ABMs usually allow wrapping operations only at the outermost level of spatial syntax: The world can be a rectangle (unwrapped), cylinder (wrapped in one dimension) or torus (wrapped in both dimensions). We allow for full compositionality of the syntax; combining and wrapping operations can be nested in any order. For instance, a trouser-shaped world can be constructed out of three rectangles A, B, C by writing $A^{\leftrightarrow} / B^{\leftrightarrow} | C^{\leftrightarrow}$.

In this section, we have defined a syntax for denoting both the current (spatially distributed) state of a CA and its topology. Departing from conventional modelling frameworks for CAs, we have made the syntax fully compositional, with every operation applicable at any level. Although our syntax deals only with certain regular shapes in two-dimensional space, generalization to other topological shapes and dimensions is possible. For instance, two-dimensional context-sensitive Lindenmayer systems, where the grid grows locally over time, for instance used for forest growth models [4], could be formalized analogously.

4 Static Semantics

It is straightforward to give a semantic interpretation for the fragment of syntax that excludes wrapping operations: The semantic domain consists of ordinary finite, rectangular matrices over L. One must merely decide how to handle the combination of incompatible shape (different heights in *beside*, different widths in *above*). In the presence of wrapping, however, things become more complicated. We shall introduce semantic interpretations stepwise in the following two subsections.

Definition 7 (World Functor). *The* ***world functor*** W *takes a local state space* L *to initial* Σ_L*-algebras* $WL = (A_L, \alpha_L)$ *and acts on singletons:*

$$Wf([a]) = [f(a)] \qquad \begin{array}{ll} Wf(x^{\leftrightarrow}) = \big(Wf(x)\big)^{\leftrightarrow} & Wf(x \mid y) = Wf(x) \mid Wf(y) \\ Wf(x^{\updownarrow}) = \big(Wf(x)\big)^{\updownarrow} & Wf(x \,/\, y) = Wf(x) \,/\, Wf(y) \end{array}$$

We write WL *for the carrier* A_L *where the initial operation is not relevant. We call an element* $x \in WL$ *a* ***world state*** *over* L.

4.1 Flat Semantics

Any world can be represented approximately as a rectangular matrix, ignoring topological details. We take the local state space L to be *pointed*, that is, L is nonempty and there is a distinguished element $\star \in L$. We define the rectangular shape and content of the interpretation of syntactic expressions by operations *wd* (*width*), *ht* (*height*) and *sl* (*select*).

$$
\begin{aligned}
wd &: WL \to \mathbb{N} & ht &: WL \to \mathbb{N} \\
wd\big([a]\big) &= 1 & ht\big([a]\big) &= 1 \\
wd(x^{\leftrightarrow}) &= wd(x) & ht(x^{\leftrightarrow}) &= ht(x) \\
wd(x^{\updownarrow}) &= wd(x) & ht(x^{\updownarrow}) &= ht(x) \\
wd(x \mid y) &= wd(x) + wd(y) & ht(x \mid y) &= ht(x) \sqcup ht(y) \\
wd(x \,/\, y) &= wd(x) \sqcup wd(y) & ht(x \,/\, y) &= ht(x) + ht(y)
\end{aligned}
$$

where $\sqcup$ is the usual *maximum* operator. The *sl* operation has dependent type, for all $x \in WL$:

$$
\begin{aligned}
sl(x) : ht(x) \times wd(x) &\to L & sl(x^{\leftrightarrow})(i,j) &= sl(x)(i,j) \\
sl(a)(0,0) &= a & sl(x^{\updownarrow})(i,j) &= sl(x)(i,j)
\end{aligned}
$$

$$
sl(x \mid y)(i,j) = \begin{cases} sl(x)(i,j) & i < ht(x), j < wd(x) \\ sl(y)\big(i, j - wd(x)\big) & i < ht(y), j \geq wd(x) \\ \star & \text{otherwise} \end{cases}
$$

$$
sl(x \,/\, y)(i,j) = \begin{cases} sl(x)(i,j) & i < ht(x), j < wd(x) \\ sl(y)\big(i - ht(x), j\big) & i \geq ht(x), j < wd(y) \\ \star & \text{otherwise} \end{cases}
$$

Note that:

1. Selection coordinates are of the form (*row*, *column*) and zero-based.
2. We write n for the set $\{k \in \mathbb{N} \mid 0 \leq k < n\}$.
3. Composition of submatrices of incompatible shapes results in padding of the smaller one with $\star$.
4. All matrices constructed from syntactic expressions are nonempty.
5. All operations are natural in L.

We write $sl^\star$ for a totalized selection operation with

$$
sl^\star(x)(i,j) = \begin{cases} sl(x) & (i,j) \in ht(x) \times wd(x), \\ \star & \text{otherwise} . \end{cases}
$$

We call expressions x, y such that $sl^\star(x) = sl^\star(y)$ **similar**, writing $x \approx y$.

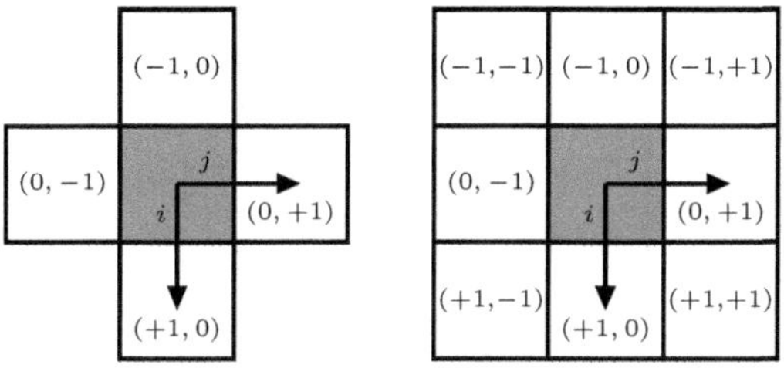

Fig. 1. Common neighborhood shapes in two-dimensional grids, with axes and chart coordinates relative to the grey center cell. Left: *von Neumann* neighborhood (4 neighbors). Right: *Moore* neighborhood (8 neighbors).

4.2 Context and Template

In a two-dimensional CA, each cell may interact only locally with a particular neighborhood. See Fig. 1 for the most common shapes; a comprehensive survey may be found in [8]. We shall use the von Neumann neighborhood shape as the running example throughout this article. Each shape may be specified by a simple data structure. The semantics of wrapping operations can then be given in terms of that structure.

Definition 8 (Context, Template). *A context is a pair* $\mathcal{C} = (C, \gamma)$ *of*

1. *a set functor* C *acting as a datatype constructor for collections of neighbor cells; for a neighborhood of* n *cells, simply put* $CX = X^n$,
2. *a distributive law* $\gamma : C^\sharp W \to WC^\sharp$, *where* $C^\sharp$ *is the* $\mathcal{C}$-**template** *functor:*

$$C^\sharp X = CX \times X \qquad\qquad C^\sharp f = Cf \times f$$

We write $c \vdash x$ *for* (c, x) *to emphasize that* $(c, x) \in C^\sharp X$ *is a template element, calling* x *the* **center**.

Such a law γ *transforms a template of worlds, namely the actual center world, surrounded by a global neighborhood of "phantom" worlds representing boundary conditions, into a world of templates of every cell embedded in its local neighborhood, thus enabling local transitions; see Fig. 2 for an example.*

The following axioms must hold for a context:

(C1) γ *preserves the shape of the center: for all* $c \in CW1; x \in W1$

$$W\dagger(\gamma_1(c \vdash x)) = x$$

where 1 *is a singleton set and* $\dagger$ *is the unique map* $\dagger : C^\sharp 1 \to 1$.

(C2) γ *does not distinguish similar neighborhoods: for all* $c_1, c_2 \in CWL; x \in WL$

$$Csl^\star(c_1) = Csl^\star(c_2) \implies \gamma_L(c_1 \vdash x) = \gamma_L(c_2 \vdash x) \ .$$

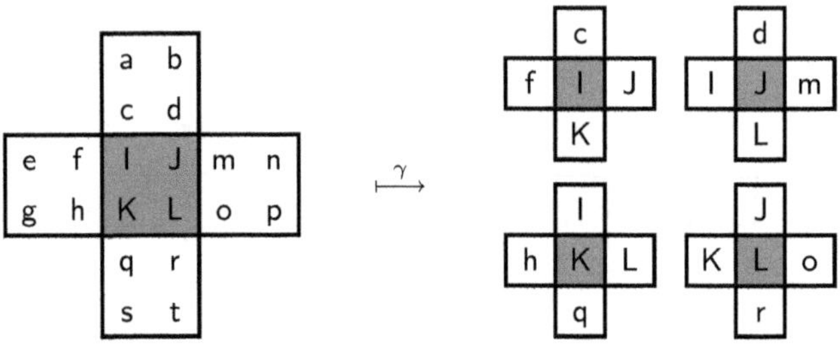

Fig. 2. Distribution of von Neumann template over a flat world. The (2×2)-shape is preserved; cf. axiom (C1).

The white part of each diagram in Fig. 1 defines a context, namely the shape C and the neighborhood distribution law γ. The center of $C^\sharp$ corresponds to the grey middle cell.

Example 1. The distributive law for the von Neumann neighborhood shape is sketched by an example application in Fig. 2. The context functor is $CX = X^4$.

For a general technique how to derive a distributive law γ from the given coordinate specifications see next subsection.

Definition 9 (Update). *Let $\mathcal{C} = (C, \gamma)$ be a context. A $C^\sharp$-algebra (L, u) is called a $\mathcal{C}$-**update** on L.*

Example 2. The CA for a diffusion system (discrete counterpart of the heat equation $\dot{u} = \alpha \Delta u$) requires the von Neumann context $CX = X^4$ and is defined by the update $d_\alpha(a, b, c, d \vdash e) = e + \alpha \cdot (a + b + c + d - 4e)$, where the parenthesis is the discrete Laplace operator. The CA for Conway's Game of Life requires the Moore context $CX = X^8$ and is defined by the update

$$v(a_1, \dots, a_8 \vdash b) = \begin{cases} 1 & \sum a_i = 3, \\ b & \sum a_i = 2, \\ 0 & \text{otherwise} \ . \end{cases}$$

Lemma 4. *Behavioral lifting for any context $\mathcal{C} = (C, \gamma)$ maps $\mathcal{C}$-updates on local states L canonically to $\mathcal{C}$-updates on world states WL.*

$$u : C^\sharp L \to L \implies W^\gamma u : C^\sharp WL \to WL$$

Definition 10 (Globalization). *Let $\mathcal{C} = (C, \gamma)$ be a context. We call the $\mathcal{C}$-update $(WL, W^\gamma u)$ the $\mathcal{C}$-**globalization** of the $\mathcal{C}$-update (L, u).*

Example 3. Reconsider Fig. 2; for the special case of linear forms u, such as the diffusion system from Example 2, the globalized operation $W^\gamma u = Wu \circ \gamma_L$ is also a rather high-level specification of a familiar array-programming technique called *convolution with a stencil/kernel.*

4.3 Regular Contexts

Distributive laws for contexts are hard to define directly in practice. The reader is invited not to be deceived by the simplicity of Fig. 2, but to try and give a complete distributive law, dealing with (nested) wrapping as well, for the von Neumann neighborhood shape.

The matter is simplified by reconsidering the relative coordinates in Fig. 1. Assume that for a context $\mathcal{C} = (C, \gamma)$, relative coordinates are given as a **chart** $\chi \in C\mathbb{Z}^2$. By adding the coordinates of the grey center cell componentwise, the absolute coordinates of neighbor cells are obtained. There are two cases:

1. The absolute coordinates of a neighbor are within the the center world: The **internal** neighbor is selected from the center.
2. The resulting coordinates are outside the the center: The **external** neighbor is selected from the appropriate context component.

For certain contexts $\mathcal{C} = (C, \gamma)$, we can define an operation sl^+ that extends the selection operation sl to external neighbors, thus unifying the cases. sl^+ operates on world templates instead of single world states and has dependent type, for all $(c \vdash x) \in C^\sharp WL$, $sl^+(c \vdash x) : \big(ht(x) \times wd(x)\big) \oplus D_\chi \to L$ where D_χ, the **domain** of χ, is the set of elements in χ, and $\oplus$ denotes the elementwise sum $X \oplus Y = \big\{(a + c, b + c) \bigm| (a, b) \in X; (c, d) \in Y\big\}$. We require that sl^+ be natural in L. Given a chart χ and an extended selection sl^+, we can implement a distributive law.

Definition 11 (Relocation). *Let C be a context functor with chart $\chi \in C\mathbb{Z}^2$. The **relocation** map is a natural transformation $\widehat{\chi}_L : WL \to WC^\sharp\mathbb{Z}^2$ defined as*

$$\widehat{\chi}_L\big([a]\big) = \big[\chi \vdash (0,0)\big]$$

$$\widehat{\chi}_L(x^\leftrightarrow) = \Big(WC^\sharp f_x\big(\widehat{\chi}_L(x)\big)\Big)^\leftrightarrow \qquad f_x(i,j) = \big(i, j \,\%\, wd(x)\big)$$

$$\widehat{\chi}_L(x^\updownarrow) = \Big(WC^\sharp g_x\big(\widehat{\chi}_L(x)\big)\Big)^\updownarrow \qquad g_x(i,j) = \big(i \,\%\, ht(x), j\big)$$

$$\widehat{\chi}_L(x \mid y) = \widehat{\chi}_L(x) \mid WC^\sharp h_x\big(\widehat{\chi}_L(y)\big) \qquad h_x(i,j) = \big(i, j + wd(x)\big)$$

$$\widehat{\chi}_L(x \mathbin{/} y) = \widehat{\chi}_L(x) \mathbin{/} WC^\sharp k_x\big(\widehat{\chi}_L(y)\big) \qquad k_x(i,j) = \big(i + ht(x), j\big)$$

where % is the usual remainder *operator. This map assigns to a world w of arbitrary state a world of templates, each containing just the* coordinates *of the cells of w relevant for a single neighborhood.*

Lemma 5. *The relocation map preserves the shape of its input: $W\ddagger\big(\widehat{\chi}_1(x)\big) = x$, where 1 is a singleton set and $\ddagger$ is the unique map $\ddagger : C^\sharp\mathbb{Z}^2 \to 1$.*

Example 4. A possible specification of von Neumann neighborhoods with $CX = X^4$, namely the one given in Fig. 1, is given as follows.

1. A chart: $\chi = \big((-1,0),(0,+1),(+1,0),(0,-1)\big)$.
   In prose, von Neumann neighbors are represented in the order *north*, *east*, *south*, *west*. Of course, every permutation is equally valid. The domain is $D_\chi = \big\{(-1,0),(0,+1),(+1,0),(0,-1)\big\}$ in any case.

2. An extended selection:

$$sl^{+}(n,e,s,w \vdash x)(i,j) = \begin{cases} sl^{\star}(n)\big(i+ht(n),j\big) & i<0, \\ sl^{\star}(e)\big(i,j-wd(x)\big) & j \geq wd(x), \\ sl^{\star}(s)\big(i-ht(x),j\big) & i \geq, ht(x) \\ sl^{\star}(w)\big(i,j+wd(w)\big) & j<0, \\ sl^{\star}(x)(i,j) = sl(x)(i,j) & \text{otherwise} \ . \end{cases}$$

where the first four cases are the subcases of external neighbors, and the last case covers internal neighbors. Note that the external cases do not overlap because of the type restriction given above.

Consider the (2×2)-torus with the flat matrix interpretation $\left(\begin{smallmatrix} a & b \\ c & d \end{smallmatrix}\right)$ defined for instance by $t = \big([a] \mid [b] \mathbin{/} [c] \mid [d]\big)^{\leftrightarrow\updownarrow}$. The relocation function $\widehat{\chi}$ for the von Neumann context yields a world $r = \widehat{\chi}(t)$ of the same shape, but populated with von Neumann contexts of coordinates in the range $wd(t) \times ht(t) = 2 \times 2$. The elements of t can be substituted for their coordinates by considering $s = WC^{\sharp}\big(sl(t)\big)(r)$, as depicted in Fig. 3.

In the above example, cells have only internal neighbors because of wrapping. In the general case, external neighbors and extended selection are required.

Theorem 2. *A chart χ and extended selection operation sl^{+} for a context functor C induce a distributive law and hence a context $\mathcal{C} = (C, \gamma)$.*

$$\gamma_L(c \vdash x) = WC^{\sharp}\big(sl^{+}(c \vdash x)\big)\big(\widehat{\chi}_L(x)\big)$$

4.4 Compositional Contexts

Definition 12. *A context $\mathcal{C} = (C, \gamma)$ is called* **compositional** *iff it admits a lifting of each syntactic operation over context formation, namely a collection of natural transformations*

$$\begin{aligned} \mathit{cosingleton}_L : &\quad CWL \to CL \\ \mathit{cohwrap}_L, \mathit{covwrap}_L : &\quad WL \times CWL \to CWL \\ \mathit{cobeside}_L, \mathit{coabove}_L : &\quad WL \times WL \times CWL \to CWL \times CWL \end{aligned}$$

such that, for all local updates (L, u) and their globalized counterparts $g = W^{\gamma}u$:

$$\begin{aligned} \big[u(\mathit{cosingleton}_L(c) \vdash a)\big] &= g(c \vdash [a]) \\ g(\mathit{cohwrap}_L(x,c) \vdash x)^{\leftrightarrow} &= g(c \vdash x^{\leftrightarrow}) \\ g(\mathit{covwrap}_L(x,c) \vdash x)^{\updownarrow} &= g(c \vdash x^{\updownarrow}) \\ g(c_1 \vdash x_1) \mid g(c_2 \vdash x_2) &= g(c \vdash x_1 \mid x_2) \qquad (c_1,c_2) = \mathit{cobeside}_L(x_1,x_2,c) \\ g(c_1 \vdash x_1) \mathbin{/} g(c_2 \vdash x_2) &= g(c \vdash x_1 \mathbin{/} x_2) \qquad (c_1,c_2) = \mathit{coabove}_L(x_1,x_2,c) \end{aligned} \tag{1}$$

Table 1. Syntax functor Σ_L

(Textual)		*singleton*		*hwrap*		*vwrap*		*beside*		*above*
(Graphic)		[-]		$\overset{\leftrightarrow}{-}$		$\overset{\updownarrow}{-}$		-|-		-/-
(Objects)	$\Sigma_L X =$	L	$+$	X	$+$	X	$+$	$(X \times X)$	$+$	$(X \times X)$
(Morphisms)	$\Sigma_L f =$	id_L	$+$	f	$+$	f	$+$	$(f \times f)$	$+$	$(f \times f)$

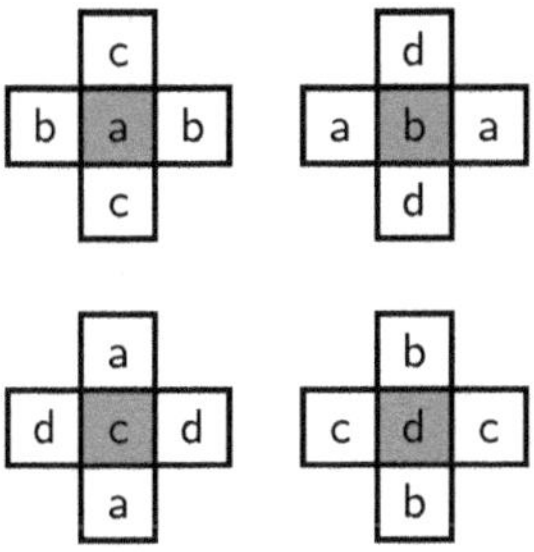

Fig. 3. Von Neumann neighborhoods on a (2×2)-torus, constructed by relocation

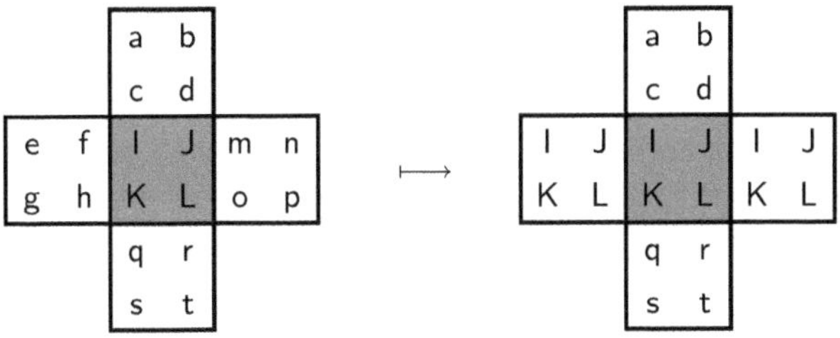

Fig. 4. *cohwrap* acting on von Neumann neighborhood of flat (2×2)-world

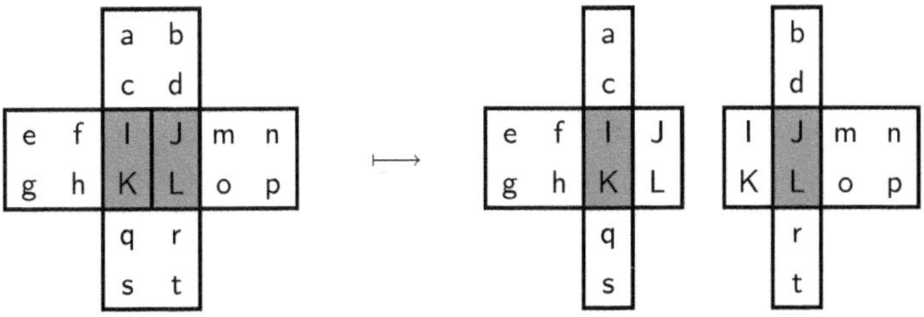

Fig. 5. *cobeside* acting on von Neumann neighborhood of flat (2×2)-world

These operations are not quite uniquely determined: Since the split operations act only on external neighbors, and only under a map $W^{\gamma}u = Wu \circ \gamma$, see equations (1), a loose specification up to similarity suffices thanks to context axiom (C2). See Figures 4, 5 for illustrations.

Space does not permit to give a full set of compositionality laws, say, for the von Neumann context here; cf. appendix.

In this section, we have formalized the context sensitivity of CAs for a variety of topologies. To this end we have introduced the notion of context and template functors to create data structures that accumulate the state of a neighborhood. We have defined local behavior of CAs (updates) as algebras of the template functor and their globalization as the behavioral lifting arising from a distributive law of context over worlds. We have addressed the problem of specifying a suitable distributive law by giving relative addresses of neighbors and a universal syntax-directed relocation procedure. Finally, we have investigated compositionality of contexts in the sense that syntactic composition can be lifted from the center of a world template over a globalized update. Together with the behavior functor to be defined in the next section, this compositionality will yield the desired spatio-temporal distributive law.

5 Behavior

Definition 13 (Behavior Functor). *Let S be a set and $\mathcal{C} = (C, \gamma)$ a context. We call the following functor the $\mathcal{C}$-**behavior** functor with **observable state** S:*

$$B^{\mathcal{C}}_S X = S \times X^{CS} \qquad\qquad B^{\mathcal{C}}_S f = \mathrm{id}_S \times (f \circ \_)$$

The functor $B^{\mathcal{C}}_S$ is the composition of the left product functor $S \times \_$ and the covariant hom-functor $\mathrm{Hom}(CS, \_)$, whose function part we abbreviate as $(f \circ \_)$ for $\mathrm{Hom}(CS, \_)(f)$. The functor $B^{\mathcal{C}}_S$ is also the signature of deterministic Moore automata with input CS (context of state) and output S.

We write $s \rhd t$ for (s, t) to emphasize that $(s, t) \in B^{\mathcal{C}}_S X$ is a behavior element. We also omit parentheses by having $\rhd$ bind less strongly than function composition $\circ$. Let (X, g) be a $B^{\mathcal{C}}_S$-coalgebra. The prose reading of an equation $g(x) = s \rhd t$ is as follows:

> The automaton specified by (X, g), when in state $x \in X$, outputs $s \in S$ and, for any input $c \in CS$, transits to state $t(c) \in X$.

Definition 14 (Update Coalgebra). *Let $\mathcal{C} = (C, \gamma)$ be a context. Let u be a $\mathcal{C}$-update on S. The structure $(S, u^{\rhd})$ where*

$$u^{\rhd}(s) = \big(s, u(\_ \vdash s)\big)$$

is a $B^{\mathcal{C}}_S$-coalgebra. Hence, the automaton specified by $(S, u^{\rhd})$, when in state s, outputs s and transits to state $u(c \vdash s)$ for any input c. In other words, update coalgebras specify automata that transduce boundary conditions to observable global state of a system. The overall behavior of the automaton is given by the unique $B^{\mathcal{C}}_S$-coalgebra homomorphism $u^{\rhd}! : S \to \Omega$.

Note that we have adopted the convention that the local state of CAs be public, formally by returning s as the output component of $u^{\triangleright}(s)$. We could introduce some privacy, and more behavioral equivalence, by outputting $o(s)$ for some observation function o instead.

In this section, we have defined a behavior functor for CAs, seen as Moore automata that consume context of state (boundary conditions) and output the current state. Updates, by construction template algebras, can be massaged into behavioral coalgebra shape. The corresponding anamorphism gives the semantic objects of automaton behavior.

6 Distributive Law

Let $\mathcal{C} = (C, \gamma)$ be a compositional context. Let L be a pointed set with default element $\star$. Let u be a $\mathcal{C}$-update on L. Our task is now to give a $\Sigma_L/B^{\mathcal{C}}_{WL}$-distributive law that encodes the update u "properly", preferrably in a generic way. Though we adopt a notation in the style of SOS, the focus is on certain distributive laws, and rule formats are considered only as a means of visualization.

Definition 15 (Rule Format). *We define a rule of the form*

$$\frac{x_1 \xrightarrow{s_1} y_1 \quad \cdots \quad x_n \xrightarrow{s_n} y_n}{k(x_1, \ldots, x_n) \xrightarrow[d]{s} y}$$

where $x_1, \ldots, x_n, y_1, \ldots, y_n, s_1, \ldots, s_n$ *are variables,* s *is an expression possibly depending on* $s_1, \ldots, s_n$*,* y *is an expression possibly depending on* $y_1, \ldots, y_n$*,* k *is a* n*-ary syntactic operation, and* d *is a* $(n+1)$*-ary function, to denote a clause of a distributive law* λ*:*

$$\lambda\big(k(s_1 \triangleright t_1, \ldots, s_n \triangleright t_n)\big) = s \triangleright l \circ (t_1 \times \cdots \times t_n) \circ d(s_1, \ldots, s_n, \_)$$
$$\textit{where } l(y_1, \ldots, y_n) = y \ .$$

The prose reading of this formula is rather involved:

> Consider n subsystems. If they are composed according to the syntactic operator k, the composite system, when in joint state $(x_1, \ldots, x_n)$,
> 1. outputs s, depending on each of the individual outputs s_i for individual state x_i of the i-th subsystem, and
> 2. for input c transits to state y, where for each y_i the post-state $t_i(c_i)$ of the i-th subsystem for state x_i and input c_i is substituted. The subinputs c_i are projected from $(c_1, \ldots, c_n) = d(s_1, \ldots, s_n, c)$.
>
> Note that Moore automata allow the computation of output *independently* of input; hence there is no vicious circle here.

Theorem 3. *If the following properties hold, then a law specified in this format is natural.*

$$\frac{}{[a] \xrightarrow[\textit{cosingleton}]{[a]} [u(\_ \vdash a)]} \qquad \frac{x \xrightarrow{s} y}{x^{\leftrightarrow} \xrightarrow[\textit{cohwrap}]{s^{\leftrightarrow}} y^{\leftrightarrow}} \qquad \frac{x_1 \xrightarrow{s_1} y_1 \quad x_2 \xrightarrow{s_2} y_2}{x_1 \mid x_2 \xrightarrow[\textit{cobeside}]{s_1 \mid s_2} y_1 \mid y_2}$$

$$\frac{x \xrightarrow{g} y}{x^{\updownarrow} \xrightarrow[\textit{covwrap}]{s^{\updownarrow}} y^{\updownarrow}} \qquad \frac{x_1 \xrightarrow{s_1} y_1 \quad x_2 \xrightarrow{s_2} y_2}{x_1 \,/\, x_2 \xrightarrow[\textit{coabove}]{s_1 / s_2} y_1 \,/\, y_2}$$

Fig. 6. Rules defining the spatio-temporal distributive law λ^u of cellular automaton with dynamics u

1. *The rules are* complete*: Every syntactic operation k is covered.*
2. *Each rule is* well-typed*:*

$$\frac{s_1, \ldots, s_n \in WL \quad t_1, \ldots, t_n : CWL \to X}{l_X \circ (t_1 \times \cdots \times t_n) \circ d(s_1, \ldots, s_n, \_) : CWL \to \Sigma_L X}$$

3. *Each rule is* syntactically natural*: $\Sigma_L h \circ l_X = l_Y \circ h^n$ for all $h : X \to Y$.*

Fig. 6 shows the rules that define λ^u. The leftmost is actually a rule scheme with one instance for each $a \in L$. The rules amount to the following equations:

$$\begin{aligned}
\lambda^u_X\big([a]\big) &= [a] \rhd \textit{singleton}_X \circ u(_ \vdash a) \circ \textit{cosingleton}_L \\
\lambda^u_X\big((s \rhd t)^{\leftrightarrow}\big) &= s^{\leftrightarrow} \rhd \textit{hwrap}_X \circ t \circ \textit{cohwrap}_L(s, _) \\
\lambda^u_X\big((s \rhd t)^{\updownarrow}\big) &= s^{\updownarrow} \rhd \textit{vwrap}_X \circ t \circ \textit{covwrap}_L(s, _) \\
\lambda^u_X\big((s_1 \rhd t_1) \mid (s_2 \rhd t_2)\big) &= (s_1 \mid s_2) \rhd \textit{beside}_X \circ (t_1 \times t_2) \circ \textit{cobeside}_L(s_1, s_2, _) \\
\lambda^u_X\big((s_1 \rhd t_1) \,/\, (s_2 \rhd t_2)\big) &= (s_1 \,/\, s_2) \rhd \textit{above}_X \circ (t_1 \times t_2) \circ \textit{coabove}_L(s_1, s_2, _)
\end{aligned} \tag{2}$$

We subscript the syntactic operations to emphasize that they are used as natural injections. It is easy to verify that these rules satisfy the naturality conditions of Theorem 3.

In this section, we have introduced a rule format similar to well-known SOS formats for our particular behavior functor. A distributive law that deals with globalization of updates in a step-wise fashion, focusing on spatio-temporal tradeoff, can be specified in that format. The generality of the rule format is not considered. The law is structured in terms of careful pairing of syntactic constructions and their context-transforming duals. It is generic in the sense that the local update affects only the singleton rule. Generalization to other syntactic functors is straightforward.

7 Equivalence

Whereas the globalization of an update reflects the array programming style of *loops*, the distributive-law encoding reflects the recursive programming style of *divide-and-conquer*. The following theorem states that both styles are equivalent.

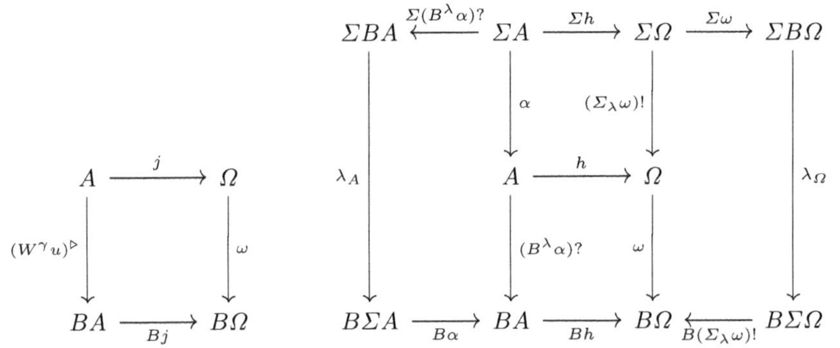

Fig. 7. Diagrams of Theorem 4. We abbreviate Σ_L, $B^{\mathcal{C}}_{WL}$, λ^u, h^u and j^u by dropping their fixed super- and subscripts.

Theorem 4. *Globalized updating of a world template is equivalent to bialgebraic semantics induced by the spatio-temporal distributive law given above:*

Let $\mathcal{C} = (C, \gamma)$ be a compositional context. Let $WL = (A, \alpha)$ be an initial Σ_L-algebra and (Ω, ω) a final $B^{\mathcal{C}}_{WL}$-coalgebra. Let u be a $\mathcal{C}$-update on L. Let $j^u = (W^\gamma u)^\triangleright!$ be the associated globalized semantics. Let $\lambda^u : \Sigma_L B^{\mathcal{C}}_{WL} \Rightarrow B^{\mathcal{C}}_{WL}\Sigma_L$ be the distributive law induced by u. Let $h^u : A \to \Omega$ be the associated bialgebraic semantics. Then $j^u = h^u$.

Proof. Show that $(W^\gamma u)^\triangleright$ forms a λ^u-bialgebra with α, that is $(W^\gamma u)^\triangleright \circ \alpha = B\alpha \circ \lambda^u_A \circ \Sigma(W^\gamma u)^\triangleright$. This can be verified by syntactic case distinction, combining the equations of compositionality (1) and distributivity (2); cf. appendix. It follows by initiality that $(W^\gamma u)^\triangleright = (B^\lambda \alpha)?$, and hence by finality that $j^u = h^u$. □

In this section, we have identified the two definition styles for a global update function, each defined by a distributive law, with two prominent programming styles for grid-based computations, namely loops and divide-and-conquer. We have demonstrated their equivalence by a straightforward bialgebraic, inductive-coinductive argument.

8 Conclusion

The class of ABMs is defined operationally as the class of program behavior that can be achieved within certain programming environments. There appears to be some consensus about what constitutes a well-behaved ABM, but only implicitly in the form of documentation standards [2], not by rigorous semantical criteria.

Environments for agent-based modelling are based on a virtual world that can be represented abstractly as a cellular automaton. We conjecture that a reasonable approximation of the notion of a well-behaved ABM can be given by requiring that its behavior be expressible as a global update with a simple neighborhood shape.

Coalgebraic formalisation of behavior has so far been largely restricted to computer science, with notable exceptions in physics. Here we showed that two-dimensional CAs and hence many ABMs can concisely be presented in this language. In a companion article (in prep.) we demonstrate how to implement the bialgebraic framework and reproduce the results of a particular ABM of ecological phenomena in the functional programming language Haskell.

In many disciplines CAs and ABMs have been used as tools for studying complex systems, i.e. systems in which the complex states can be recursively constructed from simple building blocks with persistent features. The present analysis demonstrates that this may only be part of a bigger picture: Bialgebras offer a second possibility of interpretation in which the focus is on corecursive unfolding of behavior. In another companion article (in prep.) we explore the epistemological and methodological implications of the dual viewpoints of the bialgebraic theory.

Acknowledgments. This work has been inspired by Bartek Klin's talks and writing on SOS. Jan Rutten and Milad Niqui have provided valuable feedback on earlier stages of the work, which helped considerably to shape this article.

References

1. Bartels, F.: On Generalised Coinduction and Probabilistic Specification Formats. Phd thesis. Vrije Universiteit Amsterdam (2004)
2. Grimm, V., Berger, U., Bastiansen, F., Eliassen, S., Ginot, V., Giske, J., Goss-Custard, J., Grand, T., Heinz, S., Huse, G., Huth, A., Jepsen, J., Jørgensen, C., Mooij, W., Müller, B., Pe'er, G., Piou, C., Railsback, S., Robbins, A., Robbins, M., Rossmanith, E., Rüger, N., Strand, E., Souissi, S., Stillman, R., Vabø, R., Visser, U., DeAngelis, D.: A standard protocol for describing individual-based and agent-based models. Ecological Modelling 198, 115–126 (2006)
3. Klin, B.: Structural operational semantics and modal logic, revisited. Electr. Notes Theor. Comput. Sci. 264(2), 155–175 (2010)
4. Kurth, W., Sloboda, B.: Growth grammars simulating trees. Silva Fennica 31, 285–295 (1997)
5. Kurz, A.: Logics for Coalgebras and Applications to Computer Science. Phd thesis. Ludwig-Maximilians-Universität München (2000)
6. Rutten, J.J.M.M.: Universal coalgebra: a theory of systems. Theor. Comput. Sci. 249(1), 3–80 (2000)
7. Turi, D., Plotkin, G.: Towards a mathematical operational semantics. In: Proceedings 12th International Conference on Logic in Computer Science (LICS), pp. 280–291. IEEE, Los Alamitos (1997)
8. Tyler, T.: Cellular automata, `http://cell-auto.com/neighbourhood/`
9. Trancón y Widemann, B., Hauhs, M.: Distributive-law semantics for cellular automata and agent-based models (2011), `https://www.bayceer.uni-bayreuth.de/mod/de/top/dl/96591/biabm-appendix-only.pdf` (electronic appendix)

Context-Free Languages, Coalgebraically

Joost Winter[1,⋆], Marcello M. Bonsangue[1,2], and Jan Rutten[1,3]

[1] Centrum Wiskunde & Informatica (CWI)
[2] LIACS – Leiden University
[3] Radboud University Nijmegen

Abstract. We give a coalgebraic account of context-free languages using the functor $\mathcal{D}(X) = 2 \times X^A$ for deterministic automata over an alphabet A, in three different but equivalent ways: (i) by viewing context-free grammars as $\mathcal{D}$-coalgebras; (ii) by defining a format for behavioural differential equations (w.r.t. $\mathcal{D}$) for which the unique solutions are precisely the context-free languages; and (iii) as the $\mathcal{D}$-coalgebra of generalized regular expressions in which the Kleene star is replaced by a unique fixed point operator. In all cases, semantics is defined by the unique homomorphism into the final coalgebra of all languages, paving the way for coinductive proofs of context-free language equivalence. Furthermore, the three characterizations can serve as the basis for the definition of a general coalgebraic notion of context-freeness, which we see as the ultimate long-term goal of the present study.

1 Introduction

The set $\mathcal{P}(A^*)$ of all formal languages over an alphabet A is a final coalgebra of the functor $\mathcal{D}(X) = 2 \times X^A$. Deterministic automata are $\mathcal{D}$-coalgebras and their behaviour, in terms of language acceptance is given by the final homomorphism into $\mathcal{P}(A^*)$. A language is *regular* if it is in the image of the final homomorphism from a *finite* $\mathcal{D}$-coalgebra to $\mathcal{P}(A^*)$. Or, equivalently by Kleene's theorem, if it is in the image of the final homomorphism from the set of *regular expressions*, which constitute a $\mathcal{D}$-coalgebra by means of the so-called Brzozowski derivatives.

Thus the coalgebraic picture of regular languages and regular expressions is well-understood (cf. [12] for details). Moreover the picture is so elementary that it has recently been possible [15] to generalize it to a large class of other systems, including Mealy machines, labelled transition systems, and various probabilistic automata.

In the present paper, we will develop in part a similar coalgebraic picture for *context-free* languages, which form another well-known class, extending regular languages. Our focus will be on *context-free grammars*, which constitute one of the common definition schemes for context-free languages. (Another well-known characterization is through pushdown automata, which will not be treated here.)

Because the set of all languages is a final coalgebra of the functor $\mathcal{D}(X) = 2 \times X^A$, as was mentioned above, it seems natural to try and use the same functor

⋆ Supported by the NWO project CoRE: Coinductive Calculi for Regular Expressions.

A. Corradini, B. Klin, and C. Cîrstea (Eds.): CALCO 2011, LNCS 6859, pp. 359–376, 2011.

for a coalgebraic treatment of context-free grammars and languages. We will do so in three different but equivalent ways. (i) We will, in Sect. 3, view context-free grammars (in a modified Greibach normal form) as $\mathcal{D}$-coalgebras, which we shall call grammar coalgebras; their elements will correspond to partial derivations. (ii) Next we will, in Sect. 4, define a format of behavioural differential equations with respect to the functor $\mathcal{D}$, for which the unique solutions are precisely the context-free languages. (iii) In Sect. 5, we will define context-free languages by means of generalized regular expressions, in which the Kleene star is replaced by a unique fixed point operator, and which are given a $\mathcal{D}$-coalgebra structure using a variation of Brzozowski derivatives. In all three cases, semantics is defined by the final homomorphism into $\mathcal{P}(A^*)$.

We show that the above three coalgebraic characterizations are equivalent in the following sense: a language is context-free if and only if it is in the image of the final homomorphism starting in either a grammar $\mathcal{D}$-coalgebra; or a $\mathcal{D}$-coalgebra corresponding to a (finite) system of behavioural differential equations; or the $\mathcal{D}$-coalgebra of generalized expressions with fixed point operator.

The proofs of these equivalences are not trivial, but contain few surprises, consisting of ingredients that are already present at various places in the literature. What we do see as the contribution of this paper are the three characterizations as such, together with the fact that their equivalence could be established in such an elementary fashion. We expect that this will lead to various further results, as follows.

Grammar coalgebras establish a direct correspondence between context-free grammars and context-free languages, by finality, and thus pave the way for coinductive proofs of context-free language equivalence. Furthermore, we will argue that our way of defining context-free languages through behavioural differential equations will lead to a generalization of the very notion of context-freeness to other types of systems, much in the same way as regular expressions and regular languages were generalized in [15]. A sketch of an elementary but interesting first instance hereof is given at the end of the present paper, in Sect. 6, where we will introduce the new notion of *context-free streams.* Finally, expressions with a fixed point operator are well-suited for the formulation of algebraic characterizations, which we see as yet another direction for future research.

Related Work. In contrast to regular languages, equality of context-free languages is known to be an undecidable property [6]. This may explain why not so much algebraic or coalgebraic work has been devoted to study the theory of context-free languages. The first, and only, coalgebraic treatment of context-free languages we are aware of, is presented in [5]. In this paper context-free languages are described indirectly, as the result of flattening finite skeletal parsed trees. The authors study context-free grammars as coalgebras for a functor different form ours, i.e. the functor $\mathcal{P}((A + (-))^*)$.

Algebraically, the starting point is Kozen's complete characterization of regular languages in terms of Kleene algebras, idempotent semirings equipped with a star-operation satisfying some fixed point equations [8]. In [9,2], Kleene algebras have been extended with a least fixed point operator to axiomatize fragments of the theory of context-free languages. We take a similar approach, but coalgebraic in nature and substituting the Kleene star with a unique fixed point operator. Whereas [9,2] are interested in providing solutions to systems of equations of the form $x = t$ using least fixed points, we look at systems of behavioural differential equations and give a semantic solution in terms of context-free languages (in Sect. 4) and syntactic solutions in terms of regular expressions with unique fixed points (in Sect. 5).

Regular expressions with the Kleene star replaced by a unique right-recursive fixed point operator have been studied in [17,15] for a large variety of coalgebras, including the one for deterministic automata. The observation that context-free languages can be seen as solutions to systems of equations dates back to [3].

2 Coalgebras and Deterministic Automata

In this section we give the basic definitions of coalgebras, deterministic automata and context-free grammars. A more extensive coalgebraic treatment of languages, automata and regular expressions can be found, for example, in [13,12,7].

A *coalgebra* for an endofunctor $\mathcal{F}$:**Set** $\to$ **Set** consists of a carrier set C together with a map $c{:}C \to \mathcal{F}C$. The functor $\mathcal{F}$ is usually called the *type* of the coalgebra. In this paper we will be concerned with coalgebras for structured automata [16], i.e. of type $\mathcal{D}\mathcal{F}$, for the functor $\mathcal{D} = 2 \times (-)^A$ and endofunctors $\mathcal{F}$:**Set** $\to$ **Set**. Here and in the rest of this paper, A is a finite set (in this context also called *alphabet*), 2 is the two-element set $\{0, 1\}$ and $\times$ is the Cartesian product. Sometimes we see 2 as a complete lattice with $0 \leq 1$ and join $\vee$ and meet $\wedge$ as expected. A coalgebra $(C, c{:}C \to \mathcal{D}\mathcal{F}C)$ can be interpreted as an automaton that for a given state $t \in C$ returns a pair $c(t) = \langle o(t), \delta(t)\rangle$, determining whether the state t is final (i.e. $o(t) = 1$) or not ($o(t) = 0$), and offering a structured state $\delta(t)(a) \in \mathcal{F}C$ for each input $a \in A$. Typically we will write t_a for $\delta(t)(a)$, call $o(t)$ the *output of* t and t_a the *a-derivative of* t . When confusion may arise about the coalgebra we are referring to, we will superscribe $o(t)$ and t_a with the coalgebra map c. In the case of $\mathcal{D}$-coalgebras (but not for $\mathcal{D}\mathcal{F}$-coalgebras in general), we can extend the notion of a-derivative to *word derivatives* t_w, for $w \in A^*$, by setting $t_\lambda = t$ for the empty word λ and $t_{aw} = (t_a)_w$ for $a \in A$ and $w \in A^*$.

A *homomorphism* from a $\mathcal{D}$-coalgebra (C, c) to a $\mathcal{D}$-coalgebra (D, d) is a function $f{:}\, C \to D$ preserving outputs and next states, that is, for all $t \in C$,

$$o(f(t)) = o(t) \quad \text{and} \quad f(t_a) = f(t)_a$$

(which is equivalent to the condition that $d \circ f = \mathcal{D}(f) \circ c$, where the action of the functor $\mathcal{D}$ on functions is as expected).

For example, the set $\mathcal{P}(A^*)$ of all *languages* on the alphabet A can be equipped with a $\mathcal{D}$-coalgebra map by setting for every $L \subseteq A^*$, $o(L) = 1$ if and only if $\lambda \in L$, and $L_a = \{w \in A^* \mid aw \in L\}$. This coalgebra is called *final* because for every $\mathcal{D}$-coalgebra (C, c), there is a unique homomorphism $[\![\ ]\!]_c{:}C \to \mathcal{P}(A^*)$ given by

$$w \in [\![t]\!]_c \;\text{ iff }\; o(t_w) = 1, \quad \text{for all } w \in A^*.$$

A relation $R \subseteq C \times D$ between the carriers of two $\mathcal{D}$-coalgebras (C, c) and (D, d) is called a *bisimulation* if, whenever $(s, t) \in R$, we have $o(s) = o(t)$, and $(s_a, t_a) \in R$ for all $a \in A$. Whenever there exists a bisimulation R such that $(s, t) \in R$, we say that s and t are *bisimilar* and write $s \sim t$. It holds that $s \sim t$ if and only if $[\![s]\!]_c = [\![t]\!]_d$, or, in other words, s and t are bisimilar exactly when they are mapped onto the same language.

A relation $R \subseteq C \times D$ is a *bisimulation up-to* if, whenever $(s, t) \in R$, we have $o(s) = o(t)$, and for all $a \in A$, there are $s' \in C$ and $t' \in D$ such that $s_a \sim s'$, $t_a \sim t'$, and $(s', t') \in R$. Clearly $\sim$ is a bisimulation up-to. Conversely, for every bisimulation up-to R, if $(s, t) \in R$, then $s \sim t$.

3 Context-Free Languages via Grammars

We assume the reader to be familiar with the standard notions on context-free grammars and languages, and give only the definitions and results we need in the rest of this paper. For a more comprehensive study of context-free grammars and languages, see e.g. [10].

A *context-free grammar*, or *CFG*, on a finite alphabet A is a pair (X, p), where X is a finite set of nonterminals, or variables, and $p{:}X \to \mathcal{P}_\omega((A + X)^*)$ is a function describing the production rules.[1] We use the standard notation to describe the production rules:

$$x \to t \quad \text{iff} \quad t \in p(x)\,,$$

where $x \in X$ and $t \in (A+X)^*$. Here $+$ denotes the coproduct (or disjoint union), $\mathcal{P}_\omega$ the finite power set, and $(A + X)^*$ is the set of all the strings of finite length over A and X. According to the above definition, CFGs are coalgebras for the functor $\mathcal{P}_\omega((A + (-))^*)$, and indeed a coalgebraic account of context-free grammars and context-free languages using the above functor (without the finiteness condition on the power set) is presented in [5]. There, the focus is mainly on finite skeletal parsed trees (i.e. finite strings with additional tree structure), and context-free languages are obtained after applying a flattening function. In the present paper we will depart from the above work in order to describe uniquely context-free languages in three different (but equivalent) coalgebraic forms.

In order to define the language associated to a context-free grammar, next we define the notion of derivation. Given a CFG $G = (X, p)$ and $s, s' \in (A+X)^*$, we write $s \Rightarrow s'$, and say s' is derivable from s in a single *derivation* step, whenever $s = s_1 x s_2$ and $s' = s_1 t s_2$ for a production rule $x \to t$ of G, and $s_1, s_2 \in (A+X)^*$. We say that s' is derivable from s in a single *leftmost derivation* step whenever s_1 is a (possibly empty) string of terminals in A^*. As usual, $\Rightarrow^*$ denotes the reflexive and transitive closure of $\Rightarrow$. In general, if $s \Rightarrow^* s'$, then s' is derivable from s

[1] A short note on these finiteness conditions: the finiteness conditions on both X and the powerset are required, because otherwise the set of resulting languages will be the set of *all* languages. The finiteness condition on A is essentially not required (barring some rewriting and reformulation), but it is kept here for convenience, as loosening it does not seem to lead to any significant new insights.

using only leftmost derivation steps. Therefore we can restrict our attention to leftmost derivations only.

For a CFG (X, p) and any variable $x \in X$, called the starting symbol, we define the language $L(x) \subseteq A^*$ generated by (X, p) from x by

$$L(x) = \{w \in A^* \mid x \Rightarrow^* w\}.$$

A language $L \subseteq A^*$ is called *context-free* if there exists a CFG (X, p) and a variable $x \in X$, such that $L = L(x)$.

For our coalgebraic treatment of context-free languages it will be convenient to work with CFGs with production rules of a specific form. We say that a CFG is in *weak Greibach normal form* if all of its production rules are of the form

$$x \to at \quad \text{or} \quad x \to \lambda\,,$$

where $a \in A$ is an alphabet symbol, and $t \in (A + X)^*$ is a (possibly empty) sequence of nonterminal and alphabet symbols. The main difference between weak Greibach normal form and the usual notion of Greibach normal form [4] is that here, t is not a string over X but over $(A + X)$, and hence may contain both nonterminal and alphabet symbols. Clearly, every CFG in Greibach normal form is also in weak Greibach normal form.

For every terminal symbol x, the language $L(x)$ generated by a CFG (X, p) in weak Greibach normal form is a context-free language. Conversely, every context free language L can be generated by a CFG in Greibach normal form from some nonterminal symbol [4]. Therefore CFGs in weak Greibach normal form characterize precisely the context-free languages.

3.1 A Coalgebraic Treatment of Context-Free Grammars

In this subsection we look at CFGs in weak Greibach normal form, and, for each such grammar, we define a corresponding $\mathcal{D}$-coalgebra, in the sense that the unique coalgebra homomorphism from the grammar (seen as a $\mathcal{D}$-coalgebra) to the final $\mathcal{D}$-coalgebra of all languages, maps nonterminal symbols precisely to the context-free language they generate. The key observation is that every CFG (X, p) with productions in weak Greibach normal form can be seen as a coalgebra for the functor $\mathcal{D}\mathcal{P}_\omega((A + (-))^*)$ (rather than as a $\mathcal{P}_\omega((A + (-))^*)$-coalgebra as in the previous section). More precisely, we represent the production rules by a map $p{:}X \to \mathcal{D}\mathcal{P}_\omega((A + X)^*)$, with $p(x) = \langle o(x), \delta(x)\rangle$ defined, for all terminal symbols $x \in X$, by

$$o(x) = 1 \;\text{ iff }\; x \to \lambda\,, \quad \text{and} \quad t \in x_a \;\text{ iff }\; x \to at$$

(writing as before x_a for $\delta(x)(a)$). Consider for example the grammar (in weak Greibach normal form) over the alphabet $A = \{a, b\}$ with nonterminal symbols $X = \{x, y\}$ and productions

$$x \to axa\,, \quad x \to ayb\,, \quad x \to aa\,, \quad y \to ayb\,, \quad y \to \lambda\,.$$

The language generated from x is $L(x) = \{a^n b^m a^k \mid n = m + k, n \geq 1\}$, while the language generated from y is $L(y) = \{a^n b^n \mid n \geq 0\}$. In coalgebraic form, the above productions read as follows:

$$\begin{array}{lll} o(x) = 0 & x_a = \{xa, yb, a\} & x_b = \emptyset \\ o(y) = 1 & y_a = \{yb\} & y_b = \emptyset \end{array}$$

The coalgebra associated to each CFG in weak Greibach normal form is not a proper deterministic automaton (i.e. a $\mathcal{D}$-coalgebra) because its type is of the form $\mathcal{D}\mathcal{P}_\omega((A + (-))^*)$. However, we can turn it into a deterministic automaton by embedding the nonterminal symbols X into $\mathcal{P}_\omega((A + X)^*)$ using the assignment $\eta_X{:}X \to \mathcal{P}_\omega((A + X)^*)$, mapping each $x \in X$ into the singleton set $\{x\}$ (in which x is seen as a string). In fact, we extend in a canonical manner each coalgebra $p{:}X \to \mathcal{D}\mathcal{P}_\omega((A + X)^*)$ for a CFG to what we call a *grammar coalgebra* $p^\#{:}\mathcal{P}_\omega((A + X)^*) \to \mathcal{D}\mathcal{P}_\omega((A + X)^*)$ as follows: for each finite subset $S \subseteq (A + X)^*$ we define its output value and its a-derivative by

S	$o(S)$	S_a
$\emptyset$	0	$\emptyset$
$\{\lambda\}$	1	$\emptyset$
$\{bs\}$	0	if $b = a$ then $\{s\}$ else $\emptyset$
$\{xs\}$	$o(x) \wedge o(s)$	$\{ts \mid t \in x_a\} \cup (\text{if } o(x) = 1 \text{ then } \{s\}_a \text{ else } \emptyset)$
$T \cup U$	$o(T) \vee o(U)$	$T_a \cup U_a$

Note that $\mathcal{P}_\omega((A + X)^*)$ forms a monad for the algebra of idempotent semirings with constants in A. Since $\mathcal{D}\mathcal{P}_\omega((A{+}X)^*)$ can be given an appropriate (but not trivial) algebra structure for this monad, the above inductive definition is essentially an application of the generalized determinization technique presented in [16]. More concretely, the idea of this definition is to view subsets of $(A{+}X)^*$ as languages. In fact, the above definition coincides with the coalgebra structure map of the final $\mathcal{D}$-coalgebra $\mathcal{P}(A^*)$ if we take subsets of strings in A^* (not containing nonterminal symbols). This is combined with the coalgebra map of the grammar which gives the output value and a-derivative for each nonterminal symbol, as can be seen from the fact that $p^\# \circ \eta_X = p$.

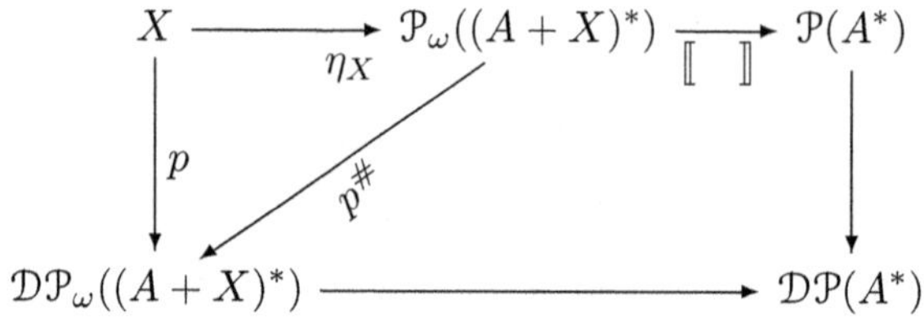

We are now ready to state our main result for this section, namely the correspondence between context-free languages and the languages associated by the final homomorphism $[\![-]\!]$ above to each nonterminal symbol of a CFG.

Theorem 1. *Let (X, p) be a context-free grammar in weak Greibach normal form over a finite alphabet A, and S a finite subset of $(A+X)^*$. For every word $w \in A^*$, we have $w \in [\![S]\!]_{p\#}$ if and only if there exists some $s \in S$ such that $s \Rightarrow^* w$.*

Proof. In this proof, we use fact that, for all $S \subseteq (A+X)^*$, if $w \in [\![S]\!]_{p\#}$, there is a $s \in S$ such that $w \in [\![\{s\}]\!]_{p\#}$, which is easily established.

We proceed by induction on the length of words w. For the empty word λ, it is easy to see that $s \Rightarrow^* \lambda$ if and only if s only consists of nonterminal symbols x that have a production rule $x \to \lambda$. Conversely, we have $\lambda \in [\![S]\!]_{p\#}$ iff $o(S_\lambda) = 1$, that is, if $o(S) = 1$: but it follows easily from the definition that this is the case iff there is a $s \in S$ consisting only of nonterminal symbols x that have a production rule $x \to \lambda$.

Assume that the inductive hypothesis holds for w, and consider the word aw. Assume $s \Rightarrow^* aw$, and consider the first term in the leftmost derivation of aw that is of the form at. Thus $s \Rightarrow^* at \Rightarrow^* aw$. By inspecting the definitions, it is easily seen that if $s \in S$, then $t \in S_a$. Furthermore, it also follows that $t \Rightarrow^* w$, and the inductive hypothesis then gives $w \in [\![S_a]\!]_{p\#}$, and hence that $o((S_a)_w) = o(S_{aw}) = 1$, from which $aw \in [\![S]\!]_{p\#}$ follows.

For the other direction, assume that $aw \in [\![S]\!]_{p\#}$. Then there must be some $s \in S$, such that $aw \in [\![\{s\}]\!]_{p\#}$, or that $o(\{s\}_{aw}) = o((\{s\}_a)_w) = 1$. We now get that $w \in [\![\{s\}_a]\!]_{p\#}$, and the inductive hypothesis now gives some $t \in \{s\}_a$ such that $t \Rightarrow^* w$. From inspecting the definitions, it is also easy to see that $s \Rightarrow^* at$. Hence, we get $s \Rightarrow^* at \Rightarrow^* aw$, which is what needed to be shown. □

It follows that a language L is context-free iff $L = [\![\eta_X(x)]\!]_{p\#}$, for some grammar coalgebra generated by a CFG (X, p), and some $x \in X$.

4 Context-Free Languages via Equations

We will now look at a characterization of context-free languages in terms of systems of behavioural differential equations, analogous to those introduced in [14]. The idea is to define context-free languages by means of equations that involve output values and derivatives for each alphabet symbol, using a simple language with only variables, choice and sequential composition. Each system of behavioural differential equations in our format has as its unique solution a language, which we prove to be context-free whenever the number of equations is finite. Conversely, for each context-free language L we will construct a finite system of behavioural differential equations with L as unique solution.

To illustrate our approach, consider the example from the previous section. A formal definition of this context-free language could be given by the following system of behavioural differential equations:

output value	a-derivative	b-derivative
$o(x) = 0$	$x_a = (x \cdot a) + (y \cdot b) + a$	$y_a = y \cdot b$
$o(y) = 1$	$x_b = 0$	$y_b = 0$

Next we present a syntax describing the format of behavioural differential equations that will be considered: let A be a finite set of *alphabet* symbols, X be a (possibly infinite) set of *variables*, and $\{o(x) \mid x \in X\}$ and $\{x_a \mid x \in X\}$ for each $a \in A$ be (syntactic) sets of symbols, representing notational variants of the variables. The variables $x \in X$ will play the role of placeholders for languages $L \subseteq A^*$, while their notational variants x_a will be placeholders for the corresponding language L_a, for each $a \in A$, and the $o(x)$ will correspond to the information whether L contains the empty string λ or not. We call a behavioural differential equation for context-free languages *well-formed* if it consists of an equation $o(x) = v$ and an equation $x_a = t$ for each $a \in A$, where $v \in \{0, 1\}$ and t is a term defined by

$$t ::= 0 \mid 1 \mid x \mid a \mid t + t \mid t \cdot t$$

where $x \in X$ and $a \in A$. (In the remainder of this paper, we will often simply write 'ab' rather than '$a \cdot b$'.) We let TX denote the set of all terms, as defined above, over the set X. A *well-formed system of equations* for X consists of one well-formed equation for each $x \in X$. Equivalently, a well-formed system of equations over X (for a fixed A) can be seen as a mapping $f{:}X \to \mathcal{D}TX$ [1] where, writing $f(x) = \langle o(x), \delta(x) \rangle$, we define $o(x)$ and, for each $a \in A$, $x_a = \delta(x)(a)$ by the values specified by the system of equations.

Before defining what a solution of a system of equations is, we need to interpret the above operations on terms as functions on languages. To this end, we transform a system of equations $f{:}X \to \mathcal{D}TX$ into a deterministic automaton $\bar{f}{:}TX \to \mathcal{D}TX$ inductively as follows:

t	$o(t)$	t_a
x	$o(x)$	x_a (as specified by f)
0	0	0
1	1	0
b	0	if $b = a$ then 1 else 0
$u + v$	$o(u) \vee o(v)$	$u_a + v_a$
$u \cdot v$	$o(u) \wedge o(v)$	$u_a \cdot v + o(u) \cdot v_a$

Again, this definition can be obtained as an application of the result in [16] since $\mathcal{D}TX$ can easily be equiped with an algebra structure for the term monad T. We can thus combine a system of equations $f{:}X \to \mathcal{D}TX$, its extension $\bar{f}$, and the final homomorphism to the coalgebra of all languages in the following diagram:

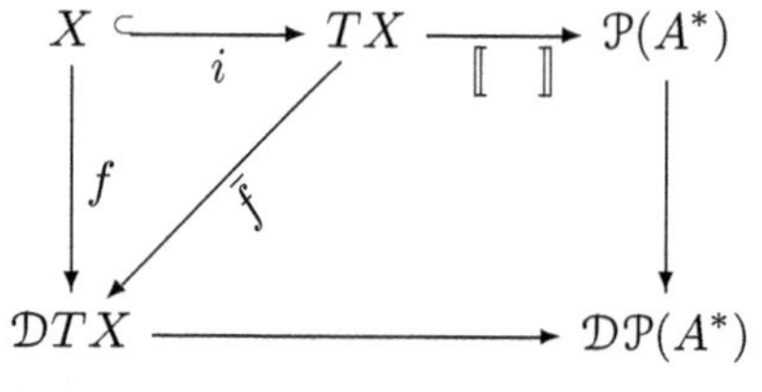

A *solution* for such a system of equations consists of a mapping of variables x to languages $L_x \subseteq A^*$ such that $L_x = [\![x]\!]$ for all $x \in X$. The above diagram basically shows that every well-formed system of equations has a unique solution. Our next goal is to prove that such a solution is a context-free language. Before we do this, however, we introduce the notion of term equivalence:

Definition 2. *We say that two terms $t, u \in TX$ are* equivalent, *denoted by $t \equiv u$, when for every well-formed system of equations (X, f), $t \sim u$ with respect to the coalgebra $(TX, \bar{f})$ generated from (X, f).*

One can easily show that the relation $\equiv$ is a congruence with respect to the sum $+$ and multiplication $\cdot$ of terms in TX, for any set X. Furthermore, TX modulo $\equiv$ forms an idempotent semiring:

Proposition 3. *For any set X and terms $t, u, v \in TX$, the following hold:*

$$\begin{array}{cc} t + u \equiv u + t & u \equiv v \Rightarrow t[u/x] \equiv t[v/x] \\ t + t \equiv t & 0 \cdot t \equiv 0 \equiv t \cdot 0 \\ 0 + t \equiv t \equiv t + 0 & 1 \cdot t \equiv t \equiv t \cdot 1 \\ t + (u + v) \equiv (t + u) + v & t \cdot (u \cdot v) \equiv (t \cdot u) \cdot v \\ (t + u) \cdot v \equiv t \cdot v + u \cdot v & t \cdot (u + v) \equiv t \cdot u + t \cdot v \end{array}$$

Equivalence between terms in TX can be extended to bisimilarity between different systems of equations on the same set of variables. This result will be convenient when proving that every context-free language is the solution of a well-formed system of equations.

Proposition 4. *If (X, f) and (X, g) are two systems of equations such that, for every $x \in X$, and every symbol a, $o^f(x) = o^g(x)$, and $x^f_a \equiv x^g_a$, then the identity relation on TX is a bisimulation up-to between the generated coalgebras $(TX, \bar{f})$ and $(TX, \bar{g})$.*

Proof. Let t be a term.

First, we have to show that $o^{\bar{f}}(x) = o^{\bar{g}}(x)$. When t is a variable, an alphabet symbol, 0 or 1, this is trivial, and when t is a compound term, this is proven by induction.

Secondly, we have to show that for all alphabet symbols a, there are terms t', t'', such that $t^{\bar{f}}_a \sim t'$, $t^{\bar{g}}_a \sim t''$, and $(t', t'') \in R$ (or, in other words, $t' = t''$). This, too, will be proven by induction, and is trivial for the base cases where t is an alphabet symbol, 0, or 1.

When t is a variable x, we have $x^{\bar{f}}_a = x^f_a \sim x^g_a = x^{\bar{g}}_a$. Taking $t' = t'' = x^{\bar{f}}_a$ then suffices.

When t is of the form $u+v$ or of the form $u \cdot v$, we will make use of the inductive assumption that the bisimilarity condition holds for u and v. But then it is easy to see that it holds for t too, making use of the fact that $\sim$ is a congruence with respect to $+$ and $\cdot$. □

We can now establish the first main result of this section, relating states of a grammar coalgebra to terms in TX.

Theorem 5. *Let A and X be finite sets. For every context-free grammar (X, p) in weak Greibach normal form and finite subset S of $(A + X)^*$ there exists a well-formed system of equations $f{:}X \to \mathcal{D}TX$ and a term $t \in TX$ such that $S \sim t$ with respect to the generated coalgebras $p^\#$ and $\bar{f}$, respectively.*

Proof. (Sketch) The system of equations f can be constructed from the grammar p as follows: $o(x) = 1$ if $\lambda \in p(x)$, and 0 otherwise; and $x_a = \sum\{\bar{s} \mid as \in p(x)\}$, where $\bar{s}$ is the obvious translation of $s \in (A+X)^*$ into a term of TX. The proof now proceeds by showing that $\{(S,t) \mid S \subseteq (A + X)^*, t = \sum\{\bar{s} \mid s \in S\}\}$ is a bisimulation up-to with respect to the generated coalgebras. □

Because for every context-free language L there exists a CFG (X, p) and $x \in X$ such that $L = [\![\eta_X(x)]\!]_{p^\#}$, it follows that *every context-free language is the solution of a well formed system of equations.*

It remains to be shown that every solution of a system of equations is a context-free language. Our approach will be to construct, for every system of equations (X, f), a context-free grammar (X, p), such that $x \sim \eta_X(x)$ for all x with respect to the generated coalgebras $\bar{f}$ and $p^\#$. To this end, we first transform our system of equations so that terms at the right hand side of all equations are in disjunctive normal form. This is possible because of the laws proven in Propositions 3 and because $\equiv$ is a congruence.

We say that a term $t \in TX$ is *conjunctive* when either $t = 1$; or when $t = a \cdot u$, where $a \in A$ and u a conjunctive term; or when $t = x \cdot u$, where x is a variable, and u is a conjunctive term. We say that a term $t \in TX$ is in *disjunctive normal form*, when either $t = 0$; or when $t = u + v$, where u is conjunctive, and v is in disjunctive normal form. Using Proposition 3, it is easy to see that for every term t, there is an equivalent (and thus bisimilar for all $(TX, \bar{f})$-coalgebras) term t' in disjunctive normal form. This implies that for every system of equations we can construct a new system of equations with the same variables and having the same solution but such that all terms on the right-hand side of the a-differential equations are in disjunctive normal form. We call a system of equations with the latter property a *system of equations in disjunctive normal form.*

We are finally ready for the other main result of this section, stating that *every solution of a system of equations is a context-free language.* Combined with the first main theorem from this section, we thus obtain the result that context-free languages are precisely the solutions of well-formed systems of equations.

Theorem 6. *For a finite set X, if (X, f) is a system of equations in disjunctive normal form, there exists a context-free grammar (X, p), such that $x \sim \eta_X(x)$ with respect to the generated coalgebras $(TX, \bar{f})$ and $(\mathcal{P}_\omega((A+X)^*), p^\#)$, respectively.*

Proof. Given a system of equations (X, f), we construct the grammar (X, p), such that

$$p(x) = \bigcup_{a \in A} \{at \mid t \text{ is a disjunct of } x_a\} \cup (\text{if } o(x) = 1 \text{ then } \{\lambda\} \text{ else } \emptyset).$$

It is easy to see that f is a translation of p in the sense of Theorem 5. Hence, it follows that $x \sim \{x\} = \eta_X(x)$ with respect to $\bar{f}$ and $p^\#$. □

5 Context-Free Expressions

In this section, we will introduce context-free expressions as an extension of regular expressions, where the Kleene star is replaced by a (unique) fixed point operator μ. We then will define a notion of Brzozowski-like derivatives for these expressions, and prove that the languages characterizable by such expressions are precisely the context-free languages. In contrast to the previous coalgebraic formalisms, this formalism gives us a *single* coalgebra of which the elements correspond exactly to the context-free languages.

Our usage of fixed point expressions with a coinductive semantics has a very similar flavour to that in [15], in which fixed point expressions are used as a characterization of regular expressions over a variety of functors. The additional expressive power obtained by the context-free expressions presented here is due to an explicit inclusion of a concatenation operator.[2] This provides an additional perspective on the treatment given here, in which 'context-freeness' is obtained by the addition of a new operator to a calculus of regular expressions[3], and may pave the way for an investigation of (1) extending this approach to other coinductively defined operators, and (2) extending this approach to a generalized notion of context-freeness for other functors.

We define the set of terms t (henceforth to be called *context-free expressions*) and guarded terms g over an alphabet A and a set of variables X as follows:

$$t ::= 0 \mid 1 \mid x \in X \mid a \in A \mid t + t \mid t \cdot t \mid \mu x.g$$
$$g ::= 0 \mid 1 \mid a \cdot t \; (a \in A) \mid g + g$$

For all closed terms t, we can describe the behaviour by defining the output value $o(t)$, and the derivative t_a for each alphabet symbol a. We do this as follows:

t	$o(t)$	t_a
0	0	0
1	1	0
b	0	if $b = a$ then 1 else 0
$u + v$	$o(u) \vee o(v)$	$u_a + v_a$
$u \cdot v$	$o(u) \wedge o(v)$	$u_a \cdot v + o(u) \cdot v_a$
$\mu x.u$	$o(u[\mu x.u/x])$	$(u[\mu x.u/x])_a$

Here $t[u/x]$, as usual, denotes the term obtained from t by replacing all *free* occurrences of x by u. Because of the guardedness conditions of terms occurring directly inside the μ operator, it is easy to see that the above specification is well-defined.

Note furthermore that we have just defined a $\mathfrak{D}$-coalgebra with the set of all closed context-free expressions as objects, and the behaviour defined above as its transition function.

[2] In [15], a translation from the familiar format of regular expressions (with concatenation) into μ-style expressions is given by means of substitution. However, this translation does not work for expressions of the type $x \cdot t$.

[3] Although this calculus does not explicitly contain the Kleene star, it can easily be expressed by means of the equality $t^* = \mu x.((t \cdot x) + 1)$.

Again, the behavioural equivalence relation $\sim$ is a congruence with respect to the sum $+$ and multiplication $\cdot$ of context-free expressions. Furthermore, the set of closed context-free expressions modulo $\sim$ forms an idempotent semiring:

Proposition 7. *For all context-free expressions t, u, v, the following hold:*

$$\begin{array}{cc}
t + u \sim u + t & u \sim v \Rightarrow t[u/x] \sim t[v/x] \\
t + t \sim t & 0 \cdot t \sim 0 \sim t \cdot 0 \\
0 + t \sim t \sim t + 0 & 1 \cdot t \sim t \sim t \cdot 1 \\
t + (u + v) \sim (t + u) + v & t \cdot (u \cdot v) \sim (t \cdot u) \cdot v \\
(t + u) \cdot v \sim t \cdot v + u \cdot v & t \cdot (u + v) \sim t \cdot u + t \cdot v \\
t[u/x] \sim u \Rightarrow \mu x.t \sim u & \mu x.t \sim \mu y.(t[y/x]) \textit{ if } y \textit{ is not free in } t \\
\mu x.t \sim t[\mu x.t/x] &
\end{array}$$

These laws can be seen as a partial (sound but not complete) axiomatization of behavioural equivalence between context-free expressions. Note that, because language equivalence of context-free languages is not semi-decidable, there cannot be any complete finitary axiomatization of behavioural equivalence.[2]

As an illustration of context-free expressions, it is easy to see that the expression $\mu x.(axb+1)$ will be mapped onto the language $\{a^n b^n\}$. As another example, consider the expression

$$\mu x.(axa + aa + a\mu y.(ayb + 1)b).$$

In the next subsection, it will become clear that this expression corresponds to the language $\{a^n b^m a^k \mid n = m + k, n \geq 1\}$ from the earlier examples.

5.1 From Systems of Equations to Context-Free Expressions

Assume we have a coalgebra generated by a system of equations, and some term $t \in TX$. From Sect. 4, we know that t is mapped by the final homomorphism to a context-free language. In this section, we will look for a context-free expression t' corresponding to t, in the sense that t and t' are mapped onto the same language, using a process of repeated substitution.

To start with, given a system of equations (X, f), we will associate with every variable x the μ-expression

$$\sigma_x := \mu x.(\sum_{a \in A} a \cdot x_a + o(x)),$$

and call it the *corresponding* or *associated* μ-expression. (As before, this notation strictly speaking does not denote a single expression, but rather a set of expressions which, by commutativity of addition, are all bisimilar.) For convenience, we also use the notation τ_x for the expression $\sum_{a \in A} a \cdot x_a + o(x)$: so $\sigma_x = \mu x.\tau_x$.

We now go on by defining the notions of *single syntactic substitutions* and *chains of syntactic substitutions*: these definitions can be seen as a formalization of the corresponding notions in [11], [17], and [15]. Given an association of expressions σ_x to variables x, an expression t' is a single syntactic substitution of t, if t' is obtained by replacing (syntactically) a *single* occurrence of a single variable x with σ_x. A chain of syntactic substitutions is a list $t_1, \ldots, t_n$ such that, for each $1 \leq i < n$, t_{i+1} is a single syntactic substitution of t_i.

We are especially interested in chains of syntactic substitutions, where the resulting expression does not contain any free variables, or only a limited set of free variables. We call such expressions closures and pseudoclosures of the original expression:

Definition 8. *We say an expression t' is a Z-*pseudoclosure *of t for a set $Z \subseteq X$ of variables, if there exists a a chain of syntactic substitutions $t_1, \ldots, t_n$ such that $t_1 = t$, $t_n = t'$ and t' only contains free variables from Z. We call a $\emptyset$-pseudoclosure simply a* closure.

As a continuation of our running example, recall the system of equations corresponding to the language $\{a^n b^m a^k \mid n = m + k, n \geq 1\}$. From this system of equations, we obtain the following assignment of expressions to variables:

$$\sigma_x = \mu x.(a(xa + yb + a) + b0 + 0) \qquad \sigma_y = \mu y.(ayb + b0 + 1)$$

From x, we obtain $\mu x.(a(xa + yb + a) + b0 + 0)$ by means of a single syntactic substitution, and another single syntactic substitution then gives us $\mu x.(a(xa + \mu y.(ayb + b0 + 1)b + a) + b0 + 0)$. This expression does not contain any free variables anymore, and therefore is a closure of x.

Some general laws about closures and pseudoclosures are easily established:

Proposition 9. *1. If u' is a W-pseudoclosure of u and v' a W-pseudoclosure of v, then $u' + v'$ is a W-pseudoclosure of $u + v$, $u' \cdot v'$ is a W-pseudoclosure of $u \cdot v$, and $\mu x.u'$ is a $W - \{x\}$-pseudoclosure of $\mu x.u$.*
2. If $t = u + v$, and t' is a W-pseudoclosure of t, then t' is of the form $u' + v'$, where u' is a W-pseudoclosure of u, and v' is a W-pseudoclosure of v'. The same fact holds if we replace $+$ by $\cdot$.

Using the previous proposition, we can establish that, for every term t, a closure t' exists. It should be noted, though, that this t' generally is not unique: for a term t, in general, many closures exist.

Proposition 10. *Given a term $t \in TX$ (that is, a μ-free term), a set of variables $Z \subseteq X$, and an assignment of expressions σ_x to variables $x \in X$ as above, there exists a Z-pseudoclosure t' of t with respect to this assignment.*

Proof. By (reverse) induction on the size of Z. If $Z = X$, the result is trivial, because every term is its own X-pseudoclosure.

Now we assume the theorem holds for $W \subseteq X$, and need to prove that, for any $x \in X$, the theorem also holds for $W - \{x\}$. We do this by induction on (μ-free) terms.

1. For terms 0, 1, and a, the result is trivial.
2. For terms $t = u + v$ or $t = u \cdot v$, the result follows from Proposition 9 and the inductive hypothesis.

3. For the variable x, we know that there must be a W-pseudoclosure u of τ_x. Then $\mu x.u$ is a $(W - \{x\})$-pseudoclosure of $\sigma_x = \mu x.\tau_x$, and hence also of x.
4. For variables $y \neq x$, assume u and v are W-pseudoclosures of τ_y and τ_x, respectively. Then $\mu x.v$ is, again, a $W - \{x\}$-pseudoclosure of x, and thus $u[\mu x.v/x]$ is a $W - \{x\}$-pseudoclosure of τ_y. Hence, $\mu y.u[\mu x.v/x]$ is a $W - \{x, y\}$-pseudoclosure (and also a $W - \{x\}$-pseudoclosure) of σ_y and y. □

With the next proposition we construct a bisimulation up-to between a coalgebra generated by a system of equations and the coalgebra of closed context-free expressions, relating every term t to all closures of it.

Proposition 11. *Given a system of equations (X, f) (yielding a corresponding expression σ_x for each variable $x \in X$), the relation*

$$R = \{(t, t') \mid t \in TX \text{ and } t' \text{ is a closure of } t\}$$

is a bisimulation up-to between the generated coalgebra $(TX, \bar{f})$ and the coalgebra of closed context-free expressions.

Proof. Say $(t, t') \in R$. It suffices to show that $o(t) = o(t')$, and that for each alphabet symbol a, there is a u_a, such that $t'_a \sim u_a$, and $(t_a, u_a) \in R$. We will do both by induction.

Showing that $o(t) = o(t')$ is immediate when $t = 0$, $t = 1$, or $t = a$, because in these cases we have $t = t'$. When $t = x$, it must be the case that t' is obtainable by a chain of single syntactic substitutions from σ_x. But, as every term obtainable by a chain of single syntactic substitutions from σ_x must be of the form $\mu x.(\sum_{a\in A} a \cdot s_a + o(x))$ for some mapping of alphabet symbols a to terms s_a, and $o(a \cdot s) = 0$ for any term s, it follows easily that $o(t') = o(x)$. When $t = u + v$, it follows that t' must be of the form $u' + v'$, where u' is a closure of u, and v' is a closure of v. We then get $o(t) = o(u) \vee o(v) = o(u') \vee o(v') = o(t')$, using the inductive assumption that $o(u) = o(u')$ and $o(v) = o(v')$. The case where $t = u \cdot v$ goes analogously.

Showing that $(t_a, t'_a) \in R$, again, is immediate when $t = 0$, $t = 1$, or $t = a$. When $t = x$, the first single syntactic substitution to obtain t' must, again, be replacing x by σ_x. As a result, t' must be of the shape $\mu x.(\sum_{a\in A} a \cdot s_a + o(x))$, where each s_a is a $\{x\}$-pseudoclosure of x_a. It follows that $u_a := s_a[t'/x]$ is a closure of x_a. But we have

$$\begin{aligned} t'_a &= (\mu x.(\sum_{a\in A} a \cdot s_a + o(x)))_a = ((\sum_{a\in A} a \cdot s_a + o(x))[t'/x])_a \\ &\sim (s_a[t'/x])_a = u_a \end{aligned}$$

so $t'_a \sim u_a$: as u_a is a closure of $x_a = t_a$, the required condition holds. The cases where $t = u + v$ and $t = u \cdot v$ are immediate from the inductive hypothesis and Proposition 9. □

Returning to our example, this proposition directly establishes that the expression $\mu x.(a(xa + \mu y.(ayb + b0 + 1)b + a) + b0 + 0)$, which is clearly bisimilar to $\mu x.(axa + a\mu y.(ayb + 1)b + aa)$, corresponds to the language $\{a^n b^m a^k \mid n = m + k, n \geq 1\}$.

The two previous propositions directly imply that, for any term in a coalgebra generated by a system of equations (and, hence, for every context-free language), we can use any closure of it as a bisimilar context-free expression. Hence, and because every term has a closure, for every context-free language we can find a context-free expression that is mapped to it by the final homomorphism:

Theorem 12. *Let L be a context-free language. There exists a context-free expression t such that $[\![t]\!] = L$.*

5.2 From Context-Free Expressions to Systems of Equations

Going in the other direction, the recipe is as follows: given a context-free expression t' in which every variable is bound by a μ-operator just once, we 'deconstruct' this expression into a system of equations (X, f), and a term t, of which t' is a closure. Then Proposition 11 implies that $t' \sim t$ with respect to the coalgebra of closed context-free expressions and the coalgebra $(TX, \bar{f})$ generated by (X, f). Hence, the final homomorphism maps t to a context-free language.

By applying a process of α-renaming, we can obtain an expression t' from any expression t such that, in t', no variable is bound twice, or, in other words, such that there are no two distinct subexpressions of t' that bind the same variable. It is easy to see that the resulting term t' will always be bisimilar to t.

Now we are able to move on to the main proposition of this section, in which for every context-free expression t' a system of equations is constructed, such that in the coalgebra generated by it, there is a term t with $t \sim t'$.

Proposition 13. *Given a closed context-free expression t, such that no two distinct sub-expressions of t are μ-expressions binding the same variable, there is a term t', such that 1) $t \sim t'$; 2) no two-subexpressions of t' are μ-expressions binding the same variable; and 3) every subexpression of t' that is a μ-expression, is of the form $\mu x.(\sum_{a \in A} a \cdot s_a + o)$, where $o \in \{0, 1\}$.*

Proof. It is easy to see that every guarded term is bisimilar to a guarded term of the form $\sum_{a \in A} a \cdot s_a + o$. When replacing the subexpressions of t that are μ-expressions by these bisimilar expressions, we obtain a new term that, by the fact that bisimilarity is a congruence, is bisimilar to the original term.

Proposition 14. *Given a closed context-free expression t', in the normal form of Proposition 13, there exists a system of equations (X, f), and a term t, such that t' is a closure of t.*

Proof. (Sketch) For any expression u', let its μ-pruning be the expression u obtained from it by replacing the outermost μ-expressions occurring in it by the corresponding variables. Because every variable is bound once, for every variable x we obtain a μ-pruning s_x of s'_x where $\mu x.s'_x$ is the μ-subexpression binding x. We then use these pruned expressions as a basis for a system of equations, and show that the original expression t' occurs as a closure of its μ-pruning t. □

As every context-free expression is bisimilar to a term in a coalgebra generated by some system of equations, it follows directly that the final homomorphism maps every context-free expression to a context-free language:

Theorem 15. *For every context-free expression t, $[\![t]\!]$ is a context-free language.*

6 Discussion

Our coalgebraic account of context-free languages in terms of grammars, systems of behavioural equations, and context-free expressions can be taken as a starting point for a generalization in at least two different and orthogonal directions. On the one hand, we can consider other languages of expressions for the functor $\mathcal{D}$ to obtain different classes of languages, and on the other hand we can generalize the notion of context-freeness to coalgebras for other functors.

As an interesting example of the first type, one could consider systems of behavioural differential equations for which the right hand side of each equation is an expression as defined by

$$t ::= 0 \,|\, 1 \,|\, x \in X \;|\, t + t \,|\, a \cdot t\,.$$

The semantic solution of such a system is given by regular languages, while the syntactic one is given by a language of expressions as studied in [17]. The corresponding notion of grammars for regular languages is then given by considering productions of the form $p{:}X \to \mathcal{D}\mathcal{P}_\omega(A^* \times X)$, i.e. *right-linear grammars* [6].

Examples of the second type of generalization will depend on the structure of the functor. Here we briefly sketch only an elementary but interesting first example. The functor $\mathcal{S}(X) = \mathbb{N} \times X$ has the set of all *streams* $\mathbb{N}^\omega$ as its final coalgebra. We note that, similar to the set of all languages, also $\mathbb{N}^\omega$ is a semiring, with elementwise addition of streams as sum, and convolution product as product. Streams can be defined by behavioural differential equations, which specify the head $\sigma(0) \in \mathbb{N}$ and tail σ' of a stream $\sigma \in \mathbb{N}^\omega$. Now let us call a stream context-free whenever it can be specified by a stream differential equation $\sigma(0) = n$ and $\sigma' = t$, where $n \in \mathbb{N}$ and t is a term of the following form:

$$t ::= n \,|\, \mathcal{X} \,|\, t + t \,|\, t \times t,$$

and where n is any natural number, $\mathcal{X}$ denotes the constant stream $(0, 1, 0, 0, \ldots)$, $+$ denotes elementwise addition of streams, and $\times$ denotes convolution product. We note that this is a straightforward variation on the approach used for behavioural differential equations for context-free languages, in Sect. 4.

As an example of a context-free stream, let γ be defined by the well-formed differential equation given by $\gamma' = \gamma \times \gamma$ and $\gamma(0) = 1$ It has as its unique solution the stream of the Catalan numbers $(1, 1, 2, 5, 14, \ldots)$, which occurs in numerous counting problems (such as the number of well-bracketed words of a specific length). The stream γ is known not to be rational, so this example nicely illustrates how the class of context-free streams extends the class of rational streams, in the same way as with languages.

Further research directions include a coalgebraic characterization of context-free languages in terms of pushdown automata [6,10], and the study of coinductive decision procedures for bisimilarity of deterministic pushdown automata, a problem that is known to be decidable [18].

Acknowledgements. We would like to thank Alexandra Silva, for valuable suggestions and discussions. Furthermore, we would like to thank the anonymous referees for their corrections and suggestions for improvements.

References

1. Bartels, F.: On Generalized Coinduction and Probabilistic Specification Formats. Ph.D. thesis. Vrije Universiteit Amsterdam (2004)
2. Ésik, Z., Leiß, H.: Algebraically complete semirings and Greibach normal form. Annals of Pure and Applied Logic 133(1-3), 173–203 (2005)
3. Ginsburg, S., Rice, H.G.: Two families of languages related to ALGOL. J. ACM 9, 350–371 (1962)
4. Greibach, S.A.: A new normal-form theorem for context-free, phrase structure grammars. Journal of the Association for Computing Machinery 12, 42–52 (1965)
5. Hasuo, I., Jacobs, B.: Context-free languages via coalgebraic trace semantics. In: Fiadeiro, J.L., Harman, N.A., Roggenbach, M., Rutten, J. (eds.) CALCO 2005. LNCS, vol. 3629, pp. 213–231. Springer, Heidelberg (2005)
6. Hopcroft, J.E., Ullman, J.D.: Formal languages and their relation to automata. Addison-Wesley Longman Publishing Co., Inc., Boston (1969)
7. Jacobs, B.: A bialgebraic review of deterministic automata, regular expressions and languages. In: Futatsugi, K., Jouannaud, J.-P., Bevilacqua, V. (eds.) Algebra, Meaning, and Computation. LNCS, vol. 4060, pp. 375–404. Springer, Heidelberg (2006)
8. Kozen, D.: A completeness theorem for Kleene algebras and the algebra of regular events. Inf. Comput. 110(2), 366–390 (1994)
9. Leiß, H.: Towards Kleene algebra with recursion. In: Kleine Büning, H., Jäger, G., Börger, E., Richter, M.M. (eds.) CSL 1991. LNCS, vol. 626, pp. 242–256. Springer, Heidelberg (1992)
10. Linz, P.: An Introduction to Formal Languages and Automata. Jones and Bartlett (1997)
11. Milius, S.: A sound and complete calculus for finite stream circuits. In: LICS 2010, Edinburgh, United Kingdom, July 11-14, pp. 421–430. IEEE Computer Society, Los Alamitos (2010)
12. Rutten, J.J.M.M.: Automata and coinduction (an exercise in coalgebra). In: Sangiorgi, D., de Simone, R. (eds.) CONCUR 1998. LNCS, vol. 1466, pp. 194–218. Springer, Heidelberg (1998)
13. Rutten, J.J.M.M.: Universal coalgebra: a theory of systems. Theoretical Computer Science 249(1), 3–80 (2000)
14. Rutten, J.J.M.M.: Behavioural differential equations: a coinductive calculus of streams, automata, and power series. Theoretical Computer Science 308(1-3), 1–53 (2003)
15. Silva, A.: Kleene Coalgebra. Ph.D. thesis. Radboud Universiteit Nijmegen (2010)

16. Silva, A., Bonchi, F., Bonsangue, M.M., Rutten, J.J.M.M.: Generalizing the powerset construction, coalgebraically. In: FSTTCS. LIPIcs, vol. 8, pp. 272–283. Schloss Dagstuhl – Leibniz-Zentrum für Informatik (2010)
17. Silva, A., Bonsangue, M.M., Rutten, J.J.M.M.: Non-deterministic Kleene coalgebras. Logical Methods in Computer Science 6(3) (2010)
18. Stirling, C.: Decidability of DPDA equivalence. Theoretical Computer Science 255(1-2), 1–31 (2001)

Preface to CALCO-Tools

The CALCO-Tools Workshop is dedicated to tools based on algebraic and/or coalgebraic principles or that are emerging from the intersection of the two approaches. These include tools developed specifically for design, checking, execution, and verification of (co)algebraic specifications, but also tools targeting different application domains while making core or interesting use of (co)algebraic techniques.

The previous CALCO-Tools editions, Bergen (Norway, 2007) and Udine (Italy, 2009), took place on the day preceding the main CALCO conference. CALCO-Tools 2011 took place on the same dates as the main CALCO conference, with no overlap between the technical programs of the two events. In this way the goal of bringing together researchers and practitioners is achieved better. My hope for the future is that the current tools presentations will inspire theoretical results of the next edition and the current theoretical results will inspire tools of the next edition.

The six papers presented at CALCO-Tools 2011 and presented in this volume were selected by the Program Committee out of nine submissions. These numbers are not spectacular but we must take into account the specificity of the workshop. The ascending evolution of the number of the presentations shows in fact that the success of the algebraic and coalgebraic techniques in practice is increasing. I strongly believe that there are still successful tools which should be presented at future editions of this workshop.

First I would like to thank the authors who submitted at CALCO-Tools 2011; their contributions ensured a high quality of the workshop. Special thanks to all Program Committee members and reviewers for their excellent work; each paper was carefully reviewed and actively discussed in the post-reviewing process. I would also like to thank the main conference chairs and organizers for their professional support.

June 2011 Dorel Lucanu

Program Committee

Paolo Baldan, University of Padova, IT
Luís Barbosa, Universidade do Minho, PT
Dorel Lucanu (chair), Al. I. Cuza University of Iasi, RO
Peter Ölveczky , University of Oslo, NO
Dirk Pattinson, Imperial College London, UK

PREG Axiomatizer – A Ground Bisimilarity Checker for GSOS with Predicates*

Luca Aceto, Georgiana Caltais, Eugen-Ioan Goriac, and Anna Ingolfsdottir

ICE-TCS, School of Computer Science, Reykjavik University, Iceland
{luca,gcaltais10,egoriac10,annai}@ru.is

Abstract. PREG Axiomatizer is a tool used for proving strong bisimilarity between ground terms consisting of operations in the GSOS format extended with predicates. It automatically derives sound and ground-complete axiomatizations using a technique proposed by the authors of this paper. These axiomatizations are provided as input to the Maude system, which, in turn, is used as a reduction engine for provided ground terms. These terms are bisimilar if and only if their normal forms obtained in this fashion are equal. The motivation of this tool is the optimized handling of equivalence checking between complex ground terms within automated provers and checkers.

Keywords: Structural operational semantics, GSOS rule format, bisimilarity, equational axiomatizations, Maude.

1 Introduction

Proving that two process terms are related by some notion of behavioural equivalence is at the heart of the equivalence-checking approach to verification. In this paper we introduce a tool named PREG Axiomatizer[1] that tackles this problem focusing on ground (*i.e.*, fully specified) terms built using operations defined using the *preg* format, a *pr*edicates *e*xtension of the *G*SOS format presented in [3]. GSOS [8] is a restricted, yet powerful, way of defining Structural Operational Semantics (SOS) for programming and specification languages in the style introduced by Plotkin in [14]. We refer the reader to [3] for the detailed description and intuition behind the *preg* rule format and the considered notion of behavioural equivalence, which is a natural extension to predicates of the classic strong bisimulation equivalence.

Building on the techniques in [2,7], we proposed in [3] a procedure to construct a finite collection of sound equations that can be used to bring any ground term to a normal form. We showed that the normal forms of two terms are equal if and

* The authors have been been partially supported by the projects "New Developments in Operational Semantics" (nr. 080039021) and "Meta-theory of Algebraic Process Theories" (nr. 100014021) of the Icelandic Fund for Research. Eugen-Ioan Goriac is supported by the doctoral grant "Extending and Axiomatizing Structural Operational Semantics: Theory and Tools" (nr. 110294-0061) of the Icelandic Fund for Research.

1 The tool is downloadable from `http://goriac.info/tools/preg-axiomatizer/`

A. Corradini, B. Klin, and C. Cîrstea (Eds.): CALCO 2011, LNCS 6859, pp. 378–385, 2011.

only if the terms are bisimilar. Given a set of actions $\mathcal{A}$ and a set of predicates $\mathcal{P}$, the normal forms we refer to are terms built according to the grammar for finite trees with predicates, namely

$$s ::= \delta \mid \kappa_P \ (\forall P \in \mathcal{P}) \mid a.s \ (\forall a \in \mathcal{A}) \mid s + s,$$

that are of the shape $t = \sum_{i \in I} a_i.t_i + \sum_{j \in J} \kappa_{P_j}$. Here the P_j's are all the predicates satisfied by t, and the t_i's are terms in normal form. The empty sum $(I = \emptyset, J = \emptyset)$ is denoted by the constant δ.

Intuitively, δ represents the process exhibiting no behaviour, $s + t$ is the non-deterministic choice between the behaviours of s and t, while $a.t$ is a process that first performs action a and behaves like t afterwards. For each predicate P we consider a constant κ_P, which denotes a process with no transitions. This process only satisfies P. A finite tree satisfies predicate P if and only if it has κ_P as a summand. We refer to predicates in $\mathcal{P}$ as *explicit predicates.* The operational semantics that captures this intuition is given by the rules of BCCSP extended with predicates. The SOS specification for this language consists of rules parameterized over all actions a and explicit predicates P:

$$\frac{}{a.x \xrightarrow{a} x}, \quad \frac{x \xrightarrow{a} x'}{x + y \xrightarrow{a} x'}, \quad \frac{y \xrightarrow{a} y'}{x + y \xrightarrow{a} y'}, \quad \frac{}{P\kappa_P}, \quad \frac{Px}{P(x+y)}, \quad \frac{Py}{P(x+y)}.$$

In [3] we showed that, for the above language, the following set of axioms [12] is sound and ground-complete for bisimilarity on the set of ground finite trees with predicates:

$$\begin{array}{ll} x + y = y + x & (x + y) + z = x + (y + z) \\ x + x = x & x + \delta = x \end{array}$$

Recall that our purpose is to find ground-complete axiomatizations like the one above for all the languages given in the *preg* format. In order to achieve this goal for operators whose rules involve negative premises, we use the *initial restriction operator* $\partial^1_{\mathcal{B},\mathcal{Q}}$ (where $\mathcal{B} \subseteq \mathcal{A}$ and $\mathcal{Q} \subseteq \mathcal{P}$ are the sets of initially forbidden actions and predicates, respectively). The semantics of $\partial^1_{\mathcal{B},\mathcal{Q}}$ is given by the following two types of transition rules:

$$\frac{x \xrightarrow{a} x'}{\partial^1_{\mathcal{B},\mathcal{Q}}(x) \xrightarrow{a} \partial^1_{\emptyset,\mathcal{Q}}(x')} \ \text{if } a \notin \mathcal{B}, \quad \frac{Px}{P(\partial^1_{\mathcal{B},\mathcal{Q}}(x))} \ \text{if } P \notin \mathcal{Q}.$$

The axiomatization of the operators $\partial^1_{\mathcal{B},\mathcal{Q}}$ is provided in [3].

Internally, PREG Axiomatizer brings the provided rule system to a "manageable" format, introducing auxiliary operators as described in [3], and afterwards performs the axiomatization itself. The tool is implemented in the Maude language [11], which has been already proven to be very useful for analyzing SOS rule formats in [13,10]. Not only did we use Maude as a programming language, but also as an equational reduction system for the generated sets of axioms.

PREG Axiomatizer is, to our knowledge, the first public tool that automatically derives sound and ground-complete axiomatizations modulo bisimilarity for GSOS-like languages. Prior to using the techniques presented in [2,3,7], one had to use ingenuity and dedicate a considerable amount of time in order to obtain axiomatizations for a language with even a limited number of operators.

The tool is generic, in the sense that the SOS specification defining the labelled transition system semantics of the process calculus is provided by the user. He or she does this in terms of *well-founded* GSOS systems, which means that these systems can be used to derive only finite labelled transition systems for the given terms (see [2] for more details). As presented in [9], the generated axiomatizations are guaranteed to be confluent, but, as a downside of our approach, only weakly normalizing. This downside is diminished by the fact that there exists a substantial decidable subclass of systems, namely the linear and syntactically well-founded ones [9], for which the generated axiomatizations are strongly normalizing. This subclass includes important languages such as CCS, CSP, and ACP.

2 Case Studies

In this section we present two scenarios involving several classic operations with their semantics extended with certain explicit predicates. Conventionally, the tool language accepts process term variables such as `X`, `X1`, `Y'`, actions like `a`, `b`, `c`, `a[0]`, `b[2]`, `c["name"]`, and predicates like `P`, `Q`, `P[1]`, `Q["prop"]`.

Example 1. Let us describe how PREG Axiomatizer is used in order to prove that "$a.(a.\kappa_\downarrow;\ b.(a.\kappa_\downarrow))$" and "$\texttt{while}\ a.b.\kappa_\downarrow\ \texttt{do}\ a.\kappa_\downarrow$" are bisimilar. Here _;_ and `while_do_` are, respectively, the sequential composition and the process loop operators (presented in [8]) extended to the *preg* format with the immediate successful termination predicate $\downarrow$ (which we choose to denote by `P` in the specification for tool consumption). In Figure 1 we present the operational semantics for these operations with the rules given both in standard notation, as well as using the syntax supported by the tool.

$$\frac{x \xrightarrow{a} x'}{x;y \xrightarrow{a} x';y} : \quad \begin{array}{c}\texttt{X -(a)-> X'}\\ \texttt{===}\\ \texttt{X ; Y -(a)-> (X' ; Y)}\end{array} \qquad \frac{x\downarrow \quad y \xrightarrow{a} y'}{x;y \xrightarrow{a} y'} : \quad \begin{array}{c}\texttt{P(X) , Y -(a)-> Y'}\\ \texttt{===}\\ \texttt{X ; Y -(a)-> Y'}\end{array}$$

$$\frac{x\downarrow \quad y\downarrow}{(x;y)\downarrow} : \quad \begin{array}{c}\texttt{P(X) , P(Y)}\\ \texttt{===}\\ \texttt{P(X ; Y)}\end{array} \qquad \frac{x\downarrow}{(\texttt{while}\ x\ \texttt{do}\ y)\downarrow} : \quad \begin{array}{c}\texttt{P(X)}\\ \texttt{===}\\ \texttt{P(while X do Y)}\end{array}$$

$$\frac{x \xrightarrow{a} x'}{\texttt{while}\ x\ \texttt{do}\ y \xrightarrow{a} y; \texttt{while}\ x'\ \texttt{do}\ y} : \quad \begin{array}{c}\texttt{X -(a)-> X'}\\ \texttt{===}\\ \texttt{while X do Y -(a)-> Y ; while X' do Y}\end{array}$$

Fig. 1. *preg* rule system for _;_ and `while_do_`

The rules involving action a are also instantiated for b. After providing this specification, the user can press the button labelled "Axiomatize" and the tool generates a Maude specification including the axioms obtained by following the procedure described in [3]. We exemplify a small part of the output which consists of the axiomatization for the `while_do_` operator:

```
 eq while X0 + X1 do X3 = while X0 do X3 + while X1 do X3 .
ceq while X do Y = a . (Y ; while X' do Y) if a . X' := X .
ceq while X do Y = b . (Y ; while X' do Y) if b . X' := X .
ceq while X do Y = k[P] if X := k[P] .
 eq while X1 do X2 = delta [owise] .
```

In order to check for the bisimilarity of the two process terms introduced at the beginning of the current example, one loads the generated specification and uses the Maude command `reduce`:

```
> reduce a . (a . k[P] ; b . (a . k[P])) ==
        while a . b . k[P] do a . k[P] .
result Bool: true

> reduce while a . b . k[P] do a . k[P] .
result PTerm: a . a . a . b . a . a . k[P]
```

We successfully used PREG Axiomatizer to further extend the operational semantics of $\_;\_$ with the predictable non-failure predicate $\neq \delta$ (which plays the role of the predicate "$\neq 0$" presented in [4]) with the rules: $\frac{x \xrightarrow{a} x' \quad y \neq \delta}{x;y \xrightarrow{a} x';y}$ $\frac{x \neq \delta \quad y \neq \delta}{(x;y) \neq \delta}$. We managed to test the property that $x;\delta$ and δ are bisimilar on various closed instantiations. It is worth noting that this property does not always hold for the initial version of $\_;\_$.

Example 2. In this example we show how we use our tool to obtain the execution tree of a network of communicating processes. This procedure is useful, for instance, when one needs to use an external model checker to verify if the communication protocol satisfies certain logical properties. Our example is based on a case study from [5].

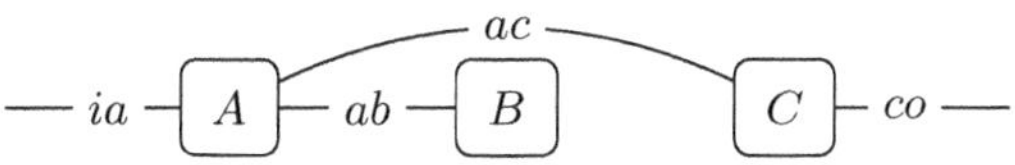

Fig. 2. Communication protocol

Consider the process network given in Figure 2 where A, B, C are the communicating processes and ia, ab, ac, co are the ports. The actions of sending, receiving, and synchronizing on the datum d over the port p are denoted by, respectively, $p!d$, $p?d$, and $p\#d$. By using these actions, the parallel composition operator $\_\|\_$, and the immediate successful termination predicate $\downarrow$, we specify the whole protocol as the term:

$$T = ia?d \, . \, (ab!d \, . \, \kappa_{\downarrow} \parallel ac!d \, . \, \kappa_{\downarrow}) \parallel ab?d \, . \, \kappa_{\downarrow} \parallel ac?d \, . \, co!d \, . \, \kappa_{\downarrow} \; .$$

We present *preg* rules for $\_\|\_$, in which $act \in \{p!d, p?d, p\#d\}$:

$$\frac{x \xrightarrow{act} x'}{x \parallel y \xrightarrow{act} x' \parallel y} \qquad \frac{y \xrightarrow{act} y'}{x \parallel y \xrightarrow{act} x \parallel y'} \qquad \frac{x \downarrow \; y \downarrow}{(x \parallel y) \downarrow}$$

$$\frac{x \xrightarrow{p!d} x' \quad y \xrightarrow{p?d} y'}{x \parallel y \xrightarrow{p\#d} x' \parallel y'} \qquad \frac{x \xrightarrow{p?d} x' \quad y \xrightarrow{p!d} y'}{x \parallel y \xrightarrow{p\#d} x' \parallel y'}$$

Fig. 3. *preg* rule system for $\_\|\_$

The input for PREG Axiomatizer consists of:

– the predicate rule

```
P(X) , P(Y)
    ===
 P(X || Y)
```

,

– all the instantiations of the first two transition rules in Figure 1 for which *act* is an action from the set $\mathcal{A} = \{ia?d,\ ab!d,\ ac!d,\ ab?d,\ ac?d,\ ab\#d,\ ac\#d,\ co!d\}$

(*e.g.*,

```
      X -(a["ia?d"])-> X'
             ===
X || Y -(a["ia?d"])-> X' || Y
```

), and

– all the instantiations of the last two transition rules in Figure 1 in which p is a port from $\{ab, ac\}$ (*e.g.*,

```
X -(a["ab?d"])-> X' , Y -(a["ab!d"])-> Y'
                   ===
      X || Y -(a["ab#d"])-> X' || Y'
```

).

We generate the process network execution tree (consisting of 582 states) by calling the command `reduce` on the specification term T:

```
> reduce ((a["ia?d"] . (a["ab!d"] . k[P] || a["ac!d"] . k[P])) ||
          a["ab?d"] . k[P]) || a["ac?d"] . a["co!d"] . k[P] .
result PTerm: a["ab?d"] . (...) + a["ac?d"] . (...) + a["ia?d"] . (...)
```

The parallel composition allows for arbitrary interleavings of the actions in $\mathcal{A}$, but it does not enforce the communication over the ports ab and ac. Hiding these ports so that other processes cannot interfere with the internal communications is desirable. This can be done with the help of the *restriction operator* $\partial_{\mathcal{B},\mathcal{Q}}$, a generalization of the initial restriction operator $\partial^1_{\mathcal{B},\mathcal{Q}}$ presented in Section 1, that preserves the imposed restrictions throughout the whole computation, not only for the first step. Forbidding independent send and receive actions over the ports ab and ac is denoted by the term $\partial_{\{p!d,p?d\,|\,p\in\{ab,ac\}\},\emptyset}(T)$. In PREG Axiomatizer we use `%%[`$\mathcal{B}$`;`$\mathcal{Q}$`]` as a syntactic notation for $\partial_{\mathcal{B},\mathcal{Q}}$:

```
> reduce %%[ a["ab?d"] a["ab!d"] a["ac?d"] a["ac!d"] ; empty ](
        (( a["ia?d"] . (a["ab!d"] . k[P] || a["ac!d"] . k[P])) ||
```

```
         a["ab?d"] . k[P]) || a["ac?d"] . a["co!d"] . k[P]) .
result PTerm: a["ia?d"] . (a["ab#d"] .  a["ac#d"] . a["co!d"] . k[P] +
                        a["ac#d"] . (a["ab#d"] . a["co!d"] . k[P] +
                                    a["co!d"] . a["ab#d"] . k[P]))
```

We also tested our tool by generating the normal form of $a^3.\kappa_\downarrow \parallel b^3.\kappa_\downarrow \parallel c^3.\kappa_\downarrow$ and obtained the same "2 page long" execution tree showed in [6], consisting of 6927 states. Maude derives this execution tree in less than 500 milliseconds on a machine with a 2.53GHz processor and 4GB of RAM.

Example 3. We now show how PREG Axiomatizer is used in order to perform equational proofs when working with predicates that have implicit behaviour. Consider, for instance, the case of the *eventual successful termination predicate* $\lightning$. It represents the extension of $\downarrow$, introduced in the previous examples, with the requirement that if $t \lightning$ holds for a term t, then $a.t \lightning$ holds for any action a.

Recall from Section 1 that our approach is based on denoting the property $\lightning$ by using the explicit process constant $\kappa_\lightning$ as a summand of the analyzed term. The above characterization of $\lightning$ is given by the axiom $a.(t+\kappa_\lightning) = a.(t+\kappa_\lightning)+\kappa_\lightning$. With this in mind, one could check, for instance, if a process t "eventually terminates" by checking if it is bisimilar to $t + \kappa_\lightning$.

In order to prove that $a.\kappa_\lightning \parallel b.\kappa_\lightning$ is bisimilar to $(a.\kappa_\lightning \parallel b.\kappa_\lightning) + \kappa_\lightning$ we need to let the tool "know" that it should treat $\lightning$ (denoted by Q in the specification) as an implicit predicate by using the operation **expandImplicit**. This operation receives a term and the set of implicit predicate names:

```
> reduce expandImplicit(a . k[Q] || b . k[Q], Q) ==
         expandImplicit(a . k[Q] || b . k[Q] + k[Q], Q) .
result Bool: True

> reduce expandImplicit(a . k[Q] || b . k[Q], Q) .
result PTerm: k[Q] + a . (k[Q] + b . k[Q]) + b . (k[Q] + a . k[Q])
```

Predicates with implicit behaviour, like $\lightning$, can only be used during the normalization process if the operators whose definition involves these predicates are given by rules that satisfy certain sanity constraints mentioned in [3]. The tool does not currently support the automated checking for those constraints, so the user needs to do it manually before using the feature presented above. The parallel composition operator does meet those constraints.

3 Discussion and Future Work

Aside from the features mentioned in Section 2, an important part of the PREG Axiomatizer engine is dedicated to checking for the conformance of specified operations and rules to the various formats presented in [3].

There are many areas in which the tool and the theory behind it can be improved. First and foremost, an important feature would be to allow the user to specify guarded recursively defined terms in order to greatly increase the complexity of the case studies our tool can handle. The most natural way to extend our approach in order to reason about the bisimilarity of such terms is to integrate the technique presented in [1], which is also based on generating complete axiomatizations for a class of GSOS languages generating regular behaviours. The main difficulty of this task will be the search for good strategies for applying the axioms and the unique fixed-point induction rule.

Another tool development direction is concerned with the ability to automatically check if the specification meets certain complex requirements. One of these requirements is, as presented in Section 1, the syntactic well-foundedness of the given system. Without this feature the user needs to be careful not to specify operators such as the *reentrant server* $!\_$, defined by the rule $\frac{x \xrightarrow{a} x'}{!x \xrightarrow{a} x' \,||\, !x}$, for which non-normalizing axioms are derived: $!x = !'(x, x)$, $!'(a.x', x) = a.(x' \,||\, !x)$. Another requirement the tool could check for consists of the sanity constraints we mentioned in Example 3.

References

1. Aceto, L.: Deriving complete inference systems for a class of GSOS languages generation regular behaviours. In: Jonsson, B., Parrow, J. (eds.) CONCUR 1994. LNCS, vol. 836, pp. 449–464. Springer, Heidelberg (1994)
2. Aceto, L., Bloom, B., Vaandrager, F.: Turning SOS rules into equations. Inf. Comput. 111, 1–52 (1994), `http://portal.acm.org/citation.cfm?id=184662.184663`
3. Aceto, L., Caltais, G., Goriac, E.I., Ingólfsdóttir, A.: Axiomatizing GSOS with predicates (2011), `http://ru.is/faculty/luca/PAPERS/extending_GSOS_with_predicates.pdf`
4. Aceto, L., Cimini, M., Ingólfsdóttir, A., Mousavi, M., Reniers, M.A.: SOS rule formats for zero and unit elements. Theoretical Computer Science (to appear, 2011)
5. Baeten, J.C.M., Basten, T., Reniers, M.A.: Process Algebra: Equational Theories of Communicating Processes. Cambridge Tracts in Theoretical Computer Science, vol. 50. Cambridge University Press, Cambridge (2010)
6. Baeten, J.C.M., Weijland, W.P.: Process Algebra. Cambridge University Press, New York (1990)
7. Baeten, J.C.M., de Vink, E.P.: Axiomatizing GSOS with termination. J. Log. Algebr. Program. 60-61, 323–351 (2004)
8. Bloom, B., Istrail, S., Meyer, A.R.: Bisimulation can't be traced. J. ACM 42, 232–268 (1995), `http://doi.acm.org/10.1145/200836.200876`
9. Bosscher, D.J.B.: Term rewriting properties of SOS axiomatisations. In: Hagiya, M., Mitchell, J.C. (eds.) TACS 1994. LNCS, vol. 789, pp. 425–439. Springer, Heidelberg (1994), `http://portal.acm.org/citation.cfm?id=645868.668513`
10. Chalub, F., Braga, C.: Maude MSOS Tool. Electron. Notes Theor. Comput. Sci. 176, 133–146 (2007), `http://portal.acm.org/citation.cfm?id=1279349.1279455`
11. Clavel, M., Durán, F., Eker, S., Lincoln, P., Martí-Oliet, N., Meseguer, J., Talcott, C.L. (eds.): All About Maude - A High-Performance Logical Framework, How to Specify, Program and Verify Systems in Rewriting Logic. LNCS, vol. 4350. Springer, Heidelberg (2007)

12. Milner, R.: Communication and Concurrency. Prentice-Hall International, Englewood Cliffs (1989)
13. Mousavi, M.R., Reniers, M.A.: Prototyping SOS meta-theory in Maude. Electron. Notes Theor. Comput. Sci. 156, 135–150 (2006), `http://dx.doi.org/10.1016/j.entcs.2005.09.030`
14. Plotkin, G.D.: A structural approach to operational semantics. J. Log. Algebr. Program. 60-61, 17–139 (2004)

PVeStA: A Parallel Statistical Model Checking and Quantitative Analysis Tool

Musab AlTurki and José Meseguer

University of Illinois at Urbana-Champaign,
Urbana, IL 61801, USA

Abstract. Statistical model checking is an attractive formal analysis method for probabilistic systems such as, for example, cyber-physical systems which are often probabilistic in nature. This paper is about drastically increasing the scalability of statistical model checking, and making such scalability of analysis available to tools like Maude, where probabilistic systems can be specified at a high level as probabilistic rewrite theories. It presents PVeStA, an extension and parallelization of the VeStA statistical model checking tool [10]. PVeStA supports statistical model checking of probabilistic real-time systems specified as either: (i) discrete or continuous Markov Chains; or (ii) probabilistic rewrite theories in Maude. Furthermore, the properties that it can model check can be expressed in either: (i) *PCTL/CSL*, or (ii) the QuaTEx quantitative temporal logic. As our experiments show, the performance gains obtained from parallelization can be very high.

1 Introduction

Statistical model checking (see, e.g., [9,11]) is an attractive formal analysis method for probabilisitic systems. Although the properties model checked can only be ensured up to a user-specified level of statistical confidence (as opposed to the *absolute* guarantees provided by standard probabilistic model checkers), the approximate nature of the formal analysis is compensated for by its better scalability, the fact that the models to be analyzed can often be known only approximately, and the interest in analyzing quantitative properties for which an approximate result within known bounds is quite acceptable.

There are many systems for which this kind of statistical model checking analysis can be very useful. For example, distributed real-time systems, including so-called cyber-physical systems, are often probabilistic in nature, both because they often use probabilistic algorithms, and due to the uncertain, stochastic nature of the environments with which they interact. Furthermore, *quality of service properties* may be as important as traditional boolean-valued properties such as safety properties. For example, in a secure communications system, *availability* of vital information may be as important as its *secrecy*, but availability may be utterly lost due to a denial of service (DoS) attack with no loss of secrecy. Suppose that such a system is hardened against DoS attacks. How should one formally analyze the effectiveness of such a hardening? What is needed is

A. Corradini, B. Klin, and C. Cîrstea (Eds.): CALCO 2011, LNCS 6859, pp. 386–392, 2011.

not a Boolean-valued yes/no answer, but a *quantitative* one in terms of the expected *latency* of messages under certain assumptions about the attacker and the network. Quantitative information may include probabilities $p \in [0,1]$, but need not be reducible to probabilities. For this reason, it is important to support statistical model checking not only of standard probabilistic temporal logics such as *PCTL*/*CSL*, but also of *quantitative temporal logics* like QuaTEx [1], where the result of evaluating a temporal formula on a path is a real number. This of course includes the case of probabilities, as values $p \in [0,1]$, and even of standard truth values, as values in $\{0,1\}$, as special cases.

This paper is about drastically increasing the scalability of statistical model checking, and also about making such scalability of analysis available to tools like Maude, where probabilistic systems can be specified at a high level as *probabilistic rewrite theories* [1], which are theories in rewriting logic [7] that may contain, in addition to regular rewrite rules, *probabilistic rewrite rules* modeling probabilistic transitions of such systems. The paper presents PVeStA, an extension and parallelization of the VeStA statistical model checking tool [10]. PVeStA supports statistical model checking of probabilistic real-time systems specified as either: (i) discrete or continuous Markov Chains; or (ii) probabilistic rewrite theories in Maude. Furthermore, the properties that it can model check can be expressed in either: (i) *PCTL*/*CSL*, or (ii) QuaTEx. As our experiments show, since statistical model checking is based on Monte-Carlo simulations, which are naturally parallelizable, the performance gains can be very high. In summary, the main contribution of this work is to parallelize the model checking algorithms and the VeStA tool itself, so that a wide range of statistical model checking analyses can be performed with high efficiency on probabilistic models such as Markov chains and probabilistic rewrite theories.

2 Efficient Parallel Statistical Analysis Algorithms

Sen et. al. [9] described an algorithm $\mathcal{A}$ based on simple hypothesis testing for statistical model checking of formulas in both: (1) *Probabilistic CTL* (*PCTL*) [5], which extends standard *CTL* by associating probability measures to computation paths of a probabilistic system and qualifying the temporal logic formulas with probability bounds, and (2) *Continuous Stochastic Logic* (*CSL*) [3,4], which further extends *PCTL* by continuous timing and qualifying temporal logic operators by time bounds. Given a probabilistic model $\mathcal{M}$, a *PCTL*/*CSL* formula $\mathcal{P}_{\bowtie p}(\varphi)$, with φ a state or path formula[1], and error bounds α and β, the algorithm $\mathcal{A}$ checks satisfiability of the formula by setting up a statistical hypothesis testing experiment such that its Type I and Type II errors are bounded, respectively, by α and β. The test is based on the sample mean of n random samples of φ computed over n Monte-Carlo simulations of the model. The algorithm uses standard statistical methods to precompute the total number n of samples needed to achieve the desired test strength (see [9] for more details).

1 We restrict our attention to non-nested probabilistic formulas here, although the algorithm of [9] can handle nested formulas as well.

To be able to express not just probabilities of satisfaction of temporal logic formulas but also *quantitative properties* such as, for example, the expected latency of a probabilistic communication protocol, *PCTL* and *CSL* have been generalized to a logic of *Quantitative Temporal Expressions* (QuaTEx) in [1], in which state formulas and path formulas are generalized to user-definable, *real-valued state expressions* and *path expressions*. In [1], Agha et. al. proposed a statistical quantitative analysis algorithm $\mathcal{Q}$ for estimating the expectation of a temporal expression in QuaTEx. Given a probabilistic model $\mathcal{M}$, an expectation QuaTEx formula of the form $\mathbf{E}[Exp]$, with *Exp* a QuaTEx state or path expression, and bounds α and δ, the algorithm $\mathcal{Q}$ approximates the value of $\mathbf{E}[Exp]$ within a $(1-\alpha)100\%$ confidence interval, with size at most δ, by generating a large enough number n of random sample values $x_1, x_2, \ldots, x_n$ of *Exp* computed from n independent Monte Carlo simulations of $\mathcal{M}$. The value returned by the algorithm as the estimator for $\mathbf{E}[Exp]$ is the sample mean $\bar{x} = \frac{\Sigma_{i\in[1,n]} x_i}{n}$. To guarantee the quality and size requirements of the confidence interval (given respectively by α and δ) for $\bar{x}$, the number n of sample values must be large enough. In general, the more accurate the estimator, the larger the number of samples required. To generate enough samples, the algorithm $\mathcal{Q}$ uses student's t-distribution to compute a $(1-\alpha)100\%$ confidence interval by iteratively generating them in batches of N samples each (with $N > 5$). Once the size of the computed interval falls below the threshold δ, $\mathcal{Q}$ halts and the sample mean $\bar{x}$ is returned (more details can be found in [1]).

In this work, we develop parallel versions $\mathcal{A}^p$ and $\mathcal{Q}^p$ of both algorithms in which the task of computing a set of n sample values for a state or path formula in *CSL* or QuaTEx is done in parallel by performing n Monte Carlo simulations in parallel. Both parallel algorithms make no assumptions about the underlying parallel architecture. For *CSL*, $\mathcal{A}^p$ assumes non-nested probabilistic formulas.

The algorithms take as input a list of available computing resources R on which the task of generating random samples is mapped. This task is first distributed as evenly as possible by determining the number of simulations m_i to be performed by each available computing resource R_i in R. In $\mathcal{A}^p$, since the total number of samples n is precomputed, m_i is simply either $\lfloor n/|R| \rfloor$ or $\lfloor n/|R| \rfloor + 1$. In $\mathcal{Q}^p$, m_i is computed as a positive integer multiple of $|R|$ and the *load factor* k, which is a parameter to $\mathcal{Q}^p$ that can be used to increase the number of simulations performed by each resource in a round. Given a verification task, the load factor k can be tuned to optimize performance, especially for lightweight simulations when the desired statistical confidence is high, as we will see in Section 4. Once m_i is determined, each resource R_i performs m_i discrete-event simulations of the model $\mathcal{M}$ and returns a list of m_i random samples. Once the samples from all resources are collected, $\mathcal{A}^p$ and $\mathcal{Q}^p$ proceed as their sequential counterparts.

3 Implementation of PVeStA

We have implemented a client-server prototype, PVeStA, of both parallel algorithms $\mathcal{A}^p$ and $\mathcal{Q}^p$, in Java, based on the Java implementation of the original algorithms in VeStA [10]. The tool, which is available for download online at `http://www.cs.illinois.edu/~alturki/pvesta`, consists of two command-line-based executable programs: (1) a client program `pvesta-client`, which implements the sequential parts of the algorithms performing simple hypothesis testing for *PCTL*/*CSL* formulas and confidence interval computations for QuaTEx expressions, and (2) a server program `pvesta-server`, which implements the role of a resource R_i that computes random samples by performing discrete-event simulations of a given model expressed as a Markov chain or as a probabilistic rewrite theory. Figure 1 presents a schematic diagram of the structure and interactions of the client and server parts of the tool.

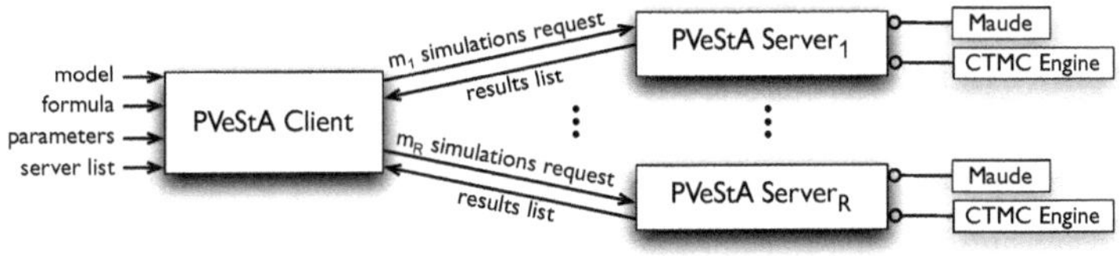

Fig. 1. Components and interactions of PVeStA

The client program first reads a list of servers R that are available for performing simulations. It then creates, using Java's managed concurrency library, a thread pool of $|R|$ *callable* computation threads, which are Java threads that implement the `Callable` interface by specifying a `run` method to be called when the thread is invoked. Each thread, which will manage simulation requests and responses with a particular server in R, is supplied with a pseudo-random seed to be used by its corresponding server to guarantee statistical independence of the simulations. The thread pool is then submitted to an *executor* object, which invokes all the threads in the pool, commencing communication with the servers in R. Upon receiving the simulations request, each PVeStA server performs the requested number of simulations using either Maude (for models expressed as probabilistic rewrite theories) or the built-in Continuous-Time Markov Chain (CTMC) engine (for CTMC models) and produces a list of sample results. The client collects all samples in an array of $|R|$ `Future` Java class objects, from which the results are extracted and then used in performing the appropriate sequential computations. For confidence interval computations, the client may need to repeat this process until enough samples are collected.

4 Experimental Evaluation

We have conducted two sets of experiments with PVeStA to evaluate the performance gains of parallelization using two different parallel architectures: (1) a high-performance computing (HPC) architecture, in which simulation tasks are distributed over different nodes in a PC cluster, and (2) a multi-core architecture, in which simulations are distributed over different processing cores within a single node. The HPC benchmarks were executed on a PC cluster consisting of 256 nodes, each of which has two (single-core) AMD Opteron 2.2GHz CPUs with 2GB of RAM. The second set of experiments was performed on a server machine having two quad-core 2.66GHz Intel Xeon processors with 16GB of RAM.

We use two examples from [10]: (1) a simplified server polling system, *Polling*, and (2) a simple tandem queuing system, *Tandem*, both expressed as continuous-time Markov chains. In addition, we use two variants of a larger case study from [2], which provides a probabilistic model in rewriting logic of the Adaptive Selective Verification (ASV) protocol [6] for thwarting DoS attacks. The first variant, denoted ASV_0, assumes a reliable communication channel, a fixed attack rate, and no message transmission delays, while the second variant, ASV_1, is slightly more realistic as it assumes a lossy channel, a variable attack rate, and random delays. Benchmarking is performed by measuring the total time required (including any additional time required for file and network I/O, thread and object management, and so on) to verify a probabilistic *CSL* formula in *Polling*, *Tandem*, and ASV_0, or a QuaTEx expectation expression in *Tandem* (load factor, $k = 100$), ASV_0 ($k = 1$), and ASV_1 ($k = 1$). The results are summarized in Table 1.

Table 1. The (average) times in seconds taken by PVeStA to complete six verification tasks using a PC cluster and a multi-core computer

	Polling (CSL)	Tandem (CSL)	Tandem (Q)	ASV_0 (CSL)	ASV_0 (Q)	ASV_1 (Q)
Simulations	16,906	16,906	46,380	1051	706	1,308
Servers			HPC Cluster			
1	6.78	9.54	17.36	494.9	770.8	1,584.3
2	2.61	4.06	8.56	248.4	385.4	798.5
4	1.24	2.01	4.26	124.2	197.1	410.5
8	0.70	1.02	2.19	62.1	103.4	221.9
12	0.59	0.77	1.53	41.4	65.3	144.3
16	0.44	0.63	1.27	31.1	52.3	116.6
20	0.42	0.56	1.14	25.1	39.4	89.9
30	0.37	0.46	0.93	16.9	26.7	63.1
60	0.38	0.43	0.82	8.7	13.7	34.2
Servers			Multi-core Computer			
1	3.83	5.53	11.26	367.7	559.7	1,167.9
2	1.70	2.60	5.43	184.5	281.1	589.5
3	1.15	1.62	3.36	122.9	189.4	396.5
4	0.86	1.24	2.53	92.3	138.7	298.3
5	0.74	1.03	2.09	74.2	113.1	243.0
6	0.66	0.86	1.84	61.8	94.5	204.5
7	0.62	0.78	1.66	53.1	85.1	181.2

As the table clearly shows, performance gains as a result of parallelization can be substantial. For example, for ASV_1, a verification task that would normally require about 27 minutes, can be completed in about 34 seconds on an HPC cluster using 60 nodes, and a 20-minute task can be done in just above 3 minutes on a multi-core machine using seven cores in parallel. In practice, several factors influence the speedups achieved by PVeStA, including the complexity of the model and the formula, and the statistical strength of the result. Figure 2(a) plots the speedups achieved against the number of servers used for HPC experiments in Table 1. We note that while performance scales almost linearly with the number of servers used for ASV_0 and ASV_1, the speedups for both *Polling* and *Tandem* begin to decelerate beyond 20 servers. This is primarily because the models *Polling* and *Tandem* are so simple that, as the number of servers increases, the time needed to generate random samples begins to be dominated by other computations in the tool. For the *Tandem*-Q experiment, which requires a fairly high statistical confidence, and thus a higher number of random samples, achievable speedups are greatly influenced by the chosen load factor k. In general, for such simple models, a higher value of k (and thus a higher number of simulations performed by a server in each round) translates into reduced processing and communication overhead and increased efficiency. For example, speedup tripled when using $k = 100$ compared with $k = 1$ for *Tandem*-Q with 60 servers. Of course, excessively high values of k result in an unnecessarily excessive number of simulations and degrade performance. Appropriate values of k can be determined by experimentation using the above ideas as guidelines. Figure 2(b), which plots speedups on a multi-core architecture, shows a similar pattern to Figure 2(a).

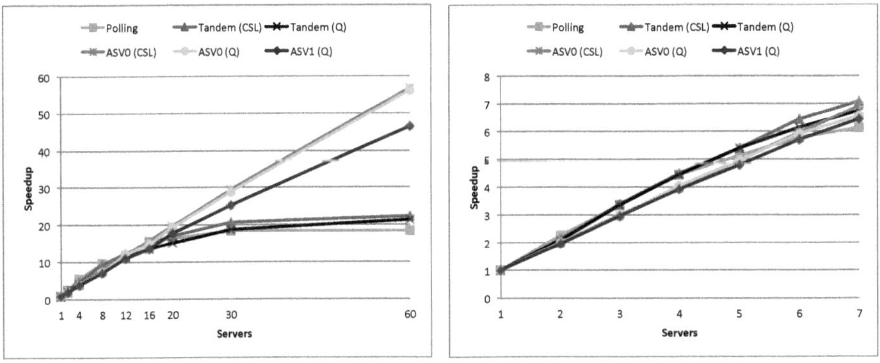

(a) HPC cluster (b) Multi-core computer

Fig. 2. The speedup using multiple PVESTA servers

5 Conclusion and Future Work

We have briefly presented parallelized algorithms for statistical model checking and quantitative analysis and described their implementations in a client-server extension of VESTA, called PVESTA. Experimental evaluation on two different parallel architectures demonstrated the expected performance gains, especially for complex probabilistic models, for which the process of computing random samples can be computationally expensive. Future work include further refining the tool to improve performance for simpler models when high parallelization factors are assumed, and extending the parallel algorithms to support parallel statistical model checking of nested *PCTL*/*CSL* formulas using ideas from [9,8].

Acknowledgments. We thank Koushik Sen, Mahesh Viswanathan, and Gul Agha for the original work on VESTA, and Gul Agha and Koushik Sen for their work on PMaude and QUATEX, which provide the basis on which PVESTA has been built. This work has been partially supported by NSF grants CNS 08-34709, CNS 07-16638, and CCF 09-05584, King Fahd University of Petroleum and Minerals, and King Abdullah University of Science and Technology.

References

1. Agha, G., Meseguer, J., Sen, K.: PMaude: Rewrite-based specification language for probabilistic object systems. Electronic Notes in Theoretical Computer Science 153(2), 213–239 (2006)
2. AlTurki, M., Meseguer, J., Gunter, C.A.: Probabilistic modeling and analysis of DoS protection for the ASV protocol. Electron. Notes Theor. Comput. Sci. 234, 3–18 (2009)
3. Aziz, A., Singhal, V., Brayton, R.K., Sangiovanni-Vincentelli, A.L.: It usually works: The temporal logic of stochastic systems. In: Wolper, P. (ed.) 7th International Conference On Computer Aided Verification, vol. 939, pp. 155–165. Springer, Liege (1995)
4. Baier, C., Katoen, J.-P., Hermanns, H.: Approximate symbolic model checking of continuous-time markov chains. In: Baeten, J., Mauw, S. (eds.) CONCUR 1999. LNCS, vol. 1664, pp. 146–781. Springer, Heidelberg (1999)
5. Hansson, H., Jonsson, B.: A logic for reasoning about time and reliability. Formal Aspects of Computing 6(5), 512–535 (1994)
6. Khanna, S., Venkatesh, S.S., Fatemieh, O., Khan, F., Gunter, C.A.: Adaptive selective verification. In: IEEE Conference on Computer Communications (INFOCOM 2008). IEEE, Phoenix (2008)
7. Meseguer, J.: Conditional rewriting logic as a unified model of concurrency. Theor. Comput. Sci. 96(1), 73–155 (1992)
8. Sen, K., Viswanathan, M., Agha, G.: Statistical model checking of black-box probabilistic systems. In: Alur, R., Peled, D.A. (eds.) CAV 2004. LNCS, vol. 3114, pp. 202–215. Springer, Heidelberg (2004)
9. Sen, K., Viswanathan, M., Agha, G.: On statistical model checking of stochastic systems. In: Etessami, K., Rajamani, S.K. (eds.) CAV 2005. LNCS, vol. 3576, pp. 266–280. Springer, Heidelberg (2005)
10. Sen, K., Viswanathan, M., Agha, G.A.: VESTA: A statistical model-checker and analyzer for probabilistic systems. In: Second International Conference on the Quantitative Evaluation of Systems (QEST), pp. 251–252 (2005)
11. Younes, H.L.S., Simmons, R.G.: Statistical probabilistic model checking with a focus on time-bounded properties. Inf. Comput. 204(9), 1368–1409 (2006)

Minlog - A Tool for Program Extraction Supporting Algebras and Coalgebras

Ulrich Berger[1], Kenji Miyamoto[2,⋆], Helmut Schwichtenberg[2], and Monika Seisenberger[1]

[1] Swansea University, Wales
[2] LMU University, Munich

Abstract. Minlog is an interactive system which implements proof-theoretic methods and applies them to verification and program extraction. We give an overview of Minlog and demonstrate how it can be used to exploit the computational content in (co)algebraic proofs and to develop correct and efficient programs. We illustrate this by means of two examples: one about parsing, the other about exact real numbers in signed digit representation.

1 Introduction

In this paper we give an overview of the interactive proof system Minlog and describe a proof-theoretic method based on realizability for developing correct programs. We particularly address inductive and coinductive proofs and the associated computation principles: iteration for initial algebras and coiteration for terminal coalgebras. Minlog is not a type-theoretic system, such as Coq or Isabelle, but based on first-order logic and has a simple mathematical (i.e. denotational) semantics. This makes it accessible to a wide range of researchers including those outside the type-theoretically minded community. Minlog is implemented in Scheme. It is an "open" system giving users full access to the code, thus inviting them to contribute to its development. Although designed as a general purpose system, most of the recent developments in Minlog are concerned with program extraction from proofs. It seems fair to say that, regarding program extraction, Minlog is the most advanced proof system. Minlog implements various methods of program extraction (realizability, Dialectica Interpretation) and extends them to classical proofs via the Friedman/Dragalin A-translation. All these techniques are refined and optimized in order improve usability and to obtain simpler programs. In addition to extracting a program from a proof, Minlog also automatically extracts a proof that the program meets its specification. In Sect. 2 we give a more detailed and technical description of Minlog and its program extraction facilities. A number of substantial case studies on program extraction have been carried out in Minlog reaching from the extraction of a normalisation-by-evaluation algorithm to the extraction of programs in

⋆ Supported by the Marie Curie Initial Training Network in Mathematical Logic – MALOA – From MAthematical LOgic to Applications, PITN-GA-2009-238381.

A. Corradini, B. Klin, and C. Cîrstea (Eds.): CALCO 2011, LNCS 6859, pp. 393–399, 2011.

constructive analysis. In this paper we present two case studies demonstrating the use of inductive and coinductive definitions in program extraction (Sect. 3) and show that they work well with Minlog's optimized extraction mechanism.

2 Program Extraction in Minlog

2.1 The Interactive Proof System Minlog

Minlog [Min, BBS+98] is an interactive proof system based on first order natural deduction calculus. It is intended to reason about higher type computable functionals, using minimal rather than classical or intuitionistic logic. Minlog implements a *theory of computable functionals*, as described in [SW11]. The underlying semantics is the Scott-Ershov model of partial continuous functionals, with *free algebras* as base types. These algebras are viewed as domains represented by Scott's information systems, whose tokens are constructor trees possibly involving the symbol $*$ ("no information"). The ideals (points, objects) of base type are consistent and deductively closed sets of tokens, possibly infinite. *Initial algebras* and *final coalgebras* are modelled by notions of totality and cototality: An ideal x is "cototal" if every constructor tree $P(*) \in x$ has a "one-step extension" $P(C\boldsymbol{a}^*) \in x$, and "total" if it is cototal and this extension relation is well-founded. Totality and cototality are instances of strictly positive inductive and coinductive definitions which are supported in general in Minlog. With every initial algebra and final coalgebra are associated operators for (co)iteration and (co)recursion. Computation is implemented efficiently via *normalization by evaluation* [BS91, BES03]. Computation is extended to proofs via the Curry-Howard correspondence. *Intuitionistic* and *classical logic* are represented by the axiom schemes $\bot \to A$ and $\neg\neg A \to A$. *Interactive proofs* are organized in a goal-directed backwards-reasoning fashion. Forward reasoning is modelled by a form of cut rule. Minlog also contains an *automated prover* for a certain fragment of (simply typed) minimal logic. Its theory (based on [Mil91]) is developed in [Sch04].

2.2 Program Extraction

One of the main motivations behind Minlog is to use it as a tool for program verification, and to exploit the proofs-as-programs paradigm for program development. Minlog's program extraction is based on Kreisel's modified realizability [Kre59]: to each formula A a type $\tau(A)$ (the type of realizers of A) and a formula $x^{\tau(A)}\ \mathbf{r}\ A$ are assigned. The formula $x\ \mathbf{r}\ A$ is to be read as "x realizes A" and intuitively means "x solves the computational problem expressed by A". Program extraction computes from a derivation d of A a term $\mathrm{et}(d)$ (the extracted program) and a proof that $\mathrm{et}(d)$ realizes A. Of particular interest are the realization of induction and coinduction by algebras and coalgebras, as well as the computational and non-computational versions of logical operators. The latter are crucial for obtaining practically useful results as they lead to drastically simplified proofs and extracted programs. The case studies on parsing (3.1) and exact real numbers (3.2) below will highlight these points. From a type-theoretic point of view, realizability collapses a dependently typed lambda-calculus (which Minlog's proof calculus is an instance of) to a simply typed lambda-calculus. The collapse happens on the level of atomic formulas, since $\tau(P\boldsymbol{t}) = \tau(P)$ where $\tau(P)$ is a simple type assigned to the predicate P, discarding the first-order terms $\boldsymbol{t}$. Minlog extends program extraction to classical proofs via a refined A-translation [BBS02] or Dialectica Interpretation (see eg. [SW11]). The theoretical foundations of program extraction from formal proofs, as implemented in Minlog, are presented in detail in [SW11]. Realizability for induction and coinduction, including applications to exact real numbers, are developed in [Ber09, BS10].

2.3 Related Work

Program extraction from proofs is also implemented in Isabelle [Isa] (for algebras) and in Coq [Coq] (cf [BBLS06] for a joint case study). The implementation in Isabelle has been modelled after Minlog's extraction. The correctness of program extraction in Coq is not based on realizability and a Soundness Theorem (as in Minlog), but on the fact that reduction of proofs is correctly simulated by reduction of extracted programs [Let03]. There exists also an experimental implementation of program extraction in Agda [Agd] (cf [Chu11]). We are not aware of substantial case studies on program extraction in these systems. Also, Minlog seems to be the only system implementing the Dialectica Interpretation and program extraction from classical proofs. We also mention RZ [BS07], a tool that computes the realizability interpretation of a mathematical statement (but does not extract programs from proofs).

3 Case Studies

3.1 Algebras for Parsing

Consider strings x, y of left and right parentheses L, R. We define inductively the predicate (grammar) S of balanced strings of parentheses by the clauses

$$\text{InitS}\colon S(\text{nil}), \qquad \text{ApS}\colon Sx \to Sy \to S(xy), \qquad \text{ParS}\colon Sx \to S(LxR).$$

The type $\tau(S)$ of realizers of S is the algebra `algS` of generation trees for S. It has one nullary constructor, cInitS, one binary, cApS, and one unary, cParS. Our goal is to prove decidability of S, i.e. $\forall_x(Sx \vee \neg Sx)$, and extract from the proof a program computing for each string x a boolean value p and a tree t: `algS` such that p decides whether Sx holds and, in the positive case, t is a generation tree for x. Note that negation, $\neg A$, is expressed in Minlog as $A \to \bot$ and disjunction, $A \vee B$, as $\exists_p((p \to A) \wedge (\neg p \to B))$. Hence the goal reads $\forall_x \exists_p((p \to Sx) \wedge ((p \to \bot) \to Sx \to \bot))$. We also consider an alternative grammar U for the same set of strings (which, in contrast to S, is deterministic)

$$\text{InitU}\colon U(\text{nil}), \qquad \text{ApU}\colon Ux \to Uy \to U(xLyR),$$

with an algebra of realizers algU with constructors cInitU (nullary) and cApU (binary). Equality of U and S is expressed by $\forall_x^{\mathrm{nc}}(Ux \to Sx)$ and $\forall_x^{\mathrm{nc}}(Sx \to Ux)$. These formulas can be easily proven by induction on Ux and Sx. The non-computational universal quantifier $\forall_x^{\mathrm{nc}}$ has the same logical meaning as the usual $\forall_x$, but it indicates that the extracted programs only operate on the generation trees for x and not on the string x itself. The variables x, y in the defining clauses for S and U are implicitly quantified by $\forall^{\mathrm{nc}}$ as well. The extracted program for the proof of $\forall_x^{\mathrm{nc}}(Sx \to Ux)$ is ([b0] and [b1,...] denote lambda-abstractions)

```
[b0](Rec algS=>algU)b0 cInitU
([b1,b2,a3,a4](Rec algU=>algU)a4 a3([a5,a6,a7,a8]cApU a7 a6))
([b1]cApU cInitU)
```

The fact that the proof is by induction on Sx is witnessed by the occurrence of the recursion operator (Rec algS=>algU) implementing (an instance of) structural recursion on algS. There is also a side induction witnessed by (Rec algU=>algU). The term above is equivalent to a program SU defined by the recursive equations

```
SU cInitS        = cInitU
SU (cApS b1 b2) = UU (SU b2) where
    UU cInitU        = SU b1
    UU (cApU a5 a6) = cApU (UU a5) a6
SU (CParS b)     = cApU cInitU (SU b)
```

The boolean value deciding whether or not Sx holds is computed as Test 0 x, the function Test being defined as a constant Test (py means parse-type):

```
(add-program-constant "Test" (py "nat=>list par=>boole"))
(add-computation-rules
 "Test 0(Nil par)"          "True"
 "Test 0(R::x)"             "False"
 "Test(Succ n)(Nil par)"    "False"
 "Test n(L::x)"             "Test(Succ n)x"
 "Test(Succ n)(R::x)"       "Test n x")
```

Note that (Nil par) denotes the empty list of parentheses and :: is the cons operation for lists. Soundness, $\forall_x^{\mathrm{nc}}(Ux \to \mathrm{Test}(0,x))$, is easy and the proof has no computational content. Completeness, $\forall_x(\mathrm{Test}(0,x) \to Sx)$, needs an inductively defined predicate State with clauses

$$\mathrm{InitState}\colon \mathrm{State}(0, \mathrm{nil}), \quad \mathrm{ApState}\colon Ss \to \mathrm{State}(n,x) \to \mathrm{State}(\mathrm{S}n, xsL)$$

and a lemma $\forall_y \forall_{n,x}^{\mathrm{nc}}(\mathrm{State}(n,x) \to \forall_s^{\mathrm{nc}}(Ss \to \mathrm{Test}(n,y) \to S(xsy)))$, proved by induction on y. Now $\forall_x \exists_p((p \to Sx) \wedge ((p \to \bot) \to Sx \to \bot))$ is proved easily[1]. The parser-term extracted from this proof is (@ is a pairing operator in infix notation, [if ...] is a generalization of the usual if-then-else allowing pattern matching on the constructors of an algebra)

[1] The third author is grateful to Makoto Takeyama for explaining his Agda code http://code.haskell.org/Agda/examples/ParenDepTac.agda to him.

```
[x0]Test 0 x0 @
 (Rec list par=>algState=>algS=>algS)x0
 ([st1,b2][if st1 b2 ([b3,st4]cInitS)])
 ([par1,x2,f3,st4,b5]
   [if par1
     (f3(cApState b5 st4)cInitS)
     [if st4 cInitS ([b6,st7]f3 st7(cApS b6(cParS b5)))]])
 cInitState cInitS
```

which corresponds to the following recursive program:

```
P x0 = Test 0 x0 @ P0 x0 cInitState cInitS   where
   P0 Nil cInitState b2              = b2
   P0 Nil(cApState b3 st4)b2         = cInitS
   P0 (L::x2)st4 b5                  = P0 x2(cApState b5 st4)cInitS
   P0 (R::x2)cInitState b5           = cInitS
   P0 (R::x2)(cApState b6 st7)b5 = P0 x2 st7(cApS b6(cParS b5))
```

Experiments (pp, nt, pt mean pretty-print, normalize-term, parse-term - Four inputs to Minlog, followed by Minlog's response):

```
(pp (nt (mk-term-in-app-form parser-term (pt "L::R:"))))
"True@cApS cInitS(cParS cInitS)"
(pp (nt (mk-term-in-app-form parser-term (pt "R::L:"))))
"False@cInitS"
(pp (nt (mk-term-in-app-form parser-term (pt "L::R::L::R:"))))
"True@cApS(cApS cInitS(cParS cInitS))(cParS cInitS)"
(pp (nt (mk-term-in-app-form parser-term (pt "L::L::R::R:"))))
"True@cApS cInitS(cParS(cApS cInitS(cParS cInitS)))"
```

3.2 Coalgebras for Exact Real Numbers

Our second case study concerns algorithms in exact real arithmetic. Whilst such algorithms have been *verified* before (see eg. [CDG06, MRE07, GNSW07, BH08]), in the present paper we show by means of an example how to *extract* them. We extract a program which for every rational number $a \in [-1,1]$ computes a signed binary digit representation, that is, a (finite or infinite) stream of digits $d_0, d_1, \ldots \in \{-1,0,1\}$ such that

$$a = \sum_i \frac{d_i}{2^{i+1}} \tag{1}$$

We let a range over abstract real numbers (we only use the properties of an ordered field) and let Qa mean that a is a rational number with absolute value ≤ 1. Our program will be extracted from a proof of the formula

$$\forall_a^{\mathrm{nc}}(Qa \to Ja) \tag{2}$$

where the predicate J is defined coinductively by the clause

$$\forall^{\mathrm{nc}}_a(Ja \to a = 0 \lor \exists^{\mathrm{r}}_b(a = \frac{b-1}{2} \land Jb) \lor \exists^{\mathrm{r}}_b(a = \frac{b}{2} \land Jb) \lor \exists^{\mathrm{r}}_b(a = \frac{b+1}{2} \land Jb)) \quad (3)$$

that is, J is the largest predicate satisfying (3). The proof of (2) proceeds by coinduction, that is, by showing that (3) holds when J is replaced by Q. The superscript r attached to the quantifier $\exists^{\mathrm{r}}$ stands for "right" and means that from a proof of a formula $\exists^{\mathrm{r}}_b A$ only the realizer of A is kept while the witness b contained in the proof is discarded. The type of realizers for J is the coalgebra of finite and infinite streams of signed digits. In our setting it is modelled as the set of cototal ideals (see Sect. 2.1) of the algebra **I** of "standard rational intervals", whose constructors are $\mathbb{I}$ (for the initial interval $[-1, 1]$) and $\mathrm{C}_{-1}, \mathrm{C}_0, \mathrm{C}_1$ (for the left, middle, right part of the argument interval, of half its length). For example, $\mathrm{C}_{-1}\mathbb{I}$, $\mathrm{C}_0\mathbb{I}$ and $\mathrm{C}_1\mathbb{I}$ should be viewed as the intervals $[-1, 0]$, $[-\frac{1}{2}, \frac{1}{2}]$ and $[0, 1]$. The cototal ideals include, for example, $\{\mathrm{C}^n_{-1}*\}_{n\geq 0}$, a "stream" representation of the real -1, and also $\{\mathrm{C}_{-1}\mathrm{C}^n_1*\}_{n\geq 0}$ and $\{\mathrm{C}_1\mathrm{C}^n_{-1}*\}_{n\geq 0}$, which both represent the real 0. Generally, the cototal ideals give us all reals in $[-1, 1]$, in the representation (1). We have formalized in Minlog a proof of (2) and extracted from it a term `neterm` of type $\iota \to \mathbf{I}$ involving the corecursion operator ${}^{\mathrm{co}}\mathcal{R}^{\iota}_{\mathbf{I}}$ associated with (3). The value of the term obtained by applying `neterm` to, say, 1/2 (in Minlog: `cGenQ(1#2)`) is an infinite ideal starting with $\mathrm{C}_1, \mathrm{C}_0, \mathrm{C}_0, \mathrm{C}_0 \ldots$ (`CIntP`, `CintZ`, ...). To compute it (again via normalization-by-evaluation, i.e., `nt`) we delay unfolding ${}^{\mathrm{co}}\mathcal{R}^{\iota}_{\mathbf{I}}$ at a fixed depth, say 5:

```
(pp (nt (undelay-delayed-corec
     (mk-term-in-app-form neterm (pt "cGenQ(1#2)")) 5)))
"CIntP (CIntZ (CIntZ (CIntZ (CIntZ (CoRec algQ=>intv)..."
```

Sources. The Minlog system is available at `www.minlog-system.de`. To run the examples download the latest version `minlog-latest.tar.gz` (SVN snapshot) and follow the installation instructions. Note that, as prerequisites, Scheme and Emacs are required. The reader new to Minlog is referred to the tutorial [CSS11] for introductory examples. The two case studies of this paper can be found in `examples/parsing/parens.scm` and `examples/analysis/ratsds.scm` together with readme files `readme-parens.txt and readme-ratsds.txt` explaining the background and how to run the case studies.

Acknowledgements. The authors would like to thank the referees for their constructive comments and criticism.

References

[Agd] Agda: `http://wiki.portal.chalmers.se/agda/`

[BBLS06] Berger, U., Berghofer, S., Letouzey, P., Schwichtenberg, H.: Program extraction from normalization proofs. Studia Logica 82, 27–51 (2006)

[BBS+98] Benl, H., Berger, U., Schwichtenberg, H., Seisenberger, M., Zuber, W.: Proof theory at work: Program development in the Minlog system. In: Bibel, W., Schmitt, P.H. (eds.) Automated Deduction. Applied Logic Series, vol. II, pp. 41–71. Kluwer, Dordrecht (1998)

[BBS02] Berger, U., Buchholz, W., Schwichtenberg, H.: Refined program extraction from classical proofs. APAL 114, 3–25 (2002)

[Ber09] Berger, U.: From coinductive proofs to exact real arithmetic. In: Grädel, E., Kahle, R. (eds.) CSL 2009. LNCS, vol. 5771, pp. 132–146. Springer, Heidelberg (2009)

[BES03] Berger, U., Eberl, M., Schwichtenberg, H.: Term rewriting for normalization by evaluation. Information and Computation 183, 19–42 (2003)

[BH08] Berger, U., Hou, T.: Coinduction for exact real number computation. Theory of Computing Systems 43, 394–409 (2008)

[BS91] Berger, U., Schwichtenberg, H.: An inverse of the evaluation functional for typed λ-calculus. In: Vemuri, R. (ed.) Proceedings 6'th Symposium on Logic in Computer Science, LICS 1991, pp. 203–211. IEEE Computer Society Press, Los Alamitos (1991)

[BS07] Bauer, A., Stone, C.A.: RZ: A tool for bringing constructive and computable mathematics closer to programming practice. In: Cooper, S.B., Löwe, B., Sorbi, A. (eds.) CiE 2007. LNCS, vol. 4497, pp. 28–42. Springer, Heidelberg (2007)

[BS10] Berger, U., Seisenberger, M.: Proofs, programs, processes. In: Ferreira, F., Löwe, B., Mayordomo, E., Mendes Gomes, L. (eds.) CiE 2010. LNCS, vol. 6158, pp. 39–48. Springer, Heidelberg (2010)

[CDG06] Ciaffaglione, A., Di Gianantonio, P.: A certified, corecursive implementation of exact real numbers. Theor. Comp. Sci. 351, 39–51 (2006)

[Chu11] Chuang, C.M.: Extraction of Programs for Exact Real Number Computation Using Agda. PhD thesis. Swansea University, Wales (2011)

[Coq] The Coq Proof Assistant, `http://coq.inria.fr/`

[CSS11] Crosilla, L., Seisenberger, M., Schwichtenberg, H.: A Tutorial for Minlog, Version 5.0 (2011)

[GNSW07] Geuvers, H., Niqui, M., Spitters, B., Wiedijk, F.: Constructive analysis, types and exact real numbers. Math. Struct. Comp. Sci. 17(1), 3–36 (2007)

[Isa] Isabelle, `http://isabelle.in.tum.de/`

[Kre59] Kreisel, G.: Interpretation of analysis by means of constructive functionals of finite types. Constructivity in Mathematics, 101–128 (1959)

[Let03] Letouzey, P.: A New Extraction for Coq. In: Geuvers, H., Wiedijk, F. (eds.) TYPES 2002. LNCS, vol. 2646, pp. 200–219. Springer, Heidelberg (2003)

[Mil91] Miller, D.: A logic programming language with lambda–abstraction, function variables and simple unification. Jour. Logic Comput. 2(4), 497–536 (1991)

[Min] The Minlog System, `http://www.minlog-system.de`

[MRE07] Marcial-Romero, J.R., Escardo, M.H.: Semantics of a sequential language for exact real-number computation. Theor. Comp. Sci. 379(1-2), 120–141 (2007)

[Sch04] Schwichtenberg, H.: Proof search in minimal logic. In: Buchberger, B., Campbell, J.A. (eds.) AISC 2004. LNCS (LNAI), vol. 3249, pp. 15–25. Springer, Heidelberg (2004)

[SW11] Schwichtenberg, H., Wainer, S.S.: Proofs and Computations. Perspectives in Logic. Assoc. Symb. Logic, Cambridge Univ. Press (to appear, 2011)

Tool Interoperability in the Maude Formal Environment

Francisco Durán[1], Camilo Rocha[2], and José M. Álvarez[1]

[1] Universidad de Málaga, Spain
[2] University of Illinois at Urbana-Champaign, IL, USA

Abstract. We present the Maude Formal Environment (MFE), an executable formal specification in Maude within which a user can seamlessly interact with the Maude Termination Tool, the Maude Sufficient Completeness Checker, the Church-Rosser Checker, the Coherence Checker, and the Maude Inductive Theorem Prover. We explain the high-level design decisions behind MFE, give a summarized account of its main features, and illustrate with an example the interoperation of the tools available in its current release.

1 Introduction

Maude [1] is a reflective declarative language and system based on rewriting logic in which computation corresponds to efficient deduction by rewriting. Several tools for the formal analysis of Maude modules have been available for a number of years. They include an inductive theorem prover [4] for equational specifications, and an LTL model checker [10], a reachability analysis tool [1], and an invariant analyzer [13] for rewrite specifications. However, many verification techniques such as the ones implemented by the tools just mentioned assume that the input module satisfies certain properties. For example, any verification task with the LTL model checker on a rewrite specification $\mathcal{R} = (\Sigma, E \cup A, R)$ assumes that $\mathcal{R}$ satisfies the so-called *executability requirements*, namely, that the equations E specifying the functional part of $\mathcal{R}$ are both ground terminating and ground Church-Rosser modulo the structural axioms A, and that the rules R specifying the concurrent transitions of $\mathcal{R}$ are ground coherent w.r.t. E modulo A. If they do not hold, then any analysis performed by the LTL model checker on $\mathcal{R}$ can in general lead to unsound and incomplete results.

Maude has been successfully used as a *metatool* in the creation of tools for verifying properties of its modules [2]. In this sense, previous work presented in [3] describes the main features of several tools concerned with the analysis of either Maude modules or of extensions of Maude. However, these tools work in isolation, making it inconvenient to switch between their execution environments and difficult to exchange data between them. In response to these limitations, we present the Maude Formal Environment (MFE), an executable and highly extensible software infrastructure within which a user can interact with several tools to mechanically verify properties of Maude modules. In MFE, tools can interoperate to discharge proof obligations of different nature without switching between different execution environments. The integration of different tools inside MFE's common environment presents the user with a consistent user interface, a mechanism to keep track of pending proof obligations, and allows the execution of several instances of each tool, among other features.

A. Corradini, B. Klin, and C. Cîrstea (Eds.): CALCO 2011, LNCS 6859, pp. 400–406.

The following tools are currently available as part of MFE: the Maude Termination Tool (MTT) can be used to prove termination of equational and rewrite specifications [5]; the Sufficient Completeness Checker (SCC) can be used to check completeness and freeness of equational specifications, and deadlock freedom of rewrite specifications [11,12]; the Church-Rosser Checker (CRC) can be used to check ground confluence and sort-decreasingness of equational specifications [7]; the Coherence Checker (ChC) can be used to check the ground coherence of rewrite specifications [8]; and the Inductive Theorem Prover (ITP) can be used to verify inductive properties of equational specifications [4,11].

Outline. In Section 2 we give a high-level overview of MFE's design features and we explain some of the commands available. In Section 3 we illustrate MFE's main functionality with a case study in which a user interacts with several of the tools in the environment. The executables, a white paper explaining MFE's design, examples, and preliminary documentation, are available at `http://maude.lcc.uma.es/MFE`. For further details on the tools available in MFE please check the given references.

2 MFE's Design and Main Features

MFE has been implemented as an extension of Full Maude [6], thus benefiting from its functionality and flexibility. Full Maude is an extension of Maude written in Maude itself that has become a common infrastructure on top of which tools can be built. In MFE, for instance, tools provided by Maude such as its LTL model checker and search command are available, as well as modules defined in Full Maude (including object-oriented ones) are directly amenable to formal verification.

MFE is highly extensible and amenable to tool interoperability given its modular design and the fact that it imposes no constraint on how each tool should model its particular domain or maintains its internal state. MFE is modeled in Maude as an interactive object-based system where tools are objects, the communication mechanism is message passing, and user interaction is available through Full Maude. Integration and interoperation of tools within MFE is module-centric, given that its main purpose is to support formal analysis of Maude modules.

The object-oriented model of MFE consists of three classes: the class `Proof` of proof objects which keep the state of specific proof requests, the class `Tool` of tool objects which manage proof objects, and a class `Controller` which inherits from the Full Maude's `DatabaseClass` and provides a centralized entry point for handling requests to the formal environment.

The `Controller` object defines the behavior of the environment and its tools with the user. The user interacts with the environment via commands which are encapsulated as messages in the object configuration. Each tool object and the controller object have a module defining the signature of the commands it can handle. The controller handles any command it can parse; since this object extends Full Maude, it handles its own commands and Full Maude ones. If the controller receives a command it cannot parse, then it delegates the message to the *active* tool (previously selected by the user). If the active tool can parse the delegated command, then it notifies the controller and handles the command. Otherwise, it will notify the failure to the controller, which in turn will output an error message of the failed command to the user.

Classes `Proof` and `Tool` define some basic functionality that can be inherited by any new tool. Class `Tool`, for example, defines a set of attributes that are convenient for supporting multiple instances of a tool and predefines some rewrite rules for managing the life cycle of proof objects. However, new tools can be added to MFE without inheriting from any of these two classes.

MFE provides the following user commands, in addition to the ones inherited from Full Maude:

`(select tool <tool-name> .)` sets `<tool-name>` as the *active* tool.
`(MFE help .)` shows MFE's help information.
`(show global state .)` shows the state of the environment.

Any command that cannot be parsed against MFE's grammar is delegated to the active tool. In this way, user interaction with any tools remains almost the same as before its integration in MFE. We refer the reader to MFE's documentation and to each tool's documentation for a detailed account of commands, restrictions, and additional examples.

In order to be flexible, MFE does not define any policy for naming tool commands. However, as a general guideline for using the environment, it is recomended that the tools provide at least the following commands, as it is the case with the tools available in MFE's current release:

`(<tool-name> help .)` shows the help information of tool `<tool-name>`.
`(show state .)` shows the state of the tool.

MFE's design supports non-trivial dependencies among tools and potential interaction complexity. For instance, a ground coherence check on a rewrite specification assumes the ground termination and ground Church-Rosser properties on its equational subspecification, and may produce a number of inductive proof obligations that could be discharged with help of ITP. Figure 1 depicts the tool-dependency graph for the current tools in MFE.

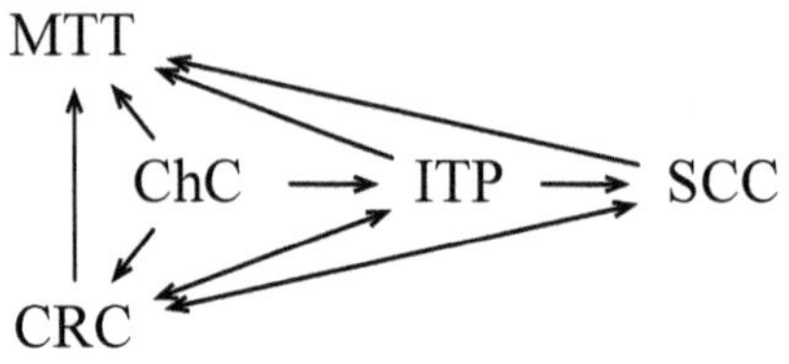

Fig. 1. Tool-dependency graph in MFE

A tool in MFE keeps track of both its pending and discharged proof obligations. It can submit proof obligations to other tools by means of a `submit` command and then be notified when these are discharged. When all proof obligations in the verification process of a module's property are discharged, the corresponding tool notifies the success result to the user or to the tool originating the verification task.

Of course, tools in general can impose constraints on its inputs. For instance, SCC does not support parametric modules but, nevertheless, proofs for such modules could be obtained by hand or by using a different tool. MFE offers a `trust` command for keeping track of proofs obtained outside MFE.

Finally, for tools which depend on external utilities not directly available from Maude such as MTT and SCC, we have extended the latest release of the Maude system with *built-in* operators associated with appropriate C++ code that interacts with the external tools. A similar extension was previously performed for the SCC [11].

3 Case Study: Ground Coherence of the Bakery Protocol

We explain how MFE can be used to prove the ground coherence of module `BAKERY`, yet another Maude specification of the Bakery Protocol. The Bakery Protocol is a classical solution by Lamport to the problem of achieving mutual exclusion between processes. The protocol is based on the procedure commonly used in bakeries where a customer is assigned a ticket number upon arrival. Here, a process is a term with sort `BProcess` built from operator `<_,_,_>` that takes a natural number with sort `Nat` as identifier, a constant term `sleep`, `wait`, or `crit` with sort `Mode` as current state, and a natural number as ticket number. A term with sort `BState` is a multiset of processes, where each process is a singleton multiset and union is denoted by juxtaposition. A term with sort `GBState` represents the system's state and it is built from operator `[[_]]` that takes a multiset of processes as argument.

The concurrent behavior of the Bakery Protocol is modeled in the `BAKERY` system module as follows.

```
(mod BAKERY is
  protecting MNAT .

  sorts Id Mode BProcess BState GBState .
  ops sleep wait crit : -> Mode [ctor] .
  op <_`,_`,_> : MNat Mode MNat -> BProcess [ctor] .
  subsort BProcess < BState .
  op __ : BState BState -> BState [ctor assoc comm id: none] .
  op none : -> BState [ctor] .
  op `[`[_`]`] : BState -> GBState [ctor] .

  var P : Mode . vars I N M : MNat . var BSt : BState .

  ---- max of the numbers assigned to processes (0 if none)
  op maxNumber : BState -> MNat .
  op maxNumber : BState MNat -> MNat .
  eq maxNumber(< I, P, N > BSt) = max(N, maxNumber(BSt)) .
  eq maxNumber(none) = 0 .

  ---- min. of the nonzero numbers assigned to processes (0 if none)
  op minNzNumber : BState -> MNat .
  op minNzNumber : BState MNat -> MNat .
  eq minNzNumber(< I, P, 0 > BSt) = minNzNumber(BSt) .
  eq minNzNumber(< I, P, s N > BSt) = minNzNumber(BSt, s N) .
  eq minNzNumber(none) = 0 .
  eq minNzNumber(< I, P, 0 > BSt, M) = minNzNumber(BSt, M) .
  eq minNzNumber(< I, P, s N > BSt, M) = minNzNumber(BSt, min(M, s N)) .
```

```
  eq minNzNumber(none, M) = M .

  rl [s2w] : [[ < I, sleep, 0 > BSt ]]
    => [[ < I, wait, s maxNumber(BSt) > BSt ]] .
 crl [w2c] : [[ < I, wait, N > BSt ]]
    => [[ < I, crit, N > BSt ]]
    if N = minNzNumber(< I, wait, N > BSt) .
  rl [c2s] : [[ < I, crit, N > BSt ]]
    => [[ < I, sleep, 0 > BSt ]] .
endm)
```

Operators maxNumber and minNzNumber operate on terms with sort BState and compute, respectively, the maximum and minimum ticket numbers in a multiset of processes, or 0 if none.

In MFE, BAKERY's executability properties can be proved in different order. For instance, we first activate CRC and use its check Church-Rosser command to first verify BAKERY's equational subspecification Church-Rosser.

```
Maude> (select tool CRC .)
The CRC has been set as current tool.
```

```
Maude> (check Church-Rosser BAKERY .)
Church-Rosser check for BAKERY
   All critical pairs have been joined.
   The specification is locally-confluent.
   The module is sort-decreasing.
```

All critical pairs are joined and consequently the specification is locally confluent. Notice also that CRC proves the equations sort-decreasing. Hence, a proof of termination would imply the ground Church-Rosser property of BAKERY's equational part.

In this case, termination of BAKERY's equational part is the only pending proof obligation. We ask the CRC to submit this proof obligation and then activate MTT.

```
Maude> (submit .)
The termination goal for the functional part of BAKERY has been submitted to MTT.
Warning: A proof of the termination of functional part of module BAKERY has not
    ↪been found.
```

MTT is not able to find a proof automatically, and we need to interact with it to go on.

```
Maude> (select tool MTT .)
The MTT has been set as current tool.
```

A proof of the termination of this specification can be found in [9]. Instead of completing the proof here, let us rely on this reference and use the trust command to tell the tool that it can assume the existence of a proof of the termination of the current module and proceed.[1]

```
Maude> (trust .)
The functional part of module BAKERY is assumed terminating.
Success: The module is therefore Church-Rosser.
```

Upon notification from MTT that a termination proof has been found, CRC notifies the user that BAKERY's functional part is Church-Rosser. Then, it only remains to prove

[1] If you are running a version of Maude that does not include the hooks to the external termination tools, you would get a message indicating that the tool cannot be used to prove the termination of the current module. You will still be able to use the trust command.

BAKERY ground coherent. We set ChC as the active tool in MFE and issue the corresponding checking command.[2]

```
Maude> (select tool ChC .)
The ChC has been set as current tool.
```

```
Maude> (check ground coherence BAKERY .)
Ground coherence checking of BAKERY
   All critical pairs have been rewritten and no rewrite with rules can happen at
       ↪ non-overlapping positions of equations left-hand sides.
   The sufficient-completeness, termination and Church-Rosser properties must
       ↪still be checked.
```

The decision procedure implemented by ChC discharges all critical pairs between equations and rules. However, this procedure requires BAKERY's functional part to be sufficiently complete, ground terminating, and ground Church-Rosser. Since the termination and Church-Rosser properties have previously been proved, ChC is notified that such proofs have been found when submitting the proof obligations.

```
Maude> (submit .)
The Church-Rosser goal for BAKERY has been submitted to CRC.
The Sufficient-Completeness goal for BAKERY has been submitted to SCC.
The termination goal for the functional part of BAKERY has been submitted to MTT.
Success: The equational theory of BAKERY does not have counterexamples for
    ↪sufficient completeness.
   However,this is under the assumption that it is ground weakly-normalizing and
       ↪ground sort-decreasing.
The functional part of module BAKERY has been checked terminating.
The module BAKERY has been checked Church-Rosser.
```

Although we already have all the pieces, we still need to select the SCC tool to complete its proof.

```
Maude> (select tool SCC .)
The SCC has been set as current tool.
```

```
Maude> (submit .)
The sort-decreasingness goal for BAKERY has been submitted to CRC.
The termination goal for the functional part of BAKERY has been submitted to MTT.
Church-Rosser check for BAKERY
   The module is sort-decreasing.
The module BAKERY has been checked sufficiently-complete.
Success: The module BAKERY is ground-coherent.
```

Thus, as desired, we conclude that system module BAKERY is (ground) coherent.

4 Future Work

More tools such as Maude's LTL and LTLR Model Checkers, Maude's Invariant Analyzer, and Real-Time Maude could be integrated in MFE. This will result in a more comprehensive environment with more features and broader applications. One could also think of MFE automatically generating the proof obligations associated to the semantics of protected and extended modules, and to that of parameterized modules.

[2] Notice that for the proof of ground coherence, assuming sufficient completeness, defined operators can be regarded as frozen (see [8]).

More ambitiously, a graphical user interface and support for distributed interoperation will enhance the user experience within MFE.

Acknowledgments. We thank the anonymous referees for their comments, and for taking the time to evaluate the manuscript and the tool. The first author was partially supported by projects TIN2008-03107 and P07-TIC-03184. The second author has been partially supported by NSF grants CNS 07-16638 and CCF 09-05584.

References

1. Clavel, M., Durán, F., Eker, S., Lincoln, P., Martí-Oliet, N., Bevilacqua, V., Talcott, C.: All About Maude - A High-Performance Logical Framework: How to Specify, Program, and Verify Systems in Rewriting Logic. LNCS, vol. 4350. Springer, Heidelberg (2007)
2. Clavel, M., Durán, F., Eker, S., Meseguer, J., Stehr, M.O.: Maude as a formal meta-tool. In: Wing, J.M., Woodcock, J., Davies, J. (eds.) FM 1999. LNCS, vol. 1709, pp. 1684–1703. Springer, Heidelberg (1999)
3. Clavel, M., Durán, F., Hendrix, J., Lucas, S., Meseguer, J., Ölveczky, P.: The Maude formal tool environment. In: Mossakowski, T., Montanari, U., Haveraaen, M. (eds.) CALCO 2007. LNCS, vol. 4624, pp. 173–178. Springer, Heidelberg (2007)
4. Clavel, M., Palomino, M., Riesco, A.: Introducing the ITP tool: a tutorial. Journal of Universal Computer Science 12(11), 1618–1650 (2006)
5. Durán, F., Lucas, S., Meseguer, J.: MTT: The Maude termination tool (system description). In: Armando, A., Baumgartner, P., Dowek, G. (eds.) IJCAR 2008. LNCS (LNAI), vol. 5195, pp. 313–319. Springer, Heidelberg (2008)
6. Durán, F., Meseguer, J.: Maude's module algebra. Science of Computer Programming 66(2), 125–153 (2007)
7. Durán, F., Meseguer, J.: A church-rosser checker tool for conditional order-sorted equational maude specifications. In: Ölveczky, P.C. (ed.) WRLA 2010. LNCS, vol. 6381, pp. 69–85. Springer, Heidelberg (2010)
8. Durán, F., Meseguer, J.: A maude coherence checker tool for conditional order-sorted rewrite theories. In: Ölveczky, P.C. (ed.) WRLA 2010. LNCS, vol. 6381, pp. 86–103. Springer, Heidelberg (2010)
9. Durán, F., Meseguer, J.: On the Church-Rosser and coherence properties of conditional order-sorted rewrite theories. Journal of Logic and Algebraic Programming Journal of Logic and Algebraic Programming (2011) (accepted for publication)
10. Eker, S., Meseguer, J., Sridharanarayanan, A.: The Maude LTL model checker. In: Gaducci, F., Montanari, U. (eds.) Proceedings of 4th International Workshop on Rewriting Logic and its Applications (WRLA 2002). Electronic Notes in Theoretical Computer Science, vol. 71 (2002)
11. Hendrix, J.: Decision Procedures for Equationally Based Reasoning. Ph.D. thesis. University of Illinois at Urbana-Champaign (2008)
12. Rocha, C., Meseguer, J.: Constructors, sufficient completeness and deadlock freedom of rewrite theories. In: Fermüller, C.G., Voronkov, A. (eds.) LPAR-17. LNCS, vol. 6397, pp. 594–609. Springer, Heidelberg (2010)
13. Rocha, C., Meseguer, J.: Proving safety properties of rewrite theories. Tech. rep. University of Illinois at Urbana-Champaign (2010), `http://hdl.handle.net/2142/17407`

WiCcA : LTS Generation Tool for Wire Calculus

Jennifer Lantair and Paweł Sobociński

ECS, University of Southampton

Abstract. We introduce the WIre CalCulus Application (WiCcA), a tool for generating and operating on labelled transition systems (LTS) that result from wire-calculus specifications. The theory behind WiCcA is explained and its uses demonstrated.

Keywords: Wire calculus, Petri nets, LTS, bisimulation, WiCcA.

1 Introduction

WiCcA is designed to simplify the study of specifications in finite wire-calculus [5] like languages, for instance the Petri calculus [6]. Such languages consist of a number of constants, each with operational semantics expressed in terms of a two-labelled transition system. Specifications are then constructed using the constants and two binary operations: tensor product '⊗' and boundary synchronisation ';', the operational semantics of each of the two operations is captured using one simple SOS rule.

Often when experimenting with specifications in such languages it is a tedious and time consuming activity to generate the operational semantics (the resulting LTS) by hand. WiCcA has been designed to automate LTS generation from such a specification, outputting a `.dot` file readable by applications associated with the GraphViz project [4].

2 Running Example

As a running example, we introduce a small process calculus that contains the constants needed to model the behaviour of a simple NAND flip-flop. These are defined in the file `flipFlop.alp` found in the `Alphabets` directory in the distribution[1], listed here in Figure 1. There are eight constant declarations: `I`, `X`, `E`, `F`, `D`, `C`, `rb` and `2N`; each has an *associated type* which is a pair of natural numbers (k, l), the intuition is that it represents a component with k wires on the left and l wires on the right. Each constant is associated with a two-labelled transition system: here the labels are words over the set of signals $\{0, 1\}$. The type of each constant determines the length of the labels: for example a constant of type $(2, 1)$ has its operational semantics given as an LTS with transitions of the form $\xrightarrow[y]{x_1 x_2}$ where $x_1, x_2, y \in \{0, 1\}$.

[1] http://users.ecs.soton.ac.uk/ps/wicca.php

A. Corradini, B. Klin, and C. Cîrstea (Eds.): CALCO 2011, LNCS 6859, pp. 407–412, 2011.

```
node I:(1,1);          node F:(1,1);          node X:(2,2);
tran I-0/0->I;         tran F-0/1->E;         tran X-00/00->X;
tran I-1/1->I;         tran F-1/1->F;         tran X-01/10->X;
                                              tran X-10/01->X;
node rb:(0,1);         node D:(1,2);          tran X-11/11->X;
tran rb-./0->rb;       tran D-0/00->D;
tran rb-./1->rb;       tran D-1/11->D;        node 2N:(2,1);
                                              tran 2N-00/1->2N;
node E:(1,1);          node C:(2,1);          tran 2N-01/1->2N;
tran E-0/0->E;         tran C-00/0->C;        tran 2N-10/1->2N;
tran E-1/0->F;         tran C-11/1->C;        tran 2N-11/0->2N;
```

Fig. 1. Example alphabet of constants (`flipFlop.alp`)

The constants represent simple digital circuit components. The intuition is that `I` corresponds to an identity wire, `rb` to ground, `E`, `F` to the two states of a one-place buffer, used to model delay, `D` and `C` to forks, `X` to a crossing of wires and `2N` to an instantaneous NAND gate. These components can be wired together using the operations of synchronisation on a common boundary ';' and tensor product '⊗'[2] , which intuitively correspond, respectively, to connecting the components side-by-side and putting them on top of each other. For details see [5,6]. Those SOS rules used within WiCcA are illustrated below:

$$\frac{P \xrightarrow[\beta]{\alpha} P' \quad Q \xrightarrow[\gamma]{\beta} Q'}{P;Q \xrightarrow[\gamma]{\alpha} P';Q'}, \qquad \frac{P \xrightarrow[\beta]{\alpha} P' \quad Q \xrightarrow[\delta]{\gamma} Q'}{P \otimes Q \xrightarrow[\beta\delta]{\alpha\gamma} P' \otimes Q'} \tag{1}$$

It is worth emphasising that *any* language that can treated by WiCcA consists of a set of user-defined constants, of which Figure 1 is a particular example. The operations ; and ⊗ are fixed and their semantics is defined equally by (1) for all such languages.

Remark 1. Constant declarations (Figure 1 contains several examples) are constrained so that both the left hand side and the right hand side of any transition declaration references a single constant. It is immediate that this restriction means that any specification handled by WiCcA leads to a finite LTS. In future work (see Section 5) we plan to relax this and consider more expressive declarations of constants, for instance by allowing arbitrary terms in the right hand sides of transition declarations.

Returning to our running example, a flip-flop can be expressed with the following term:

```
I o rb o rb o I ; I o D o D o I ;
                  I o E o I o I o E o I ; 2N o X o 2N ; C o C   (2)
```

[2] In WiCcA synchronisation on common boundary and tensor product are represented, respectively, with ASCII characters ';' and 'o'.

It is included as the file `flipflop.des` in the `Designs` directory. By drawing the appropriate symbols from left to right, it is clear that the above term corresponds to the following circuit diagram:

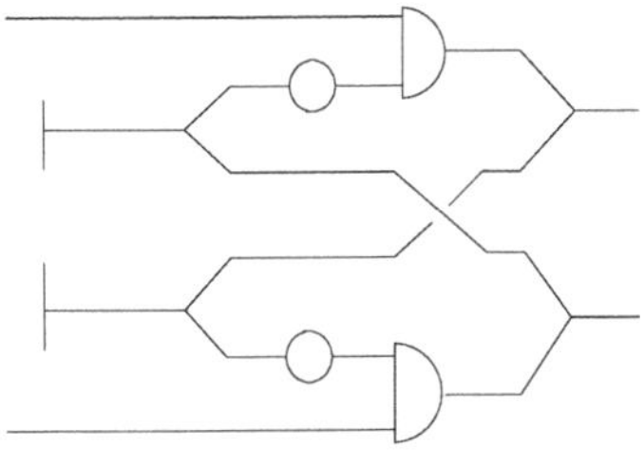

Fig. 2. Circuit diagram of (2)

Extracting the operational semantics by hand is tedious, so WiCcA generates a `.dot` file which can be used to display the LTS of (2). Figure 3, generated by WiCcA and typeset by GraphViz, illustrates the behaviour of (2):

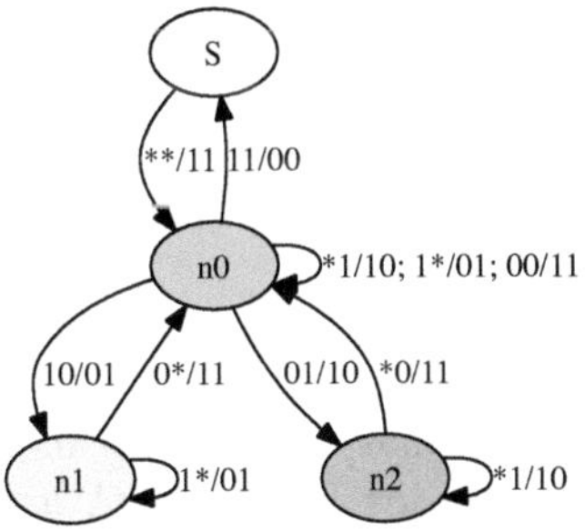

Fig. 3. LTS of the Figure 2

Some labels in Figure 3 contain the symbol '$*$' that represents all possible labels from within the alphabet $\{0, 1\}$. Where more than one '$*$' is present all possible combinations are implied. The states correspond to the following terms:

$$\textbf{S}: C[t_{E,E}] \qquad \textbf{n0}: C[t_{F,F}] \qquad \textbf{n1}: C[t_{F,E}] \qquad \textbf{n2}: C[t_{E,F}]$$

Where $t_{X,Y} = \texttt{I} \circ X \circ \texttt{I} \circ \texttt{I} \circ Y \circ \texttt{I}$ and $X, Y \in \{\texttt{E}, \texttt{F}\}$, the context C represents the remainder of (2).

Figure 3 captures the expected correct behaviour of a model NAND flip-flop, as well as behaviour on the normally omitted "restricted" or "don't care" input combinations. States **n1** and **n2** are the two stable states, corresponding to the two logical states of a flip-flop. When in state **n1**, as long as the first input remains high, the output is `01`. The state may be changed to `n2` with the input `01`, note that this requires two "clock ticks". The input `11` is normally described as "no change", and indeed, when in either of the stable states this combination does not result in a state change. However, our LTS exposes a problem with the asynchronous nature of the circuit in that if `11` is applied during a state-change (either from **n1** to **n2**, or from **n2** to **n1**) before the circuit reaches a stable state, oscillating behaviour (between **n0** and **S**) results; indeed such behaviour can be observed in physical circuit implementations laid out according to to Figure 2. The "restricted combination", or "don't care", input to a NAND flip-flop is `00`. Our labelled transition system demonstrates that this results simply in an output of `11` at state `n0`.

3 Features of WiCcA

WiCcA is written in Java [3], whilst the interpreter is written in JavaCC [2], a compiler compiler. Using Java and JavaCC ensures that WiCcA is platform independent. WiCcA processes scripts that contain all the required information to generate, reduce and compare a user's LTS. The basic information within a script is the alphabet of constants, the specification, operations to be performed on the LTS and file names for saving the generated LTS. Currently supported operations are reduction w.r.t. bisimilarity and weak bisimilarity and bisimilarity checking of two specifications. A simple example script, which may be found in the distribution of WiCcA in the `WICCA.jar` directory, is shown below:

```
input flipFlop.alp
v1 eval E
save v1 -> testFlipFlopDesign
```

Here `E` is the (very simple) specification and `v1` is a script variable that is assigned to the LTS generated from the specification. The `save` command generates a `.dot` file (the GraphViz file format) and writes it to the file name provided as the second argument. In the `.dot` file, WiCcA assigns randomly generated colours to each node and writes the graph of the generated LTS to a new file; bisimilar nodes are always assigned the same colour.

Reduction w.r.t. bisimilarity is implemented within WiCcA using the basic partition refinement algorithm of Kanellakis and Smolka [7,1]. Reduction w.r.t. weak bisimilarity depends on the user first declaring a `empty` symbol: labels constructed entirely using this symbol are considered to be silent. Weak reduction is then obtained by first performing transitive closure on the transitions in the obvious way, followed by standard partition refinement.

We now give an example of reduction w.r.t. bisimilarity. Continuing with the running example alphabet (see Table 1) two new components are added:

```
 node LD:(1,2);
tran LD-0/00->LD;
tran LD-1/10->LD;
tran LD-1/01->LD;
```

```
 node LC:(2,1);
tran LC-00/0->LC;
tran LC-10/1->LC;
tran LC-01/1->LC;
```

Fig. 4. Additional components for the Alphabet (`flipFlopExpanded.alp`)

The file `flipFlopExpanded.alp` may be found in the directory `Alphabets`. Components `LD` and `LC` represent simple switches. Using these switches we construct the specification for a pair of switches with one-place buffers in the center:

$$\texttt{LD ; E o E ; LC} \tag{3}$$

The term (3) is included as `switchs.des` in the `Designs` directory and corresponds to the following circuit diagram:

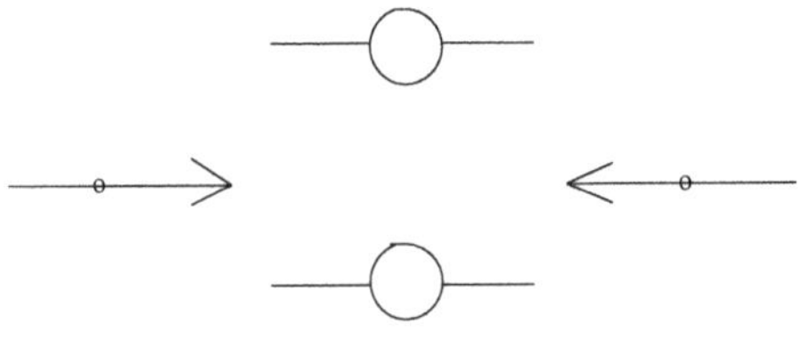

Fig. 5. Circuit diagram of (3)

The following LTS were generated by WiCcA from (3), showing the statespace before and after reduction:

The states of Figure 6 correspond to the following terms:

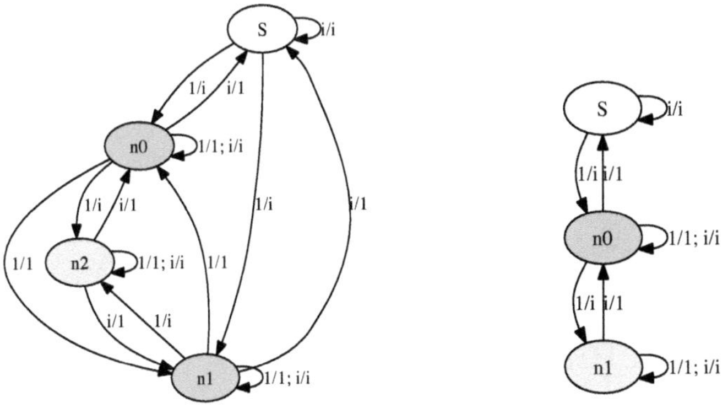

Fig. 6. Behaviour of (3) before and after reduction

S: $C[t_{E,E}]$ **n0**: $C[t_{F,E}]$ **n1**: $C[t_{E,F}]$ **n2**: $C[t_{F,F}]$

where $t_{X,Y} = X \texttt{ o } Y$ with $X, Y \in \{\texttt{E}, \texttt{F}\}$ and context C is the rest of (3). The two bisimilar nodes correspond to situations where the signal has been stored in one of the two buffer components—the precise identity of the component (upper or lower) is invisible to the environment.

4 Running WiCcA

WiCcA is equipped with a simple GUI that allows user-friendly access to some features. An initial development version of WiCcA is available for download[3]. Installation instructions and FAQ are available at the site.

5 Future Work

WiCcA is currently in active development and, at the time of writing, supports the basic functionality outlined in this paper. In particular, the tool does not currently support circuit diagrammatic notation, an important feature of wire calculus. Moreover only finite wire calculus expressions are modelled; this is a consequence of how constants are defined (see Remark 1). In particular, recursion is not supported at this stage. We plan to address both these points in future work, in particular:

(i) take advantage of the graphical potential of the wire calculus, both by displaying syntactical specifications as circuit diagrams (for example, as done by us manually in Figures 2 and 5), as well as eventually allowing users to directly edit circuit diagrams; from the users' point of view this would mean bypassing syntactic specifications entirely;
(ii) consider a more expressive subset of the wire calculus that includes restricted forms of recursion. In those cases we plan to use symbolic techniques to study the resulting LTS.

References

1. Kanellakis, P.C., Smolka, S.A.: CCS expressions, finite state processes, and three problems of equivalence. In: Proceedings of the Second Annual ACM Symposium on Principles of Distributed Computing, PODC 1983, pp. 228–240. ACM, New York (1983)
2. Metamata. Javacc, `http://javacc.java.net/`
3. Sun Microsystems. Java, `http://www.java.com/en/`
4. AT&T research and the open source community. Graphviz, `http://www.graphviz.org/`
5. Sobociński, P.: A non-interleaving process calculus for multi-party synchronisation. In: ICE 2009 (2009)
6. Sobociński, P.: Representations of Petri net interactions. In: Gastin, P., Laroussinie, F. (eds.) CONCUR 2010. LNCS, vol. 6269, pp. 554–568. Springer, Heidelberg (2010)
7. Sokolsky, O., Cleaveland, R.: Equivalence and preorder checking for finite-state systems (1990)

[3] http://users.ecs.soton.ac.uk/ps/wicca.php

SHACC: A Functional Prototyper for a Component Calculus

André Martins[1], Luís S. Barbosa[1], and Nuno F. Rodrigues[2]

[1] DI - CCTC, University of Minho, 4710-057 Braga, Portugal
[2] DIGARC - IPCA 4750-810 Barcelos, Portugal

Abstract. Over the last decade component-based software development arose as a promising paradigm to deal with the ever increasing complexity in software design, evolution and reuse. SHACC is a prototyping tool for component-based systems in which components are modelled coinductively as generalized Mealy machines. The prototype is built as a HASKELL library endowed with a graphical user interface developed in *Swing*.

Keywords: software composition, Mealy machines, prototyping.

1 Introduction

SHACC is a HASKELL -based prototype for a calculus of state-based components framed as generalised Mealy machines detailed in [1,2]. A typical example of such a state-based component is the ubiquitous *stack*. Denoting by U its internal state, a stack of values of type P is handled through the usual

$$\mathsf{top} : U \longrightarrow P, \quad \mathsf{pop} : U \longrightarrow P \times U \quad \text{and} \quad \mathsf{push} : U \times P \longrightarrow U$$

operations. A 'black box' view, however, hides U from the stack environment and regards each operation as a pair of input/output ports. For example, the top operation becomes declared as $\mathsf{top} : \mathbf{1} \longrightarrow P$, where $\mathbf{1}$ stands for the nil (or unit) datatype. The intuition is that top is activated with the simple pushing of a 'button' (its argument being the stack private state space) whose effect is the production of a P value in the corresponding output port. Similarly typing push as $\mathsf{push} : P \longrightarrow \mathbf{1}$ means that an external argument is required on activation but no visible output is produced, but for a trivial indication of successful termination. Such 'port' signatures are grouped together in the diagram below. Combined input type $\mathbf{1} + \mathbf{1} + P$ models the choice among three functionalities (top, pop and push, in this order), of which only one takes input of type P.

$$\begin{cases} \mathsf{pop} : & \mathbf{1} \longrightarrow P \\ \mathsf{top} : & \mathbf{1} \longrightarrow P \\ \mathsf{push} : & P \longrightarrow \mathbf{1} \end{cases}$$

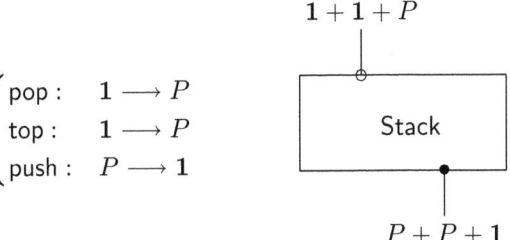

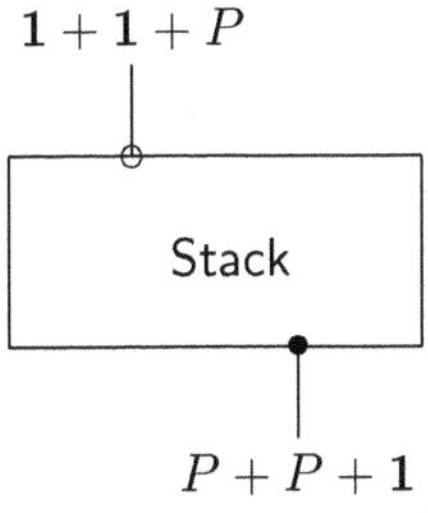

A. Corradini, B. Klin, and C. Cîrstea (Eds.): CALCO 2011, LNCS 6859, pp. 413–419, 2011.

Component Stack encapsulates a number of services through a public *interface* providing limited access to its internal *state space*. Furthermore, it *persists* and *evolves* in time, in a way which can only be traced through observations at the interface level. One might capture these intuitions by providing an explicit semantic definition in terms of a function $[\![\mathsf{Stack}]\!] : U \times I \longrightarrow (U \times O + \mathbf{1})$, where I, O abbreviate $\mathbf{1} + \mathbf{1} + P$ and $P + P + \mathbf{1}$, respectively. The presence of $\mathbf{1}$ in its result type indicates that the overall behaviour of this component is *partial*: in a number of state configurations the execution of some operations may fail. Function $[\![\mathsf{Stack}]\!]$ describes how Stack reacts to input stimuli, produces output data (if any) and changes state. It can also be written in a curried form as

$$\overline{[\![\mathsf{Stack}]\!]} : U \longrightarrow (U \times O + \mathbf{1})^I$$

that is, as a *coalgebra* $U \longrightarrow \mathsf{T}\, U$ for functor $\mathsf{T}\, X = ((X \times O) + \mathbf{1})^I$.

The Stack example illustrates the basic elements of a semantic model for state-based components: *a)* the presence of an *internal state space* which evolves and persists in time, and *b)* the possibility of *interaction* with other components through well-defined interfaces and during the overall computation. This favours adoption of a *coalgebraic* modelling framework: components are inherently dynamic, possess an observable behaviour, but their internal configurations remain hidden and should be identified if not distinguishable by observation. The qualificative 'state-based' is used in the sense the word 'state' has in automata theory — the internal memory of the automaton which both constrains and is constrained by the execution of component operations. Such operations are encoded in a functor which constitutes the (syntax of the) component interface. Building on such a representation, reference [1] developed a calculus of component composition. The experimental tool SHACC presented here provides a HASKELL based prototyper for this calculus.

Outline. The following section provides a brief overview of the calculus and an example. The prototyping tool is described in section 3.

2 A Components' Calculus

Given a collection of sets I, O, ..., acting as component interfaces, a component taking input in I and producing output in O is specified by a pointed coalgebra

$$\langle u_p \in U_p, \overline{a}_p : U_p \longrightarrow \mathsf{B}(U_p \times O)^I \rangle$$

where u_p is the initial state, B a strong monad capturing the component behaviour model (*e.g.*, partiality, as above, or non determinism or ...), and the coalgebra dynamics is given by $a_p : U_p \times I \longrightarrow \mathsf{B}\ (U_p \times O)$. This definition means that the computation of an action in a component will not simply produce an output and a continuation state, but a B-structure of such pairs. The monadic structure provides tools to handle such computations. Unit (η) and multiplication (μ), act, respectively, as a value embedding and a 'flatten' operation to reduce nested behavioural effects. Strength, either in its right (τ_r) or left (τ_l) version, caters for context information.

References [1,2] introduce a small set of component combinators and study their properties. Their implementation in SHACC is parametric on the component behaviour discipline encoded in a monad B.

Components with compatible interfaces (for example, $p : I \longrightarrow K$ and $q : K \longrightarrow O$) can be composed sequentially as

$$p \,;\, q \;=\; \langle \langle u_p, u_q \rangle \in U_p \times U_q, \overline{a}_{p;q} \rangle$$

where $a_{p;q} : U_p \times U_q \times I \longrightarrow \mathsf{B}(U_p \times U_q \times O)$ is detailed as follows [1]

$$\begin{aligned} a_{p;q} &= U_p \times U_q \times I \xrightarrow{\times r} U_p \times I \times U_q \xrightarrow{a_p \times \mathsf{id}} \\ &\mathsf{B}(U_p \times K) \times U_q \xrightarrow{\tau_r} \mathsf{B}(U_p \times K \times U_q) \xrightarrow{\mathsf{B}(a \cdot \times r)} \\ &\mathsf{B}(U_p \times (U_q \times K)) \xrightarrow{\mathsf{B}(\mathsf{id} \times a_q)} \mathsf{B}(U_p \times \mathsf{B}(U_q \times O)) \\ &\xrightarrow{\mathsf{B}\tau_l} \mathsf{BB}(U_p \times (U_q \times O)) \xrightarrow{\mathsf{BB}a^{\circ}} \\ &\mathsf{BB}(U_p \times U_q \times O) \xrightarrow{\mu} \mathsf{B}(U_p \times U_q \times O) \end{aligned}$$

HASKELL monadic technology provides all the ingredients for a direct implementation of this definition, suitably parametric on a strong monad b. Each component is represented by a monadic function from pairs of state-input values to b-computations of state-output pairs. The HASKELL definition of each combinator in the calculus follows closely the corresponding mathematical construction, as illustrated below for sequential composition. Computation proceeds through Kleisli composition. Note, finally, that in order to guarantee state persistence (and propagation of state values) the implementation of SHACC resorts to HASKELL state monad which is suitably combined with monad b capturing the underlying behavioral model.

```
seqCompostion :: Strong b =>
    ((u,i)-> b (u,k)) -> ((v,k)-> b (v,o))
     -> ((u,v), i) -> b ((u,v),o)

seqCompostion p q = mult . (fmap (fmap assocl)). (fmap lstr).
                    (fmap (id >< q)) . (fmap xl).
                    rstr . (p >< id) . xr
```

The identity of sequential composition is component $\mathsf{copy}_K = \langle * \in \mathbf{1}, \overline{a}_{\mathsf{copy}_K} \rangle$, where $a_{\mathsf{copy}_K} = \eta_{1 \times K}$. The monoidal structure is expressed as bisimulation equations:

$$\begin{aligned} \mathsf{copy}_I \,;\, p \sim p \sim p \,;\, \mathsf{copy}_O \\ (p \,;\, q) \,;\, r \sim p \,;\, (q \,;\, r) \end{aligned}$$

[1] The definition resorts to standard isomorphisms, such as associativity (a) and exchange ($\times r : A \times B \times C \rightarrow A \times C \times B$, $\times l : A \times (B \times C) \rightarrow B \times (A \times C)$), as well as to natural transformations $\tau_r : T \times - \Longrightarrow T(\mathsf{id} \times -)$ and $\tau_l : - \times T \Longrightarrow T(- \times \mathsf{id})$ denoting right and left monad strength.

Parallel composition, denoted by $p \boxtimes q$, corresponds to a synchronous product: both components are executed simultaneously when triggered by a pair of legal input values. Note, however, that the behavioral effect, captured by monad B, propagates. For example, if B can express component failure and one of the arguments fails, product fails as well. Two other tensors capture other forms of component aggregation: *external choice* $\boxplus$ and *concurrent* $\circledast$ composition. When interacting with $p \boxplus q : I + J \rightarrow O + R$, the environment chooses either to input a value of type I or one of type J, which triggers the corresponding component (p or q, respectively), producing the relevant output. In its turn, concurrent composition combines choice and parallel, in the sense that p and q can be executed independently or jointly, depending on the input supplied.

A *wrapping* mechanism $p[f, g]$ which encodes the pre- and post-composition of a component with a function is defined as a combinator which generalises the renaming connective found in process algebras. Moreover, any function $f : A \longrightarrow B$ can be lifted to a component whose interfaces are given by their domain and codomain types. Formally, a function $f : A \longrightarrow B$ gives rise to component $\ulcorner f \urcorner = \langle * \in \mathbf{1}, \overline{a}_{\ulcorner f \urcorner} \rangle$ *i.e.*, a coalgebra over $\mathbf{1}$ whose action is given by currying $a_{\ulcorner f \urcorner} = \mathsf{B}(\mathbf{1} \times B) \cdot (\mathsf{id} \times f)$.

Finally, generalized interaction is catered through a sort of "feedback" mechanism on a subset of the inputs. This is defined by a combinator, called *hook*, which connects some input to some output wires and, consequently, forces part of the output of a component to be fed back as input. Formally, the *hook* combinator $- \lnot_Z$ maps each component $p : I + Z \longrightarrow O + Z$ to $p \lnot_Z : I + Z \longrightarrow O + Z$.

3 The SHACC Tool

The SHACC was developed as a *proof-of-concept* prototype for the component calculus proposed in [2,1]. It allows the (interactive) definition of state-based components through the set of combinators available in the calculus: Figure 1 illustrates the application of the *hook* combinator to link a user-specified number of ports with opposite polarity.

The definition of a new, base component is directly made in HASKELL . A specific strong monad B is chosen to model the envisaged behavioral effect. The code below corresponds to a Stack component, where B is instantiated to HASKELL `Maybe` monad to capture partiality.

```
stack  (xs, ("Push", Just a))   = Just ( a:xs, ("Push", a))
stack  (xs, ("Pop", Nothing))
       |(xs == [])    = Nothing
       | otherwise    = Just ( tail xs, ("Pop", head xs))
stack  (xs, ("Top", Nothing))
       |(xs == [])    = Nothing
       | otherwise    = Just (  xs, ("Top", head xs))
```

In a subsequent step the component's interface is created from a suitable annotation in the source code. For this example:

```
@Input: (( 1:Pop + 1:Top) + P:Push)
@Output:(( P:Pop + P:Top) + 1:Push)
```

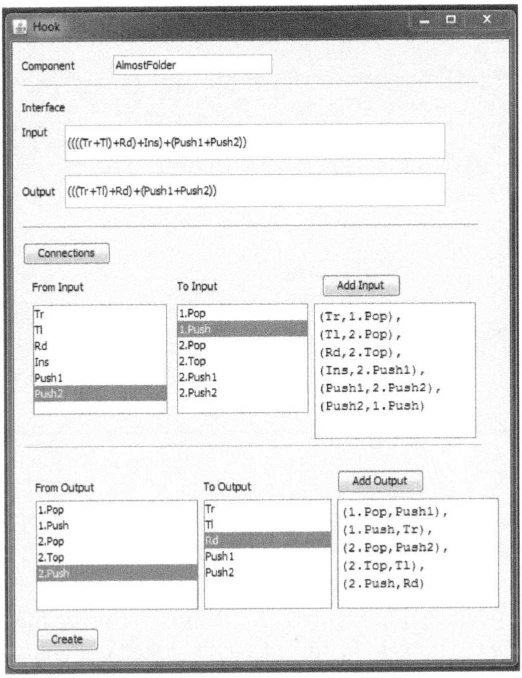

Fig. 1. Linking ports through the *hook* combinator

where `Pop`, `Top` and `Push` are introduced as labels for the component's available services.

Figure 1 refers to an example from the SHACC library, in which a virtual version of a paper folder is built through the combination of two stacks modelling, respectively, the folder left and right piles.

The Folder component provides ports corresponding to the operations *read*, *insert* a new page, *turn a page right* and *turn a page left*. Its construction requires first an adaptation to be performed on each instance of the Stack component. This is needed, for example, to hide the *top* operation on the left stack whereas renaming the *top* on the right as the Folder *read* operation. In a second stage, both stacks are put together through the $\boxplus$ combinator and, finally, suitable feedback loops are established, through the *hook* operator, to connect ports. This ensures, for example, that the left turn of a page is achieved through a *pop* performed on the right stack connected to a *push* on the left one. Formally, this amounts to the following expression in the component calculus (see [4] for a detailed discussion)

$$\mathsf{Folder} \;=\; ((\mathsf{LeftS} \boxplus \mathsf{RightS})[wi, wp]) \Lsh_{P+P}$$

where $\mathsf{RightS} = \mathsf{Stack}[id + \nabla, id]$ and $\mathsf{LeftS} = \mathsf{Stack}[i_2 + Id, (id + !_{P+1}) \cdot a_+]$.

A crucial ingredient in defining Folder is to suitably wrapp the two underlying Stack components so that the intended output-input ports are effectively connected. Formally this is achieved through the *wrapping* combinator, as in the specification of LeftS and

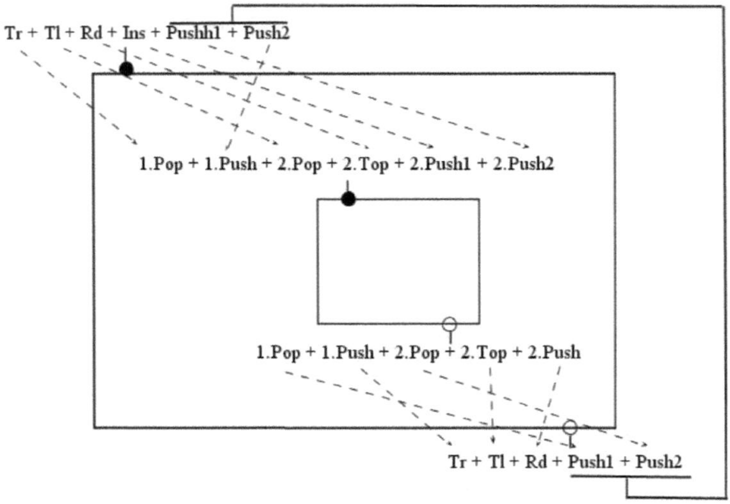

Fig. 2. Linking output to input ports

RightS. The effect is depicted in Figure 2. In SHACC, however, the user has the option of manually selecting the ports to be linked, as illustrated in Figure 1.

SHACC allows both the (interactive) definition of this sort of component expressions and their execution in a simulation mode. Actually, once components are defined either from scratch (*i.e.*, by providing the corresponding HASKELL code directly) or by composition of other components, SHACC offers an environment for testing by simulation. The *Run* window in the tool offers two simulation modes: a *free* mode in which, if the component's behaviour model allows, execution may lead to 'disaster' (*e.g.*, by violation of port pre-conditions on a *partial* component), and a *safe* mode in which the effect of a port operation is foreseen and eventually precluded. Component testing, on the other hand, can be made in a purely interactive way, running event by event, or by executing a whole sequence of events specified through a regular expression and supplied to the tool. Figure 3 illustrates the tool execution mode.

The box labelled *State* in Figure 3 shows the initial value of the component's state. Box *Operation*, on the other hand, accepts the component service to be called. On

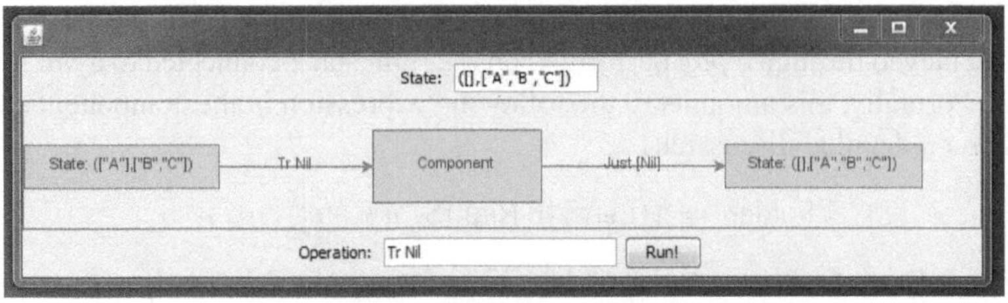

Fig. 3. Component prototyping in SHACC

executing a service from the component's interface SHACC displays three boxes representing the component state before, during and after service completion.

The SHACC tool is composed of a HASKELL combinator library and a graphical user interface developed in Swing. The choice of HASKELL was motivated by its expressiveness and extensibility, which provides an ideal means to support domain specific languages. Most important was the direct encoding in HASKELL 's 'monadic technology' of the entire component representation and manipulation. In particular, one resorted to HASKELL 's readily available *state monad* implementation, for storing the internal state of components being executed. This, together with HASKELL 's specific *do notation* for monadic type values manipulation, greatly reduced the effort of implementing the prototyper and its different execution modes.

Another important implementation detail, again resorting to monadic technology, is error detection and handling in a way which conforms to the underlying behavioral model. according to the execution mode). As already explained, in the Folder example above this resorts to the native `Maybe` data type, which forms an instance of the monad class, thus allowing for error propagation and detection at each specific point of component execution.

Finally, integration with *Swing*, to provide a user-friendly interface, proved effective.

Availability. SHACC is available from `shacc.wetpaint.com`. For the underlying calculus see references [2,1,4]. A refinement theory for this sort of component models is documented in [5,3].

References

1. Barbosa, L.S.: Towards a calculus of state-based software components. Journal of Universal Computer Science 9(8), 891–909 (2003)
2. Barbosa, L.S., Oliveira, J.N.: State-based components made generic. In: Peter Gumm, H. (ed.) CMCS 2003. Elect. Notes in Theor. Comp. Sci., vol. 82.1. Elsevier, Amsterdam (2003)
3. Barbosa, L.S., Oliveira, J.N.: Transposing partial components: An exercise on coalgebraic refinement. Theor. Comp. Sci. 365(1-2), 2–22 (2006)
4. Barbosa, L.S., Sun, M., Aichernig, B.K., Rodrigues, N.: On the semantics of componentware: A coalgebraic perspective. In: He, J., Liu, Z. (eds.) Mathematical Frameworks for Component Software, pp. 69–117. World Scientific, Singapore (2006)
5. Meng, S., Barbosa, L.S.: Components as coalgebras: The refinement dimension. Theor. Comp. Sci. 351, 276–294 (2005)

Author Index

GPSR Compliance

The European Union's (EU) General Product Safety Regulation (GPSR) is a set of rules that requires consumer products to be safe and our obligations to ensure this.

If you have any concerns about our products, you can contact us on ProductSafety@springernature.com

In case Publisher is established outside the EU, the EU authorized representative is:

Springer Nature Customer Service Center GmbH
Europaplatz 3
69115 Heidelberg, Germany

Batch number: 09473985

Printed by Printforce, the Netherlands